Case Studies in Environmental Science
SECOND EDITION

LARRY UNDERWOOD
Northern Virginia Community College

Australia • Canada • Mexico • Singapore • Spain • United Kingdom • United States

Publisher: Emily Barrosse

Publisher: John Vondeling

Developmental Editor: Gabrielle Goodman

Production Manager: Susan Shipe

Art Director: Caroline McGowan

Cover Designer: Jacqueline LeFranc

Production Service: Dartmouth Publishing, Inc.

Copyright © 2001 by Thomson Learning, Inc.
Thomson Learning™ is a trademark used herein under license.

ALL RIGHTS RESERVED. No part of this work covered by the copyright hereon may be reproduced or used in any form or by any means — graphic, electronic, or mechanical, including photocopying, recording, taping, Web distribution, or information storage or retrieval systems — without the written permission of the publisher.

For information about our products, contact us:
Thomson Learning Academic Resource Center
1-800-423-0563
http://www.wadsworth.com

For permission to use material from this text, contact us by
Web: http://www.thomsonrights.com
Fax: 1-800-730-2215
Phone: 1-800-730-2214

Printed in the United States of America
10 9 8 7 6 5 4 3 2

ISBN 0-03-031582-4

Preface

Environmental science often finds itself embroiled in controversy. The science itself is not controversial. Among scientists there is general agreement as to its findings and principles. But when environmental science brushes against economics and politics, controversies arise. Inevitably, environmental science becomes embroiled in environmental issues.

These are some of the most contentious issues society faces. Often they involve economic interests, preservationists, one or more user groups, politicians, regulatory agencies (local, state, and/or federal), and the press in a process that comes to resemble a barroom brawl. What starts out as a minor disagreement escalates into an all-consuming issue that divides communities and defies reason. Environmental science is often caught in the middle.

In the pages that follow, we examine 12 such issues. Each one is an example of how a specific group of people in a specific region faced a specific issue. These may or may not be the most important issues facing society. What is important is that they illustrate how a community has dealt or is dealing with an environmental controversy. In every case progress has been made, but no case sees complete resolution. Indeed, a characteristic of these issues is that they tend never to go away completely. Intensities ebb and flow as decisions are made and events take their course, but the issues are always there, waiting somewhere in the background. The price for involvement in environmental issues is constant vigilance.

Must this always be so? What would it mean to satisfactorily resolve an environmental issue?

If resolution is the goal, keep these points in mind when dealing with these and similar issues:

1. There are no good guys and no bad guys, just differing points of view. In fact, even differing viewpoints are often not that far off from each other. Most developers see the value of a clean environment and most preservationists admit the importance of a strong economy. They differ mainly in the order they value economic versus environmental factors. How can we get combatants in these issues to be sensitive to the opinions of others?

2. The role of environmental science is somewhat limited, and as scientists, we cannot solve all problems or answer all questions. For example, what would be the environmental impacts if large numbers of species become extinct in a given area? Environmental science can answer this one. How much money should we spend to save endangered species? This is a question that environmental science cannot answer by itself. The answer is political, economic, perhaps moral, but not

scientific. In general, what kinds of questions should scientists answer and which ones should we avoid?

3. Our goal should be "win/win" solutions. Too often issues become "win/lose." If the environment wins, the economy loses. If the economy wins, the environment loses. How often do we hear, "A particular environmental policy will cost too much; thousands of jobs will be lost." Must it always be this way? What are the economic benefits of clean, healthy environments or healthy populations of endangered plants? Often, it's a matter of attitude.

4. These issues showcase real people dealing with real-life issues. Related issues are being dealt with by other people in other regions. What are the related issues in your community? How will you contribute? Do you wish to enflame, inform, ignore, or resolve?

The 12 issues that constitute *Case Studies in Environmental Science* are organized regionally—at least one issue is included from each of the major regions of North America. Each starts with a commentary that introduces the issue, including its importance, relevance, and seriousness. The introductory text is followed by a series of readings that provide additional background and address the specific issue and region. At the end of each unit, a series of questions guides the reader through the intricacies and complexities of the issue.

Case Studies in Environmental Science is more than a textbook. An integral part of the learning system is an extensive Web site authored and maintained by Kevin R. Henke, University of Kentucky, Lexington. The site can be found at <www.harcourtcollege.com/lifesci/envicases2>. This Web site and its links are powerful additions to the text, allowing exploration of issues in greater detail with access to the most up-to-date information on any environmental topic. Using the same organization as this book, the site supplements each unit with coverage of the case study issue in all regions of the United States and Canada. Resources for each unit enable instructors and students to go beyond the readings via links to other Web sites and critical thinking questions. The links contain a diverse wealth of environmental information on national as well as local issues. This information includes both archived, historical materials (such as the Dust Bowl of the 1930s) and detailed discussions on today's headlines. Furthermore, the Web site contains practical suggestions for becoming more environmentally proactive, including how to save energy and more effectively recycle wastes at home. Links will be periodically reviewed to ensure that they are still active. Inactive links will be replaced with new links on related topics.

Environmental science and involvement in environmental issues are exciting and necessary aspects of modern life. For many of you, it will be your life's passion. Enjoy!

LARRY UNDERWOOD
Woodbridge, Virginia
April 2000

Contents

Unit One   1

Northeast: Land-Use Issues

INTRODUCTION: Future Use of the Woodbridge Research Facility 1

READINGS:
    Reading 1: Mason Neck Addition 5
    Reading 2: Recommended Reuse for the Woodbridge Research Facility 6
    Reading 3: Minority Report Letter from the Woodbridge Reuse Committee to
                the Secretary of the Army 17
    Reading 4: Comprehensive Conservation Plan and Environmental Assessment
                for Occoquan Bay National Wildlife Refuge 19

QUESTIONS FOR DISCUSSION 25

Unit Two   26

Northeast: Control of Wildlife Populations

INTRODUCTION: Management of White-Tailed Deer in Pennsylvania 26

READINGS:
    Reading 1: The White-Tailed Deer: A Keystone Herbivore 29
    Reading 2: Attitudes of Pennsylvania Sportsmen Towards Managing White-Tailed
                Deer to Protect the Ecological Integrity of Forests 36
    Reading 3: White-Tailed Deer Research/Management 42
    Reading 4: Urban Deer Contraception: The Seven Stages of Grief 49
    Reading 5: Pennsylvania Deer Statistics 1982–1998 54

QUESTIONS FOR DISCUSSION 56

Unit Three   57

Canada: Alternate Energy Resources

INTRODUCTION: James Bay Hydroelectric Project, Canada 57

READINGS:

    Reading 1:   The La Grande Complex Development and its Main
                        Environmental Issues 60

    Reading 2:   James Bay and Northern Québec Agreement and
                        Subsequent Agreements 64

    Reading 3:   Economic and Social Development of the Aboriginal Communities 69

    Reading 4:   Native People and the Environmental Regime in the James
                        Bay and Northern Québec Agreement 75

QUESTIONS FOR DISCUSSION 90

Unit Four   91

Great Lakes: Water Resources

INTRODUCTION: Great Lakes Cleanup 91

READINGS:

    Reading 1:   A Strategy for Virtual Elimination of Persistent Toxic Substances 94

    Reading 2:   Comments on the *Canada–United States Strategy for the Virtual Elimination of
                        Persistent Toxic Substances in the Great Lakes Basin* 116

    Reading 3:   Pollution, Thirst for Cheap Water Threaten Great Lakes, Study Says 122

    Reading 4:   Protecting Lake Superior: A Community-Based Approach 124

QUESTIONS FOR DISCUSSION 133

Unit Five   134

Southeast: Biodiversity and Nonindigenous Species

INTRODUCTION: Controlling Nonindigenous Species in Florida 134

READINGS:

    Reading 1:   Why We Should Care and What We Should Do 137

    Reading 2:   Fido-Munching Toads Take Florida by Leaps and Bounds 141

    Reading 3:   Florida: A Wacky Wild Kingdom 142

    Reading 4:   Management in National Wildlife Refuges 143

    Reading 5:   Plant Management in Everglades National Park 146

    Reading 6:   Management on State Lands 152

    Reading 7:   Florida's Biological Crapshoot with Carp 157

    Reading 8:   Will Beetle Bombs Recapture Everglades—Or Overrun It? 158

    Reading 9:   Ecological Effects of an Insect Introduced
                        for the Biological Control of Weeds 159

QUESTIONS FOR DISCUSSION 162

Unit Six   163

Midwest: Soil Conservation

INTRODUCTION: Avoiding Another Dust Bowl 163

READINGS:

    Reading 1: Conserving the Plains: The Soil Conservation Service
                    in the Great Plains 166

    Reading 2: The Great Plains Conservation Program, 1956-1981:
                    A Short Administrative and Legislative History 173

    Reading 3: Summary Report: 1997 National Resources Inventory 184

    Reading 4: The Buffalo Commons, Then and Now 224

    Reading 5: Comment on the Future of the Great Plains: Not a Buffalo Commons 226

    Reading 6: The Bison Are Coming.. 228

QUESTIONS FOR DISCUSSION 230

Unit Seven   231

Northwest: Wildlife Management

INTRODUCTION: Managing Timber in the Pacific Northwest.............................. 231

READINGS:

    Reading 1: The Spotted Owl Controversy and the Sustainability
                    of Rural Communities in the Pacific Northwest 234

    Reading 2: Having Owls and Jobs Too 242

    Reading 3: The Birds—The Spotted Owl: An Environmental Parable 245

    Reading 4: Timber Owners Cut a Deal to Preserve Wildlife Habitat 251

    Reading 5: U.S. Considers 80% Increase in Sierra Logging 254

QUESTIONS FOR DISCUSSION 255

Unit Eight   256

California: Solid Waste Disposal

INTRODUCTION: Solid Waste Disposal in Sonoma County................................ 256

READINGS:

    Reading 1: The 3 R's of Solid Waste and the
                    Population Factor for a Sustainable Planet 259

    Reading 2: Sonoma County Waste Management
                    Agency 1992-1997 Progress Report................................... 262

    Reading 3: Why Recycle? 275

    Reading 4: Buy Recycled—On Earth Day and Every Day,
                    Shop the Recycled Way 277

    Reading 5: Earth Pledge: Twelve Things You Can Do Today to Fight
                    Global Warming and Environmental Destruction 278

Reading 6:	Elder Earthkeepers Honored with Gifts From the Earth	281
Reading 7:	What is Bay Area Creative Re-Use?	282
Reading 8:	Environmentalism Begins at Home: Green Building Represents a Vital Earth-Friendly Action	283
Reading 9:	Helping Business Donate/Sell Waste Material: Sonoma County's Material Exchange, www.recyclenow.org/SonoMax	284

QUESTIONS FOR DISCUSSION ... 285

Unit Nine   286

Northeast: Acid Precipitation

INTRODUCTION: Minimizing Acid Precipitation in the Adirondacks ... 286

READINGS:

Reading 1:	EPA 1998 Compliance Report: Acid Rain Program	289
Reading 2:	Emissions Trading of Sulfur Dioxide: The U.S. Experience	300
Reading 3:	Acid Rain: A Continuing National Tragedy	305
Reading 4:	A Washington, D.C. Press Conference	309
Reading 5:	Public Service Announcement Campaign a Ringing Success	310

QUESTIONS FOR DISCUSSION ... 311

Unit Ten   312

Southwest: Air Quality

INTRODUCTION: Improving Air Quality in Los Angeles ... 312

READINGS

Reading 1:	Smog City Case Studies	315
Reading 2:	Cleaner Fuels and Cleaner Cars	321
Reading 3:	The False Promise of Electric Cars	338
Reading 4:	New Vehicles Now Less Efficient Than Junked Ones	343

QUESTIONS FOR DISCUSSION ... 344

Unit Eleven   345

Southeast: Wildlife Management

INTRODUCTION: Managing Bluefin Tuna in the Western Atlantic ... 345

READINGS:

Reading 1:	Bluefin Tuna in the West Atlantic: Negligent Management and the Making of an Endangered Species	348
Reading 2:	Historic Rationale, Effectiveness, and Biological Efficiency of Existing Regulations for the U.S. Atlantic Bluefin Tuna Fisheries: A Report to the United States Congress	353

QUESTIONS FOR DISCUSSION ... 382

Unit Twelve 383

Northwest: Toxic Wastes and Pollution

INTRODUCTION: What If There Were Another Oil Spill in Prince William Sound, Alaska?....... 383

READINGS:
- Reading 1: The Two Faces of the Exxon Disaster 386
- Reading 2: 1999 Status Report on the *Exxon Valdez* Oil Spill........................390
- Reading 3: 2000 Status Report on the *Exxon Valdez* Oil Spill....................... 397

QUESTIONS FOR DISCUSSION 408

unit 1

Northeast: Land-Use Issues

Introduction
Readings:
Reading 1: Mason Neck Addition
Reading 2: Recommended Reuse for the Woodbridge Research Facility
Reading 3: Minority Report Letter from the Woodbridge Reuse Committee to the Secretary of the Army
Reading 4: Comprehensive Conservation Plan and Environmental Assessment for Occoquan Bay National Wildlife Refuge
Questions for Discussion

Future Use of the Woodbridge Research Facility

Introduction

In urban and suburban communities, it is often difficult to establish and retain natural areas. Open land is at a premium. Often, as soon as a tract becomes available, competing, incompatible uses vie for its control. Citizens get caught up in what soon becomes local controversy. Their needs and wants frequently conflict. Citizens want open land for passive or active recreation and to provide habitat for wildlife, but they need jobs, services, and sources of revenue to support local governments. Land-use decisions are often long, tedious, and contentious.

Prince William County (PWC) in Northern Virginia is caught up in just such an activity. In recent years, PWC has been a bedroom community for Washington, D. C., located just 20 miles (32 km) away. Without an industrial base, PWC is finding it difficult to squeeze enough tax revenues from citizens to provide for their needs. To broaden its commercial tax base and to provide citizens with local jobs, the county is doing what it can to lure businesses and other economic interests to the area.

In 1991, PWC saw an opportunity to obtain a sizable tract of land from the federal government. To save federal tax dollars, the U.S. government asked the Base Realignment and Closure Commission to identify bases no longer essential to military needs. Bases so designated were to be closed and the land they occupied either transferred to other federal agencies, given to state or local governments, or sold on the open market. To minimize the economic impact on local communities, local governments were asked to establish Reuse Committees, made up of local citizens and interested parties, to study alternatives and make recommendations to the Army.

The Woodbridge Research Facility (WRF) was determined to be such a site. Located in PWC's southwest corner, this top-secret U.S. Army laboratory was designated for closure. Nearly 600 acres (240 hectares) of potentially prime real estate fronting the Potomac River was suddenly up for grabs.

Background

In many respects, the WRF is a unique and important site. Its potential economic value is unquestioned. It lies less than one mile (1.6 km) south of the "Route 1 Corridor" (Figure 1), a strip development running through the county made up of numerous small- to medium-sized businesses serving the region. Immediately adjacent, to the northwest of the facility, is an industrial complex. Perhaps more important, immediately northeast of the facility a major commercial development (Belmont Development) is planned, consisting of new homes, a hotel and conference center, a marina, an aquarium, and a golf course. To county land planners, the additional acreage would create jobs and expand the commercial tax base.

The area also had surprising environmental values. First settled in the 1700s, the facility had been farmed until 1950, when approximately 600 acres (240 hectares) were transferred to the U.S. Army. In 1970, the facility became part of the Harry Diamond Laboratory. For the next 24 years, research at the WRF was top secret. The exact nature of much of this research. the site's value as a wildlife habitat. Part of this research involved generating huge electromotive forces, similar to those generated in atomic bomb blasts, and studying their effects on military equipment. This involved arrays of huge antennas that could focus energy on equipment.

Perhaps surprisingly, the army's research was not incompatible with wildlife. The top-secret nature of the research kept the general public at bay for 40 years. Four main buildings, where human activities were concentrated, were situated in a central compound of approximately 11 acres (4.5 hectares). A system of roads radiated from the central compound linking areas outside the compound. Habitat diversity is high. At least 20 different vegetative communities have been identified. A variety of wetland communities cover approximately 300

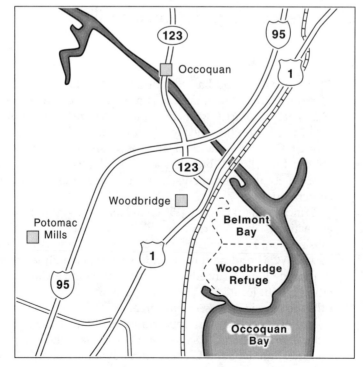

Figure 1

acres (120 hectares). These include tidal marshes, freshwater swamps, seasonal wetlands, and a pond. Open woodland habitat comprises another approximately 170 acres (70 hectares) of upland meadows. For more than 40 years, the Army mowed these areas annually for security and related reasons, but otherwise they were preserved. As a result, surprisingly diverse communities arose within the grassland ecosystem.

Understandably, the Army's interest lay in areas other than analysis of habitat and wildlife. Ecologically, the area was virtually unstudied. Starting in 1993, an ad hoc community of interested citizens documented some of the facility's diversity. More than 620 species of plants have been identified and catalogued. Many are found nowhere else in the county or region. The area is also attractive to birds; 220 species have been identified. Additionally, the grasslands support large populations of mice and voles, which are important to over-wintering raptors. Ospreys nest on the facility each summer, and bald eagles feed and rest there daily. More studies are needed.

The Problem

No sooner was base closure announced than various interests began to vie for control of part or all of the land. Base Realignment and Closure regulations gave the Army three options: (1) transfer the base to another federal agency, (2) transfer it to a state or local government agency, or (3) sell the land on the open market. The Army was instructed to both close the base and minimize local economic impact.

The Fish and Wildlife Service (FWS) voiced an early interest in the land. The WRF land would join two units of the Mason Neck Wildlife Refuge, which the FWS manages. Some local environmentalists sided with the FWS. They reasoned that any region growing as fast as PWC needs to preserve as much of its natural heritage as it can.

Local educators and scientists saw an opportunity to develop the facility into a regional environmental education center. Particularly appealing was the facility's combination of buildings, roads, and outstanding habitat values. Here was an area uniquely suited for visits and study by school children, older students, and adults.

Citizens interested in economic development stressed WRF's possible contribution to local tourism.

The Library of Congress asked for a portion of the facility on which to build warehouses to store excess books and documents. Public opposition to this request grew swiftly. At the urging of the Virginia Congressional Delegation, the library found an alternate, less environmentally sensitive area and withdrew its proposal.

Into this milieu of simmering and conflicting interests stepped the Reuse Committee. For more than two years, meetings and hearings were held to determine the will of the community and to sort through various possibilities. There was general agreement on the need for and value of a regional environmental education center. The wetland and woodland areas along with the existing buildings and roads could well serve this purpose, they reasoned. That left more than 100 acres (40 hectares) of upland meadows, along the northern border of the facility, immediately adjacent to the proposed Belmont Development. This land, among other uses, would allow for the expansion of the proposed golf course.

Unfortunately for the process, this was the very habitat that was most important to over-wintering raptors and that supported many of the rarest plants. Here were the seeds for protracted controversy.

On May 16, 1994, the Committee submitted its recommendations. Eight days later, four members of the committee issued a minority report, disagreeing with the majority view and recommending alternative actions. For several months after the release of the two sets of conflicting recommendations, no significant action was taken officially by the county, the army, the FWS, or private citizens. But behind the scenes, there was considerable activity. Most significant, several local and national environment groups urged Congress to take action and settle the issue. In August 1994, at the urging of Congresswoman Byrne (D-VA) and Senator Robb (D-VA), with the concurrence of Senator Warner (R-VA), the following section was added to the pending Military Appropriations Bill:

- Sec. 128. Land Transfer, Woodbridge Research Facility, Virginia
- (a) Requirement of transfer.—Not withstanding any other provision of law, the Secretary of the Army shall transfer, without reimbursement, to the Department of the Interior, a parcel of real estate consisting of approximately 580 acres [230 hectares], comprising the Army Research Laboratory Woodbridge Facility, Virginia, together with any improvements thereon.
- (b) Use of Transfer Property.—The Secretary of the Interior shall use appropriate parts of this real property for (1) incorporation into the Mason Neck Wildlife Refuge and (2) work with the local governmental and the Woodbridge Reuse Committee to plan any additional usage of the property, including an environmental education center: Provided, that the secretary of the Interior provided appropriate public access to the property.

In September 1994, President Clinton signed legislation that transferred the entire facility to the U.S. Fish and Wildlife facility. They renamed the facility Occaquan Bay National Wildlife Refuge (OBNWR).

Taking over and managing such a facility is neither easy nor straightforward. The first task or the FWS was to develop a facility management plan that would recognize the needs and desires of a highly politicized general public, the habitat needs of species the refuge is intended to preserve, and the federally mandated mission statement of the agency. The following readings provide additional background.

"Mason Neck Addition"
Spencer S. Hsu

Prince William Weekly, August 15, 1994, p.D05. © 1994 *The Washington Post.* Reprinted by permission.

A proposal to incorporate a surplus Army post into the Mason Neck Wildlife Refuge got the approval of Congress this month, making it likely that birdwatchers and hikers will soon tread on land once used for secret testing associated with nuclear weapons.

Transfer of the 580-acre Woodbridge Research Facility, formerly known as the Harry Diamond Laboratories, in Eastern Prince William County was included in a military appropriations bill that now will go to President Clinton, who is expected to sign it.

The bill calls for the Army to turn over the land, which borders the Potomac River, to the U.S. Fish and Wildlife Service on Sept. 30.

Over the years, the site had a variety of Defense Department roles. Once it was a secret radio listening post; later it was an electronic testing ground, where certain effects of nuclear blasts were simulated. The tests did not involve explosions or radioactivity and did not harm the environment, according to Army officials.

The land is on a peninsula at the confluence of the Potomac and Occoquan rivers, less than 15 minutes south of the Capital Beltway, and it harbors 200 animal species and 300 kinds of plants. It is one of the most lush habitats and bird marshes in the Washington area, said Dennis Shiflett, spokesman for the Virginia Wildlife Federation.

J. Frederick Milton, manager of the Mason Neck refuge, said the land will be added to the 2,200 acres the Fish and Wildlife Service already manages on the Fairfax County side of the Occoquan.

It would be opened to the public before the end of the year, allowing dawn-to-dusk access for boating, fishing, hiking and other recreation, he said.

Sens. Charles S. Robb (D-Va.) and John W. Warner (R-Va.) led the legislative move, ending a struggle that began in 1991 when the post was slated for closing. Last year, Congress killed an attempt by the Library of Congress to use the land for warehousing books and records.

This spring, Prince William County tried to save some of the land for development, and a consultant's study foresaw creation of 1,000 jobs and more than $500,000 in annual tax revenue if development were permitted.

The legislation provides for keeping the land in its natural state, although it allows for environmental education programs that would create some jobs.

reading 2

"Recommended Reuse for the Woodbridge Research Facility, Prince William County, Virginia"

Woodbridge Reuse Committee, May 16, 1997, selected pages.

1. Executive Summary

Following announcement of the Army's decision to abandon and dispose of the Woodbridge Research Facility (formerly the Harry Diamond Lab), the Prince William County Board of Supervisors indicated their interest in the reuse of this important resource. To pursue this, the Board created the Woodbridge Reuse Committee, which was charged with exploring and recommending the most appropriate reuse for the site.

The WRC in Phase I of the reuse planning effort, after extensive analysis and discussion, *recommends the Woodbridge Research Facility site be used for a combination of:*

- *Environmental protection and enhancement;*
- *Educational, institutional, and cultural uses; and,*
- *Commercial (tax paying, job supporting) development...*

... provided criteria and procedures are established in Phase II to assure the uses and activities are made compatible.

This WRC recommendation has evolved from a process that involved the insight of the members and the organizations they represent, the input from two citizens meetings, and analysis of the applicable local and regional plans. This process resulted in the formulation of goals, development and evaluation of alternatives, and the recommended conceptual reuse plan andimplementation strategy.

The basic goals supporting the recommended reuse are:

- Protect and enhance the natural environment located on the WRF, including:
    — shoreline
    — wetlands
    — uplands
    — flora/fauna
- Encourage public access and enjoyment of the natural features, with appropriate management to prevent damage or degradation
- Create programs to allow the natural environment to be used for environmental education, research and other related activities (arts, education, etc.)
- Allow and encourage the use of the remainder of the site for commercial use (tax paying and employment generating)

Based on these general concepts, more specific reuse proposals have been developed for the sites four sub-areas—the compound, the wetlands, the shoreline, and the uplands. These are described as follows:

The Compound–The buildings and land can be used or redeveloped to accommodate the resource management activities, the educational, institutional and cultural activities, and private commercial uses. This includes accommodations for overnight visitors.

The Wetlands–The wetlands should be protected as a natural resource, and be used for environmentalresearch and education. Maximum public access, consistent with the objective of preserving and enhancing the natural eco-system, is recommended for walking paths or trails for observing wildlife and other passive activity. Provision will also be made to continue the regional Potomac Trail.

The Shoreline–The shoreline should be made available for fishing, canoeing, public boat launch, and mooring for research vessels related to the educational and institutional activity.

The Uplands–Up to 50% of the 100 acre uplands, excluding the compound (that is, less than 10% of the total acreage at the WRF), can be used for commercial (tax paying, employment generating) uses, particularly for environmentally oriented businesses and institutions. The remainder of the uplands will remain undeveloped nad programs will be initiated to sue these areas to enhance the natural systems. The uplands areas adjoining the golf course and hotel/marina proposed by the adjoining Belmont Center may be the location for private development to benefit from the values of these amenities.

All of the recommended uses will be able to take advantage of the nearby VRE/AMTRAK station, reducing dependency on private auto access.

These recommended uses are designed to accommodate the views of the major constituencies interested in the site—those interested in protecting the environment, those interested in use for educational, institutional and cultural activities, and those interested in economic development. Accommodating increased activity and additional development, without threatening the eco-system, will require:

- the establishment of carefully crafted performance criteria for the design, use, and management of all activities to be accommodated; and,
- procedures for enforcing the performance criteria, both for inclusion in any transfer agreement and after transfer have taken place.

Assuming the Army will accept the reuse recommendations, the next steps will be negotiating the performance criteria, evaluating the institutional capabilities of available organizations and/or creating a Local Redevelopment Authority, and securing the necessary funding. Because no institution or organization has been identified that has the charter, or existing resources to accommodate or support all of recommended uses, the evaluation of institutional capabilities will be an important element of the next phase of the process.

Inclusion of commercial tax paying and employment generating uses will give this portion of the site considerable market value, which, according to the BRAC policy, can provide the Army and the County significant proceeds (in a 60-40 split). A portion of these funds can possibly be used to support the environmental management, educational, institutional and cultural activities. The economic analysis of the recommended reuse estimates the market value of the land, at the .30 FAR, to be almost $4 million. This analysis also estimates the economic benefits to Prince William County to be an annual real estate tax of over $750,000 and 2400 jobs. This will also generate an estimated $3.9 million in real estate value.

These recommendations are the result of extensive effort, analysis and negotiations by the Woodbridge Reuse Committee, in their effort to respond to the responsibility given them by the County, and their responsibility to the citizens of the adjoining community, the County, and the region. The Woodbridge Reuse Committee is now prepared to initiate Phase II to assure the optimum use of this valuable asset—The Woodbridge Research Facility.

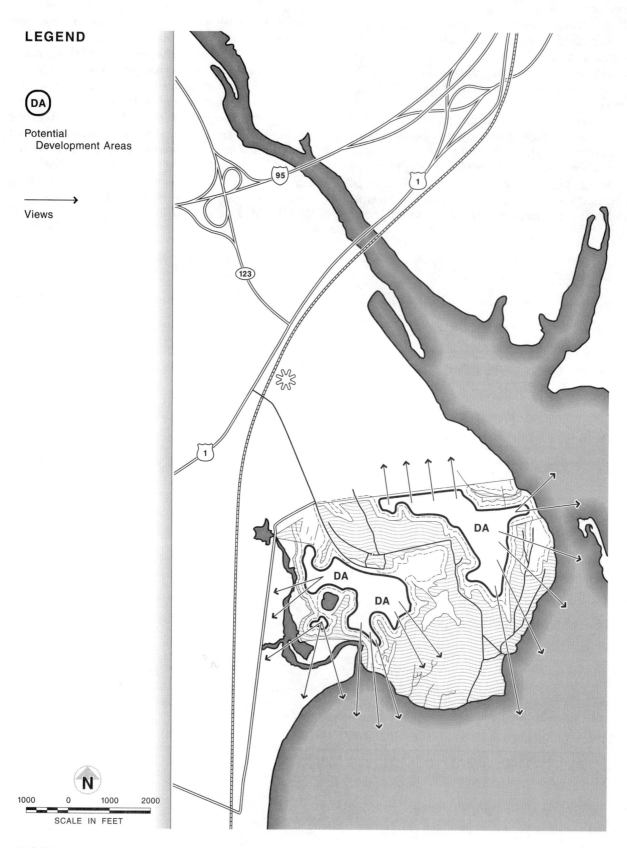

Exhibit 3.1–Site Potential Development Areas

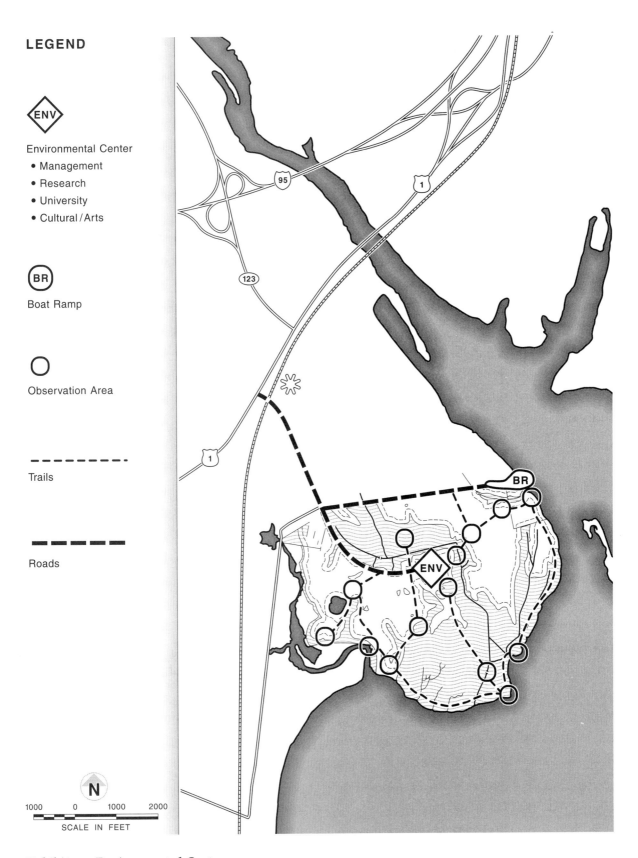

Exhibit 3.2–Environmental Center

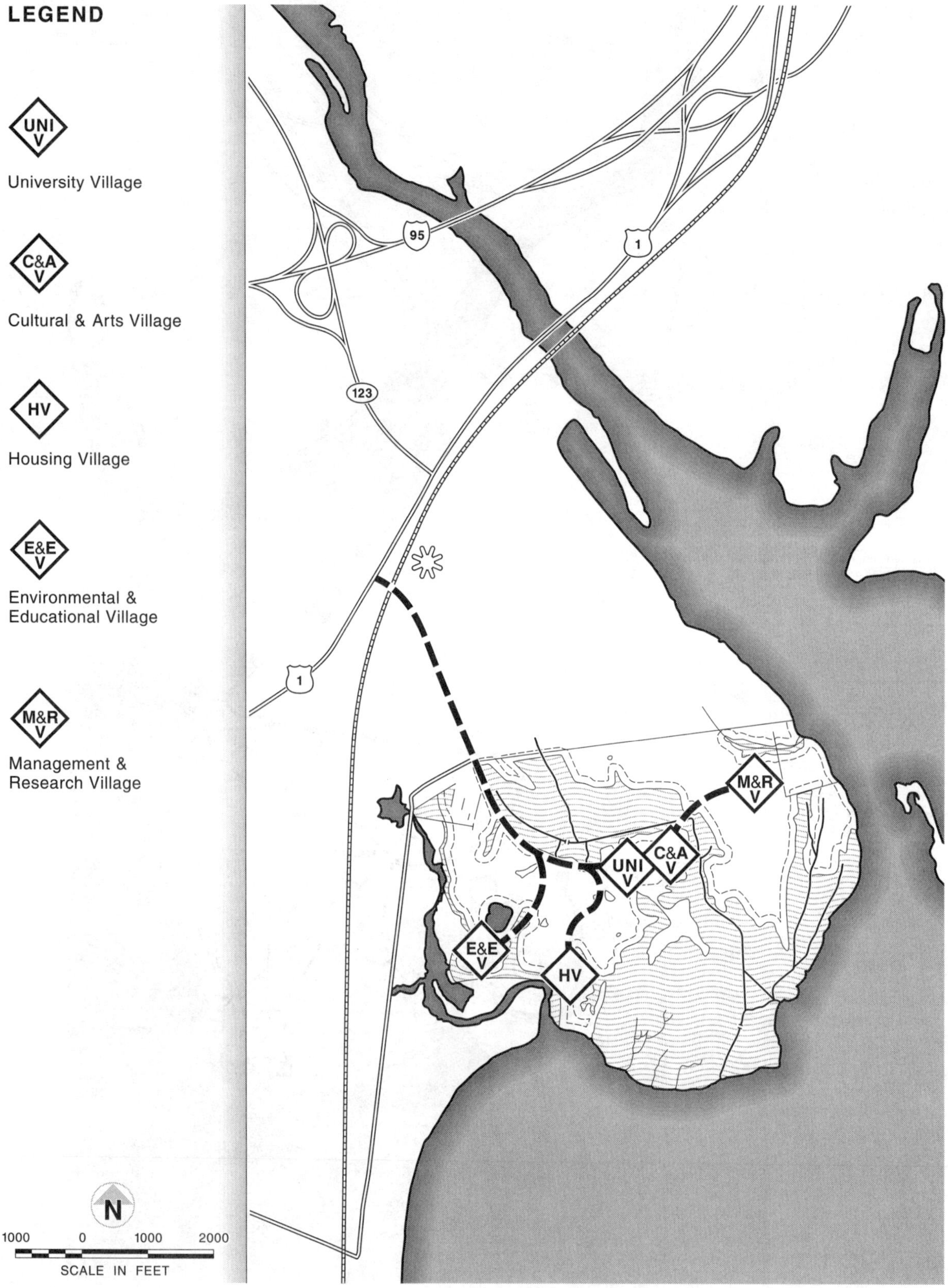

LEGEND

◇ UNI V — University Village

◇ C&A V — Cultural & Arts Village

◇ HV — Housing Village

◇ E&E V — Environmental & Educational Village

◇ M&R V — Management & Research Village

SCALE IN FEET: 1000 0 1000 2000

Exhibit 3.3–Green Villages

10 Unit One: Northeast: Land-Use Issues

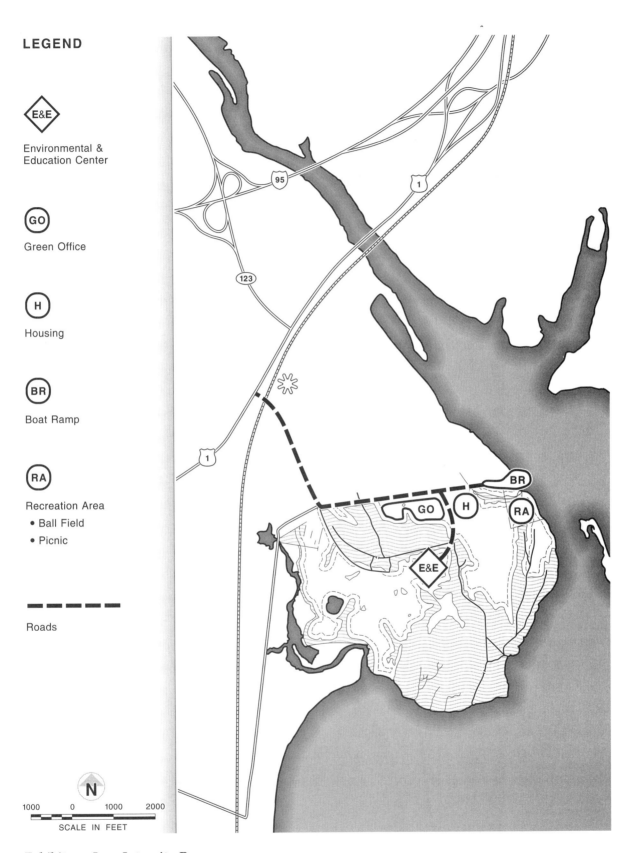

Exhibit 3.4–Low Intensity Reuse

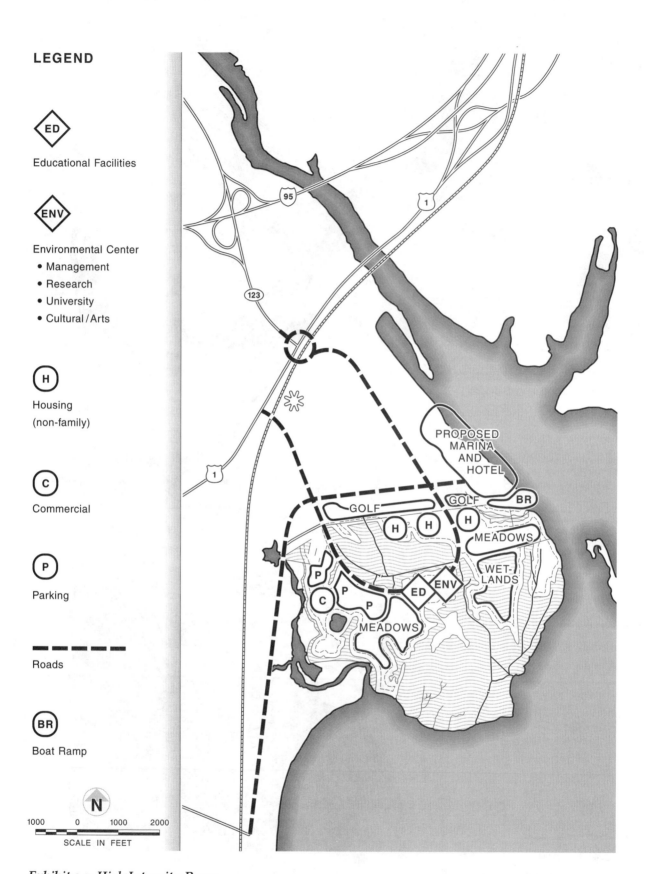

Exhibit 3.5–High Intensity Reuse

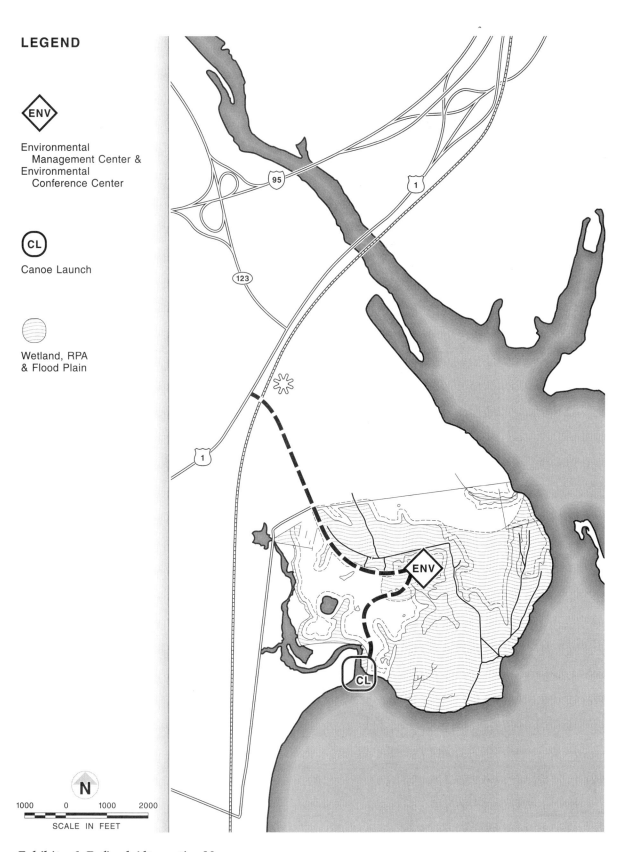

Exhibit 3.6–Refined Alternative No. 1

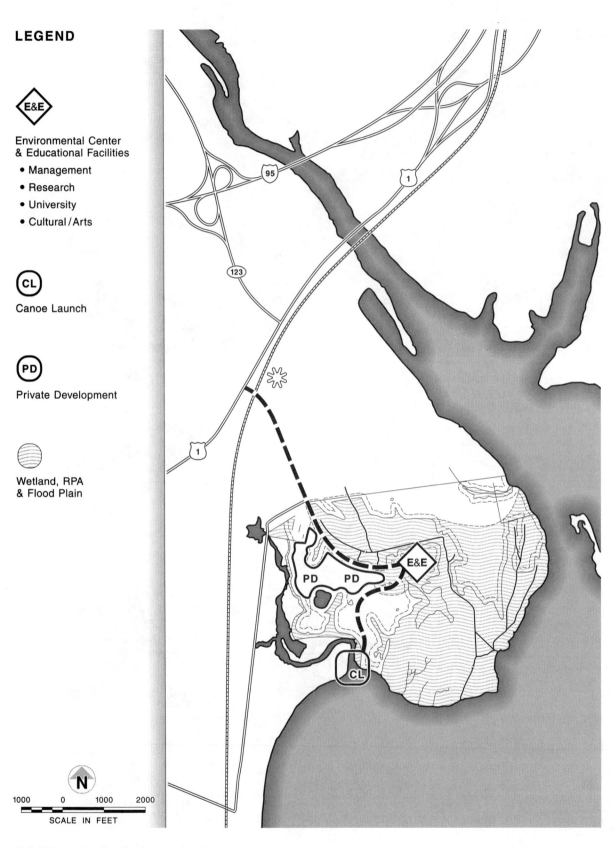

Exhibit 3.7–Refined Alternative No. 2

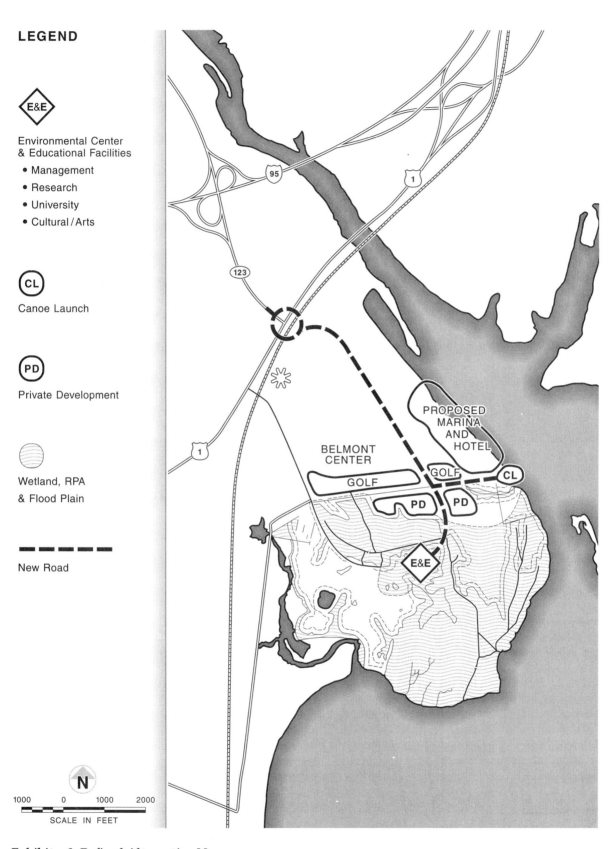

Exhibit 3.8–Refined Alternative No. 3

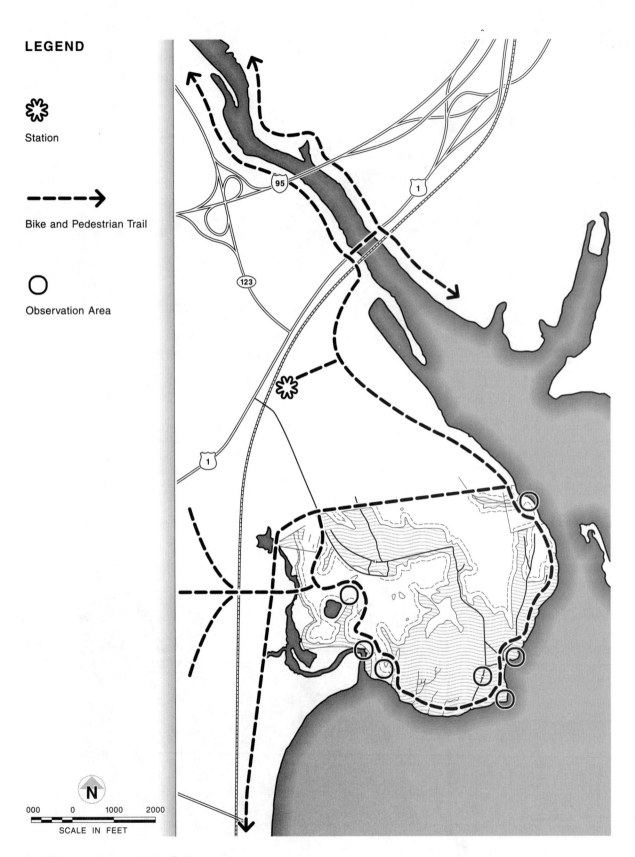

Exhibit 3.9–Potomac Trail Connections

Minority Report Letter from the Woodbridge Reuse Committee to the Secretary of the Army, May 24, 1994

12823 Mill Brook Court
Woodbridge, VA 22192

May 24, 1994

The Honorable Togo West
Secretary of the Army
600 Army Pentagon
Washington, D.C. 20310

Dear Secretary West,

As members of the committee charged with recommending alternative future uses of the Harry Diamond Laboratories' Woodbridge Research Facility (WRF), we wish to advise you of our strong disagreement with the Concept Reuse Plan which was delivered to your staff last week. It was not our intention to produce a minority report. We had hoped that the Plan would fully and fairly address the essential land issues and the strong public sentiment about WRF future uses. Absent that, we are compelled to place on the record our fundamental dissatisfaction with the way the reuse study has been managed and with the product.

The Reuse Plan may appear to represent a systematic analysis of reuses, but it does not. From our vantage point the Committee has not discharged its basic obligation to the broad community affected by the WRF's disposition and, in our opinion, has not properly executed its contract with the consultant preparing this reuse plan. We believe you will find the Committee has not fulfilled the terms specified in the Office of Economic Assistance grant to study alternative uses. The Plan does not reflect a conscientious effort by the Committee to develop concrete alternatives (who, what, where, when, why, how), evaluation criteria, rigorous cost/benefit analysis and selection of a clearly defined preferred reuse proposal. The resulting construction of a vague range of possibilities without specific form or substance is not what should or could have been produced with the available time and public funds.

The Committee had ample opportunity to define and evaluate specific, detailed alternatives but chose not to do so. Our sense is that certain members of the Committee and County staff chose not to address who should control the site because that would have required directly considering whether an important wildlife area should be turned over to a local government or the Department of the Interior. There is no question that the County Government's goal for this committee was to secure Prince William County control of all or at least the supposedly suitable development areas of the WRF. This has been its consistent, but not publicly examined, position for some time (Resolution: 8 Sep 92). Not explicitly stated in this Plan, but made clear in a related staff paper (Lawson: 10 May 94) is that foremost among the "essential elements" of this Plan is the provision for local control of the WRF through conveyance of all of the site to Prince William County. The local agency available to manage this site would be the Prince William County Park Authority. The attached paper (Pokorny: 16 May 94), made available to the Committee, suggests the Park Authority's strength is planning and managing active recreational facilities. It lacks experience with either refuges or nature centers. The Committee was told there is not a single natural resource person on the Park Authority's staff.

We believe the Plan omits any specific development option for the WRF's open natural areas because the Plan's drafters clearly recognized the strong public distaste for other than refuge-related uses there. Nevertheless, despite an absence of information on who would develop or what would be developed, the Committee majority allocated up to one-half of the property's uplands for development. Despite an absence of rationale for commercial development on these uplands, the Committee accepted an unsubstantiated conclusion that, eventually, something could be built there which in some unspecified fashion could be compatible with the remaining natural areas and also yield extraordinary local employment and tax benefits.

We disagree with the assertion in the Plan that no organization has been identified which has "the charter or existing resources to accommodate or support" appropriate future uses of this site. During the course of the Committee's work, the Interior Department's U. S. Fish and Wildlife Service made a compelling case that it has the mission, authority, experience and resources to manage the refuge and the related public, environmental education\training activities that are appropriate for this special site.

We are sorry that the Committee majority did not choose to cooperate with the Department of the Interior on alternative future plans for this facility. That explicit decision (Committee vote, 25 Apr 94) exacerbated the already potentially competitive nature of this process. By ignoring the repeated encouragements of your staff (Owens: 31 Jan 94) and others to work with Interior, the Committee majority missed the opportunity to draft realistic plans for cooperative uses. The result is wasted public funds and this seriously flawed plan.

We fully support the rationale behind this Administration's effort to return economic vitality to communities adversely effected by base closures. We are sure you recognize, however, that the Woodbridge Research Facility's closure, because of scale and other circumstances, will have no such adverse effects (ARL: Final EIS, Nov 93, 3–225). This closure represents an opportunity, not a problem. The governing factor in this closure should be how best to preserve and enhance the natural resource that has flourished under the Army's long custody. We hope this Plan has not obscured the probability that, without development on the WRF's uplands, refuge–related activities on this site will generate more economic return for the local community than the Army's present uses.

We regret that the majority of the Committee appears not to have recognized or accepted the special value to the county and region of all of this site's wildlife areas, but we know that you will understand the testimony of the many individuals and organizations confirming the unquestionable ecological worth of both the WRF's uplands and wetlands and their delicate interrelationship.

We are most disappointed at the Committee majority's lack of regard for the public's opinions on this matter. Public information was scarce. Neither of the public meetings was widely publicized. It appears that none of the individuals promised notice of the second meeting by the Chair were, in fact, notified. Finally, those speaking for preservation at the public meetings were characterized as "a minority" (when, in fact, they were a substantial majority) and "outsiders" (although the great majority were local area residents).

We believe it is important to address the issue of what constitutes an outside interest in this matter. We understand the Defense Department seeks representation and input from all constituencies affected by a base closure, regardless of jurisdictional boundaries. In the case of this public natural resource, the interests of potential beneficiaries throughout the greater Washington DC area need to be taken into account. Certainly residents of Fairfax County communities immediately adjacent to the WRF are much affected, undoubtedly more affected, than many Prince William County residents, e.g., the Federation of Lorton Communities (Resolution: 8 Mar 94) represents residents directly opposite the WRF and much closer to it than is the Prince William County seat of government. Regrettably, we observed that some on the Committee chose to accord lesser importance to comments and opinions from beyond Prince William's borders, in some cases beyond the borders of the Woodbridge District. We believe you will agree that this is not a sufficiently broad perspective for charting the future course of a regionally important public resource.

We are confident that you will recognize the importance of assuring full and permanent protection for all of this property. The scarcity and unique diversity of its wildlife habitats make the Woodbridge Research Facility a resource of great educational and economic value to the entire National Capital Area. We trust your decision will be based not on the interests reflected in this Reuse Plan, but on the clear, broad, long–term interests of the general public.

We would be pleased to meet with you or your staff to clarify any issues related to the Reuse Committee's proceedings or the Concept Reuse Plan.

Respectfully,

James Waggener   Charles Creighton
Robert Maestro   Dennis Shiflett

reading 4

"Comprehensive Conservation Plan and Environmental Assessment for Occoquan Bay National Wildlife Refuge"

U.S. Fish and Wildlife Service, December 1997, selected pages.

I. Introduction and Background

Purpose of and Need for Action

This Comprehensive Conservation Plan has been prepared for the Occoquan Bay National Wildlife Refuge in Woodbridge, Prince William County, Virginia. Its purpose is to identify what role the refuge, with its biological resources, existing facilities, and educational opportunities, will play in support of the mission of the National Wildlife Refuge System, and how it will address community expectations for public use. The plan outlines intended management direction and expectations to guide operations of the site following transfer from the U. S. Army to the U.S. Fish and Wildlife Service, and for up to 15 years thereafter.

The 654-acre parcel of land formerly known as the Woodbridge Research Facility and the Marumsco National Wildlife Refuge is located near the confluence of the Occoquan and Potomac Rivers, tributaries to Chesapeake Bay. The research facility, which served as an Army communications and research center for several decades, closed its operations in September 1994 under the Base Realignment and Closure Act (BRAC). Local initiative and support led to the signing of legislation by President Clinton in September 1994, authorizing transfer of the entire facility to the U.S. Fish and Wildlife Service. As a prerequisite to accepting transfer of the Woodbridge Research Facility, the Service required that an "EPA-approved cleanup plan" be prepared by the Army. To that end, the Army prepared the document entitled "BRAC Cleanup Plan, Revised Version III." That document describes the current environmental condition of the property, and presents the Army's plan and schedule for taking appropriate environmental clean up and restoration actions.

The refuge will be managed as one of three refuges comprising the Potomac River National Wildlife Refuge Complex, located in Prince William and Fairfax Counties, Virginia. Mason Neck NWR, established in 1969 under the authority of the Endangered Species Act, was the Nation's first national wildlife refuge for bald eagles. Featherstone NWR, established in 1970, is located along the western shoreline of Occoquan Bay, south of Occoquan Bay NWR. Marumsco NWR, the freshwater marsh on Marumsco Creek, was carved from the Woodbridge Research site, and established by transfer from the Department of Defense to the Service in 1972 for its "particular value in carrying out the national migratory bird management program." The Occoquan Bay NWR will rejoin the former Woodbridge Research Facility land with that of Marumsco NWR.

As a classified Army site, the Woodbridge Research Facility has long been closed to the public. Mowed and cleared for electronics testing, the now open land contains a diversity of grassland and wetland plant species unusual in the heavily developed Potomac region. Its diverse habitats support a correspondingly high number of wildlife species, particularly migrant land and waterbirds. The Service will manage the land to provide early successional habitats and appropriate wildlife-dependent recreational opportunities, to educate visitors on the results and benefits of habitat management for wildlife, and for the enjoyment and benefit of people. Within the 10-acre, fenced compound in the center of the refuge are four buildings formerly used as research and testing facilities. Interest in these buildings is for their reuse as environmental education facilities.

The Service prepared this plan for Occoquan Bay NWR to:

- Provide a clear statement of the desired future conditions for habitat, wildlife, facilities, and people;
- Ensure that management of the refuge reflects the policies and goals of the National Wildlife Refuge System;
- Ensure the compatibility of current and future uses of the refuge;
- Provide long-term continuity and direction for Refuge management; and
- Provide a basis for operation, maintenance, and development of budget requests.

Mission of the U.S. Fish and Wildlife Service

"...provide Federal leadership to conserve, protect, and enhance the Nation's fish and wildlife and their habitats for the continuing benefit of the American people."

The Service has primary responsibility for migratory birds, endangered species, anadromous and interjurisdictional fish, and certain marine mammals. The Service also manages the National Wildlife Refuge System, the world's largest collection of lands set aside specifically for the protection of fish and wildlife populations and habitats. Over 510 national wildlife refuges provide important habitat for native plants and many species of mammals, birds, fish, invertebrates, amphibians, and reptiles. They also play a vital role in preserving endangered and threatened species. Refuges offer a wide variety of recreational opportunities, and many have visitor centers, wildlife trails, and environmental education programs. Nation-wide, over 29.5 million visitors annually hunt, fish, observe and photograph wildlife, or participate in interpretive activities on national wildlife refuges.

Mission of the National Wildlife Refuge System

"...to administer a national network of lands and waters for the conservation, management, and where appropriate, restoration of the fish, wildlife and plant resources and their habitats within the United States for the benefit of present and future generations of Americans."

On October 9, 1997, President Clinton signed organic legislation for the development and operation of the National Wildlife Refuge System. With respect to the System, it is the policy of THE UNITED STATES OF AMERICA that:

(A) each refuge shall be managed to fulfill the mission of the System, as well as the specific purposes for which that refuge was established;

(B) compatible wildlife-dependent recreation is a legitimate and appropriate general public use of the System, directly related to the mission of the System and the purposes of many refuges, and which generally fosters refuge management and through which the American public can develop an appreciation for fish and wildlife;

(C) compatible wildlife-dependent recreational uses are the priority general public uses of the System and shall receive priority consideration in refuge planning and management;

(D) when the Secretary determines that a proposed wildlife-dependent recreational use is a compatible use within a refuge, that activity should be facilitated, subject to such restrictions or regulations as may be necessary, reasonable and appropriate.

Chesapeake Bay/Susquehanna River Ecosystem Priorities

The Occoquan and Potomac Rivers significantly contribute to the Chesapeake Bay. The Chesapeake Bay watershed covers a basin of 64,000 square miles, encompassing portions of Delaware, Maryland, Pennsylvania, New York, Virginia and West Virginia. Waters from this expansive landscape flow into the largest estuary in the United States. The watershed contains an array of habitat types that support thousands of different species of fish and wildlife. The challenge to all stewards of such a diverse watershed is finding a way to ensure that all of its parts are considered in making decisions that affect the natural and human resources of the area. The following priorities are the framework for Service efforts and management in the Chesapeake/Susquehanna watershed.

Endangered Species Resource Priority–Protect, monitor and restore threatened and endangered species, and candidate species facing immediate or serious decline.

Wetlands Resource Priority–Protect and restore vegetated palustrine and riverine wetlands with emphasis on the seven areas identified in Recent Wetlands Status and Trends in the Chesapeake Watershed (Tiner 1994): Southeastern Virginia, Virginia Piedmont, Maryland Eastern Shore, Western Delaware, Virginia Upper Coastal Plain, Virginia Blue Ridge/ Appalachians, Northeastern Pennsylvania.

Interjurisdictional Fish Resource Priority–Restore and maintain self-sustaining populations of interjurisdictional/anadromous species (American shad, hickory shad, river herring, striped bass, and Atlantic sturgeon), coastal migratory fishes identified in the Atlantic Coastal Fisheries Cooperative Management Act of 1993, and those specifies for which the Fisheries Management Workgroup of the Chesapeake Bay Program has developed fishery management plans.

Non-Game Birds Resource Priority–Reverse the decline of migratory bird populations identified in Migratory Nongame Birds of Management Concern in the Northeast (Schneider and Pence 1992) including grassland species and other migrant Neotropical birds.

Waterfowl and other Migratory Game Birds Resource Priority–Restore waterfowl populations to 1970's levels by the year 2000 as identified in the North American Waterfowl Management Plan and the Chesapeake Bay Waterfowl Policy & Management Plan.

Legal Mandates

Administration of National Wildlife Refuges is governed by various Federal laws, Executive Orders, and regulations affecting land and water use as well as the conservation and management of fish and wildlife resources. Policies of the Service guiding all aspects of refuge administration are stated in its primary management documents and in the Service Manual.

The National Wildlife Refuge System Improvement Act of 1997 mandates the development of a comprehensive conservation plan for all units of the National Wildlife Refuge System, compliant with the National Environmental Policy Act and its implementing regulations. It recognizes wildlife-dependent recreation as priority public uses of refuge land.

Management is further guided by the Fish and Wildlife Act of 1956 and the National Wildlife Refuge System Administration Act that authorizes the Secretary of the Interior to permit any uses of a refuge "…whenever it is determined that such uses are compatible with the major purposes for which such areas were established."

The Refuge Recreation Act of 1962 requires that any recreational use of refuge lands be compatible with the primary purposes for which a refuge was established and not inconsistent with other previously authorized operations.

The National Historic Preservation Act of 1966 provides for the management of historic and archaeological resources that occur on any refuge. Other legislation, such as the Endangered Species Act, the North American Wetlands Conservation Act, and particularly the National Environmental Policy Act (NEPA) all provide guidance for the conservation of fish and wildlife and their habitats.

II. Refuge Management—Goals and Objectives

Refuge Purpose

The purpose of Mamrumsco NWR, which becomes part of Occoquan Bay National Wildlife Refuge, is for the "particular value in carrying out the national migratory bird program." (16 U.S.C. §667b: An act authorizing the transfer of certain real property for wildlife, or other purposes).

Woodbridge Legislation (H.R. 4453) applicable to Occoquan Bay NWR states: "(b) The Secretary of the Interior shall use appropriate parts of this real property for (1) incorporation into the Mason Neck National Wildlife Refuge and (2) work with the local government and the Woodbridge Reuse Committee to plan any additional usage of the property, including an environmental education center: Provided, that the Secretary of the Interior provide appropriate public access to the property."

Considering the above purposes as they relate to the management of Occoquan Bay NWR, they are interpreted and defined as follows:

The Purposes of Occoquan Bay NWR are:
1. As a refuge and breeding area for migratory birds, interjurisdictional fishes, and endangered species.
2. As an outdoor classroom to provide the public with educational opportunities relating to fish and wildlife resources; and
3. For other compatible recreational uses including: fishing, wildlife observation, interpretation and wildlife photography.

Refuge Vision Statement

Occoquan Bay NWR is envisioned to be a key refuge in the National Widlife Refuge System. The important grassland and wetland habitats are important to the Nation's wildlife in this highly urbanized area. Furthermore, the variety of habitat types accessible to refuge visitors and the refuge's proximity to the Nation's capitol provide unparalleled opportunities to demonstrate the role of national wildlife refuges, particularly the benefits of habitat management for wildlife.

Natural Resources–The refuge is managed for primary benefit of migratory birds and threatened or endangered species, with an emphasis on early successional habitats and wetland habitats. Habitat management is an active and interactive program which also serves as the focus for the education programs.

Visitor Use–Within an urban setting, Occoquan Bay NWR demonstrates the importance of the natural world to the human quality of life, and the human role in preserving and enhancing wildlife habitat. Local communities enthusiastically identify the area as a destination for wildlife-oriented public use that enhances the quality of life in the Potomac area. As a result of visiting the refuge, the public gains an appreciation of the co-existence of urban and natural areas. The refuge is a showcase for the Service and other resource partners for environmental education and resource management. A flexible and dynamic learning environment is created in a natural setting. Clean, safe, accessible, wildlife-compatible, and high quality experiences for diverse audiences, within the carrying capacity of the refuge, are provided.

Environmental Education (EE)–In collaboration with many partners, a wide range of innovative, stimulating, general public and environmental education programs and activities is provided. EE is the process of integrating environmental concepts and management with the educational activities of the Service. Activities such as wildlife resource programs, interpretation, outdoor classrooms, and educational assistance are provided as educational activities. When these activities deal with environmental concerns, incorporate basic ecological concepts, or focus on the role of humans in the ecosystem, they become forms of environmental education. Occoquan Bay activities are designed to promote an awareness of the basic ecological foundations for inter-relationships between human activities and the natural system. The primary objectives of the environmental education effort in the Service are to conserve and enhance our fish and wildlife resources, and to motivate citizens to learn the role of management in the maintenance of healthy ecosystems so they can effectively support wildlife conservation.

Facilities–The refuge provides safe, high-quality facilities and visitor opportunities for both Service and non-Service programs, primarily for those activities not available in nearby areas.

Refuge Goals and Objectives

The broad goals of Occoquan Bay NWR support the direction of the Service and the Chesapeake Bay Ecosystem Priorities. These goals step down the stated "Refuge Purposes of Occoquan Bay NWR" into management direction. These goals aided in the selection of the "Preferred Alternative" and the development of this final Comprehensive Conservation Plan. Each goal is supported by the measurable, achievable objectives with specific strategies and tasks needed to accomplish them. Objectives are intended to be accomplished in a 10- to 15-year time frame. Actual implementation may vary as a result of available funding.

The following are Occoquan Bay NWR goals and objectives. Accompanying each objective are its strategies, both long term and annual activities, and projects that are means to achieve the refuge objectives. Staffing and funding needed to accomplish the strategies are outlined.

Wildlife and Habitat Management

Goal I: Maintain, restore, and enhance grassland and wetland habitats to support a diversity of plants and animals.

Objective I: The refuge will maintain approximately 290 acres in grassland habitat in a variety of successional stages to maximize the potential habitat for the greatest diversity of breeding and migratory bird species.

Strategies:
- For the first 4 to 5 years, the refuge grassland acreage will be managed at approximately one-third each of 1-year, 2-year, and 3-year growth.
- Design and implement an inventory monitoring program to identify specific use of each phase of growth by breeding and migrating birds.
- To better identify plant and wildlife responses to prescribed fire management, a prescribed burn plan for small (<20-acre) sites will be implemented which will monitor and evaluate changes in plant composition as compared to areas mowed. The State of Virginia and Blackwater NWR have expressed interest in being involved in developing and implementing the plan.
- In year 4 or 5, utilize data collected from the above monitoring programs to develop a management plan that identifies the proportion of habitat in each successional stage, the method of habitat manipulation. The plan will reevaluate habitat management objectives for each unit. The plan will consider area use by migrating and breeding birds, numbers of species, species of concern, and habitat response to type of manipulation.

Objective 2: The refuge will maintain approximately 180 acres in wetland habitat in the current mix of wetland types for migratory bird species.

Strategies:
- Implement a water quality monitoring program and work with Belmont Development to ensure adequate water quantity flow to the refuge.
- Design and implement a survey of migratory bird use and nesting in the various wetland types.
- Evaluate the benefits of restoring the natural water regime to the NW are of the refuge where natural water flow has been redirected through ditching.
- Utilize data collected to develop a management plan that identifies the proportion of habitat maintained in each wetland type. The plan will reevaluate management objectives and will consider area use by migrating and breeding birds, numbers of species, species of concern, and methods for controlling water flow.

Objective 3: Encourage research that will provide needed data for improved management of Occoquan Bay and other units of the National Wildlife Refuge System.

Strategies:
- Fill biologist position to coordinate biological research.
- Design and implement the following biological surveys:
    1. Use and response by migratory birds to successional stages of grasslands.
    2. Use by migratory birds of each wetland type.
    3. Responses in plant composition to grassland mowing and burning regimes.
    4. Inventory of invasive species.
    5. Water quality and quantity monitoring.
    6. Monitoring of beaver population and effects on vegetation.
    7. Monitoring of deer population and effects on vegetation.
    8. Bald eagle use of refuge (reinstate monitoring).
    9. Changes in wildlife use along trails relative to visitation numbers.
    10. Changes in fish productivity relative to water quality in Marumsco Creek.
    11. Water quality effects upon aquatic communities.
    12. Monitoring and assessment of interjurisdictional fish species.
- Develop a volunteer program for local researchers, students, etc. to collect the biological data required above.
- Confirm and expand existing vegetation cover type maps.

- Obtain adequate computer equipment and training to operate the Potomac River NWR Complex habitat and wildlife databases.

Goal II: Prevent and control invasive species that impact native plant and animal communities.

Objective 4: The refuge will provide optimum conditions for migratory birds by maintaining the whitetail deer population withing the habitat carrying capacity.

Strategies:
- Survey deer population, work with the State of Virginia game biologist to evaluate effects of deer on vegetation and identify potential methods for managing the population.
- In FY98, begin deer population management on the refuge.

Objective 5: The refuge will maintain desired wetland diversity by evaluating the impact of beaver activity on wetland structure, composition, and water flow through refuge and by identifying and implementing potential methods for managing the population.

Strategy:
- Work with the Virginia Native Plant Society to locate the invasive plants. Methods of control or elimination will be identified by FY2000.

Goal III: Provide habitat and protection for federally listed threatened or endangered species.

Objective 7: Ensure that Bald eagles are protected on the refuge.

Strategies:
- Monitor Bald eagle use on the refuge. Minimize disturbance through adjustments in visitor access.
- Enhance eagle perching opportunities in areas of lower boat disturbance.

Objective 8: Provide habitat that supports State-listed rare species, species of Service management concern, and globally rare species.

Strategy:
- Monitor use of the refuge by the above categories of species, working with VA Division of Natural Heritage; review species' use and needs; and determine how management can support these species.

Visitor Use

Goal IV: A public that values fish and wildlife resources, understand events and issues related to these resources, and acts to promote fish and wildlife conservation.

Objective 9: Visitors will (a) know that wildlife can benefit from active management, (b) feel it is important to protect land for wildlife, and (c) actively support wildlife conservation.

Strategies:
- Within two years, evaluate and identify space, equipment, programs, and messages needed to achieve visitor use objectives.
- Identify opportunities to increase awareness of the NWRS, such as NWR week activities.
- Provide interpretation of wildlife, habitats, and habitat management techniques and benefits to wildlife.
- Develop a volunteer program for refuge programs and on-site interpretation.

Goal V: A public that values and supports the National Wildlife Refuge System.

Objective 10: Provide effective wildlife and ecosystem-based education.

Strategies:
- Assist Prince William County and other surrounding counties with refuge-related information for environmental educational curricula.
- Design and implement a training course for teachers and educators desiring to utilize the refuge.

Objective 11: Increase awareness of the NWRS and its benefits to wildlife and people.

Strategies:
- Develop volunteer program to help collect data and provide visitor assistance and interpretation.
- Cooperate in development of school curricula based on the habitats and management techniques employed on the refuge.
- Incorporate key refuge messages into teacher training to serve as a prerequisite to use of the refuge as an outdoor classroom for EE programs.
- Within two years identify other tools, space requirements, opportunities, and partnerships to increase awareness of the NWRS through Environmental Education.

Objective 12: Expand Refuge outreach opportunities.

Strategies:
- Support the activities of the Trust or other non-profit organization to function as its cooperating association and/or a "Friends of Potomac River NWR."
- Identify opportunities to increase visibility of the Service including special events, partnership activities and implementing strategies outlined in the Regional 100 on 100 campaign.

Goal VI: The provision of opportunities for high quality, compatible wildlife-dependent recreation and environmental education related to habitat and wildlife management and the historic/cultural significance of the Occoquan Bay NWR.

Objective 13: Provide opportunities for visitors to view and photograph wildlife.

Strategies:
- Upon transfer of the site to the FWS, implement Phase 1 of the public access including auto and walking routes.
- Develop and implement a monitoring program that measures change in wildlife use relative to visitation numbers in a zone of 50 ft. along trails through various habitat types. Information to be used for trail management, seasonal closures, and habitat management.
- Redirect access away from areas of active contamination clean-up.
- Incorporate responsible wildlife viewing etiquette into public and environmental education.
- Design demonstration habitat manipulations in the vicinity of public trails to inform visitors of the value of habitat management and to increase use of those areas by wildlife for improved viewing opportunities.
- Within five years, provide wildlife viewing and photography opportunities.

Objective 14: Minimize impact to sensitive wildlife and plant species on the refuge.

Strategies:
- Use existing roads and trails wherever possible.
- Review final placement of all facilities for impacts to sensitive species.
- Enact seasonal closures of trails or areas of the refuge, or manage visitation numbers, as determined by monitoring program data.

Objective 15: Provide limited opportunities for fishing on or near the refuge.

Strategies:
- Evaluate within two years, two river sites and the pond for disabled accessible fishing opportunities. Work with Belmont Development to explore the potential fishing opportunities associated with the marina or other shorefront construction.
- Pursue access to Featherstone NWR to develop high quality fishing opportunities.
- Based on evaluations and current fish contamination issues, develop an educational fishing event on or near the refuge.

- Work with State, Federal, County, and concerned citizens to improve the water quality of Marumsco Creek as a fisheries nursery area.

Objective 16: Identify history of human impacts to the site, opportunities created or lost, and future opportunities.

Strategy:
- Interpret human modification of vegetation, alteration of waterways, restoration opportunities to better understand the role of habitat management.

Administration

Goal VII: Efficient administration of functions that support and pursue the vision for the refuge.

Objective 17: Develop the operational capability to accomplish the objectives of the comprehensive plan.

Strategies:
- Create a welcoming facility that provides efficient administration and maintenance space, equipment necessary to support the programs identified, and staff to develop the programs and maintain high levels of public and educational involvement in the refuge activities.
- Within 2 years, working with partners, define the requirements of and expectations for the welcoming facility.

Objective 18: Ensure the health and safety of all refuge users, including staff.

Strategies:
- Monitor contamination clean-up from health, safety, and welfare standpoints.
- Adjust public access as required by contaminate clean-up activities.

Objective 19: Implement projects in a manner sensitive to the cultural resources of the site.

Strategy:
- Review construction projects with the Regional Historic Preservation Officer for their sensitivity to cultural resources.

Questions for Discussion

1. How can communities deal productively with issues such as disposition of the Woodbridge Research Facility? Often factions polarize unrelentingly. It is important to remember that each sees itself as the good guy and the other side as the enemy. How can leaders avoid entrenched positions?

2. One technique in resolving disputes is to try to understand the opposition's viewpoint. In this issue, one side wants habitat protection and environmental education and the other wants economic development and local control. Can the two sides reach a compromise? Are there any potential economic benefits for the community from wildlife preservation or environmental education? Are there any development possibilities that would be compatible with wildlife and habitat?

3. This exercise lends itself to role playing. Key persons to portray could be:

   - County executives, anxious to solve the area's development needs.
   - Members of the committee representing both economic and environmental interests, charged with formulating recommendations.
   - Developers, eager to add to the Belmont Development.
   - FWS personnel, eager to add to their wildlife refuge.
   - Preservationists, eager to establish parks and save wildlife habitat.
   - Members of the community representing development, environmental, and education interests.

4. How will the FWS manage the OBNWR? What are their overall management goals? How will each goal be implemented?

5. Who were the winners and losers in this issue? Could another solution have provided a compromise satisfactory to all sides? Can a wildlife refuge provide any economic benefits to a county or region?

Going Beyond the Readings

Visit our Website at <www.harcourtcollege.com/lifesci/envicases2>
to investigate land-use issues in additional regions of the United States and Canada.

unit 2

Northeast: Control of Wildlife Populations

Introduction
Readings:
Reading 1: The White-Tailed Deer: A Keystone Herbivore
Reading 2: Attitudes of Pennsylvania Sportsmen Towards Managing White-Tailed Deer to Protect the Ecological Integrity of Forests
Reading 3: White-Tailed Deer Research/Management
Reading 4: Urban Deer Contraception: The Seven Stages of Grief
Reading 5: Pennsylvania Deer Statistics 1982–1998
Questions for Discussion

Management of White-Tailed Deer in Pennsylvania

Introduction

Managing white-tailed deer (*Odocoileus virginianus*) is often exceptionally contentious and difficult. Everyone, it seems, who has any interest in the outdoors has an opinion on these animals. Management of wildlife in the United States is generally the responsibility of special state agencies. Frequently, these agencies find themselves carefully picking their way among interest groups whose goals and agendas are difficult to reconcile.

Some preservationists and hunters believe that there cannot be too many white-tailed deer. Their message to managers is to do whatever is necessary to enhance deer populations. Although they agree on this fundamental point, they agree on little else.

Preservationists want white-tailed deer, and indeed all wildlife, left alone. "Let nature take its course," the die-hards cry. They delight in seeing large numbers of nearly tame animals at every turn. They want populations large, healthy, and contented.

Hunters also want large numbers of white-tailed deer, but for different reasons. They want long hunting seasons, large bag limits, and big trophies. Perhaps no other game animal is as important to American hunters as white-tailed deer. Nationwide, 80 to 90 percent of hunting involves this species. Businesses such as sporting-goods stores, guides, and hotels and motels cater to and side with hunters. These interests recognize tremendous economic potential—billions of dollars annually—from white-tailed deer hunting. They reason that more deer means more hunting and more profit.

Not everyone wants large numbers of white-tailed deer. Farmers worry about crops. Indeed, farmers can lose two-thirds of a given crop to white-tailed deer. Nursery operations are particularly hard hit; young sprouting plants are tempting food when deer populations are high and food is limited.

Foresters, too, often see white-tailed deer as pests. Commercial forests must be periodically harvested and then reforested. White-tailed deer eat young trees, retarding reforestation efforts. If deer populations are too high, reforestation fails completely. Some species of trees are more susceptible to browsing than others; thus white-tailed deer can adversely affect tree diversity. Forests with few species are more susceptible to insects and diseases. In economic terms, high numbers of white-tailed deer can mean low profits.

White-tailed deer are sometimes problems for motorists. Nothing changes a person's attitude toward deer management faster than hitting one at 60 mph (100 km/h). This is not a trivial problem. Pennsylvania motorists are involved in more than 40,000 deer–auto accidents each year. Human lives are lost. Cars are totaled. Car repairs cost hundreds of dollars. In Pennsylvania such statistics adversely affect auto insurance rates.

Farmers, foresters, and motorists urge managers to hold white-tailed deer within manageable numbers. That translates to "numbers low enough that my resources are not affected."

State managers find themselves in the middle of these groups. Their first responsibility is intelligent resource management. This means stable populations that will stay healthy into the future. Doing so requires considerable technical skill. Managers need to know a great deal about ecological relationships, population dynamics, animal diseases, animal nutrition, statistics, computer modeling—the list could go on and on. Technical knowledge is not enough. Managers also need to somehow satisfy the varied interests of a critical general public. Modern management practices fall into three general areas: (1) manage populations, (2) manage habitat, and (3) manage people. The first two are often considerably easier than the last.

Background: Biology of the White-Tailed Deer

The white-tailed deer is perhaps the most widely studied animal in the world. White-tailed deer are unusually widespread throughout North and Central America. They range from the Arctic to the tropics, generally preferring open forest. They also roam deserts and swamps, in grasslands and in deep forest, along seashores and at high elevations. They are a wilderness species that can tolerate suburbs and even large urban parks.

White-tailed deer are as varied in food preferences as they are in habitat choices. They eat more than 100 species of plants. In the East, acorns and dogwood fruit are favorite foods. If these foods are not available, they will eat almost any other plant, from seaweed to mushrooms to evergreen boughs. They are a bane to farmers, who can hardly plant a crop the deer won't eat. White-tailed deers' favorite crops seem to be corn, soybeans, and young trees. They have an uncanny ability to choose nutritious food. Given a choice, they will eat fertilized rather than unfertilized green beans.

White-tailed deer breed in the fall. Under favorable conditions, they are prodigious reproducers. When the environment is right, female fawns will breed when only six or seven months old, and 10 to 15 percent will have twins or triplets. Under adverse conditions, females shut down reproduction, fail to ovulate, and thus save themselves until conditions improve. The white-tailed deer population has generally been increasing during the last century. In precolonial days, the total population may have numbered 40 million. Excessive hunting in the nineteenth century reduced numbers continent wide to around 500,000 in 1908. Today, the total white-tailed deer population approaches 30 million.

The Problem

Perhaps nowhere in the mid-Atlantic region have white-tailed deer been managed more successfully than in Pennsylvania. All of the aforementioned interest groups are well represented in the state. Many are politically active and vociferous. Furthermore, white-tailed deer are unusually important to the state. Deer hunters currently exceed one million—one of the largest populations in the United States. Hunting is an important industry in the state, but so are farming and forestry.

The state has a long history of managing white-tailed deer. Pennsylvania enacted its first Game Law in 1721 to protect these deer. In 1895 the state legislature created the Board of Game Commissioners, whose major responsibility is managing white-tailed deer. In the ensuing century, the numbers of white-tailed deer in Pennsylvania followed North American trends, from dangerously low to unacceptably high.

Current management efforts focus on reducing the overall population to sustainable levels. Hunting has been the method of choice in controlling deer. Success requires close cooperation between managers and hunters. During the late 1980s it appeared that the partnership was working, but in the early 1990s something apparently changed. The following readings illustrate some problems faced by the Pennsylvania Game Commission in managing white-tailed deer.

"The White-Tailed Deer: A Keystone Herbivore"

Donald M. Waller and William S. Alverson

Wildlife Society Bulletin, 1997, 25(2): 217–226. Reprinted by permission.

Issues Surrounding Deer Management

During the last 3 centuries, sweeping manipulations of habitat for agriculture, silviculture, and, to a lesser degree, game management have improved and expanded habitat for white-tailed deer (*Odocoileus virginianus*) across much of the landscape in the eastern United States. For most of this century, wildlife managers sought to protect and enhance populations of deer. With the specter of extirpation still haunting their memory, wildlife managers worked hard in the early 20th century to devise and enforce bag limits, short hunting seasons, and buck-only hunts in order to protect the recovering herds. As they professionalized, wildlife managers were quick to follow Leopold's (1933) suggestion that the way to manage game is to manage habitat. For white-tailed deer, this meant favoring edge and early successional habitats by creating gaps and grassy openings in regions dominated by mature forest. Clear-cuts, in particular, continue to be promoted for their immediate production of slash for browse and their ready succession to shade-intolerant species such as aspen (*Populus* spp.) that provide good summer browse, at least for a few years (e.g., Masters et al. 1993, Johnson et al. 1995).

Nearly a decade ago, we warned wildlife professionals and conservation biologists about the ecological consequences of overabundant deer populations (Alverson et al. 1988). Since then, deer populations and their ecological and economic impacts appear to have increased and worsened. As we approach the next millennium, it behooves wildlife managers to contemplate what consequences will result from their actions in their own professional lifetimes. We hear more each year about the high costs of crop and tree-seeding damage, deer–vehicle collisions, and nuisance deer in suburban locales (Conover et al. 1995). Beyond these substantial economic costs, however, we face new and often vexing issues regarding the ecological costs of overabundant deer. In some cases, it appears that these consequences will extend over decades and perhaps even centuries. This makes it even more important now than it was a decade ago that wildlife managers assume responsibility and take action to minimize the ecological effects of chronically overabundant deer populations.

The wildlife management profession has begun to respond to the economic and ecological impacts of overabundant deer. In 1995, Wisconsin's Bureau of Wildlife elected to scrutinize the impacts of its own deer management policies via a comprehensive Environmental Assessment (VanderZouwen and Warnke 1995). This was the first attempt, that we are aware of, to seriously consider the broad range of ecological and environmental impacts pertaining to a state's deer management policy. Partly in response to this assessment, the Wildlife Bureau also began to institute new hunting regulations in some areas to ensure that more, especially more female, deer were killed (e.g., their 1996 "Earn a Buck" program). This, however, has proved to be a major challenge, both because it is difficult to adjust hunter effort as more areas become off-limits to hunting and because hunters favor a tradition and management they see as contributing to, rather than diminishing, their prospects for personal hunting success. Many hunters remain skeptical about the seriousness of deer impacts, or at least the need to reduce deer densities in their own areas (Diefenbach et al. 1997). Thus, in addition to facing the irony of having done their job too well, wildlife managers must now muster the effort to document, and publicize, the negative consequences of overabundant deer if they are to effectively influence hunter effort, which is their primary tool for adjusting deer density.

While managers understand that they can boost deer and other game populations by manipulating habitat, they have been slower to acknowledge the converse, i.e., that managing for abundant deer brings reciprocal effects for their habitats. If our current deer densities adversely affected only a few particularly sensitive species, or if these effects only occurred intermittently (during peaks in deer abundance), or locally (say, in deer yards), then the ecological issues they posed could be addressed via focused and proximate efforts, or perhaps even dismissed as not being a major management issue. Allocating large efforts to document and ameliorate such scattered impacts would appear misguided and wasteful, and traditional approaches to herd management would appear well justified. If instead, however, current deer densities substantially affect many species, and if their impacts are geographically widespread and chronic, then wildlife managers face a different and more serious set of issues (Garrott et al. 1993). In particular, they face the immediate need to accurately monitor and assess the range and nature of these impacts, and, simultaneously, to reset management goals (and perhaps redesign management techniques) to substantially reduce the severity and scale of these impacts.

Here, we review evidence for the contention that chronically high densities of white-tailed deer are having multiple, and often substantial, deleterious ecological impacts across many regions. To structure this review, we specifically consider whether deer are acting as a "keystone" herbivore to substantially alter ecological communities (Paine 1969). We define a keystone species as one that: (1) affects the distribution or abundance of many other species, (2) can affect community structure by strongly modifying patterns of relative abundance among competing species, or (3) affects community structure by affecting the abundance of species at multiple trophic levels. Power et al. (1996) added that keystone species were expected to have disproportionately large impacts on communities. The concept of a keystone species was originally applied to carnivores that affected the relative abundance and competitive interactions among their prey; now, however, most ecologists accept the idea as pertaining to species on any trophic level (Hunter 1992, Paine 1995). With this in mind, we briefly review past and current efforts to assess the nature and severity of the ecological impacts of deer. Considering our 3 criteria for a keystone species, we discuss the impacts of deer on tree seedlings (criteria 1 and 2 above), herbaceous plants (criteria 1 and 2), and species on higher trophic levels (criterion 3). Because we and others have already reviewed much of the older literature elsewhere (Alverson et al. 1988, Warren 1991), we concentrate here on more recent results. While these data remain far from comprehensive, we conclude that ample evidence exists to publicly acknowledge the substantial risks posed by sustaining high deer densities. We therefore conclude by discussing the larger management issues these results raise.

Effects on Trees and Shrubs

Wildlife biologists and foresters have known for many years that deer can strongly affect the absolute and relative abundance of woody species (e.g., Leopold et al. 1947, Webb et al. 1956). Such research in trees is both straightforward (involving the tabulation of size or age classes) and of practical importance, given the economic value of trees. Indeed, these effects are so widespread that forestry textbooks have routinely mentioned deer browse as a problem in regenerating particular species (e.g., oaks [*Quercus* spp.]) for years (e.g., Allen and Sharpe 1960).

Some of the best information on tree impacts comes from the Allegheny National Forest in northwestern Pennsylvania, where foresters have long been concerned that high deer densities depress the regeneration of several valuable hardwood species to well below acceptable stocking levels (Harlow and Downing 1970, Marquis 1974, Marquis 1981, Tilghman 1989). Marquis (1975) noted that this region has been heavily browsed since recovery of the deer herds in the 1930s. Suppression or elimination of palatable seedlings and saplings results in a slow but steady conversion of the stand to less-palatable species such as American beech (*Fagus grandifolia*), which is thus given a competitive advantage. Whitney (1984:403) concluded that deer are "one of the more important determinants of forest structure in the Allegheny Plateau over the past 50 years." At very high deer densities and under certain conditions, the seedlings and saplings of all tree species are eliminated and stands with park-like, grass and fern-dominated understories emerge. Such conditions appear doubly troubling for tree seedlings in that the ferns themselves interfere with the germination, growth, and survival of desirable tree seedlings (Horsley and Marquis 1983), and so may extend the indirect effects of deer browsing on tree regeneration.

Quantitative studies of the effects of deer on the regeneration of tree species and resulting changes in forest composition are becoming common in other areas as well. Robertson and Robertson (1995:68) studied Pennypack Wilderness, a 324-ha natural area northeast of Philadelphia, and concluded that "the striking lack of regeneration by species destined for position in the canopy presages significant structural and compositional shifts in the forest as existing canopy trees die." In suburban areas like this, adverse effects of deer browsing are compounded by the invasion of woody and herbaceous exotic plants and direct human impacts. In a comprehensive assessment of the impacts of deer browsing on forests in central Illinois, Strole and Anderson (1992:141) noted that "deer took a disproportionately large amount of browse from relatively uncommon species" such as white oak (*Quercus alba*) and shagbark hickory (*Carya ovata*). Similarly, in a 10-year study of upland beech-maple (*Fagus-Acer*), lowland ash-ellm (*Fraxinus-Ulmus*), and young pin oak (*Q. palustris*) forests in Ohio, Boerner and Brinkman (1996:309) concluded that "deer browsing was more important than environmental gradients or climate factors in determining seedling longevity and mortality." In studying eastern white pine (*Pinus strobus*) at the southern limit of its range in southwestern Wisconsin, Ziegler (1995) noted that although pines have been present for >12,000 years with self-replacement occurring through 1948 (McIntosh 1950), regeneration failures attributable to browsing are now occurring in several stands.

Slow-growing conifers like eastern hemlock (*Tsuga canadensis*) appear particularly sensitive to the sustained effects of deer browsing. Hough (1965) examined changes over a 20-year interval in a 1,620-ha virgin hemlock-hardwood stand in the Allegheny Mountains and noted that deer arrest typical patterns of succession by eliminating the advance regeneration of hemlock seedlings. Similarly, in the upper Midwest, Anderson and Loucks (1979) examined demographic profiles on hemlock and sugar maple in mixed stands and concluded that differential browsing by deer was favoring sugar maples over hemlocks in the smaller size classes. In a follow-up study, Anderson and Katz (1993:203) inferred from exclosure data that exclosure "periods as long as 70 years may be required for shade-tolerant trees to achieve a size class distribution characteristic of all-aged forests." Working in the Porcupine Mountains in Upper Michigan, Frelich and Lorimer (1985) also documented dramatic differences in hemlock regeneration between sites with higher and lower deer densities and predicted that hemlock would be essentially eliminated from stands if browsing levels did not change.

Our own recent research on stands of eastern hemlock and northern white cedar (*Thuja occidentalis*) in northern Wisconsin and western Upper Michigan has confirmed the significant role that deer play in affecting patterns of seedling recruitment (Waller et al. 1996; Fig. 2). We are testing various hypotheses (see Mladenoff and Stearns 1993) that might explain regional patterns of failed regeneration of these 2 ecologically important trees. The presence of small hemlock seedlings at most of our study sites suggests that poor seed production or unsatisfactory sites for initial germination are not limiting seedling recruitment. Other factors traditionally thought to influence seedling abundance, such as stand composition, light, and substrate, also affect the probability of initial seedling establishment. However, numbers of seedlings in larger size classes are for the most part independent of local abiotic conditions and stand composition, but vary strongly across land ownerships in parallel with local deer abundance, as estimated by pellet counts or the browse observed on sugar maple saplings. Results from our experimental exclosures confirm that regional differences in deer abundance account for most of the differential in growth between hemlock seedlings in and outside of these exclosures (Alverson and Waller 1997).

Like hemlock, northern white cedar is a slow-growing, late-successional conifer that often fails to regenerate in northern forests. In response to these failures, the U.S. Forest Service declared a regional moratorium on cutting this important timber species. Conspicuous browse lines, evident in many stands, suggest that deer are an important factor in these failures. Like hemlocks, small cedar seedlings occur disproportionately on "nurse" logs or stumps but suffer high mortality and show few associations with this substrate when older (Scott and Murphy 1987). In a detailed study of a swamp forest, Blewett (1976) concluded that deer were the major limiting factor. We also observe that the size distributions of cedar seedlings differ greatly among nearby ownerships that differ conspicuously in deer density.

Canada yew (*Taxus canadensis*), mountain maple (*Acer spicatum*), yellow birch (*Betula alleghaniensis*), and mountain ash (*Sorbus* spp.) all decreased in apparent response to increasing deer densities in northern Wisconsin (Balgooyen and Waller 1995), as expected

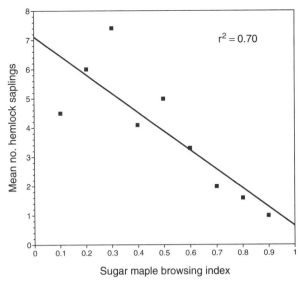

Fig 2. Declines in the abundance of eastern hemlock (Tsuga canadensis) seedlings with increases in local deer browsing. These data reflect the mean abundance of 30–99 cm hemlock seedlings within 14 ¥ 21-m study plots averaged over all study sites within a given browse index class. The browse index at each site was computed from the average proportion of sugar maple (Acer saccharum) twigs between 20 and 180 cm high observed to be browsed by deer. Data from 1996 field season (Rooney and Waller unpubl. data).

given deer browse preferences (Dahlberg and Guettinger 1954, Stiteler and Shaw 1966, Williamson and Hirth 1985). The effects of deer on shrubs such as hobblebush (*Viburnum alnifolium*) are also well documented (Hough 1965, Whitney 1984). Canada yew, an understory shrub of northeastern forests, has long been known to be favored by deer as browse (Beals et al. 1960, Foster 1993). Allison (1990a,b) documented that deer browsing directly affected vegetative cover by yew. He determined also that deer indirectly affected yew's abundance by eating male cones, thereby causing pollen limitation and reducing seed production and reproductive success. *Torreya taxifolia,* a closely related, rare shrub endemic to northern Florida, already experiencing rapid population decline due to fungal infections (Schwartz et al. 1995), is suffering additional losses to antler rubbing by locally abundant deer populations (M. Schwartz, Illinois Nat. Hist. Surv., Urbana, pers commun.).

Effects on Herbaceous Plants

Most of the plant diversity within our forests exists not as trees but rather as herbaceous understory species. Most of these herbaceous plants never grow above the "molar zone" of browse susceptibility and thus are subject to the threat of repeated grazing. In addition, herbaceous plants constitute the bulk of deer summer diets (87%; McCaffery et al. 1974). Finally, whereas the early abundance of tree seedlings and saplings is often discernible through the legacies they subsequently write in the canopy, deer grazing may obliterate herbaceous species without a trace, thereby making it difficult to infer the importance of this factor. Thus, we need careful studies to determine how and where deer may be changing the relative abundances of our native herbs. Although there are few studies on the impacts of deer on herbaceous species, the recent work reviewed here suggests that deer may have substantial and pervasive effects on many herbaceous communities.

Miller et al. (1992) recently compiled reports from both the literature and phone interviews of deer herbivory on 98 rare species. Surprisingly high proportions of rare lilies (40%), orchids (39%), and dicots (56%) were reported to be adversely affected by deer. (They received no reports of adverse effects on rare graminoids.) Many of these accounts were disturbingly dramatic, including the loss of all flowering stems of Loesel's twayblade (*Liparis loeselii*) to grazing in 1984 at an important site for this species in Kentucky, and all flowering stems of the only known population of white fringeless orchid (*Platanthera integrilabia*) in 1989. There were also many accounts of local extirpations caused by deer browsing. Miller et al. (1992) report that such impacts have grown so severe that many conservatory preserves now routinely fence rare or valuable herbs on their property, copying the practice of foresters and orchardists.

Augustine (1997) also noted strongly selective deer preferences for lilies and other monocots in 4 old-growth remnant forests in southeastern Minnesota. He noted that species of Trillium and Uvularia (both Liliaceae) were favored even after they became scarce. He concluded that deer population densities (which in his study area ranged from 10 to 50/km^2) were a crucial issue affecting the conservation and restoration of forest remnants. In Illinois, Anderson (1994) studied the response of *Trillium grandiflorum* to deer browse and proposed that its height could be used to indicate local browse pressure. Inferring from these data, he recommended a density of <4–6 deer/km^2 to retain viable populations of this and similarly palatable herb species. Such selective herbivory also implies that favored species may be overgrazed even when other species show few signs of grazing.

Detailed historical comparisons can provide insight into the dramatic impacts of deer on herbaceous community structure. For example, Rooney and Dress (In Press) observed that hemlock and hemlock-beech stands in Heart's Content, an old-growth forest in northwestern Pennsylvania, lost 59% and 80% of their ground flora species, respectively, between 1929 and 1995. They attributed these losses to the direct and indirect effects of deer herbivory and recommended reducing fern abundance, erecting an exclosure, and reintroducing extirpated species to protect this valuable virgin forest remnant. Drayton and Primack (1996) and Leach and Givnish (1996) noted similarly catastrophic losses from the local floras of Middlesex Fells, outside Boston, and prairie fragments in southern Wisconsin, but inferred no conclusions regarding the role of deer. It is also conceivable that deer could indirectly threaten the persistence of herbaceous species by removing flowers or flowering individuals and so reduce pollen availability and seed set (as with yew) or by reducing populations of pollinators and thus pollinator service (Kearns and Inouye 1997).

In the absence of historical information, inferences about the impacts of deer on herbs are usually drawn by comparing areas known to differ in deer density. Exclosures provide an experimental method to create appropriate controls, but if they were not established before deer became numerous, then they may be of little use as herbs may have already been extirpated. Perhaps more useful, then, are the 'natural experiments' provided by local or regional variation in ambient deer densities. Rooney (1998), for example, compared the density, size, and reproductive condition of a susceptible lily, Canada mayflower (*Maianthemum canadense*), growing on both high and low boulders in the Allegheny National Forest, reasoning that boulders ≥2 m high would act as refuges from deer browsing. He found significantly higher densities (3 times higher than on low boulders), larger plants, and more frequent flowers on high boulder tops in this species, although a second, control, species not susceptible to browse (*Oxalis montana*) showed no such differences.

Balgooyen and Waller (1995) exploited regional variation in deer densities in northern Wisconsin and the Apostle Islands of Lake Superior to infer both shorter-, and longer-term effects of deer browsing on woody and herbaceous communities. We noted declines in several woody plants, overall herbaceous species diversity, and specific declines in wild sarsaparilla (*Aralia nudicaulis*), Canada mayflower, and bluebead lily (*Clintonia borealis*). Whereas the impacts on woody species appeared to be reversible in this study, adverse effects on overall herbaceous diversity persisted for >30 years, with Clintonia apparently being extirpated from Madeline Island. Historical effects also were complex in that reductions in deer density allowed recovery of yew on some islands to the point where this coniferous shrub shaded out some herbs. In aggregate, however, both the current and historical effects of deer were strongly negative. Together, these data implied that densities <4.5 deer/km^2 in this region were most compatible with retaining a full complement of herbaceous species. This study was unusual in that it provided a glimpse of how long deer impacts might persist.

Thus, deer have substantial impacts on both particular herbaceous and woody species and overall plant community structure (Fig. 3). For those areas where they still persist, both trillium and bluebead lily serve as convenient indicators of the intensity of deer herbivory. Further long-term monitoring of these and other species could give us a much clearer view of these effects, particularly if steps were taken to assess variation in the intensity and duration of herbivory. Experimental exclosures and reintroductions coupled with models of the long-term effects of shifts in competitive relationships and community structure would also further our understanding. However, the species-specific nature of these responses implies that we may

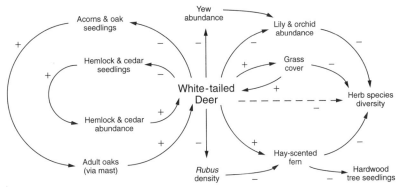

Fig 3. Documented and inferred interactions of white-tailed deer with various plant species in forests of the northeastern and midwestern United states. Further interactions with species on other trophic levels are not shown (but see Ostfeld et al. 1996). For sources, see text.

need a considerable body of research to fully assess the range of responses by herbs (and woody plants) to deer herbivory.

Effects on Other Trophic Levels

Our last criterion for determining if white-tailed deer might be considered a keystone herbivore was that their impacts extend to other trophic levels. Such cascading trophic interactions have been found to be important in aquatic ecosystems (Carpenter 1988).

In his experimental study of the effects of contrasting deer densities on bird populations in Pennsylvania, deCalesta (1994) found that intermediate-canopy-nesting birds declined 37% in abundance and 27% in species diversity at higher deer densities. Five species dropped out at densities of 14.9 deer/km^2 and another 2 disappeared at the highest density of 24.9 deer/km^2. In contrast, such deer densities did not consistently affect the diversity or abundance of small mammals at these sites (deCalesta, unpubl. rep.).

McShea and Rappole (1992) noted multiple effects at other trophic levels, which prompted them to label deer a keystone species. Casey and Hein (1983) documented losses of 3 bird species in a wildlife-research preserve stocked with high densities of deer, elk (*Cervus canadensis*), and Mouflon sheep (*Ovis aries*). The New Hampshire Natural Heritage Inventory Program similarly reported that deer depressed populations of the Karner Blue Butterfly (*Lycaeides melissa samuelis*), a federally endangered species, by browsing on its lupine host plants (reported in Miller et al. 1992). While expecting deer effects primarily on ground-nesting birds, McShea (1997) found multiple effects on bird species nesting at several levels in the forest, apparently reflecting complex interactions with squirrels (*Sciurus* spp.) and acorn crops. By competing with squirrels and other frugivores for oak mast, deer further affect many other species. For example, squirrel abundance affects rates of bird-nest predation, whereas mice influence the abundance of gypsy moths (Ostfeld et al. 1996). In reviewing these complex interactions in eastern deciduous forests, Ostfeld et al. concluded (1996:327) that "the complex interactions we have uncovered suggest that it is likely to be difficult or impossible to manage simultaneously for multiple uses (both recreational and industrial) of the forests."

These multiple trophic level interactions may extend to include our own species. Elevated deer populations are thought to contribute to the incidence of Lyme disease in humans via the complex interactions of deer ticks (*Ixodes*), white-footed mice (*Peromyscus leucopus*), and spirochaetes (*Borrelia*) that affect the survival and transmission of the disease (Wilson et al. 1985, 1988, 1990; Deblinger et al. 1993; Conover et al. 1995; see also http://www.lymenet.org).

The Deer Dilemma

The accumulating body of evidence briefly described here clearly points toward the conclusion that white-tailed deer have reached, and sustained, densities across much of the eastern, northern, and southern United States sufficient to cause manifold and substantial ecological impacts. Furthermore, deleterious impacts on biotic communities have been noted for more than half a century (Leopold 1946). As noted by Power et al. (1996), keystone species are not always of high trophic status, often exert their effects through ecological interactions other than direct consumption, and often only in particular contexts. Impacts of keystone species may also require decades, or centuries, to be fully manifest (Terborgh 1986). By any and all of the criteria we use to define them, white-tailed deer must be accepted as a keystone herbivore in eastern deciduous forests at this time.

Despite numerous studies, over many years and areas, we are only beginning to appreciate in aggregate the biological costs of deer browsing. It is already clear, however, that they are quite large—probably much larger than most wildlife managers expected—and cannot be ignored. Most investigators who have tested for impacts of deer browsing have found them. Many of these impacts have been more dramatic than expected and often more indirect and subtle than expected.

Deer browsing clearly favors some species, e.g., hay-scented ferns (*Dennstaedtia punctilobula*) while eliminating others, e.g., Canada yew, often with cascading effects as these intermediary species compete strongly with tree seedlings and herbs. Whereas we have evidence already for these particular interactions (Fig. 3), we can hardly imagine all the ways in which deer browsing will affect patterns of forest diversity once significant shifts in canopy composition start to occur in response to prolonged failures of regeneration of dominant tree species.

Despite the fact that pernicious deer effects on some plants, birds, and mammals have been manifest for decades, we have been slow in recognizing these effects and appreciating their magnitude. This explains, but only in part, why we have been slow to acknowledge the scale of these impacts and to incorporate this knowledge into more aggressive herd management. With inadequate monitoring, it has been easy to remain (blissfully) unaware of many deer impacts. Yet foresters, orchardists, and other landowners suffering economic losses have been acutely aware of these deer impacts. They often have chosen, however, to address these impacts via proximal solutions, such as browse-proof tubes around individual seedlings and fencing of small areas. Unfortunately, these methods are expensive, limited in extent of application, and do not protect nontarget species.

Taking a larger view, it is evident that our view of deer ecology has centered on Odocoileus virginianus itself rather than the web of species it affects. For example, whereas studies of deer forage and range condition are common, studies of collective and cumulative ecological impacts are not. Likewise, we have found it difficult to develop a perspective that transcends local conditions and recent history. We myopically tend to see the elevated deer populations of recent decades as the biological norm, without reference to the conditions under which forest species evolved and adapted over far longer periods of time. Furthermore, the changes deer bring, although chronic and pervasive, are often slow in developing and subtle in appearance. Our ability to detect these subtle changes is impaired by the lack of natural or actual experimental controls for contrast as our landscapes have become increasingly homogenized and subject to high deer populations.

All of this complicates the job of wildlife managers considerably. An originally simple professional mission to accept responsibility and funds to protect the deer herd and provide a steady supply for the hunt has become far more complex. Deer management has forcefully injected wildlife managers into the unexpected and more difficult role of managing habitats and ecosystems. If managers accept responsibility (and income) for managing deer herds, then they must also accept the corresponding duties to understand deer impacts and ameliorate the adverse biological side effects of high deer densities when and where those occur.

Proposals

We propose 2 courses of action to address the increasing responsibilities of deer managers. The first is to commit firmly to expanding research and monitoring of deer. No manager should be expected to set target deer densities without basic and continuing information on the direct and indirect impacts of management actions. Such monitoring will be especially effective if we can devise efficient and reliable indicators capable of serving as "early warning signs" of impending ecological change. Many of the studies reviewed above are aimed at providing such indicators and could be extended and cross-calibrated. Monitoring efforts should also focus on those elements of the ecosystem likely to be most sensitive to high deer densities, including species such as those whose populations are lowered or eliminated (e.g., orchids) and those whose populations may be inflated (e.g., *Ixodes* and *Borrelia*) so as to affect other species. By necessity, such monitoring should be broad in taxonomic scope because birds, mammals, herpetiles, and invertebrates, as well as herbaceous and woody plants, may be affected. In short, accurate and convenient indicators of ecological impacts should be incorporated into routing management activities.

To accomplish all this, we propose that some reasonable fraction of license revenues (approx 5–10%) specifically be dedicated to devising and implementing comprehensive monitoring programs, including the associated research such programs require. It might also be possible to permit more use of Pittman–Robertson Act funds for this kind of monitoring and research. Although such programs will not be simple to establish, they are clearly needed if we are to have the accurate and timely information we need to pursue adaptive management (Holling 1978, Walters 1986, Nielsen et al. 1997). Fortunately, an increasing cadre of wildlife and conservation biologists appear qualified and eager to participate in these efforts.

Establishing and funding an expanded program for monitoring deer impacts would greatly enhance the information base for managers (as well as providing convincing evidence to hunters and the general public of the extent and severity of such impacts). These efforts must be accompanied, however, by an enhanced ability to use this information to set and adjust deer densities across our contemporary landscapes. Therefore, the second course of action we recommend involves extending our abilities to raise or lower deer populations at various scales. This is, of course, a necessary skill if we are to pursue adaptive management in response to changes in conditions over space and time. It is not yet clear, however, which management practices will best allow managers to manipulate and control deer densities, even locally and temporarily. Although traditional programs for deer management served effectively to protect recovering deer herds and satisfy hunter de-mand, they have not yet served effectively to correct the excesses of this success. Further experimentation is needed to assess how effectively, and at what scales, deer densities may be reduced by increasing deer kill or by running conventional habitat management "in reverse." We currently lack the technical means to assess population densities at intermediate scales, e.g., townships. We also find that the primary tool for adjusting deer densities (i.e., hunter effort) is heavily constrained by local restrictions on hunting and hunter preferences.

Most state programs assess and manage deer densities over the rather broad geographic scales included in deer management units. Target deer densities are often set at about half the estimated carrying capacity in order to maximize deer population growth rates and, thus, the yield of harvestable animals. Managers often seek to reach such targets throughout each deer management unit and often uniformly among units as well. The homogenous deer densities such management brings, however, may not permit sensitive plant and animal species to escape, even locally or temporarily, from lethal deer-density conditions. At the same time, pervasive changes in contemporary landscapes in favor of edge and disturbed habitats also serve to elevate and homogenize deer densities. Such conditions may well represent a radical departure from historical and prehistorical conditions when deer densities presumably were lower and varied widely in response to landscapes dominated by late-successional species, with more continuous canopy cover, and higher densities of top carnivores.

Sixty years ago, wildlife professionals followed the path blazed by Aldo Leopold, embracing the power of habitat manipulation as a fundamental tool in game management. In the years since then, our awareness of ecological interactions, ecological losses, and our uncertainty have expanded greatly (Alverson et al. 1994). At the same time, our concepts of "wildlife" have broadened to include many more of the organisms sharing a given habitat. The challenge facing wildlife managers today and in the 21st century is to produce and encourage game production in some areas of the landscape without reducing the diversity of wildlife in the broadest sense that share this landscape. These broad conservation goals were enunciated clearly by Leopold in his later years, culminating in "The Land Ethic" (Leopold 1949:259). We join him in asking: "Can management principles by extended to wildflowers?"

Acknowledgments. We thank R. Warren for the invitation to prepare this piece and T. Rooney and R. Warren for their helpful suggestions on an earlier draft of this paper. Support for our field work on hemlock and cedar has come from the NSF (DEB BSR-9000102), the USDA Forest/Rangeland/Crop Ecosystems program (Award #93-00648), and the W.F. Vilas Trust Estate.

"Attitudes of Pennsylvania Sportsmen Towards Managing White-Tailed Deer To Protect the Ecological Integrity of Forests"

Duane R. Diefenbach, William L. Palmer, and William K. Shope

Wildlife Society Bulletin, 1997, 25(2): 244–251.

Abstract   We analyzed 13 years of white-tailed deer (*Odocoileus virginianus*) population data for Pennsylvania and conducted a mail survey of Pennsylvania hunters to compare trends in deer populations to the opinions of hunters on deer management. We used deer harvests, estimates of deer population parameters, and forest inventories, for 1982–1994, to monitor deer densities relative to overwinter carrying capacities (deer-density goals). The overwinter deer density that Pennsylvania forests could support without adversely affecting tree regeneration declined, statewide, from 23 to 21 deer/259 ha (1 mi.^{2}) of forest between 1978 and 1989. In contrast, deer densities statewide peaked at 34 deer/259 ha of forest in 1987 and declined to 29 deer/259 ha of forest by 1994 (41% above goal). The decline in deer densities was the result of harvest regulation changes designed to increase the antlerless harvest. However, a 1995 hunter survey revealed 66% of hunters believed deer populations were too low in the area where they did most of their hunting. The majority (44%) of hunters agreed that antlerless permits should be reduced, and 19% believed they should be eliminated. The majority of hunters agreed that controlling deer populations was necessary (87%), that deer populations should be kept in balance with natural food supplies (89%), and that deer affected plant and animal communities (56%). The majority disagreed that damage to Pennsylvania forests by deer was a problem (57%) or that deer caused serious conflicts with other land uses (44%). Support from sportsmen for programs to reduce deer populations and protect forest ecosystems will require that hunters understand the adverse ecological effects of too many deer on forest communities. In the meantime, support for Pennsylvania Game Commission management recommendations will be needed from other special-interest groups, including farmers, foresters, conservationists, the timber industry, and the scientific and conservation communities.

White-tailed deer (*Odocoileus virginianus*) are a dominant species of forested ecosystems in the eastern United States. In high densities deer may change the tree species composition of forests (Tilghman 1989) and the species diversity and abundance of herbaceous understories (Hough 1965), and affect other species of birds and mammals (Healy et al. 1987, Brooks and Healy 1989, DeCalesta 1994). Deer were nearly extinct in Pennsylvania by 1900, but populations recovered as protective laws were enforced and areas that had been logged for timber regenerated (Kosack 1995). Deer benefited from habitat fragmentation, elimination of native predators (Noss and Cooperrider 1994:204–205), and management activities of state wildlife agencies. The interspersion of woodland and cropland has been especially beneficial. In 1995, Pennsylvania averaged 30 deer/259 ha of forested habitat following the hunting season (Pa. Game Comm., Harrisburg, unpubl. data).

Deer are an economically important game species in Pennsylvania, and deer hunting results in >$245 million annually in retail sales and $122 million in wage earnings (Southwick Assoc., Int. Assoc. Fish and Wildl. Agencies, unpubl. rep., 1994). The Pennsylvania Game Commission (PGC) is responsible for managing the state's birds and mammals, including the white-tailed deer (34 Pa. Consolidated Statutes). Traditionally, this has meant protecting and managing deer to maximize hunter recreation, although in recent years the agency has had to address problems caused by deer-human conflicts (Kosack 1995). Each year, >40,000 deer are reported killed on Pennsylvania highways (PGC, Harrisburg, unpubl. data). Witmer and deCalesta (1992) reported that the average repair bill (in 1993 dollars) was $1,100–2,200 for a deer–vehicle collision, which, when multiplied by 40,000 road-killed deer, equaled $44–88 million annually. Also, suppressed timber regeneration, damage to farm crops, and conflicts with suburban and urban landowners are problems the PGC is trying to resolve. Estimates of crop damage and damage to urban landscapes are unavailable, but Marquis (1981) estimated that annual timber losses to deer browsing within Pennsylvania's Allegheny hardwood forest may be $367 million per year.

The PGC's deer management program has several goals: (1) provide a sustained harvest of deer for hunters, (2) keep deer populations in balance with their natural food supplies, (3) alleviate conflicts between deer and humans in urban and suburban areas, and (4) minimize damage to farm crops. The PGC has been addressing these goals through several programs since the late 1970s. First, biologists in the Bureau of Wildlife Management estimate population density of deer annually by collecting data on hunter harvest, reproductive rates, sources of mortality, and age structure of the population. Second, farmers are permitted to shoot deer that may be causing crop damage, to receive subsidies for fencing to exclude deer from cropland, and to enroll in a program that affords hunters extended hunting opportunities. Third, the PGC has implemented extended hunting seasons and liberal bag limits for deer in counties with limited suitable land for hunting and special areas, such as military installations and prisons. Fourth, the PGC issues permits to municipalities with deer problems if traditional hunting techniques are not effective in reducing deer populations. Fifth, the PGC conducted research to estimate deer densities that would not adversely affect tree regeneration and to ensure that deer populations can be sustained long term. This research was used to establish deer-density goals for each deer management unit (i.e., county) in Pennsylvania.

The PGC does not have deer management goals that explicitly address biodiversity issues. Noss and Cooperrider (1994:205–207) recommended that forest-reserve management in highly fragmented landscapes should include controlling overabundant herbivores. The PGC established overwinter carrying capacities based on the effect of deer browsing on tree regeneration (W. E. Drake and W. L. Palmer, final rep., Pa. Game Comm., Harrisburg, 1986). However, deer densities that allow for tree regeneration may not be adequate to protect endangered plants (e.g., Bratton and White 1980, Miller et al. 1992) or other species of wildlife (e.g., DeCalesta 1994). Although current deer density goals may not protect other species from the adverse effects of deer, no data are available to justify lowering current deer density goals.

Although the PGC has many deer programs, not 1 goal has been attained to the satisfaction of any special-interest group. The PGC's failure is the result of conflicting objectives among Pennsylvania residents. Some people want lower deer densities to reduce deer–vehicle collisions, crop damage, and urban–suburban deer problems, but many sportsmen believe there are too few deer to hunt. Also, the composition of forests in Pennsylvania is shifting from seedling–sapling timber, which supports in highest density deer, to older-aged timber. The changing composition of Pennsylvania's forests necessitates lower deer densities to protect tree regeneration in many traditional hunting areas.

Access to land for hunting confounds the objectives of managing deer populations. Only about 15% of Pennsylvania's land area is public and open to hunting.

Most of the remaining property is private land. The PGC has an incentive program for landowners to keep their property open to public hunting, but this land only amounts to an additional 1.2 million ha. Hunting is the most effective means of controlling deer populations, but restricted access to land by hunters is an increasing problem in urbanized areas. In northeastern Pennsylvania, the amount of land posted against hunting and trespassing is increasing because of housing development.

In 1995, individual sportsmen, sportsmen groups, and legislators petitioned the PGC to reduce or eliminate antlerless licenses to allow the deer population to increase. This outcry was the consequence of actions by the PGC, beginning in the late 1980s, to reduce deer populations to protect forest regeneration. In response, in part to political pressure to reduce antlerless harvests, we conducted a survey of deer hunters to assess their opinions and knowledge about deer management issues in Pennsylvania.

We analyzed deer population and forest habitat trends in Pennsylvania during 1982–1994, and present results of a 1995 survey of Pennsylvania hunters. We suggest that many sportsmen do not accept current PGC deer management and that sportsmen neither understand nor recognize the adverse ecological effects of deer upon forested ecosystems.

Study Area

Pennsylvania has 6.88 106 ha of forest, which comprise 59% of the state's land area (Alerich 1993). Southeastern Pennsylvania is in the Coastal Plain and Piedmont regions of the Atlantic Coast, an area primarily of farmland or urban centers (e.g., Philadelphia). The Ridge and Valley Region encompasses much of south-central Pennsylvania and extends from the southern part of the state along the Maryland border into northeastern Pennsylvania. The Ridge and Valley Region is dominated by oak (*Quercus* spp.) and hickory (*Carya* spp.) forests on the ridges. There is farmland in the valleys. To the north and west of the Ridge and Valley Region are the Allegheny Plateaus. These areas are primarily northern hardwoods, typically black cherry (*Prunus serotina*), sugar maple (*Acer saccharum*), and red maple (*Acer rubrum*), but also substantial amounts of oak–hickory forests. Both northwestern and northeastern Pennsylvania were glaciated, with the northwestern corner of the state primarily in farmland and forest lowland associated with Lake Erie.

We present deer densities and forest stand composition for 6 regions of Pennsylvania: northwest (NW), southwest (SW), northcentral (NC), southcentral (SC), northeast (NE), and southeast (SE). Region boundaries were based on political units of similar size, but they corresponded, for the most part, to the physiographic regions described above.

Methods

Habitat Conditions and Deer-Density Goals

We classified U.S. Forest Service inventory data for 1978 and 1989 in Pennsylvania into seedling–sapling, pole, and saw timber (Alerich 1993). We included noncommercial timber in the saw-timber class because it was unlikely to be harvested and therefore was similar to saw timber in terms of food availability for deer. Research conducted by the PGC (W. E. Drake and W. L. Palmer, final rep., PGC, Harrisburg, 1991; Tzilkowski et al., final rep., PGC, Harrisburg, 1994) estimated the forest could support overwinter deer densities of 60 deer/259 ha in seedling–sapling timber, 5 deer/259 ha in pole timber, and 20 deer/259 ha in saw timber without adversely affecting tree regeneration. Thus, deer density goals for each Pennsylvania county were calculated by summing the product of ha of each forest stand type by the corresponding overwinter carrying capacity.

Population Monitoring and Management

We estimated county-level deer densities after the hunting season annually from 1982 to 1995 using consistent data-collection methods and population-estimation procedures (W. L. Palmer and W. K. Shope, Annu. white-tailed deer rep., PGC, Harrisburg, 1994). Successful hunters were required to return a postage-paid, business-reply card with the date and location of each deer harvested. By checking >40,000 harvested deer taken to meat processors, we estimated the percentage of harvested deer actually reported, and the age structure of the harvest. Also, during the spring, road-killed females were aged and their reproductive condition (number and sex of fetuses) recorded.

We estimated deer populations using a modification of the change-in-ratio (CIR) procedure described by Shope (1978). The CIR technique was appropriate, because about 75–80% of adult males and about 30% of adult females were harvested each year. We used the modified CIR procedure to generate the previous 7 populations of adult bucks and adult does. Then, we generated between-season adult survival rates and minimum fawn populations. Fawn survival rates were estimated by dividing postseason fawn estimates by the number of preseason 18-month-old deer in the subsequent population estimate. We obtained preseason estimates of adults using forward projections with 5-year-average survival rates calculated for the adults and fawns. We estimated the preseason fawn population by multiplying the number of adult does by the maximum reproductive rate, and postseason population estimates by subtracting estimated legal harvests.

After estimating deer densities and predicting future harvests, we estimated the number of antlerless licenses necessary to bring about desired changes in the deer populations for each county. Antlerless allocations were designed to increase (or decrease) deer populations to the desired levels within a specified number of years (usually 3–5 yrs).

The PGC is managed by a Board of Commissioners comprised of 8 governor-appointed Pennsylvania residents, who approve all seasons and bag limits. Deer density goals for each county were approved by the commissioners in 1979. Antlerless allocations were presented to the commissioners, who also reviewed public input on the recommended allocations, and either accepted or modified the antlerless allocation.

Hunter Opinion Survey

In 1995, we mailed self-administered, mail-back questionnaires to 2,000 randomly selected purchasers of the 1,130,090 general hunting licenses sold during the 1993–1994 license year. This license permitted a hunter to harvest an antlered deer during the regular, rifle deer season, and purchase 1 or more licenses to hunt deer in the archery, muzzleloader, and antlerless deer seasons.

The 4-page questionnaire contained 66 questions. The first part of the questionnaire asked in which deer-hunting seasons the respondent had participated, and what was his or her success. The second part of the questionnaire presented 45 statements to which respondents were asked to indicate their level of agreement. We used a 5-point scale for respondents to indicate their level of agreement (strongly agree, agree, undecided, disagree, or strongly disagree). These statements presented possible regulation changes, and assessed the respondent's agreement with current regulations. Also, we included statements to assess respondents' opinions about deer populations, the effects of deer on forests, deer conflicts with human activities, and possible management actions to address specific deer problems. Responses to these statements were grouped into 3 categories: agree, undecided, and disagree. The questionnaire then asked respondents to rate (poor, fair, good, excellent, don't know) 5 areas of the agency's deer management program (archery regulations, urban–suburban deer management, antlerless allocations, muzzleloader seasons, crop-damage program), and overall deer management. The questionnaire ended with 6 demographic questions (e.g., marital status, education, etc.).

The initial survey was mailed in late January 1995, and we mailed a reminder postcard 1 week after the first mailing to increase response rates. If we received no response after 2–3 weeks following the first mailing, we mailed a second copy of the questionnaire. We used

TABLE 1. Forest inventories (ha) by stand type,[a] and deer density goals for Pennsylvania, 1978 and 1989.

Region	1978 Forest inventory (ha) of stand types			Deer density goal (deer/259 ha)	1989 Forest inventory (ha) of stand types			Deer density goal (deer/259 ha)
	Seedling-sapling	Pole	Saw		Seedling-sapling	Pole	Saw	
NW	255,113	249,674	549,595	26.2	136,233	218,594	686,864	22.0
SW	276,351	279,718	500,903	26.4	222,997	281,790	584,301	24.4
NC	227,400	651,381	1,040,916	19.7	187,256	580,157	1,110,068	19.4
SC	156,694	267,027	530,429	22.3	146,852	284,380	554,257	21.8
NE	286,452	455,837	514,630	23.6	183,630	460,240	634,806	20.5
SE	129,240	130,017	300,697	25.6	91,167	179,486	327,892	21.5
State	1,331,253	2,033,658	3,437,172	23.3	968,137	2,004,650	3,898,190	21.2

[a] Adapted from Alerich (1993).

techniques described by Dillman (1978) to maximize the response rate to the survey. We took a sample of nonrespondents and contacted them by phone to assess nonresponse bias. We conducted chi-squared tests of homogeneity to assess differences between respondents and nonrespondents.

Results

Habitat Conditions and Deer Density Goals

In 1989, 59% of Pennsylvania (6.88 106 ha) was forested. This amount increased <1% between 1978 and 1989 (Table 1). However, seedling–sapling stands, which support the highest overwinter density of deer, declined 27%, pole timber stands changed little (<2%), and saw timber area increased 13%. The result of this change in forest composition was that the estimated overwinter carrying capacity for deer declined. Statewide, the deer density goal declined from 23 deer/259 ha of forest in 1978 to 21 deer/259 ha of forest in 1989, and regionally the deer density goal decline ranged from 0.3 deer/259 ha of forest in north-central Pennsylvania to 4.2 deer/259 ha of forest in northwestern Pennsylvania (Table 1).

Population Monitoring and Management

Deer population densities peaked in the late 1980s, but after the 1994 hunting season they were still above densities observed in the early 1980s (Fig. 1). The antlered harvest, which is a function of population size and hunter effort (because no restriction is placed on number of licenses sold), increased and then stabilized over the last 5–6 years in most regions (Fig. 2). Antler-less harvests are regulated by the number of antlerless permits issued in each management unit: the correlation (r) between the allocation and antlerless harvest was 0.94 (12 df, $P < 0.001$). Consequently, antlerless harvests peaked in the early 1990s, when the PGC increased the number of antlerless permits to reduce the deer population (Fig. 3). In addition, in 1988, the PGC eliminated the harvest restriction of 1 deer per hunter and permitted hunters to purchase ≤3 antlerless licenses, if available. Also, hunters were permitted ≤4 antlerless licenses, if

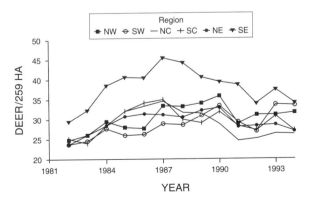

Fig 1. Density of deer (deer/259 ha) for each region of Pennsylvania, 1982–1994.

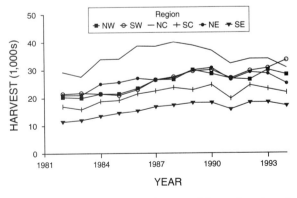

Fig 2. Numbers of antlered deer harvested for each region of Pennsylvania, 1982–1994.

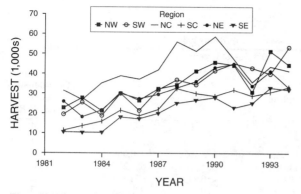

Fig 3. Numbers of antlerless deer harvested for each region of Pennsylvania, 1982–1994.

available, in 6 counties surrounding Philadelphia (SE region) and Pittsburgh (SW region) because these counties were recognized as having high deer densities and limited areas suitable for hunting.

Hunter Opinion Survey

After 2 mailings, the response rate was 72% (1,375 responses) after excluding 108 undeliverable questionnaires. We attempted to contact 146 nonrespondents by phone, but 80 (55%) could not be reached, 31 (21%) refused to respond, and 35 (24%) responded. We detected no important differences (c^2 tests) between the answers of respondents to the mail survey and those nonrespondents who responded when contacted by phone.

Most (97%) of Pennsylvania's 1.1 million hunters participated in ≥1 deer seasons. Most (61%) deer hunters thought deer populations where they hunted were too low, 35% considered them about right, and 4% thought deer populations were too high. Consequently, 44% supported reducing antlerless allocations (37% disagreed), and 19% favored eliminating antlerless allocations altogether (63% disagreed).

Most (87%) hunters agreed that controlling deer populations was necessary, and 13% disagreed or were undecided (Table 2). Besides the low support for antlerless seasons, 24% of respondents believed there were not enough deer unless some deer were starving to death each year. Most (56%) respondents agreed that deer have the ability to affect plant and animal communities, but only 51% of those believed deer cause serious problems involving other land uses. Likewise, 89% agreed that keeping deer in balance with natural food supplies was necessary, but only 24% of those hunters believed that deer damage to forests in Pennsylvania was a problem.

Discussion

One hundred years ago the few deer remaining in Pennsylvania existed primarily in the north-central region of the state. By 1916, however, the director of the PGC was promoting regulated, antlerless seasons to control deer populations (Kosack 1995), and by the 1930s the PGC was trying to convince sportsmen that antlerless seasons would not decimate deer populations and that food scarcity was limiting deer weights and numbers of deer in many parts of the state (Luttringer 1931). In 1994, deer inhabited every county in the state, 395,000 deer were harvested by hunters, >42,000 were killed on highways, and nearly 2,000 were shot because of crop damage. Statewide, Pennsylvania averaged >29 deer/259 ha following the 1994 hunting season. PGC research indicated that, to prevent adverse effects on tree regeneration, the statewide average for overwinter deer densities should not exceed 21 deer/259 ha.

Researchers have documented the effects of deer on Pennsylvania forests and songbird communities

TABLE 2. Agreement of 1,357 hunters to statements presented in a mail-survey questionnaire on deer management, Pennsylvania

Statement	Percent of respondents		
	Agree	Undecided	Disagree
Controlling deer populations is necessary.	87	6	8
Keeping deer populations in balance with natural food supplies is necessary	89	4	7
Deer damage to forests in Pennsylvania is a problem.	21	22	57
Deer have the ability to affect plant and animal communities	56	24	20
Deer cause serious conflicts with other land uses, such as forestry, farming, highways, and other development.	37	19	44
We don't have enough deer unless some are starving to death each year.	24	20	56
Allocations for antlerless permits should be eliminated in the county where I hunt.	19	19	63
Allocations for antlerless permits should be reduced in the county where I hunt.	44	19	37

(Tilghman 1989, DeCalesta 1994). Tilghman (1989) recommended deer densities ≤18 deer/259 ha of forest. However, deer densities after the 1994 hunting season in northwestern and north-central Pennsylvania were 31 and 26 deer/259 ha of forest, respectively. The consequences of these deer densities are evident, especially in northern Pennsylvania. Tree regeneration has been affected by deer browsing. In recent years some private, industrial landowners harvested timber only where fencing had been erected to exclude deer from tree seedlings for 4–6 years (T. Eubanks, Int. Paper Co., Coudersport, Pa., pers. commun.). In mature or old-growth forest stands, ferns typically are the dominant understory vegetation. In the northern hardwood forest, regeneration is dominated by black cherry, and herbaceous-vegetation diversity is reduced (Hough 1965, Tilghman 1989, Witmer and DeCalesta 1992). In some areas even brambles (Rubus spp.) fail to grow unless deer-exclusion fences are erected (J. P. Dzemyan, PGC, Smethport, pers. commun.).

The results of our survey were particularly troubling because many hunters opposed reducing deer populations. Clearly, hunters lacked knowledge about the interaction of deer and their habitat and the effects of too many deer on forested ecosystems. Pennsylvania hunters apparently did not recognize the deleterious effects of current deer populations on forest ecosystems in the state. Consequently, attempts by biologists to reduce deer populations in Pennsylvania were rejected. In 1995 biologists recommended that >750,000 antlerless licenses be issued for the 1995–1996 hunting season. Because of public outcry at PGC meetings and letters to state legislators, antlerless allocations were reduced by the Board of Commissioners to 656,000 licenses. Consequently, deer populations were expected to increase in northern Pennsylvania and remain stable or increase slightly in southern counties. In the northern counties where deer population increases were expected, only a small proportion of forested habitat is in seedling–sapling timber (Table 1). In these areas deer densities have already seriously reduced forest regeneration and plant species diversity.

Attempts to protect biological diversity must address the management of dominant species, such as white-tailed deer (Garrott et al. 1993). Deer affect all forested habitats in Pennsylvania, and thus, may exert as great an effect as human disturbances (e.g., timber harvesting, livestock grazing) on plant and animal communities. In fact, because deer are such an important, managed game species their effects on forest ecosystems could be considered an indirect human disturbance.

Several factors must be addressed if deer managers in Pennsylvania intend to protect forest biological diversity. Foremost, hunters need to understand the effects of deer on forest communities and to support deer density goals that will protect forest ecosystems. Deer densities observed in the 1980s demonstrate that Pennsylvania can support many deer. However, the adverse effects of deer are not acknowledged or recognized by deer hunters. Moreover, managing deer to protect forest regeneration is sound management and will ensure sustained deer harvests. Deer densities that reduce forest regeneration result in habitat that supports fewer deer in the long-run. Thus, we believe that the management of deer populations at currently established goals will increase the likelihood of sustained harvests because important food resources will be protected from excessive browsing.

The conflicting goals of people wanting fewer deer (e.g., farmers, landowners) versus those wanting more deer (i.e., hunters) have created controversy for decades. This type of conflict erupted when New Jersey tried to establish an antlerless deer season in the late 1950s (Tillett 1963). In contrast to deer management 30 years ago, Pennsylvania has better estimates of deer populations and harvests and a greater knowledge of the adverse effects of too many deer on forested ecosystems. Unfortunately, there was little difference in sportsmen's opinions about deer populations and management than those described by Tillett (1963).

Scientists and conservationists must play a greater role in providing comment on PGC management goals, strategies, and actions. Also, these groups must play a greater role in public education and communication efforts in the future. Management recommendations for Pennsylvania's deer herd are based on the best scientific information available, but these recommendations must also have public support (Gilbert and Dodds 1992). Legislators and governor-appointed commissioners of the PGC must recognize that biologist's recommendations have scientific and public support. In addition, more research is needed on the effects of deer on forest ecosystems because the PGC's current deer density goals are based on tree regeneration for timber harvesting. We assume that if adequate tree regeneration occurs, other species of plants and animals will benefit as well. However, research providing empirical data to support this assumption is needed.

Acknowledgments. We thank the many Pennsylvania Game Commission personnel who have collected biological data on the white-tailed deer population. The manuscript benefited greatly from the suggestions of R. C. Boyd, C. W. DuBrock, J. D. Hassinger, R. J. Pedersen, and an anonymous referee.

"White-Tailed Deer Research/Management"

Pennsylvania Game Commission. Bureau of Wildlife Management, August 5, 1999.

Abstract

We used data on deer reproduction, sex and age of harvested deer, license numbers of successful hunters, and reported harvests to estimate 1998 and 1999 deer populations by management unit. Wildlife Conservation Officers (WCOs) also conducted winter deer mortality surveys along preselected routes in their respective districts. Our 1998-99 winter deer density of 34 deer/mi^2 of forest land was about 13% higher than in 1997-98. The 1998-1999 winter deer loss index of 0.15 deer/mile was well below previously recorded losses. We projected a preseason deer population of 1,401,911 for 1999. The deer population and allocation models indicated more than 1,121,000 antlerless licenses were needed just to stabilize the growing deer population in all units. Since 1.121 million licenses far exceeds demand, and the necessary harvest could not be achieved in a 3-day season, the Executive Office and senior staff developed several options with longer season length to increase the efficiency of antlerless licenses. Options required the sale of unsold licenses, and incorporated the historical maximum license sales for each county. The 3-6-12 option required 782,000 licenses, and kept a 3-day antlerless season in some counties while increasing the season to 6 or 12 days in other counties. A second option, called the 6-12 option, required 761,850 licenses, and was for a 6- or 12-day season for all counties. The 3-day season would be the traditional season, the 6-day season would extend the traditional season through Saturday, and the 12-day season would have a concurrent antlered and antlerless season during the second week of the traditional antlered deer season, and run through the following week with 6 days of antlerless deer hunting. The objective of all options was to harvest 301,000 antlerless deer, and stabilize the population in all counties. The Commissioners did not consider the extended season options, and they approved the same antlerless allocations used in 1998, totaling 797,200 licenses excluding special regulations counties. Special regulations counties were allocated 93,500 antlerless licenses. A proposal to allow the sale of up to 2 antlerless licenses per hunter also failed. We recommend changing the antlerless season to include a Saturday and/or a holiday to expand hunting opportunities for individuals who may have conflicts with the traditional Monday through Wednesday season. We also recommend increasing the number of days of antlerless hunting to 6 or 12 days to increase individual success. Increasing participation opportunity and hunter success will reduce the number of licenses needed to regulate deer harvests. Both actions may improve hunter support of antlerless deer hunting. In addition, we recommend that hunters be permitted to purchase and use all unsold licenses, and that flintlock hunters be required to have an antlerless license to harvest an anterless deer during the flintlock season.

Objective

To determine deer population sizes and harvest recommendations by management unit.

Procedures

To obtain data on reproduction by age class, WCO's examined female deer killed by various causes from 1 February through 31 May 1998. They recorded location (county, township, and proposed deer management unit), date killed, cause of death and number of embryos for each doe on a form attached to a deer jaw envelope. They also removed one side of the lower jaw from each deer for age determination. Jaws were forwarded to wildlife biologists who made the age assignments in July 1998. Personnel in the Bureau of Automated Technology Services (BATS) processed the reproductive data and provided summary reports for the state and each county.

During the 1998 antlered and antlerless rifle seasons, 30 data collection teams examined deer in assigned areas. Each team spent at least three days during each season collecting ages, sexes, counties of harvest, and hunting license numbers from harvested deer found in butcher shops and other locations. Deer teams determined deer ages using tooth wear and replacement (Severinghaus 1949).

BATS personnel input and processed data from 1998–99 deer harvest report cards submitted by hunters and the biological collections by the deer teams. BATS also provided a PC download for population analysis. For each county the download included: the reported antlered harvest, the reported antlerless harvest, reporting rates, age and sex breakdowns of the harvest, reproductive data, combined reported regular three-day antlerless rifle and antlerless archery harvests, and the total antlerless rifle and archery harvests. We used the download data in DEERPOP and PROJECT software (Shope pers. commun.) to estimate 1998 and project 1999 county deer populations. Besides estimating populations, we used PROJECT to develop antlerless allocation recommendations for 1999.

In late March and early April, WCOs conducted winter deer mortality surveys in their assigned districts. Each WCO walked three 1.5-mile routes along stream bottoms to locate possible winter losses. They recorded the sex and age of all dead deer found and submitted the data to us for analysis. We converted their data to a deer/mile index and compared it with previous winter loss indexes to decide if we needed to adjust any projected county estimates for excessive winter losses.

Findings

WCOs provided usable reproductive data from 1,748 females examined during the 1998 prefawning season. The 1998 sample was 24% smaller than in 1997. Most of the decrease was due to an increase in the number of road-kill deer contractors from which no reproductive data were collected. Twenty-seven (27%) of the female fawns, 90% of female yearlings, and 92% of the adult females were pregnant. Pregnant fawns averaged 1.28 embryos/doe, pregnant yearlings 1.63 embryos/doe, and pregnant adults 1.78 embryos/doe. The average reproductive rates for pregnant and barren fawns, yearlings, and adults were 0.35, 1.46, and 1.63 embryos/doe, respectively. The average reproductive rate for all females was 1.04 embryos/doe.

We estimated a 1998–99 statewide winter density of 34 deer/mi^2 of forested habitat. This density was about 13% higher than the 1997–98 winter density (Table 1). The statewide winter deer population was 62% higher than the agency goal of 21 deer/mi^2.

Statewide, WCOs found 0.15 dead deer/mile on winter survey routes in 1999. In most counties, winter losses were well below the high losses recorded in 1978 (Table 2). Because winter losses were nominal, we made no adjustments in projected population estimates for 1999.

We projected a preseason state population of 1,401,911 (54 deer/mi^2 of forest land) for the 1999 fall hunting season. This figure does not include counties with special regulations. Projected county densities (excluding counties with special regulations) ranged from lows of 20, 24, 31, 32, and 32 deer/mi^2 of forest land in the counties of Cameron, Clinton, Pike, Elk, and Monroe, respectively, to highs of 116, 115, 112, 108, and 101 deer/mi^2 of forest land in the counties of Washington, Montour, York, Lehigh, and Greene, respectively. The lowest projected rates of population increase from postseason 1998 to preseason 1999 was 33% in Cameron, Clinton, Sullivan, and Elk counties.

The highest projected rates of population increase were 71–94% in Greene, Washington, Lawrence, and Beaver counties (Table 3).

The deer population model indicated that about 1.121 million antlerless licenses must be issued in 1999 under the current season format just to hold populations stable statewide. If we were to move all county deer populations to goal by 2003 using the present format, as stated in the PGC Strategic Plan, the number of antlerless licenses needed would be much higher. This allocation question was not even discussed by Bureau of Wildlife Management staff in 1999 for 3 major reasons. First, it was politically unacceptable; second, an allocation of this size, even with the sale of surplus licenses, would far exceed demand; and third, the season length is too short to attain the necessary harvest. Senior management staff concluded that the best situation attainable for antlerless allocations for 1999 was to hold each county population stable. Because the statewide allocation needed to hold each county stable under the current 3-day format could not be sold, we needed an alternative to reduce the allocation, yet still achieve the desired harvest to stabilize each county population. That alternative was to lengthen the antlerless season, thus making antlerless licenses more efficient. We could then issue fewer permits to achieve the desired harvest.

We had limited information on 6- and 12-day antlerless seasons from previously extended antlerless seasons and special regulations areas. We also knew what the maximum sales for antlerless licenses were during years with the sale of surplus licenses. At a meeting of senior management staff and Bureau of Wildlife Management staff, it was decided to propose an antlerless allocation to the Commissioners with each county having a 3-, 6-, or 12-day antlerless season. The allocation would be designed only to hold each county population stable, i.e., even with the longer season, the resulting harvest would be the same as with the traditional 3-day season. This decision was made to allow the recently chartered Deer Management Working Group another year to make recommendations concerning antlerless deer seasons to the Commissioners that are acceptable to all stakeholders. The 3- and 6-day season would begin at their traditional times. The 12-day season would begin on the second Monday of antlered season and have 6 days of concurrent antlered and antlerless season and the final 6 days of antlerless deer only hunting.

In summary, a 3-day season would required 1,121,000 licenses to harvest 301,000 antlerless deer. A 6-day statewide season would require 866,400 antlerless licenses to harvest the same number of deer; and a 12-day statewide season would require 595,850 antlerless licenses to harvest the same number of deer. Bureau of Wildlife Management staff developed 2 modifications from the varying season lengths into a 6-12 option and a 3-6-12 option. Under the 6-12 option, 41 counties would have 6-day seasons, and 20 counties would have 12-day seasons. Under the 3-6-12 option, 9 counties would have 3-day seasons, 32 counties would have 6-day seasons, and 20 counties would have 12-day seasons. The 6-12 option would require an antlerless allocation of 761,850 licenses, and the 3-6-12 option would require an allocation of 782,000 licenses.

Each of the allocations would require the sale of unsold licenses to reach their full effectiveness. It is also important to realize that all options are designed to harvest the same number of antlerless deer (301,000), and that this harvest would only stabilize the population in each county. This allocation would not move any county population toward its Commission-approved density goal, and no progress would be made toward reaching the deer management objective as stated in the Commission-approved strategic plan.

This proposal with different options to harvest 301,000 antlerless deer and stabilize the population in all counties was submitted to the Commissioners in April 1999. The Commissioners did not approve any part of the proposal, and accepted a proposal to revert to last year's county allocation totaling 797,200 antlerless licenses, excluding special regulations counties. Special regulations counties had 93,500 antlerless licenses allocated. This approved allocation was designed to manage 1998 (not 1999) deer populations with the sale of surplus license populations. Many counties had increased significantly since 1998. This approved allocation has no specific deer management objective.

The Commissioners also failed to reinstate the sale of surplus licenses for counties other than special regulations counties and the southwest region counties. Thus, attainment of an adequate antlerless deer harvest will not be possible due to lack of an adequate antlerless license allocation, and the restriction of antlerless licenses to 1 per hunter. In special regulations counties, hunters will be permitted to purchase licenses until the entire allocation is sold. In all southwest region counties except Allegheny, hunters can purchase a second antlerless license. Allegheny County is a special regulations county.

For the approved allocation to be effective in meeting the strategic plan objective, the Commissioners must reinstate the regulation permitting hunters to purchase and use unsold antlerless licenses in counties outside of the special regulations areas. Because of the restriction on sale of antlerless licenses, sales are expected to be similar to the 660,200 sold in 1998. Due to this restriction in license sales, the statewide overwintering deer population increased 13% from January 1998–January 1999. Without the surplus sale regulation in place, antlerless deer harvests in most if not all county units of the state will be insufficient to maintain control of the deer population. Predominately agricultural counties in the southern and western areas of the state will be especially hard hit by a failure to adopt this proposal. Much of the higher allocation that the model generated for 1999 was a result of the Commissioners eliminating the sale of surplus licenses in 1997 and 1998 (except for southwest region counties in 1998). Elimination of the sale of surplus licenses left many unsold licenses in important agricultural areas of the state. Consequently, inadequate antlerless harvests occurred in those areas, and populations rose sharply.

Hunters and some commissioners may be pleased with the increases in deer abundance. However, we have already been shown the lesson of rapid deer increases and deer overpopulation in Pennsylvania. If everyone with a stake in deer management knew the history of deer in the northcentral counties, they would understand why biologists stress the need to manage deer within the capacity of the land to sustain them over the long term. Unfortunately, the range, damage inflicted decades ago in the northcentral and northeast counties has never been permitted to heal. It still cannot heal, and given the restrictions in antlerless allocation and subsequent harvests, the result can only be further damage to the deer range. Stakeholders who do not understand this neglect the principles of deer management and deer biology, including deer reproduction and especially the impacts of deer to their environment. In recent years, populations in the southern half of the state have grown rapidly, and are being carried at levels far beyond what the forested land can support for long periods of time. And the same effects of overbrowsing and range deterioration by deer that we already experienced in the northcentral counties are being seen in the southern counties. We have already learned the lessons of carrying too many deer in the northcentral and northeastern parts of Pennsylvania. Our deer management program needs to address these problems before they occur in the southern half of the state, and to allow the range to recover across other parts of the state.

Because antlerless allocations have been restricted in recent years, statewide we are currently carrying 62% more deer than our Commission-approved goal. Pennsylvania is now carrying more deer than ever before. This fall, a projected 1.4 million deer will be available when the archery season opens in October. Because of the low antlerless allocation and restriction of surplus license sales, populations will likely again increase significantly in 2000. Due to the projected abundance of deer, and subsequent reproduction, we cannot hold populations stable in any county unit with the current antlerless season format. To stabilize or reduce populations toward Commission-approved goals, we will have to move to some alternative format with a longer rifle season for antlerless deer, and permit hunters to purchase antlerless licenses until allocations are sold out. The only other solution possible is a severe reduction in populations due to a harsh winter. If this were to occur, there would have to be mass starvation of deer and accompanying habitat degradation caused by starving deer. The real losers in the years following a catastrophic winter loss will be deer, other species of wildlife, habitat quality essential to all wildlife, and ultimately deer hunters. This surely is not a responsible way to manage Pennsylvania's deer and other wildlife resources.

Large antlerless allocations are difficult to sell. Moving to a 6- and/or a 12-day season would increase license efficiency and thus decrease the number of licenses we need to issue. This reduction in allocation, however, is in relation to a 3-day season. Because of recent population increases, it will still require high allocations, even with a longer season format.

With two exceptions, we are currently obtaining all deer statistics for proposed deer management units (DMUs) that have been obtained for county-based units. The exceptions are success rates for antlerless deer hunters, and estimating populations. We will be working with BATS to create additional computer programming that will calculate success rates. Population estimates cannot be made for management units because we need 7 years of data to do the modeling. Although some reduction in confidence is expected until 7 years of data are acquired, 5 years of DMU data are needed to mitigate variation from year to year in weather, food supply and distribution of hunter pressure. We need long-term averages to overcome short-term fluctuations. We currently have DMU data for 1995–1998. We need data from the 1999 hunting season to have 5 years of deer data to estimate populations. If the Commission grants approval to the proposed DMUs, we can make new DMU antlerless allocations in the year 2000.

Recommendations

Many hunters may be excluded from hunting antlerless deer during the traditional three-weekday gun season because of school or work commitments. Therefore, we suggest changing the present season format to one that would include Saturdays and/or holidays. About 82% of the respondents to the 1993 PSU deer hunter survey indicated that they hunted deer sometime during the previous 5 years. Current annual participation is running about 57%. Therefore, we believe there may be more hunter interest in antlerless deer hunting than current license sales indicate.

We also recommend increasing the number of days to 6 or 12 for gun hunters to pursue antlerless deer. This would increase individual success. Because of recent increases in deer populations and projected increases this year, we cannot stabilize or reduce most if not all deer populations without increasing the number of antlerless deer hunting days. Our hunter surveys suggest that hunters react more positively to programs they can participate in and that success has a positive influence on their attitudes. Therefore, increasing both opportunity and success could improve hunter support for antlerless deer hunting.

For antlerless deer licenses, we recommend that hunters be permitted to purchase and use all unsold licenses. We also recommend that the Commission include the muzzleloaders hunters in the antlerless license system.

Literature Cited

Severinghaus, C. W. 1949. Tooth development and wear as criteria of age in white-tailed deer. *J. Wildl. Manage.* 13:195–216.

TABLE 1. County forest statistics, winter deer density goals, and estimated winter density trends from the winter of 1994–95 through the winter of 1998–99 for Pennsylvania. Special regulations countries are excluded.

Country	% Forest	mi² of forest land				Goal[b]	Winter deer density estimates				
		Seedling sapling	Pole timber	Saw timber	Total		94—95	95—96	96—97	97—98	98—99
Adams	33	33	41	99	173	24	41	41	40	50	58
Armstrong	54	98	43	214	355	29	37	37	45	44	52
Beaver	48	33	60	117	210	22	37	36	34	39	36
Bedford	72	172	212	342	726	25	25	32	30	31	29
Berks	35	40	85	175	300	21	44	41	56	49	60
Blair	64	59	113	166	338	22	31	40	36	41	40
Bradford	59	127	269	280	676	22	28	28	31	37	42
Butler	50	75	110	212	397	23	26	33	42	42	47
Cambria	64	52	116	271	439	21	31	28	28	29	33
Cameron	94	20	86	266	372	19	23	17	19	15	15
Carbon	75	67	114	105	286	23	30	33	32	21	27
Centre	76	104	304	429	837	20	25	26	27	27	29
Clarion	61	91	85	194	370	26	33	28	41	42	41
Clearfield	74	145	305	398	848	21	38	39	37	33	37
Clinton	87	33	275	464	772	16	19	17	18	18	18
Columbia	53	29	102	126	257	19	33	35	34	39	46
Crawford	48	42	158	285	485	18	35	31	35	33	39
Cumberland	35	17	87	90	194	17	29	32	27	34	37
Dauphin	50	51	85	129	265	23	27	22	22	20	27
Elk	91	64	137	552	753	21	30	29	23	21	24
Erie	47	100	49	224	373	29	28	33	30	30	36
Fayette	61	74	114	292	480	23	32	26	28	26	33
Forest	93	50	43	304	397	23	36	33	29	32	39
Franklin	44	77	40	219	336	27	34	36	45	34	34
Fulton	69	34	91	177	302	20	23	30	31	30	30
Greene	56	44	111	169	324	20	35	50	45	50	59
Huntingdon	75	94	210	353	657	21	24	33	36	39	40
Indiana	61	100	160	243	503	23	34	32	36	33	39
Jefferson	61	21	74	308	403	19	35	36	42	39	37
Juniata	66	18	80	161	259	18	28	31	37	29	34
Lackawanna	68	59	105	147	311	23	32	34	30	23	32
Lancaster	13	0	11	114	125	19	49	29	48	49	57
Lawrence	42	24	43	84	151	22	24	17	21	23	28
Lebanon	34	18	26	78	122	23	31	26	26	31	38
Lehigh	29	12	20	68	100	22	37	39	52	52	66
Luzerne	66	60	273	253	586	17	33	30	29	26	33
Lycoming	77	85	310	559	954	19	28	29	27	23	24
McKean	81	90	237	485	812	20	26	26	26	25	30
Mercer	39	35	62	166	263	22	31	36	35	37	40
Mifflin	72	35	56	205	296	22	23	25	27	29	32
Monroe	76	38	178	245	461	18	22	24	25	17	22
Montour	27	9	0	27	36	30	55	50	57	55	72
Northhampton	34	29	18	80	127	27	29	30	39	47	51
Northumberland	50	45	78	105	228	23	22	22	26	23	26
Perry	64	10	92	253	355	17	27	34	38	30	37
Pike	82	42	149	260	451	19	20	23	27	20	22
Potter	86	73	202	652	927	20	22	19	23	24	31
Schuylkill	71	110	295	146	551	20	32	26	31	34	37
Snyder	51	18	76	75	169	18	26	26	30	31	33
Somerset	64	157	238	294	689	24	34	30	29	29	29
Sullivan	86	18	139	230	387	16	21	23	23	20	27
Susquehanna	65	114	134	283	531	25	29	36	45	34	37
Tioga	66	103	305	352	760	19	26	27	31	30	38
Union	68	6	79	129	214	16	24	26	27	27	26
Venango	72	26	111	348	485	19	32	23	36	25	34
Warren	79	62	109	527	698	21	29	27	30	30	31
Washington	50	132	113	182	427	28	34	46	50	49	67
Wayne	66	54	154	272	480	20	28	39	39	30	38
Westmoreland	51	137	98	283	518	28	32	41	40	39	48
Wyoming	62	47	82	118	247	23	28	29	34	31	30
York	27	9	55	180	244	18	47	52	48	51	69
Total	59	3,738	7,740	15,051	26,529	21	29	30	31	30	34

[a]Forest statistics are based on 1989 U.S. Forest Service inventory data for Pennsylvania.
[b]Goals are based on 60 deer/mi², 5 deer/mi², and 20 deer/mi² for seedling/sapling, pole, and sawtimber stands, respectively.

TABLE 2. Dead deer found on wintery survey routes in 1999 and dead deer found/mile surveyed in 1999 and 1978 in Pennsylvania.

Country	1999 Miles	1999 Dead deer	Dead deer/mile 1999	Dead deer/mile 1978
Adams	10.00	1	0.10	0.33
Allegheny	10.90	3	0.28	0.15
Armstrong	8.70	5	0.57	0.11
Beaver	7.25	5	0.69	0.00
Bedford	14.70	3	0.20	1.35
Berks	15.10	1	0.07	0.00
Blair	16.50	5	0.30	4.00
Bradford	18.90	2	0.11	0.81
Bucks	9.00	0	0.00	
Butler	10.50	3	0.29	0.09
Cambria	9.20	0	0.00	2.18
Cameron	4.50	0	0.00	13.60
Carbon	15.00	3	0.20	0.13
Centre	18.50	0	0.00	3.35
Chester	11.50	1	0.09	0.00
Clarion	10.50	0	0.00	1.88
Clearfield	14.50	1	0.07	5.17
Clinton	10.50	0	0.00	0.87
Columbia	11.75	4	0.34	0.83
Crawford	27.50	2	0.07	0.33
Cumberland	9.50	0	0.00	0.55
Dauphin	11.75	2	0.17	1.67
Delaware	1.50	1	0.67	
Elk	9.65	2	0.21	1.86
Erie	15.70	1	0.06	0.08
Fayette	12.00	3	0.25	0.00
Forest				0.42
Franklin	11.10	8	0.72	0.29
Fulton	4.60	0	0.00	0.75
Greene	9.00	4	0.44	0.83
Huntingdon	10.50	0	0.00	0.95
Indiana	12.00	0	0.00	2.16
Jefferson	11.10	6	0.54	1.00
Juniata	5.80	0	0.00	2.67
Lackawanna	11.10	0	0.00	2.24
Lancaster	16.20	0	0.00	0.00
Lawrence	4.50	0	0.00	0.33
Lebanon	4.60	0	0.00	
Lehigh	6.00	2	0.33	0.00
Luzerne	14.80	3	0.20	0.78
Lycoming	34.20	5	0.19	0.70
McKean	16.50	7	0.42	1.23
Mercer	9.50	0	0.00	0.00
Mifflin	6.50	0	0.00	0.77
Monroe	9.50	1	0.11	4.10
Montgomery	10.00	1	0.10	0.14
Montour	4.50	0	0.00	0.00
Northhampton	5.90	2	0.34	
Northumberland	4.50	0	0.00	1.67
Perry	4.50	0	0.00	1.01
Philadelphia	4.50	5	1.11	
Pike	10.00	1	0.10	4.33
Potter	21.50	4	0.19	3.69
Schuylkill	9.00	1	0.11	0.74
Snyder	5.55	0	0.00	0.63
Somerset	19.00	7	0.37	3.93
Sullivan	4.50	1	0.22	0.75
Susquehanna	10.50	0	0.00	3.97
Tioga	29.00	1	0.03	4.17
Union	4.50	0	0.00	1.09
Venango	10.50	3	0.29	0.38
Warren	13.00	0	0.00	2.10
Washington	10.25	4	0.39	0.29
Wayne	18.50	0	0.00	16.42
Westmoreland	15.50	1	0.06	3.03
Wyoming	4.50	0	0.00	0.00
York	23.00	0	0.00	
1998 Total	742.10	115	0.15	
1978 Totals	686.05	1,330	1.94	

TABLE 3. County deer population densities (deer/mi^2 of forest land) and projected rates of population increase from postseason 1997 to preseason 1998. Special regulations countries are not included.

Country	1998 deer densities		1999 projected preseason density	% Population increase
	Preseason	Postseason		
Adams	77	58	95	64
Armstrong	78	52	84	62
Beaver	56	36	70	94
Bedford	42	29	45	55
Berks	86	60	97	62
Blair	54	40	61	53
Bradford	61	42	70	67
Butler	70	47	77	64
Cambria	47	33	51	55
Cameron	18	15	20	33
Carbon	35	27	39	44
Centre	38	29	41	41
Clarion	58	41	67	63
Clearfield	51	37	56	51
Clinton	22	18	24	33
Columbia	63	46	73	59
Crawford	60	39	64	64
Cumberland	51	37	60	62
Dauphin	38	27	42	56
Elk	31	24	32	33
Erie	54	36	60	67
Fayette	46	33	52	58
Forest	55	39	60	54
Franklin	46	34	50	47
Fulton	43	30	47	57
Greene	87	59	101	71
Huntingdon	54	40	62	55
Indiana	59	39	64	64
Jefferson	57	37	60	62
Juniata	46	34	51	50
Lackawanna	41	32	47	47
Lancaster	80	57	92	61
Lawrence	41	28	54	93
Lebanon	53	38	58	53
Lehigh	92	66	108	64
Luzerne	43	33	47	42
Lycoming	31	24	33	38
McKean	40	30	45	50
Mercer	62	40	65	63
Mifflin	43	32	46	44
Monroe	28	22	32	45
Montour	104	72	115	60
Northhampton	71	51	82	61
Northumberland	38	26	42	62
Perry	52	37	57	54
Pike	29	22	31	41
Potter	40	31	45	45
Schuylkill	51	37	55	49
Snyder	44	33	49	48
Somerset	43	29	45	55
Sullivan	34	27	36	33
Susquehanna	51	37	55	49
Tioga	50	38	58	53
Union	38	26	39	50
Venango	49	34	56	65
Warren	46	31	50	61
Washington	94	67	116	73
Wayne	51	38	55	45
Westmoreland	72	48	79	65
Wyoming	42	30	44	47
York	93	69	112	62
Totals	49	34	54	59

"Urban Deer Contraception: The Seven Stages of Grief"
Jay F. Kirkpatrick and John W. Turner, Jr.

Wildlife Society Bulletin, 1997, 25(2):515–519. Reprinted by permission.

Until about 1990, deer contraception was based on techniques so impractical that few people paid attention to it. The public knew little, and what state fish and wildlife agencies knew, they ignored. The media didn't know the technology existed, and regulatory agencies, like the U.S. Food and Drug Administration (FDA), had no policies that applied to it.

With the advent of immunocontraception and demonstrations in the early 1990s that a contraceptive vaccine could be delivered to wild horses (*Equus caballus*; Kirkpatrick et al. 1997), captive exotic species (Kirkpatrick et al. 1996a,b), and white-tailed deer (*Odocoileus virginianus*; Turner et al. 1996), the subject rapidly emerged from obscurity to become immensely controversial and then sensational. Today, while the science of wildlife immunocontraception moves quickly along, its actual application to white-tailed deer is bogged down in a morass of social and political turmoil. It is clear that the science has outstripped our social and political capacities to deal with it. Thus, the overriding question for the authors—and probably for many others—is why is there so much controversy, irrationality, and even hostility surrounding deer contraception? We will try to explore this subject in the context of our own experiences with immunocontraceptive trials in free-roaming populations. The story that follows is a distillate of our experience in about 30 communities and parks and involving hundreds of people. We have made the names in this story fictitious, but the content is, unfortunately, real.

The Call

The adventure begins, usually, when a group of residents from the Town of East Overshoe calls one of us and begs for help in saving their urban deer from a planned hunt or cull. These are generally nice people who dislike the killing of animals in general, and in their backyards in particular. The first and most consistent characteristic we notice about them is that they have absolutely no legal authority to do anything about the deer. For many years we consented to come to their communities to discuss the subject in a public forum.

The Town Meeting

The town meeting is a reliably consistent phenomenon. Its participants include (1) those who want to save "their" particular deer, (2) those who object to hunting in general, (3) those who object to management of any kind, (4) those who hate deer for eating their shrubbery or defecating on their lawns, or who believe that the deer will give them Lyme disease or wreck their cars, (5) some township and county officials who want to be reelected, (6) at least 1 representative from the state fish and wildlife agency, (7) some shotgun hunters, (8) some bow hunters, (9) a representative from either an animal-rights or an animal-welfare organization, and (10) the media.

We take about 30 minutes to present our talk on state-of-the-art deer contraception. After that, almost everyone ignores us and the real show begins. We mostly watch and listen. The discussion begins with a review of the evidence that there are too many deer. The deer eat the blue flowers of those gardeners who grow blue flowers, and their lawns are scattered with deer feces. Numerous testimonies are given on the number of deer–vehicle collisions in the area, the number of cases of Lyme disease in town, and the process by which deer have destroyed the town's only remaining 1-ha woodlot (the town's other woodlot was flattened the previous year for a mini-mall). Finally, the estimate of deer living in East Overshoe is given as somewhere between 500 and 2,000.

The town's bird-watching club provides data (the only real data other than our report that finds its way into this meeting) indicating that over the past 15 years the number of bird species has declined by 64% in East Overshoe. They blame it on the condition of the woodlot and, therefore, the deer. They do not point out that in the last 15 years, suburban development has reduced the extent of forest in East Overshoe by 87%.

There is an immediate rebuttal by the deer lovers. Those who grow red flowers, which the deer don't eat, don't seem to have any trouble, and the droppings on their lawns go unnoticed or, they explain, are easily removed by raking. While these people acknowledge there have been some deer–vehicle collisions, they attribute the accidents to a lack of reflectors and signs, and to people who drive too fast. This group doesn't have much to say about the woodlot, largely because they value the deer more than the trees. They estimate the deer population to be somewhere around 45 animals.

A passionate speech is now made by an animal-rights representative, who, disliking any form of hunting, cites the moral and ethical dimensions of killing animals and argues, "Let nature take its course." This is followed by a more calm and reasoned speech by an animal-welfare representative (who is ignored by the animal-rights representative) about the need for more tolerant attitudes about urban wildlife, the risks of wounding, and the dangers of using lethal weapons within the town boundaries. The animal-rights representative jumps up at this point and makes it clear that immunocontraception is the means by which all sport hunting can be ended, and an audible stir occurs in the hunter groups. This last comment is made despite the description we provided earlier of deer contraception—a process that requires getting to within 30–40 m of each deer and darting it several times over a 5-year period. At this point the town officials become very quiet, and the media representatives begin to realize that they have before them the makings of a really interesting story.

Now it's the hunters' turn. The shotgun hunters speak of sport, recreation, a wasted resource, and the safety of 00 buckshot. Some speak on behalf of the needy and on how a hunt can feed the hungry. Next the bowhunters speak and give similar testimony. Both hunter groups are concerned about the effects of the contraceptive drugs. "What will happen," they ask, "if someone eats a treated deer?" But just minutes before we have explained that it is totally digestible protein, like the meat itself, and won't pass through the food chain. The hunter groups estimate the deer herd to be about 10,000.

One of the deer lovers who reads quite a bit, points out that 00 buckshot is really not all that safe and that although bow hunting may be described by some as recreation, it is not a management tool. Another deer lover cites anecdotes about wounded deer with arrows sticking out of them running about the community.

The state fish and wildlife agency representative reminds everyone that the state is the only legal entity that has the right to make decisions about deer management. This person is concerned about what the FDA thinks and about the dangers of darts lying about in the city park (although he does not seem to have the same concern about broadhead arrows buried under the same grass). He is courteous, and for his reward, invokes the wrath of almost everyone in the group. When pressed, he admits that he has no idea how many deer are in East Overshoe and no systematic studies to document the health of the town's woodlot. His position is that "too many deer are too many deer" regardless of how many there are.

Each group accuses the other groups of either inflating or deflating the number of deer in town, and when questioned by us, not a single group has any data to support its numbers. Finally, the town–county officials speak up. "Who will pay for contraception?" they ask. The state agency representative makes it clear that the state won't, and the hunters and deer haters agree vociferously. The deer lovers claim that they can raise the necessary money. The mayor asks who will do the work, and the city manager makes it clear that the town won't. The deer lovers indicate that they can find people who will do the work, probably for nothing. We point out that, public intuition to the contrary, it requires significant training and skill to successfully dart wild deer with contraceptives. One of the more thoughtful hunters points out that he has read an article, written by a computer population modeler, reporting it would be impossible to control deer with contraception. We acknowledge the reference but point out that the study employed population data collected 2,000 miles away from a herd of deer living a few miles from the Arctic Circle. The relevance of site-specificity to deer contraception in East Overshoe is pondered by the group for 12 seconds.

Next, a heated discussion of the health of the deer ensues. The deer lovers see only healthy deer, the deer haters see only sickly deer. It soon emerges that during the past year about 100 deer were killed on the local highways, but no one has examined these deer to see if they were healthy or not. The state fish and wildlife agency representative suggests that 50 deer be killed to assess their health. A deer lover points out that these deer will not be healthy after they are killed.

Another deer lover—let's call him Ned—who is an avid reader begins to question the state fish and wildlife agency representative—let's call him Joe. "Do you have nonhuntable deer populations in the state?" he asks. Reluctantly, Joe admits that there are nonhuntable populations here and there. Next, Ned asks him if the state is responsible for the state's wildlife. "Yes," Joe replies. "All the wildlife?" Ned asks. "Yes," responds Joe. "Even the nonhuntable wildlife?" presses Ned. "Yes," replies Joe, knowing exactly where this discussion is going. Ned puffs on his pipe for effect, pauses, and then asks what the state is doing about managing the nonhuntable deer populations. Poor Joe, who has very little to do with making policy and who at that moment wants very much to be somewhere else, says quietly, "Nothing."

At this point the deer lovers try to capitalize on the "nothing" response. The pitch of voices rises and emotions are running high, but all that happens is that all the arguments we have just described begin a second round, and, much later, a third round. The meeting finally ends after exchanges become hostile and insults frequent, without decisions by anyone with legal authority to act on the problem.

For many years we actually were foolish enough to keep accepting invitations to come back to these meetings, which occurred several times a year. Our time and energy resources were used up, and the same old arguments occurred over and over again. In most instances decisions were deferred, in some it was decided to cull deer, and in just a few cases it was decided to try contraception. As we grew wiser, we decided not to return to East Overshoe until someone with the legal authority to do something had actually decided to try contraception. For the sake of this essay, let's assume that the East Overshoe officials determined that there were more deer lovers than deer haters on this particular election year, and that it was in their best interest to try contraception.

The Media: Part I

The newspaper reporters and television people absolutely loved the town meeting. For weeks after the meeting, the papers and television screens are filled with reports of the heated arguments and close-ups of grimaces and snarls. Soon the "Letter to the Editor" sections of the newspapers begin to fill up with indignant letters from all sides. The local sportswriters make fun of the idea of deer contraception and air out all the old questions: "Will men grow breasts if they eat meat from one of these treated animals?" or "How do you get the condoms on the bucks?" or "How will sport hunting survive if this becomes a reality?" In most cases, they disparage us personally (we have grown to philosophically enjoy some of the more vitriolic remarks—after all, we don't live in these communities). Few of the published "facts" regarding the science of deer contraception are correct; the media focuses on the interpersonal conflicts rather than substantive issues. Much of this printed matter is sent to us by residents of East Overshoe, and occasionally we even get a tape or two from the local broadcasts regarding the conflicts. If we had not been at the meeting ourselves, we would hardly recognize what these stories are about. In general, the media merely inflames the issues and offers nothing constructive to the community in the way of education.

The Proposal

We carry out the early field projects ourselves, but know that the deer issue is much bigger than us. So, we involve The Humane Society of the United States (HSUS). Part of their role is to sponsor an Investigational New Animal Drug (INAD) exemption, which provides for FDA authorization to conduct field studies with deer using the contraceptive vaccine. What the FDA wants for each deer field study is a proposal that explains in detail how the project will be carried out, who will be conducting the research, how data will be collected, and so forth. The HSUS provides East Overshoe officials with a generic proposal to aid them with writing this document. But no one in East Overshoe has the time to do this or understands how to copy the generic proposal, and someone at HSUS ends up writing the proposal. During the course of this exercise, several heated arguments break out in East Overshoe between officials and residents, over who will do the work, where exactly will the work be carried out, and who will pay for the project. Finally it is decided that several municipal park employees will do the work on their off hours and that several philanthropists in town will provide much of the funding. The battle over the precise site of the work is more intense, because everyone with 3 deer in their backyard wants the project to be carried out with "their" deer. The proposal is written and is sent to FDA, which reviews it and sends back a letter full of comments. This causes immediate confusion within the community. Has the FDA approved the project? No, the FDA has merely "commented" upon the project, setting conditions and recommending changes. Unless the FDA specifically says it cannot go ahead, it can go ahead, because there is already an INAD exemption for the vaccine, permitting research in animals. No one in the community understands that "approval" by the FDA is reserved for the manufacture and marketing of commercial drugs, and has nothing to do with our contraceptive vaccine.

After FDA has commented, the proposal next goes to the state fish and wildlife agency. In this agency, several heated discussions are held behind closed doors. One group passionately opposes the project because they fear that 3 park employees with dart guns will bring an end to hunting in America. Another group opposes the project largely because they dislike HSUS. Fortunately, cooler heads prevail and some people, who understand that deer contraception and sport hunting are separate and unrelated issues, approve the project. In fact, several of these people are relieved that they do not have to authorize either a public hunt or a bait-and-shoot program in a town park rimmed with people who already dislike the idea of killing semi-tame deer in downtown East Overshoe. Some don't even care if contraception works; they are just thankful that the problem has been dumped into someone else's lap.

The Project

This is the only part of the entire saga that goes fairly well. The park employees are sent off to Washington, D.C., where they are trained by HSUS. They begin to learn how to box-trap deer and how to chemically immobilize deer for the purposes of eartagging. They work with an array of capture guns and darts. In class they learn how the vaccine works, strategies for delivering the vaccine, regulatory issues, and even the importance of a good education program to keep the public informed. Soon after, they begin their actual work.

In the few field projects we have conducted thus far, deer contraception works fairly well, with fertility reductions of 72–86% (Kirkpatrick et al. 1996, McShea et al. 1997). Urban deer are reasonably accessible, and using bait stations we average a time investment of about 1 person-hour per deer. A dart that leaves a visible dye mark on the treated animal at the same time it injects the vaccine has been developed and used with some success (R. Naugle, HSUS, Gaithersburg, Md, pers. commun.). It has been possible to deliver the contraceptive vaccine to a significant proportion of the wild deer, and contraceptive efficacy has been good enough to stop herd growth and bring about a small decline after only 3 years of treatment (B. Underwood, Natl. Biol. Surv., Syracuse, N.Y., pers. commun.). Our lessons have been that obstacles to deer contraception are social and political, not a lack of science. And, while the scientific dimensions of a project move smoothly ahead (in a relative sense), the social and political dimensions do not.

The Media: Part II

Now that park employees are out there in the park, darting deer with the vaccine, the media rushes forth to record the event. Reporters try to tag along after the researchers, stumbling over logs and frightening the deer. It's worse with the television personnel, who follow along with a camera man, a sound man, and a correspondent. The whole scene must be awesome to a previously unmanaged deer. Few deer get darted when this retinue of followers are present, and that only means that they stay longer. However, most remarkably, and despite the fact that only 3 deer have been treated in East Overshoe's park so far, the media declares the project a success!

Epilogue

Although the project has moved along with reasonable success, some people in the community think it is moving too slowly and should be expanded to other areas. They begin harassing town officials. After receiving no satisfaction at the town hall, they propose creating a private foundation that will take over the project, and they demand that HSUS provide the vaccine to them. HSUS politely refuses, citing FDA obligations. The disgruntled residents inquire as to where they might buy blowguns. Everyone ignores them and they become more hostile.

A group of hunters, still upset that the project went forward at all, hires an attorney. The attorney writes us a letter demanding that we send him every paper ever published on the subject. Our own legal counsel tells us to ignore the letter. We never hear from them again, but several articles appear in hunting magazines. They claim that contraception won't work and that if it did, the gene pool would be ruined. The next article tells how to kill the biggest bucks. Another group that opposes the contraceptive project contacts the media and tells them about 4 deer that have died mysterious deaths, probably from the vaccine they received. The 4 deer, 3 of which were males that never received the vaccine, died of rat poison that someone had put out on their property. This has already been confirmed by the state's wildlife laboratory, but the media never bothered to ask.

An animal-rights advocate writes a letter to the local newspaper, pointing out once again, that this contraceptive project, which by now has treated 76 deer over 2 years, will be the instrument to end all sport hunting in the Western Hemisphere. This stimulates a barrage of letters from the hunting community to the local newspaper. Incidentally, each new publicity cycle engages a whole new cadre of people who has never heard of wildlife contraception.

The state fish and wildlife agency in the neighboring state, which abhors the idea of deer contraception, contacts the fish and wildlife agency in East Overshoe's state and asks them to stop the project. They cite some new evidence, based on the genetics of inbred chickens, that this "mass immunization" of 76 deer will change the nature of the country's deer herd. They suggest that only healthy animals will respond to the vaccine and soon 20 million North American deer will be unhealthy. Meanwhile, the average life expectancy of deer hunted in their state is about 1.5 years.

The East Overshoe forester, who is responsible for the 1-ha woodlot, complains that while the growth of the deer herd has been stopped in only 2 years, his forest is in no better shape. In reality it took hundreds of deer about 20 years to create the damage, and the woodlot is also used by horseback riders, picnickers, dog walkers and dirt bikers. He insists that the problem be solved overnight. No one considered the prospect that reducing the number of deer may not help the forest at all, because relatively few deer can keep the debilitated forest from regenerating. Someone from a federal agency who knows about these things suggests that East Overshoe either eliminate deer or put a fence around the forest.

By the waning days of the project, 86 deer have been treated and the growth of the herd has slowed or stopped and in some cases the population is beginning to decline. The cost of the scientific effort has been about $50 per deer, but when the costs of the phone calls, town meetings, copying charges, peoples' time, postage, attorney's fees, travel to and from town meetings, and coffee and donuts for the various meetings is calculated, the cost reaches $17,000 per deer. Deer hunting goes on as usual in other portions of the state, and plans have been drawn up to convert one-half of the 1-ha woodlot to yet another mini-mall.

As we prepare to leave East Overshoe we take stock; we are merely researchers trying to solve a problem. The deer are in this fix because people put them there. We suburbanized their historic habitat. Then we built up humanity all around them, so they couldn't get out even if they wanted to. We owe them a solution. Working together, we can find it; the solution will most certainly be a compromise. But until we put aside our egotism, territorialism, and defensiveness and sort through the facts as a focused, interdisciplinary team, all of us and the deer will suffer.

We can't extend this story much further; it is at the point where we find ourselves now. What the future will hold is anyone's guess. One thing is certain though; we will probably spend more time conducting research on contraception in wild horses, zoo animals, and even elephants (Fayrer–Hosken et al. 1997). Although none of these 3 endeavors are apolitical, the problems they arouse pale when compared to the turmoil and emotions aroused by deer contraception.

reading 5

Pennsylvania Deer Statistics 1982–1998

Pennsylvania Game Commission, Bureau of Wildlife Management, Spring 1999

The purpose of this publication is to summarize important data collected by the Game Commission to provide trend information for Pennsylvania's deer herd. Each year, personnel collect data on the proportion of hunters who return harvest report cards, age structure of the population, reproductive rates, sex ratios, habitat information, number of deer killed by dogs, the number poached, and the number destroyed for crop damage. These data are used to estimate deer harvests and populations.

For each county, we present three graphs depicting important population and harvest statistics. **First, population density is presented as number of deer per square mile of forested habitat following the hunting season.** The population density goal is the density of deer that can be supported during the winter without adversely affecting forest regeneration: 60 deer/mi2 in seedling/sapling habitat; 5 deer/mi2 in pole timber; and 20 deer/mi2 in saw timber. In special regulation counties (Allegheny, Bucks, Chester, Delaware, Philadelphia, and Montgomery County), prior to 1992, we set the population goal at 50% below carrying capacity. Since 1992, the population goal in special regulation counties has been 5 deer per square mile of forested habitat. In non-forested habitats the population goal is 0 deer per square mile. Note that the scale for the population density graph is the same for all counties so that graphs for different counties can be compared directly.

Second, harvest of antlered and antlerless deer is presented per square mile of forested habitat. The harvest of antlered deer is primarily a function of deer abundance and hunger effort, and the harvest of antlerless deer is primarily a function of the number of permits issued. By checking license numbers on harvest tags of field-checked deer against the license numbers on report cards returned to the Game Commission, we can determine the proportion of hunters who report their deer harvest. In turn, this information can be used to adjust the reported harvest to reflect the actual harvest of deer. For example, if we receive 100 harvest report cards, but only 50% of the field-checked deer were reported, we know that 200 deer were harvested. Note that the scale of the vertical axis for this graph is different for each county.

Third, we present hunter reporting rates of harvested deer. By law, hunters must report the harvest of a deer to the Game Commission. Currently, only about 50% of the hunters comply with this regulation. Knowing the number of deer harvested each year is important information for managing the resource. We depend on sportsmen to provide harvest data. Successful deer management begins with the cooperation of sportsmen.

DEER HARVEST
Pennsylvania, 1982-98

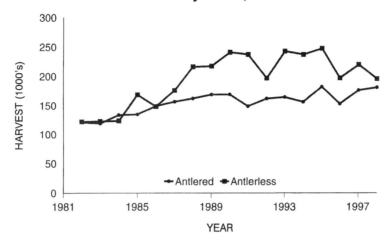

DEER DENSITY
Pennsylvania, 1982-98

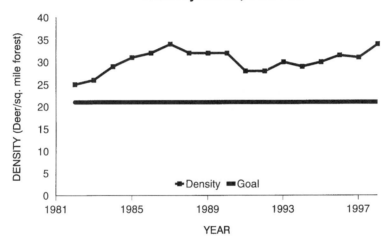

REPORTING RATE FOR HARVEST REPORT CARDS
Pennsylvania, 1982-98

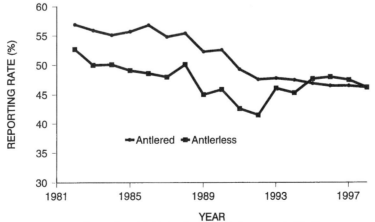

Reporting Rate = Proportion of field-checked deer that are reported to Game Commission (>30,000 deer field-checked per year).

Questions for Discussion

Read all of the readings before answering the questions.

1. What options are available for managing the white-tailed deer in the United States? Discuss techniques of managing numbers (chiefly through regulated hunting), managing habitat (reforestation, regulating timber harvests), and managing people (education). How can each of these be done?

2. In the paper by Diefenbach, Palmer, and Shope, hunting was the only method mentioned for regulating deer numbers. The paper by Kirkpatrick and Turner is a somewhat tongue-in-check evaluation (exposé?) of the use of birth control methods to control deer populations. According to one of the authors (Kirkpatrick), some of the details were overstated (costs per deer probably didn't really approach $17,000 per deer), but all incidents reported were experienced by the authors. How might implementation of these methods be facilitated? What is the role of science in this issue? What role would you play if a similar issue arose in your community?

3. Another possible method of reducing deer numbers might be reintroduction of predators. Both wolves and mountain lions formerly lived in Pennsylvania. How do you suppose various interest groups would respond to this proposal?

4. Notice how interest groups overlap and conflict in complex ways. Hunters and preservationists may come together to push for more deer, but disagree on bag limits and length of season. Farmers and hunters both want long hunting seasons and large bag limits, but disagree on what is the most desirable population level. Farmers who lease land to hunters might disagree with with other farmers who use land more traditionally. Where do ecologists, who want balanced ecosystems, fit in?

5. The three graphs for the "Pennsylvania Deer Statistics 1982–1998" show some interesting trends. How would you describe recent trends in deer numbers (expressed as density) and deer harvests? Are these data conflicting? How can both be correct? The reporting rate for harvesting deer has steadily decreased since 1985. What does this mean? What recommendations could you make for the Pennsylvania Game Commission to reverse this trend? How does the information contained in these graphs compare to graphs presented in the Diefenbach, et al. paper?

6. According to the report, what would be the optimal deer population in Pennsylvania? Are actual numbers moving toward target numbers? Why or why not? Specifically, how do the game biologists propose to bring these numbers into agreement?

7. Is deer population control a widespread problem in North America? How do other states, provinces, and regions manage their deer populations?

8. This topic lends itself to class debates. Indeed, it may not be possible to avoid them.

Going Beyond the Readings

Visit our Website at <www.harcourtcollege.com/lifesci/envicases2> to investigate wildlife overpopulation issues in additional regions of the United States and Canada.

Canada: Alternate Energy Resources

Introduction

Readings:

Reading 1: The La Grande Complex Development and its Main Environmental Issues

Reading 2: James Bay and Northern Québec Agreement and Subsequent Agreements

Reading 3: Economic and Social Development of the Aboriginal Communities

Reading 4: Native People and the Environmental Regime in the James Bay and Northern Québec Agreement

Questions for Discussion

James Bay Hydroelectric Project, Canada

Introduction

An uneasy peace has settled over Eastern Canada. Warily, Native Americans, developers, government authorities, and environmentalists live out an ill-defined truce. Hydro-Québec, one of the world's largest electric utility companies, has no immediate plans to construct more dams on rivers flowing into James Bay. Water built up behind such dams flows through turbines and generates electricity, which Hydro-Québec is in the business of selling. Phases I and II supply Montreal, one of Canada's largest and fastest-growing metropolitan areas, with much of its energy.

Hydro-Québec's proposal to build Phase III is currently on hold. Powerful political and economic interests in Québec Province and elsewhere hope the delay is only temporary. Their reasoning is that the whole region needs or will soon need abundant, inexpensive energy to carry industrial development into the next century. Excess energy can be sold at a profit to neighboring energy-hungry regions, such as the northeastern United States. Québec has rivers—lots of them—and needs a firm, reliable economic base. The power of Québec's rivers can answer these needs; that is the intended purpose of Phase III.

But the project has equally powerful opponents. Cree Indians and Inuit natives (Eskimos) have used these rivers and the lands they drain for thousands of years. They see the rivers as their people's lifeblood. Harnessing them, they fear, will harness the people and bring a cherished lifestyle to an untimely end. Those concerned with the environment lament loss of habitat. Huge dams turn forests, meadows, and bogs into areas unsuited for moose, songbirds, and beaver. Potential environmental impacts, they warn, may affect water and air quality well beyond Québec.

Background

The James Bay region is not an easy area in which to work. Only about 300 miles (480 km) north of the Great Lakes, it is directly linked to the Arctic Ocean through the expansive Hudson Bay. The region was covered with glaciers until only about 5000 years ago, and is fiercely cold in winter. Soils are thin, poor, and support a straggly forest of mainly black spruce. Ground cover is a more diverse community of lichen ("caribou moss"), mosses, and hardy perennial wildflowers. This is taiga, an ecosystem that supports more than 200 bird, 40 mammal, and 20 fish species. If the region is poor biologically, it is rich hydrologically. Nearly one-third of Canada's fresh water flows into James Bay.

Hydro-Québec's dream to convert the energy of these flowing waters into electricity is partially fulfilled. Construction of Phase I started in May 1972. Four major dams and three generating stations were constructed along La Grande River, nearly 620 miles (1000 km) north of Montreal. The Eastmain River to the south and the Caniapiscau River to the north and east were diverted into the La Grande River watershed to ensure water flow and uninterrupted energy flow during dry seasons and years. The magnitude of the project was staggering. Never had anything so huge been attempted so far north in pristine wilderness. During peak construction, 18,000 workers were employed. Workers need amenities such as eating and sleeping quarters, hospitals, and work spaces, so five new towns were carved out of the wilderness, linked by 930 miles (1500 km) of new roads. A network of five power lines comprised of nearly 12,000 towers and 18 relay stations channeled electricity to Montreal. In December 1985, at a cost of more than $13 billion (Canadian), the system was built and functioning, with a capacity to generate more than 10,000 megawatts (MW) of electricity.

Construction of Phase II started two years later. It called for construction of five more generating stations to increase capacity an additional 5,000 MW. This phase involved relatively little modification to the river systems. Construction was completed in 1996.

Hydro-Québec has worked hard to assuage the concerns of native peoples and to minimize environmental impacts. In 1975, after extensive negotiations among Hydro-Québec, Québec Province, the government of Canada, and associations representing Cree and Inuit natives, the James Bay and Northern Québec Agreement was signed. It directly addressed numerous native settlement claims, provided payment of $225 million (Canadian) to the natives, established an extensive housing program, and set up native-controlled administrative infrastructures throughout the region.

The Problem

With a project so large, environmental impacts were, of course, inevitable. Complicating the picture was the fact that when the project started, relatively little was known about ecological aspects of the region. Hydro-Québec was confident that impacts could be minimized. Throughout the 1970s, the James Bay territory was considered to be a vast natural laboratory. During Phase I, more than 200 scientists conducted studies and inventoried biological resources. In 1973, Hydro-Québec and federal and provincial governments signed the Biophysical Agreement, which defined responsibilities for the studies. The Société d'énergie de la Baie James (SDBJ), a subsidiary of Hydro-Québec, assumed overall responsibility for coordinating research and monitoring functions. Their goals were threefold: "Ensure the quality and biological productivity of the milieux affected by construction; permit Native people to carry on their traditional activities; and re-establish an attractive environment near frequented areas." (The La Grande Complex, Phase I, page 53). In 1975, the La Grande Complex Remedial Works Corporation (SOTRAC) was set up to study, plan, and carry out corrective work and programs designed to minimize impacts on traditional native uses of the land. Follow-up programs continue today.

And what of Phase III, officially known as the Grande-Baleine complex, more familiarly referred to as the Great Whale River project? It would create more dams and three additional generating stations 150 miles (240 km) farther north on rivers draining into Hudson Bay. To Hydro-Québec, Phase III is a logical extension of Phases I and II. Building on their experiences on the last 20 plus years, construction should proceed smoothly with minimal disruptions to natives and nature. But Crees and Inuit are opposed. They sold one river, realized questionable benefits, and are being asked to sacrifice another. How many rivers can a people sell and still retain their identity? Environmental interests question the assumption that impacts can be and have been kept minimal. Phase III construction would occur where taiga transitions into tundra—a different ecosystem with new sensitivities. Have previous impacts, critics ask, been truly minimal or glossed over? Has anything in previous studies been missed?

For the moment there is peace. For decades opponents fought. Battles were won and lost on both sides. Warily they rest. But nobody thinks peace will last.

reading 1

"The La Grande Complex Development and its Main Environmental Issues"

Hydro-Québec, Ltd. © 1995 by Hydro-Québec. Reprinted by permission.

Project Context

The James Bay territory lies between the 48th and 55th north parallels and covers 350,000 km² (135,187 sq. mi), an area the size of Germany or about two-thirds of France. The territory, which includes all or part of the drainage basins of six large rivers, is home to some 11,000 Cree Indians, belonging to eight communities scattered across the region and the community of Whapmagoostui located just outside its northern boundary.

The Hydroelectric Development

The basin developed by the La Grande hydroelectric complex is contained within this territory. Covering nearly 177 000 km² (68,366 sq. mi.), it encompasses the drainage basin of La Grande Rivière and waters diverted from the upper basins of the rivière Caniapiscau to the east and the rivière Eastmain to the south. The two-phase project comprises nine generating stations, and required the impounding of seven reservoirs with a total area of 13,577 km² (6130 sq. mi.), including 11,505 km² (4444 sq. mi.) of land surface that was flooded. Total capacity under both phases of the complex will exceed 15,800 MW.

The Grande-Baleine project, with an installed capacity of 3212 MW, is planned for an area north of the La Grande complex. It calls for one control reservoir created out of an enlargement of Lac Bienville, three generating stations, each with its own reservoir, and the diversion of most of the Petite rivière de la Baleine. The reservoirs and diversion lakes would total 3391 km² (1310 sq. mi.) in area, including 1667 km² (644 sq. mi.) of flooded land.

The total land area flooded by the La Grande complex development represent 6.5% of the La Grande Rivière drainage basin. Before the development, nearly 15% of the drainage basin was occupied by water.

An asphalt road more than 700 km (435 mi.) long connects Matagami in the south to Radisson in the north and then continues on to the village of Chisasibi. A secondary gravel road about 600 km (373 mi.) long extends east to the Caniapiscau reservoir. Further south, another secondary access road runs about 100 km (62 mi.) to a few substations and the Cree community of Nemaska. Since 1993, a northern road connects Chibougamau to the main James Bay road by way of the Albanel substation. At the request of Crie communities, permanent roads linking these two villages to the main Matagami-Radisson road were completed in 1995.

Environmental Monitoring and Research

Since the project began, regular monitoring has been conducted on the main parameters of the new ecosystems in the developed basins, such as water quality, fish, and mercury levels in the flesh of fish. Other studies have looked at the habitats of certain terrestrial wildlife and waterfront species, and Native land use for traditional activities as well as recreational and tourist use by non-Native visitors. The projects' social and economic impacts on the Native peoples are also the subject of research. These studies are carried out by government agencies, Hydro-Québec and its subsidiary, Société d'énergie de la Baie James, along with a number of university research centers. The results of this environmental follow-up and research are used in assessing the impacts of future projects and optimizing mitigative measures.

Overall Situation

In search of new sources of energy to meet Québec's needs in the 1980s, Hydro-Québec looked to the vast James Bay territory. The hydroelectric potential of its river basins is considerable. This potential, as described in Information Sheet No. 1 on the environment prior to the La Grande complex development, stems from the impermeable nature of the Canadian Shield rock formations that constitute this region, the characteristics of its topography and hydrology which are the legacy of past glacial events, and present climatic conditions.

"In 1965, Hydro-Québec resumed studies, begun the preceding decade, of the five major rivers flowing into James Bay. Initially, attention concentrated on the development of three rivers in the southern part of the James Bay territory, the Nottaway, Broadback and Rupert. This hydroelectric development, the NBR complex, included the diversion of the upper reaches of the Nottaway and Broadback rivers into the Rupert, on which would be built seven powerhouses.

"In 1970, studies of the James Bay rivers intensified as it became more apparent that hydroelectricity was still the most economical way of satisfying future growth in demand for power. Completion of Churchill Falls generating station and the last installations in the

Manicouagan-Outardes complex by Hydro-Québec would satisfy Québec's needs to the end of the 1970s. The James Bay project was intended to meet needs beyond that date.

"The James Bay development was officially launched in April 1971. In July of that year, the National Assembly of Québec passed the *James Bay Region Development Act,* covering a territory of 350,000 km2, or 20% of Québec. Under this Act, the Société de développement de la Baie James (SBDJ—James Bay Development Corporation) was created, and was given the mandate to promote the development and exploitation of natural resources in the territory and to see to the administration and management of that territory in order to promote its development.

"The *James Bay Region Development Act* specified, however, that development of the hydroelectric resources would be entrusted to a subsidiary, the Société d'énergie de la Baie James (SEBJ—James Bay Energy Corporation), which was incorporated in December 1971 for the specific purpose of developing the hydroelectric potential of the rivers on the Québec side of James Bay. SEBJ, half of whose shares were owned by Hydro-Québec, became a wholly owned subsidiary of Hydro-Québec in 1978, although its initial objective was unchanged.

"On the basis of studies already completed and following additional studies, it was decided in the spring of 1972 that there were definite advantages in beginning the hydroelectric development of the James Bay territory in the northern sector, on the La Grande Rivière, including the diversion of portions of adjacent rivers, rather than in the southern sector with the NBR complex. A number of factors supported this choice, the main ones being the lesser environmental impact of the La Grande Rivière hydroelectric complex and its economic advantages. These were due in particular to favorable geological conditions and to the profile of the river, which allowed its potential to be developed with a small number of medium-head power stations. In January 1974, after the technical, economic and environmental studies had been reviewed by Hydro-Québec, SDBJ and SEBJ, the optimum setup for the La Grande complex was selected."[1]

La Grande, Phase 1

Phase 1 of the hydroelectric development of the La Grande complex began in 1973 and was completed in 1985. It includes three generating stations built under SEBJ supervision: La Grande-2, La Grande-3 and La Grande-4, which are now operated by Hydro-Québec. Each of these powerhouses has its own reservoir. Additional inflows come from the reservoirs of the diverted Caniapiscau, Eastmain and Opinaca rivers. The Caniapiscau reservoir constitutes the head component in the La Grande complex. It provides interannual regulation for the complex, to which it transfers a mean annual flow of around 790 m^3/s. The installed capacity of the three generating stations is 10,282 MW (see Table 1), for annual generation of 62.4 TWh and a utilization factor between 60% and 80%. The area of the reservoirs created in Phase 1 totals 11,335 km^2; their drawdown ranges from 7 m to 12 m.

The diversions of the Eastmain and Opinaca rivers (EOL diversion) and the Caniapiscau (Laforge diversion)

TABLE 1 Main Characteristics of the La Grande Complex (Phase 1)

Project	Level (m) Maximal	Level (m) Minimal	Active storage (millions of m^3)	Reservoir area (km^2) (maximum level)	Number of generating units	Installed capacity (MW)	Annual energy (TWh)	Design flow[1] (m^3/s)	Net rated head (m)
La Grande-2	175.3	167.6	19,365	2,835	16	5,328	35.2	4,300	137.2
La Grande-3	256.0	243.8	25,200	2,420	12	2,304	12.6	3,260	79.2
La Grande-4	377.0	366.0	7,160	765	9	2,650	14.6	2,520	116.7
Caniapiscau	535.5	522.2	39,070	4,275	—	—	—	1,130	—
EOL	215.8	211.8	3,395	1,040	—	—	—	1,980	—
Total			**94,190**	**11,335**	**37**	**10,282**	**62.4**		

[1](Excerpted from La Grande Rivíré: A Development in Accord with its Environment—SEBJ, 1988.)

to the La Grande Rivière doubled the latter river's energy potential. The EOL diversion led to a 90% reduction in flow in the Eastmain, while the Laforge diversion reduced the flow of the Caniapiscau by 40%, measured at the rivers' respective mouths. As a result of these substantial hydraulic inflows, the mean annual flow of La Grande Rivière, at its mouth, rose to 3,400 m^3/s from 1,700 m^3/s, and its mean winter flow was increased eightfold.

La Grande, Phase 2

The installations plan for Phase 2 of the La Grande complex calls for the construction of six generating stations: La Grande-1, La Grande-2-A, Laforge-1, Laforge-2, Brisay and, eventually, Eastmain-1 (see Table 2). Work on Phase 2 began in 1987. La Grande-1 and Laforge-2 generating stations were scheduled for commissioning in 1995 and 1996, respectively. The other three stations are already in operation. Negotiations pertaining to Eastmain-1 powerhouse are still under way between the Native people and Hydro-Québec. Altogether, the six generating stations should add 5.594 MW to the total installed capacity of the complex, along with annual output of 21.1 TWh with a utilization factor of around 60%.

Only three new reservoirs would be required: the La Grande 1, Laforge 1 and Eastmain 1 reservoirs. The other powerhouses use reservoirs already included in Phase 1. The area of these new reservoirs is 2242 km^2, and their drawdown fluctuates between 1.5 m and 13 m.

Phases 1 and 2

Once the nine powerhouses in both phases are built, the total installed capacity of the La Grande complex would be approximately 15,784 MW, with annual generation of 83.1 TWh. The developed drainage basin of La Grande Rivière (phases 1 and 2) covers an area of 176,800 km2. When development of the La Grande complex is complete, the total reservoir area will be 15,876 km2. Taking into account natural bodies of water that already existed, the area of land flooded by the reservoirs equals 6.5% of the total area of the La Grande Rivière drainage basin.

To carry the generated electricity to the load centres in the south, six energy transmission lines were built: five 735-kV lines and one 450-kV line. This system, which runs more than 10,000 km in total length, serves Québecers mainly, and also allows the export of surplus electricity to the United States.

Environmental Issues

When development of the hydroelectric potential of the James Bay territory was initiated, a federal-provincial task force made up of nationally recognized authorities was formed to evaluate the environmental conse-

TABLE 2 Main Characteristics of the La Grande Complex (Phase 2)

Project	Level (m) Maximal	Level (m) Minimal	Active storage (millions of m^3)	Reservoir area (km^2) (maximum level)	Number of generating units	Installed capacity (MW)	Annual energy (TWh)	Design flow[1] (m^3/s)	Net rated head (m)
La Grande-2-A	175.3[1]	167.6[1]	(19,365)[1]	(2,835)[1]	6	1,998	2.2	1,620	138.5
La Grande-1	32.0	30.5	98	70.2	12	1,368	7.5	5,950	27.5
Eastmain-1[2]	283.0	274.0	4,210	624	4	640	2.8	1,120	63.0
Laforge-1	439.0	431.0	6,857	1,288	6	838	4.5	1,613	57.3
Laforge-2	481.0	479.5	390	260	2	304	1.8	1,200	27.4
Brisay	535.5[3]	522.2[3]	(39,070)[3]	(4,275)[3]	2	446	2.3	1,130	37.5
Total			11,555	2,242.2	32	5,594	21.1		

[1] Refers to La Grande 2 reservoir, already included in Phase 1.
[2] Negotiations for this project are under way with the Crees.
[3] Refers to Caniapiscau reservoir, already included in Phase 1.

quences of this development. Since many aspects of this region were unknown, the group therefore recommended that the La Grande complex development project be considered a vast natural laboratory, where multidisciplinary research and studies could be conducted to determine how the ecological processes would be modified by the development scheme. Subsequently, following this group's recommendation, the overall environmental issues of this hydroelectric development were closely monitored during the construction phase by the project developer, the Société d'énergie de la Baie James, and then by Hydro-Québec once the facilities were in operation. As both corporations have enshrined environmental protection and enhancement in their environment policies, they have acted to ensure the greatest possible harmony in integrating the hydroelectric complex into its environment.

The environmental issues, in other words the main environmental concerns related to the various technical components of the development of the La Grande complex as previously described, result mainly from changes in the hydrological regimens of the rivers involved and in the land areas affected by the construction of the facilities and work camps as well as the infrastructure. Indirectly, these physical changes have an impact on wildlife resources and habitats and on the users of these resources, in particular the Native people.

The environmental issues and themes discussed in these information sheets take into account the environmental concerns of Québecers as a whole as well as those expressed by a number of groups of ecologists, academics and others, both in Canada and abroad. These concerns relate principally to the anticipated impacts of the development of the La Grande hydroelectric complex on the natural and human environments involved.

reading 2

"James Bay and Northern Québec Agreement and Subsequent Agreements"

Hydro-Québec, Ltd. © 1996 by Hydro-Québec. Reprinted by permission.

James Bay and Northern Québec Agreement

The James Bay and Northern Québec territory is part of the domain formerly granted to the Hudson's Bay Company in 1670 by King Charles II of England, a domain then called "Rupert's Land" and ceded to Canada in 1870. On two occasions thereafter, the Canadian Parliament enlarged the area of the Province of Québec with portions of this same territory: the provincial boundaries were moved north to the Rupert River, near the 52nd parallel, in 1898, and then, in 1912, as far as Hudson Strait.

The **Québec Boundaries Extension Act, 1912**, specified "that the province of Québec will recognize the rights of the Indian inhabitants in the territory above described to the same extent, and will obtain surrenders of such rights in the same manner, as the Government of Canada has heretofore recognized such rights and has obtained surrender thereof. . . "

When the decision was made in 1971 to develop the hydroelectric potential of the James Bay drainage basin, very little was known about this region in biophysical and human terms, and environmental concerns were still in an early stage. The development of the La Grande complex was likely to bring about physical and biological changes that could affect the Aboriginal peoples' way of life. Understanding this region and its inhabitants, in order to determine the impacts of the proposed developments, constituted a major issue (see Information Sheet No. 2). Another issue consisted in determining, formulating and implementing measures able to mitigate the negative impacts and enhance the positive impacts. Furthermore, the development of the hydroelectric resources of Northern Québec raised the question of the ancestral rights of the approximately 5000 Crees, 3500 Inuit and 400 Naskapis then living in the James Bay and Northern Québec region, territory covering an area of 1 million km^2.

Québec honored its obligation to recognize and restore the rights of the territory's Aboriginal people by signing, in 1975, the **James Bay and Northern Québec Agreement** (JBNQA) with the Crees and Inuit of Québec and, in 1978, the Northeastern Québec Agreement (NQA) with the Naskapis of Québec. In signing these agreements, the Crees, Inuit and Haskapis of Québec exchanged their ancestral rights, of indefinite scope and nature, for clear, defined rights, lands, financial compensation (see Tables 1 and 2) and other benefits.

TABLE 1 James Bay and Northern Québec Agreement (in millions of $)

	Compensation			Mitigative Measures	Total
	Canada	Québec	Hydro-Québec	Hydro-Québec	
Québec Inuit	13.8	48.8	30.4		93.0
Québec Crees	20.7	72.7	45.4	30.0	168.8

TABLE 2 Northeastern Québec Agreement (in millions of $)

	Compensation			Total
	Canada	Québec	Hydro-Québec	
Naskapis	1.585	5.065	3.000	9.650
Crees	0.037	0.112		0.150
Inuit	0.037	0.112		0.150

Philosophy and Principles

"This Agreement has enabled us to accomplish two great tasks to which the government committed itself. It enables us to fulfill our obligations to the aboriginal peoples who inhabit our North, and to affirm finally Québec's presence throughout its entire territory.

"In undertaking the negotiations with the aboriginal peoples, we have followed two guiding principles, two principles of equal importance. The first is that Québec needs to use the resources of its territory, all its territory, for the benefit of all its people. The use of these resources must be reasonably planned. The future needs of the people of Québec must be anticipated.

"The second principle is that we must recognize the needs of the aboriginal peoples, the Crees and the Inuit, who have a different culture and a different way of life from those of other peoples of Québec."[1]

The provisions of the JBNQA cover 30 chapters. The Crees and Inuit gain financial compensation, lands and defined rights in several areas such as local autonomy, harvesting of wildlife resources and pursuit of traditional activities, economic development, administration of justice, health, social services, education and environmental protection. The Naskapis of Québec obtained similar rights under the NQA. Certain rights concerning wildlife harvesting and the environmental regime were included in the JBNQA through an amendment to that Agreement (Complementary Agreement No. 1).

A central element in these two agreements is the land regime. The James Bay territory was subdivided into three categories of land. Category I lands (14,022 km^2) were allocated to each of the Cree and Inuit communities and the Naskapi community, for their exclusive use. The villages are located on Category I lands.

The Cree and Naskapi Category I lands are subdivided into IA and IB lands. Lands in Category IA were assigned by Québec to the federal government for the benefit of Crees and Naskapis who wished to remain under the jurisdiction of the federal government; the Cree villages and the Naskapi village are located on these lands. The federal government passed the Cree-Naskapi (of Québec) Act, which replaces the Indian Act as far as local administration on these lands is concerned. Ownership of Category IB lands was handed over to land corporations made up of members of each of the communities. These corporations are set up as municipalities under Québec jurisdiction and run by the council that administers IA lands. The Inuit villages were established as municipalities under jurisdiction of the Québec government. Ownership of the Category I lands of each of the Inuit communities was assigned to land corporations whose members are the Inuit beneficiaries of the JBNQA in each of these communities.

Category II lands (155,736 km^2) adjoin Category I lands. They are public lands where the signatory Aboriginal peoples have exclusive hunting and fishing rights. Category III lands (896,242 km^2) are the remaining public lands, where Aboriginal people are entitled to pursue their harvesting activities under the hunting, fishing and trapping regime set forth in the JBNQA. Throughout most of these lands, the Agreement's beneficiaries enjoy the exclusive right to trap, as well as exclusive harvesting rights for certain wildlife species.

The JBNQA has consequently led to the rational organization of the territory of James Bay and Northern Québec (1,066,000 km^2, or 410,000 sq. mi.). It has also allowed the orderly development of the region, while complying with the commitments made to the Aboriginal peoples.

Hydroelectric Developments

Chapter 8 of the JBNQA contains all of the technical provisions, remedial measures and obligations related to the development of *Le Complexe La Grande (1975)* as conceived at the time of signing the JBNQA.

Since its construction was under way when the JBNQA was signed, the parties agreed to exempt *Le Complexe La Grande (1975)* from the environmental regime set forth in the JBNQA. However, to ensure the protection of the biophysical and human environments, the Agreement provided for the creation of three bodies with Aboriginal members and targeting specific objectives.

First of all, the Environmental Expert Committee, established before the signing of the JBNQA to advise SEBJ on its environmental protection program, was confirmed in this role by the Agreement. The JBNQA allowed the Crees and Inuit to name one representative each to this Committee.

It was difficult to predict the consequences which the La Grande complex would have on traditional wildlife harvesting activities (see information sheets no. 2 and 15). To this end, a second agency, the La Grande Complex Remedial Works Corporation (SOTRAC), was formed. The board of directors of this joint body is made up of an equal number of representatives of the Crees and SEBJ. SOTRAC's mandate covered everything connected with the Cree's hunting, fishing and trapping activities. A budget of $30 million was granted to SOTRAC to study, plan and implement remedial measures during construction and operation of the complex, with a view to mitigating the negative impacts of *Le Complexe La Grande (1975)* on the Crees' hunting, fishing and trapping activities.

A third body, the Caniapiscau-Koksoak Joint Study Group (GECCK), was set up to study the impacts of the Caniapiscau diversion, downstream from the diversion

---

[1] Excerpt for the opening remarks made by John Ciaccia, member of the National Assembly for Mount Royal and special representative of Premier Robert Bourassa in the James Bay negotiations, on November 5, 1975 at the opening of the standing Parliamentary Committee of the National Assembly of Québec on Natural Resources convened to examine the Agreement with the James Bay Crees and the Inuit of Québec, prior to its signature. Unoffical translation of remarks which were made in the French language.

point, on fish and wildlife north of the 55th parallel, and particularly on the fish harvest by the people of Kuujjuaq. It was also meant to recommend remedial measures to be carried out by SEBJ. The municipal corporation of Kuujjuaq could name one member to this group. Following the signature of the NQA, the Naskapis also were able to appoint one representative, and GECCK become responsible for studying the possibility of partially regulating the flow of water in the Caniapiscau drainage basin. Upon examination, this measure proved unnecessary from an ecological standpoint.

Chapter 8 further contains a number of technical provisions concerning the aims and principles of certain remedial measures in the reservoirs and diversion zones, such as deforestation and flow control. It also lists specific commitments aimed at improving living conditions in some Cree villages, such as drinking water and power supply. Finally, Hydro-Québec and SEBJ undertook to apply preferential means to enable the Crees to obtain competitive jobs and contracts on Le Complexe La Grande (1975), commensurate with their abilities.

The possibility of developing the Nottaway, Broadback and Rupert rivers (NBR complex) and the Grande-Baleine complex is also explicitly mentioned, and the chapter contains a brief description of these developments. It states that these two projects, along with any substantial change or addition to Le Complexe La Grande (1975), are considered future projects subject to the environmental regime set forth in the JBNQA, but only as far as their ecological impacts are concerned.

Amendments to the JBNQA and New Agreements

A clause in Section 8 of the JBNQA permits Hydro-Québec and SEBJ to conclude agreements covering mitigative, remedial and compensatory measures for future projects. In addition, the JBNQA allows the signatory parties concerned to amend the provisions of the Agreement.

Accordingly, between 1978 and 1993, 12 amendments were made to the JBNQA (Complementary Agreements No. 1 to 12). Hydro-Québec and SEBJ signed five new agreements with the Québec Crees for additions and changes to the La Grande complex (1975). Following GECCK's studies, another agreement was reached with the Inuit, to obtain a release regarding the impacts of the Caniapiscau diversion on the use of wildlife resources north of the 55th parallel. These six subsequent agreements led to the incorporation of five amendments to the JBNQA (Complementary Agreements No. 4, 5, 7, 9 and 11).

Chisasibi Agreement

(Complementary Agreement No. 4)

This agreement was signed on April 14, 1978. It covered two main subjects: the relocation of the Crees of Fort George and the change in site of La Grande-1 generating station. The Société de développement de la Baie James and the governments of Canada and Québec are also signatories of this agreement.

The Crees of Fort George living on le de Fort-George (also referred to as Governor's island) had expressed the desire to move their village to the mainland, to a new site called Chisasibi (see Information Sheet No. 14). The Agreement permitted this relocation and specified its terms and conditions. It stated that the Canadian government and SEBJ would allot $50 million to new housing, to replace the letters of commitment and obligation originally drawn up for the village of Fort George (see Table 3).

Complementary Agreement No. 4 permitted the construction of a revised version of La Grande-1, at

TABLE 3 Subsequent Agreements (in millions of $)

Agreement	Compensation	Mitigative Measures (measures, studies, infrastructures)			Total
	Hydro-Québec	Canada	Québec	Hydro-Québec	
Chisasibi (Crees)		10.0		40.0	50.0
Sakami Lake (Crees)	8.0			17.5	25.5
La Grande (1986) (Crees)	97.0			15.0	112.0
Mercury (1986) (Crees)			4.4	12.4	16.9
Opimiscow (Crees)	50.9[1]			25.0	75.9
Kuujjuaq (1988) (Inuit)	34.5			14.0	48.5

mile 23 on the La Grande Rivière, instead of the version planned at mile 44. The Chisasibi Agreement also provided for remedial measures in connection with La Grande-1, Revision 1.

Sakami Lake Agreement

(Complementary Agreement No. 5)

This agreement was signed on July 4, 1979. Its purpose was to provide for remedial measures and other benefits for the Cree community of Wemindji, following the rise in the level of Lac Sakami.

The JBNQA stipulates that, for environmental reasons, the maximum level of Lac Sakami must not normally exceed the maximum level officially recorded in past years. The work necessary to comply with this obligation proved to entail considerable expense, without ensuring compliance, however. Complementary Agreement No. 5 permits the level of Lac Sakami to rise to a level that must not normally exceed 613 feet above average sea level.

This complementary agreement specifies various remedial and compensatory measures. SEBJ contributed $8 million to a development fund to improve living conditions for the community of Wemindji. SEBJ further agreed to carry out $17.5 million worth of infrastructure work, for the benefit of the community of Wemindji (see Table 3). Finally, SEBJ undertook to carry out remedial work in the Lac Sakami area related to the diversion of the Eastmain and Opinaca rivers, wildlife and the use of this region by the Crees.

The Agreement set up the Sakami Eeyou Corporation, mandated to plan and implement the compensatory measures laid out by the Agreement and to see to the well-being of the Cree community of Wemindji. This agency also manages the development fund.

La Grande (1986) Agreement

(Complementary Agreement No. 7)

This agreement was signed on November 6, 1986. Its objective was to determine remedial and mitigative measures for the La Grande-1 (1986), La Grande-2-A, Brisay and Radisson-Nicolet-Des Cantons energy transmission line projects, community benefits, economic measures and other measures in favor of the James Bay Crees. Complementary Agreement No. 7 amends the description of *Le Complexe La Grande (1975)* to include these projects.

The Agreement also provided for the creation of the James Bay Eeyou Corporation to ensure a more efficient structure for governing relations between Hydro-Québec and the James Bay Crees. The assets, rights, interests and obligations of SOTRAC, established under the JBNQA, were then transferred to the Eeyou Corporation.

A mitigative measures fund was constituted (SOTRAC 1986), comprising the balance of SOTRAC's funds—$17 million—plus an additional $15 million paid by Hydro-Québec under the present agreement (see Table 3). The Eeyou Corporation, which succeeded SOTRAC, administers this fund in the manner prescribed for SOTRAC under th JBNQA or for any new mitigative measure approved by majority vote of the directors.

Hydro-Québec contributes $50 million to the Cree Community Fund and $45 million to the Cree Economic Assistance Fund, both to benefit the Cree communities of Québec.

The James Bay Eeyou Corporation performs SOTRAC's previous functions, manages the funds mentioned above and acts as a permanent forum to handle more efficiently matters affecting the James Bay Crees and Hydro-Québec. The board of directors of the James Bay Eeyou Corporation is made up of 20 Cree members and four members appointed by Hydro-Québec.

Under the heading of mitigative measures, Hydro-Québec also undertook to carry out work such as facilitating access to the north bank of the La Grande Rivière by building a road, supplying the community of Chisasibi with a reliable water intake, facilitating use of Hydro-Québec's infrastructures by the Crees and setting up, in cooperation with SEBJ, the Environmental Expert Committee to advise Hydro-Québec and SEBJ, with two members designated by the James Bay Eeyou Corporation. Additional studies were also planned on the long-term overall impact of hydroelectric projects, including mercury, which ws the subject of a separate agreement.

The Agreement contains provisions concerning the connection of five Cree villages to Hydro-Québec's power grid, as well as clauses on the training and employment of Crees for operation of *Le Complexe La Grande (1975)*.

Mercury (1986) Agreement

The Mercury Agreement was signed on November 6, 1986, at the same time as the La Grande (1986) Agreement. The object of this agreement was to institute a mercury program to reduce health risks and provide for remedial measures that would enable the Crees to carry on their harvesting activities and preserve their way of life.

This program is managed by the seven-member James Bay Mercury Committee. The Cree Regional Authority (CRA) and Hydro-Québec each designate two members, while the Québec government and the Cree Board of Health and Social Services of James Bay each name one. The seventh, non-voting member, the

Chairman, is appointed by the Québec government on recommendation of the CRA and Hydro-Québec.

The program involves studying, evaluating and formulating remedial measures in three distinct areas—health, environment, and sociocultural and economic activities—so that the Crees can continue their harvesting activities (see Information Sheet No. 7).

Subject to prior consultation with the Mercury Committee, the Cree Board of Health and Social Services of James Bay carries out the health-related part of the program. Hydro-Québec looks after the rest of the program, which covers the other two areas, namely environment and the sociocultural and economic aspect. An amount of $16.9 million was allocated to implementing the mercury program from 1987 to 1996, inclusive (see Table 3).

Kuujjuaq (1988) Agreement

(Complementary Agreement No. 9)

This agreement was signed on October 21, 1988. Its object was to agree upon community and individual benefits, economic measures and other measures in favor of the Inuit, in order to fulfill the obligations set forth in the JBNQA with respect to the diversion of the rivière Caniapiscau.

Complementary Agreement No. 9 amends the JBNQA: the Inuit discharge Hydro-Québec and SEBJ from all impacts due to the diversion of the Caniapiscau. However, this discharge does not apply to the effects which might be caused by the production of methyl-mercury north of the 55th parallel as a result of *Le Complexe La Grande (1975)* or any other hydroelectric development.

Under the Agreement, Hydro-Québec pays $48 million, distributed over seven different funds, for remedial measures, compensation, studies and development to benefit the Inuit (see Table 3). Some of the funds are meant to compensate for damage or inconveniences caused by the drop in water level and declines in fish catches, while others are designed to promote the socioeconomic development of the Inuit of Kuujjuaq and of Québec. Funds are also granted for remedial work.

Opimiscow Agreement

(Complementary Agreement No. 11)

This agreement was signed on January 8, 1993. It covers the following projects: Laforge-1, Laforge-2, 3rd Le Moyne-Tilly line, 2nd La Grande-2-A–Radisson line, 12th transmission line, the series capacitor project and the series compensation project for the northwestern grid (Abitibi, Albanel, Chibougamau and Némiscau substations).

Complementary Agreement No. 11 consequently amends the description of *Le Complexe La Grande (1975)*.

The purpose of this agreement is to specify additional environmental and remedial measures for the projects, protect the Crees' way of life, facilitate the implementation of these projects, enhance the Crees' community development and provide for other benefits for them.

Payments spread over 50 years are allocated to four funds for the Cree's benefit. These payments represent $50 million in 1992 dollars (see Table 3). There are three community funds: one for the benefit of the Cree Nation, one benefiting the Crees of Chisasibi and the third for the Crees of Wemindji. A fourth fund, the Indoho Fund, was set up to promote the Crees' traditional activities and mitigate the projects' impacts on them.

Lastly, a remedial measures fund was established for carrying out remedial and environmental measures during project construction. Hydro-Québec contributes $25 million to this work and other remedial measures.

The Agreement provides for the creation of the La Grande Complex Remedial Works Corporation (1992), also known as the Opimiscow-Sotrac Company. An equal number of representatives of the Crees and Hydro-Québec sit on the board of directors. The Company administers the Indoho Fund, studies, plans, evaluates and authorizes the work and program financed by the remedial measures fund, and ensures cooperation between the Crees and Hydro-Québec on the projects.

Under the Agreement, an implementation committee is established to see to the application of the Agreement and the existence of a permanent forum to handle disputes.

***

Altogether, under all the agreements, the Crees of Québec will have received $295 million as compensation and $154 million in mitigative measures, for a combined total of $449 million.

For the Inuit, compensation amounts to $128 million, and mitigative measures, $14 million, for a combined total of $142 million.

Conclusion

Hydro-Québec considers it important to establish a dialogue with the peoples concerned, so that their concerns and expectations may be taken into account. They must also be involved in designing and implementing the mitigative measures. This type of approach, formalized in a negotiated agreement, allows the peoples concerned to quickly become familiar with the new conditions of their environment and the project elements likely to generate impacts. Furthermore, it makes it possible to manage the impacts through remedial measures and compensation funds. All these elements of an agreement ensure a project's integration into the affected communities.

reading 3

"Economic and Social Development of the Aboriginal Communities"

Hydro-Québec, Ltd. © 1996 by Hydro-Québec. Reprinted by permission.

Cree Communities

Impact of the Income Security Program

The impact of the La Grande complex hydroelectric development on the use of wildlife resources (see Information Sheet No. 15), which comprises a physical and biological dimension, must be placed in the context of the Income Security Program for Cree hunters and trappers (ISP) established in 1975 under the terms of the James Bay and Northern Québec Agreement (JBNQA). The purpose of this program is to preserve the traditional way of life by providing financial and other assistance to families or individuals that hunt or trap on a more or less continuous basis (at least 120 days a year). Whereas in 1971, 600 families took part in hunting, fishing and trapping activities on a regular basis, the number of units benefiting from the program consistently hovered around 1200 throughout the eighties. This corresponded to over one-third of the Cree population at the time.

All of the studies conducted show that this program, instituted in all the Cree communities, has made it possible to preserve a way of life that seemed in jeopardy, or at least slow its decline, which had been under way since before 1975. Some signs observed in the Aboriginal villages that do not benefit from the program indicate that traditional activities seem less vigorous there, despite the introduction of subsidy programs (such as one paying the costs of air transport to remote hunting grounds). Between 1978 and 1995, the number of beneficiaries of the Program ranged from 2704 to 3740 persons, out of a total population of approximately 11,000 in 1995. The popularity of the Program reflects the persistence and strength of the values associated with the traditional way of life (see graphs on the next page).

However, a downward trend can be seen from year to year since 1984. This phenomenon seems mainly linked to a decrease in the number of children registered, despite the growth in the Cree population. In 1994–95, for example, the beneficiaries represented only 24.1% of the overall Cree population. Some of the new generation, raised in villages and more highly educated, apparently shows less interest in a way of life oriented toward hunting, fishing and trapping. Wildlife harvesting activities seem to be becoming more "professional," moreover: they are practiced less and less by the family, and more and more by individuals on their own. In addition, the 18–30 and over 50 age brackets are proportionally better represented among ISP beneficiaries than in the Cree population as a whole.

The program has also affected the main objective of spending time in the bush—apart from its strictly financial importance—and even contributed to restoring its original objectives. Increasingly, the purpose of spending time in the bush is to feed the trapper's family and immediate circle; income derived from the sale of furs forms only a marginal proportion (about 2%) of total income generated by the pursuit of traditional activities. Furthermore, ISP participation seems inversely proportional to the number of jobs available. When unemployment increases among young people or adults, they tend to register more for the ISP. The Program's popularity thus partly reflects, when other sources of income (such as temporary jobs) start to dry up, the precariousness of a social stratum—made up of the young, the old and the less educated—for whom the ISP represents a kind of social safety net.

The income provided by the ISP has also made it possible to bring more efficient hunting and trapping technologies, such as snowmobiles, into general use. These vehicles, which had already been introduced before the signing of the JBNQA, allow greater distances to be covered quickly, and are used to transport game. In addition, the hunter is no longer obliged to change camp periodically in order to remain close to his traps and sources of subsistence. It is now possible to build a permanent, more comfortable camp. However, some observers consider that the program has turned the traditional hunter into a "harvest manager," and thus helped alter the very meaning of traditional activities. The dispenser of resources is no longer the "Master of the Animal" who captures the latter with his gun or net, but actually the Administration which issues the cheques hundreds of kilometres away, in Québec City.

One fact nevertheless seems clear: traditional activities no longer carry the weight they used to, in terms of income. Even though hunters' incomes more than tripled between 1971 and 1981, according to Salisbury (1986), their percentage in relation to the communities' total income decreased to 43% from 61%, while salaries increased in proportion to 52% from 23%. This gap seems to have widened even further over the past 15 years. As Jean-Jacques Simard (1996) explains, "starting in 1981, the stratum of wage earners made, in general, nearly twice as much money as that of regular hunters (1.7), and this gap even grew slightly in 1989 (1.9)". Moreover, Simard estimated that the average annual income of a family of five engaging primarily in subsistence activities was $28,000 in 1989, while the income of an equivalent family living mainly from wages amount-

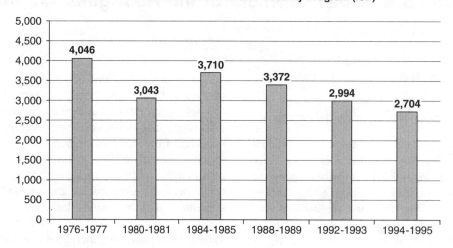

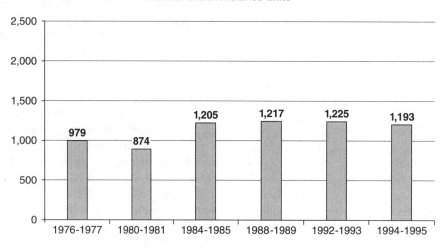

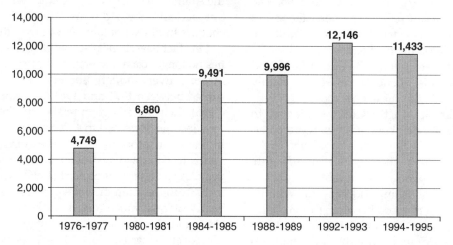

70 Unit Three: Canada: Alternate Energy Resources

ed to $47,800, putting the income of the former family on a par with that of the Canadian lower middle class.

Socioeconomic Development of the Villages

This strong growth in overall wages, a sign of a consolidation of the job market, undoubtedly represents one of the most significant impacts of the project and the Agreement. The jobs offered previously were precarious and seasonal. A large proportion of them are now permanent and much better paid. Most often provided by one of the many agencies set up in the wake of the Agreement (or by the band councils), these jobs are concentrated mainly in the public service sector which the transfer of government services has enabled to develop on the local and regional levels. This devolution of powers—combined with the payment of benefits—and its consequences on the occupational structure, in the form of higher-paid jobs, seem to have differentiated the development of people covered by the Agreement from that of other Aboriginal peoples, living in remote or isolated regions, who are not subject to the JBNQA. It is important to point out, at the same time, as was demonstrated by Simard (1996), that all these trends were already under way when the JBNQA was signed. The Agreement did accelerate them, however, or at the very least gave them new impetus.

Furthermore, the extremely rapid growth in population, in the order of 25% over 10 years, increased the needs for housing and infrastructures. This factor, coupled with the economic spinoffs of the contracts granted by SEBJ and Hydro-Québec to Cree businesses, fostered the expansion of the construction sector in the villages. Phase II of the La Grande complex, for example, generated wage spinoffs in the Cree communities, mainly in Chisasibi, totalling $7 million in current dollars, resulting from job site hiring (see Information Sheet No. 11). The growth in public services and in the construction sector in the villages in turn led to the development of commercial activities. Interest earned from compensation paid under the terms of the Agreement allowed several businesses to be started up, including an airline (Air Creebec) and the Cree Construction Company. The presence of these businesses played a large part in increasing the amount of the contracts awarded to Aboriginal companies during construction of Phase II of the La Grande complex ($272 million from 1987 to 1994) in comparison with Phase I ($3.8 million). In addition, while the contracts awarded in Phase I mainly involved deforestation work, those in Phase II were much more diversified: construction, maintenance and upgrading of roads and buildings, supply of services, etc. Mitigative measures also constituted a major source of spinoffs in the Cree villages.

This expansion of salaried employment has increasingly given rise to a distinction between families living from wages and those living from subsistence activities. This differentiation has actually culminated in the appearance of social strata, which some commentators are even referring to as social classes. Already, a gulf seems to be developing between the education level of the children of hunters and those of wage earners, according to Salisbury. The emergence of this system of social stratification, detectable as far back as the early seventies (when three-quarters of job incomes were monopolized by one-third of the families), appears to have been accentuated by the signing of the various agreements (see Information Sheet No. 3). According to Proulx (1996), the top of the pyramid is occupied by the families of politicians or managers, followed by those of wage earners and traditional hunters registered in the ISP. At the bottom of the social ladder, a certain proportion of the population does not seem to have access to either salaried employment or the ISP.

Sociopolitical Development and its Social Costs

To all intents and purposes, the JBNQA led to the creation of a new level of government, controlled and run by the Québec Crees, which provides them with considerable latitude in terms of political autonomy. The construction of the new road network furthered the development of exchanges between previously isolated communities as well as the emergence of a sense of belonging which some people do not hesitate to describe as national. These changes have resulted in the establishment of an impressive number of consultation and decision-making bodies and processes for the relatively small number of Crees who live in the James Bay territory (11,237 residents at September 16, 1996).

These structures take up the energies of a large proportion of the population. According to many observers, they have also set in motion a process of bureaucratization of Cree society which has been superimposed on the old ways of solving conflicts and networks of social solidarity. This evolution has entailed significant social costs which are not unlike those seen in advanced industrial societies: a certain marginalization of young people, along with a growing precariousness in their living conditions; tensions between band councils and regional authorities, which are typical of modern political structures; as well as various forms of stress experienced by individuals torn between many different cultural trends and responsibilities, which they are nonetheless managing to integrate.

The relocation of the village of Fort George, situated on the island of the same name, to the Chisasibi site represents another major source of impacts. The decision to relocate was based on the possible erosion of the island and on the many long-term benefits (such as connection to the road network) which the move offered the village community. While approved by the residents of Fort George in a consultation process, this relocation did not go entirely smoothly, like many other projects of the same type carried out elsewhere in the world. Despite a distinct improvement in quality of life in the new village, several dozen residents refused to move to Chisasibi, and some are still living on the island. The older residents who moved seem to have experienced a sense of loss.

For people faced with very rapid demographic and social development, their preresettlement life has sometimes come to symbolize an ideal world, marked by greater family and community solidarity, which has apparently disintegrated since then, leading to a whole range of social problems, in the view of some residents. To partially offset these problems, traditional celebrations are held every summer on the site of the old village.

Finally, the development of the La Grande complex also brought about the construction, in the midst of Cree lands, of a new village—Radisson—inhabited by the workers in charge of operations at the La Grande complex. Substantial commercial ties quickly formed between Radisson and Chisasibi.

Inuit Communities

The Inuit population, made up of 7,863 residents (at September 16, 1996), is spread over 14 villages located on the coasts of Hudson Bay, Hudson Strait and Ungava Bay. Several dozen Inuit also live in the village of Chisasibi.

The economic, social and cultural impacts occurring in Inuit villages are similar in many ways to those observed in Cree villages. The Inuit villages have also gained from urban reconstruction or renovation programs. Some of the benefits ($34.5 million) paid to the community of Kuujjuaq, as compensation for the reduction in flows in the Koksoak, were used to build community facilities. For several reasons, researchers have emphasized the impacts of the JBNQA on the cooperative movement, which they had viewed as the main factor in emancipating the Inuit and which is still considered by some just as valid an alternative model today, ensuring greater economic and political autonomy.

The signing of the JBNQA, which extinguishes territorial rights, is perceived by the two villages that refused to ratify the Agreement as a factor marginalizing the self-determination aspirations embodied by the cooperative movement. According to some researchers, the JBNQA has led to considerable bureaucratization, financed by the transfer payments made by the various departments of the Québec government; increased dependence on powers established in the south, and as a corollary, a loss of control which undermines the development of a sense of responsibility as well as the definition of traditional roles; a more folkloric approach to subsistence activities, which are now subsidized; development of the service sector to the exclusion of much else; etc.

The partial relocation of the Inuit population of Kuujjuarapik north to the site of Umiujaq (near Lac Guillaume-Delisle) also had major social repercussions. A portion of the population wanted to move closer to its hunting grounds, located further north, and also return to a way of life considered more traditional. However, a large part of the population of Kuujjuarapik, which was using lands located south of this community, refused to move. The relocation therefore led to a split in this community, along with a new, minority status for the Inuit of Kuujjuarapik in relation to the Crees of the neighboring community of Whapmagoostui.

The Inuit of Kuujjuarapik believe, moreover, that they could have obtained more category I and II lands near their village if the creation of Umiujaq had not been considered at the time of negotiating the JBNQA. In addition, departures were more numerous among the older people and young people under 15 years of age, with a resulting impact on the demographic structure of the two populations. It seems that the teenagers were sometimes reluctant to move, feeling less inclined to return to lands and a way of life that were relatively foreign to them. Nor did they want to leave a village offering more services.

The Naskapi Community

The Northeastern Québec Agreement and, more specifically, the Income Security Program (ISP) which it instituted enabled the Naskapis to return to their traditional activities (caribou hunting, craft production, etc.), which were becoming marginal after the community settled in Schefferville. In terms of employment, this agreement mainly sought to promote the integration of the Naskapis into the mining industry, which was still thriving in Schefferville at the time of signing, and many of them were trained to this end. However, the Iron Ore mine, the main employer in the region, closed its doors in 1981, ending this dream.

The Agreement conferred considerable powers on the Kawawachikamach Band Council, as well as rights over a territory. Budgets were also granted to open a clinic, run a school and set up a police force. The Band Council quickly became the principal employer, thus leading, here as elsewhere, to pronounced development of the service economy, which seems to benefit women primarily (generally those with the most education). According to some researchers, the Agreement made it possible to largely offset the sizable economic losses caused by the mine's closing, but at the cost of a dependence by the band on government programs, and by the residents on the band. Here as well, the only solution appears to lie in a diversification of the economy.

The creation of the village of Kawawachikamach,

located 15 km northeast of Schefferville, also represents an old dream for the Naskapis, who until then had often had to live alongside other ethnic groups (Inuit, then Montagnais or Innu). The band's relocation to the new village, in 1983, was marked by initial difficulties—such as a lack of commercial and recreational infrastructures, as well as some public services—which did not seem to have been fully overcome several years later. This could be partly explained by the relatively small size of the Naskapi population (630 residents, at September 16, 1996).

The absence of these services seems to have handicapped the village's economic development, and may be related to a series of problems experienced by the young people in subsequent years. The new site of the village also isolated them from Schefferville, the centre of the region's social life.

CREE COMMUNITIES (September 26, 1995)

	Residents*	Categorie I lands		Categorie II lands	
		sq mi	sq km	sq mi	sq km
CHISASIBI	2,832	505.6	1,309.6	6,536.0	16,928.2
EASTMAIN	450	189.0	489.5	1,384.0	3,584.5
MISTISSINI	2,386	533.0	1,380.4	6,896.0	17,860.6
NEMISCAU	459	59.0	152.8	784.0	2,030.6
WASKAGANISH	1,442	303.0	784.8	3,947.0	10,222.7
WASWANIPI	985	231.0	598.3	2,949.0	7,637.9
WEMINDJI	970	198.0	512.8	2,634.0	6,822.0
WHAPMAGOOSTUI	591	122.0	316.1	1,895.3	4,908.8
OUJÉ-BOUGOUMOU	506	—	—	—	—
TOTAL	10,621	2,140.6	6,544.3	27,025.3	69,995.3

The Naskapi Band of Québec living in
KAWAWACHIKAMACH: 599   142.0   326.3   1,600.0   4,144.0

INUIT COMMUNITIES (September 26, 1995)

	Residents*	Categorie I lands		Categorie II lands	
		sq mi	sq km	sq mi	sq km
AKULIVIK	336	215.3	557.7	2,003.9	5,190.9
AUPALUK	111	243.2	629.8	1,559.7	4,039.7
CHISASIBI	55	17.4	45.1	0	0
INUKJUAK	1,089	215.4	557.8	3,043.7	7,883.2
IVUJIVIK**	139	202.7	524.9	1,766.9	4,576.3
KANGIQSUALUJJUAQ	539	243.2	629.8	2,119.7	5,490.1
KANGIQSUJUAQ	447	234.2	606.7	2,000.7	5,181.9
KANGIRSUK	341	243.0	629.5	1,878.2	4,864.6
KUUJJUAQ	1,250	243.2	629.8	3,428.7	8,880.4
KUUJJUARAPIK	474	5.8	15.0	234.7	608.0
UMIUJAQ	243	220.5	571.0	3,113.7	8,064.5
PUVIRNITUQ**	1,054	241.9	626.6	3,278.9	8,492.4
QUAQTAQ	233	224.9	582.4	1,612.2	4,175.7
SALLUIT	905	241.6	625.7	2,707.7	7,013.0
TASIUJAQ	176	243.2	629.8	1,428.7	3,840.3
TOTAL	7,392	3,035.5	7,816.5	30,231.4	78,301.0

* Aboriginal living within the territory as defined by the James Bay and Northern Québec Agreement and complementary Agreements (JBNQA, 1975) and by the Northeastern Québec Agreement (NEQA, 1978)

** Inuit Community which did not sign the JBNQA (lands, undelimited).

Map of the hydroelectric development and Aboriginal communities in Northern Québec

"Native People and the Environmental Regime in the James Bay and Northern Québec Agreement"

Evelyn J. Peters

Arctic, 1999, 52 (4):395–410. © The Arctic Institute of North America.

Abstract A major objective of the Cree and Inuit in signing the 1975 James Bay and Northern Québec Agreement was to protect the environment and thus secure their way of life based on harvesting activities. The main elements of the federal, provincial, and Agreement environmental protection regimes are compared with respect to principles derived from the growing literature on indigenous peoples and environmental assessment. The Agreement contained pioneering provisions for environmental assessment; yet those provisions have not met many of the expectations of the Native people. Part of the dissatisfaction derives from the Agreement itself; some sections are vague and difficult to translate into practices; the advisory committee structures are not well suited to Native cultures; and the right to develop is woven throughout the sections on environmental protection. However, failures and delays in implementing the Agreement have also contributed to this dissatisfaction. These issues have implications for the negotiation strategies of other groups.

Introduction

Despite assumptions about their inevitable passing, indigenous hunting, fishing and trapping economies have survived in many areas of the North American continent. The flexibility of indigenous economies has enabled them to avoid or accommodate frontier activity and intrusion by withdrawing to more and more marginal lands (Brody, 1983; Kayahna Area Tribal Council, 1985). However, there are limits beyond which hunting economies and frontier development become irreconcilable. Introducing a collection of articles on the geography of indigenous struggles, Stea and Wisner (1984:5) described the most recent phase in a history of encroachment on indigenous peoples' land:

> Having spent the better part of one hundred years pacifying their indigenous peoples and allocating to them what was then perceived as the very worst land . . . the major industrial powers are suddenly finding that these areas . . . over enormous energy and mineral resources. Thus the dominant industrialists buttress these nation-states' own attempts [to incorporate marginal lands] with notions of "energy crisis" and "strategic minerals." Oil, coal, and uranium are the chief foci, although in some places surface and even underground water has begun to be re-evaluated by capital as potentially strategic.

This phase may result in the destruction of those indigenous hunting economies that have survived.

Maintaining a subsistence economy is seen as a decisive factor in indigenous peoples' struggles for cultural survival (Berger, 1985; Jull, 1988; Simon, 1992; Brascoupé, 1993; Nuttall, 1994). Indigenous peoples internationally have used a variety of strategies to combat the threat that uncontrolled economic development poses for their economies and ways of life (Stea and Wisner, 1984; Burger, 1987). In Canada, the possibility of enhancing their ability to maintain a subsistence economy has been a major impetus for indigenous peoples' attempts to assert sovereignty over their lands and their lives through land-claim negotiations and self-government arrangements (Peters, 1989). While some arrangements have existed for more than two decades, there is almost no work that evaluates their appropriateness in protecting environments for the subsistence use of indigenous peoples (Young, 1995 is an exception).

In Canada, the James Bay and Northern Québec Agreement (hereafter, the Agreement) was signed in 1975. While the Agreement was the product of an out-of-court settlement rather than a land-claim settlement, it became an important precedent in subsequent land-claim negotiations. A major Cree and Inuit objective in negotiating the Agreement was to secure their way of life based on harvesting activities (Feit, 1980, 1988; Brooke, 1995). The process of implementing the Agreement has helped to define its status in Canadian law. This experience may be of interest to other indigenous groups attempting to design mechanisms for protecting their lands and economies.

After a brief background to the negotiation of the Agreement and to Cree and Inuit objectives with respect to environmental protection, this paper outlines the main elements of the environmental protection regimes for Cree and Inuit lands under the Agreement. The Cree protest against the Great Whale hydroelectric project is described to illustrate some issues involved in implementing provisions of the Agreement. While the

Inuit did not oppose the project in the same way, the Cree experience provides an important context for the analysis. The final section evaluates the environmental protection regime in the Agreement and the appropriateness of Agreement, federal, and provincial regimes for protecting indigenous economies, using principles derived from the growing literature on environmental assessment and indigenous peoples. The paper relies mainly on materials produced by the Cree and Inuit and individuals working for their organizations: as the question here is whether provisions in the Agreement meet the needs of the indigenous residents, their views count the most.

Negotiating the Agreement

In 1971, the Québec government began to construct the James Bay Hydroelectric Project in an area not yet ceded by the Native peoples, and which was still used by them in their traditional hunting pursuits (Richardson, 1975; LaRusic, 1979; Salisbury, 1986). The land in question had been transferred by the federal government to Québec under the 1912 Boundary Extension Act, with the condition that the province obtain surrender of Native interests in the area prior to development. When the province failed to negotiate, the Cree and Inuit instituted legal proceedings against Québec and the James Bay Development Corporation in 1972.

Mr. Justice Malouf of the Québec Superior Court granted the Cree and Inuit a hearing and, after receiving the testimony of Inuit and Cree hunters about their continuing use and occupation of these lands, accorded an interlocutory injunction against the hydroelectric development in process. The Malouf decision was suspended a week later pending a final decision by the Québec Court of Appeal on a permanent injunction. Québec's fear that in the end the Cree might be awarded substantial damages and the inability of the Cree and Inuit to stop construction during court proceedings provided the impetus for both sides to negotiate an out-of-court settlement. An agreement-in-principle was reached on 15 November 1974. On 21 November 1974, the Québec Court of Appeal denied the Cree and Inuit a permanent injunction, putting an end to litigation on that issue. A final agreement was completed in November 1975.

Québec's intention in negotiating the Agreement was to affirm its presence and jurisdiction over the territory and to open the area for economic development (Ciaccia, 1976: xi–xxiv). Hydroelectric development of Québec's northern rivers was a long-standing objective of Robert Bourassa, Québec's Premier during many of the events described here (Bourassa, 1985). In an attempt to prevent aboriginal people from blocking regional development, Québec insisted that the province retain final control over authorization of economic development in the territory referred to in the Agreement, except on lands reserved for aboriginal communities.

Unable, in the face of Québec's objectives, to gain jurisdiction over the whole territory, Cree and Inuit negotiators followed a number of strategies to protect the land for subsistence activities. They negotiated a detailed hunting regime that allowed them continuing access to wildlife resources on most of the lands of the territory referred to in the Agreement (Section 24), and they obtained financial support programs for Cree and Inuit harvesters (Sections 29 and 30). The focus of this paper, however, is on Sections 22 and 23 in the Agreement, which attempt to protect the environment for harvesting activities.

When the Agreement was negotiated, there were few precedents or models for environmental protection. Where they existed at all, federal and provincial environmental and social protection regimes were weak, with no explicit social component, limited provisions for public involvement, and no general obligation to conduct impact assessments and reviews. Yergeau (1988:296) reviewed the environmental regime in the Agreement a decade after it was signed:

> The Agreement is a pioneering document in the field of environmental protection, in particular with respect to the criteria governing environmental and social impact statement. . . .[T]hat it stated all this in 1975 was evidence of an unheard-of boldness far in advance of any other legislation in the world dealing with the environmental impacts of development.

In 1980, Feit (1980:12) demonstrated a cautious optimism about the effectiveness of the environmental regime outlined in the Agreement.

> The combination of these distinctive principles, obligatory consultations, special representations and mechanisms for inputs into consultations, was thought to sufficiently constrain the exercise of governmental authority that it would have to respect and, in part, serve Cree interests and concerns. The compromises involved in accepting this more limited control over development activities and social and environmental protection were hard ones for the Cree, and it is still too soon to tell whether or not these provisions will be adequate.

With a decade, Feit's evaluation had changed.

> The threat to the hunting economy posed by relatively unregulated industrial development of the region pinpoints the. . . failure of the agreement process to effectively resolve conflicts over resource control and economic development. Large-scale industrial development projects are continuing on

Cree lands... [T]he failure to adequately regulate development is a major future threat to the revitalized hunting sector. These threats demonstrate... that the agreement process was unable to resolve fundamental conflicts between the interests of Cree and those of wider economic and political institutions of the capitalist economy or the liberal democratic state. (1989:96-97)

Nevertheless, the sections of the Agreement referring to environmental protection played a major role in the Cree protest against the Great Whale River hydroelectric project, which was put on hold in 1994. Moreover, the federal and provincial governments and the Cree and Inuit have been negotiating implementation issues for more than two decades (Peters, 1980; Brooke, 1995; Penn, 1995). Thus it is difficult to know whether the problem lies with specific provisions of the Agreement, or with the ways in which those provisions have (or have not) been put into effect. The following paragraphs attempt to evaluate these issues.

Main Components of the Social and Environmental Protection Regime

Land Regime

The Agreement distinguishes three main classes of land (Categories I–III) with respect to allocation of title, resources, interest, and jurisdiction. Provincial insistence on opening up the area for economic development limited the size of the land base over which the Cree and Inuit could negotiate ownership or control.

Category I lands correspond to the locations of Cree and Inuit villages and their peripheries (Figs. 1 and 2). Category IA lands are areas transferred by Québec to Canada for the exclusive use and benefit of the Cree bands. Category IB lands are areas whose ownership has been transferred by Québec to Cree landholding corporations. In the case of the Inuit, Category I lands are areas whose ownership Québec has transferred to Inuit Village Corporations. Category I lands are sufficient in size for community sites with some buffer between the community and adjacent development. However, such areas provide only very limited protection for the hunting economy. Moreover, Category I lands are subject to fairly extensive expropriation powers of the federal and provincial governments.

Category II lands adjoin Category I lands. They are lands under provincial jurisdiction on which the Cree and Inuit have exclusive rights of harvesting and outfitting. Category II lands may also be expropriated by Canada or Québec for development.

Category III lands cover the balance of the territory referred to in the Agreement. Native people have exclusive rights to trap and create commercial fisheries for some species on Category III lands. These lands are areas of joint use by Native and non-Native peoples, but Native people are to be subject to a minimum of control or regulation with respect to hunting and fishing, and any controls imposed on them must be decided by a Hunting, Fishing, and Trapping Coordinating Committee on which Native people are represented. (The harvesting regime for the Agreement is found in Section 24 of the Agreement. A full description of this

TABLE 1. Enviromental protection bodies under the Agreement.

Committee	Membership	Section
Reviewing and Formulating Laws and Regulations for Environmental Protection		
South of the 55th Parallel		
James Bay Advisory Committee on the Environment (JBACE)	Four representatives appointed by each of the Cree Regional Authority (CRA), Quebec, Canada, plus the Chair of the Hunting, Fishing and Trapping Coordinating Committee	22.3
North of the 55th Parallel		
Kativik Environmental Advisory Comittee (KEAC)	Three representatives appointed by each of the Kativik Regional Government (KRG), Quebec, Canada	23.5
Setting Guidlines for Environmental and Social Impact Assessment		
South of the 55th Parallel		
Evaluating Committee (COMEV)	Two representatives appointed by each of the Cree Regional Authority (CRA), Quebec, Canada	22.5
North of the 55th Parallel		
Kativik Environmental Quality Commission (KEQC)	Four representatives appointed by the KRG, five by Quebec	23.3
Federal Review Committee North (FRC-North)	Two representatives appointed by the KRG, three representatives by Canada	23.4
Evaluating and Reviewing Impact Assessments		
South of the 55th Parallel		
Provincial Review Committee (COMEX)	Two representatives appointed by the CRA, three by Quebec	22.6
Federal Review Committee (FRC-South)	Two representatives appointed by the CRA, three by Canada	22.6
North of the 55th Parallel		
Kativik Environmental Quality Commission (KEQC)	Four representatives appointed by the KRG, five by Quebec	23.3
Federal Review Committee North (FRC- North)	Two representatives appointed by the KRG, three representatives by Canada	23.4

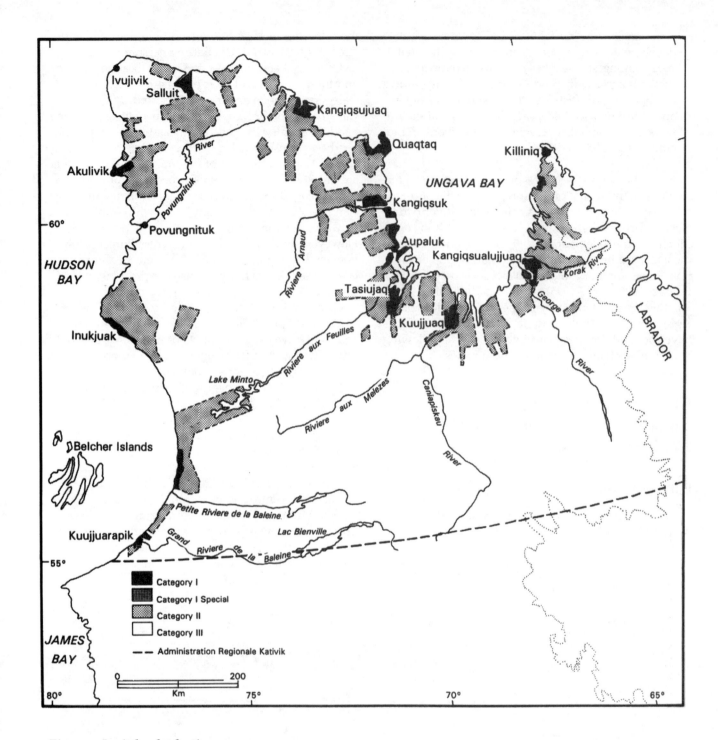

Figure 1 Inuit land selection, 1975

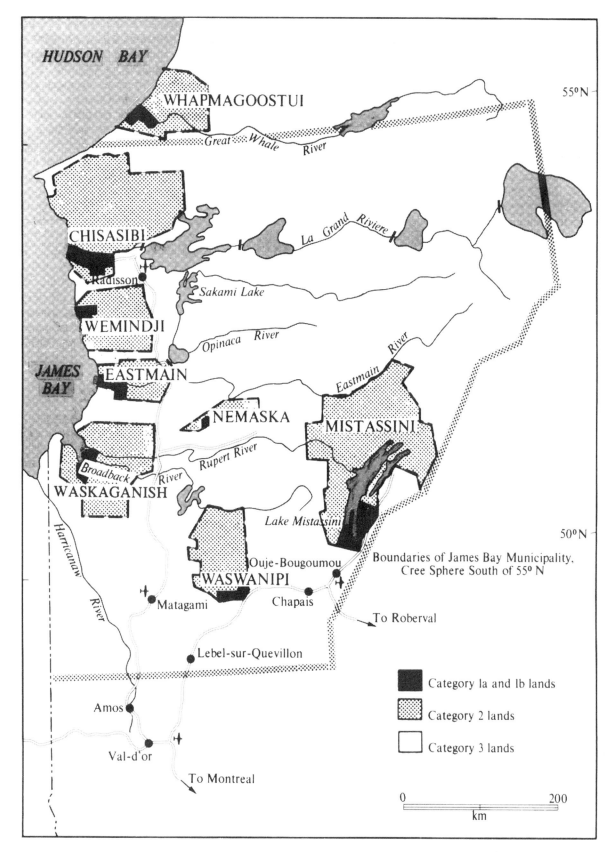

Figure 2 Cree land selection, 1975

regime is beyond the scope of this paper, but see Brooke, 1995.) While the Cree and Inuit have little power to restrict non-renewable resource or economic development on Category II and III lands, and while the Agreement explicitly gives Québec the right to economic development on these lands, Québec's rights are subject to the environmental regimes set out in the Agreement.

Environmental Protection Regimes

Sections 22 and 23 of the Agreement set out the processes and criteria to be employed in assessing the social and environmental effects of developments south and north of the 55th parallel, respectively. The Great Whale River hydroelectric development, described in the following section, would have been located in both areas.

The James Bay Advisory Committee on the Environment (JBACE) and the Kativik Environmental Advisory Committee (KEAC) were created to facilitate Cree and Inuit input into all aspects of decision making about the management of development in the area referred to in the Agreement (Table 1). These committees were established as consultative bodies to responsible governments on issues concerning the environmental regime and the formulation of laws relating to the environment. Their responsibility is to review existing and proposed development-related legislation and regulations (including environmental impact assessments) that affect Cree and Inuit environments and recommend environmental laws, regulations, and other measures to responsible governments. Federal and provincial governments are required to fund a secretariat for the committees, and the committees can call on expert advice if necessary.

The bodies formed to deal specifically with impact assessment processes are also described in Table 1. Federal and provincial governments fund staffing requirements for these bodies and pay for experts required in their deliberations. When a new development is proposed, the review bodies make recommendations to the appropriate Administrator about the need for an environmental impact statement and its nature and extent. The Administrator makes the decision and, if appropriate, issues guidelines for the assessment to the proponent. There is no provision for public consultation at this stage. The review bodies evaluate the environmental impact statement submitted by the proponent and, except for the Kativik Environmental Quality Commission (KEQC), recommend to the Administrator under what conditions a development may proceed. The KEQC can decide whether or not a development may proceed, and the Québec Administrator must obtain the approval of the Lieutenant Governor in Council if he or she does not wish to follow the decision of the KEQC. During the evaluation and review stage, the review bodies may hold public consultations or invite written comments concerning the environmental impact statement.

The Agreement indicates that a Cree of Kativik local government "Administrator" is responsible for decisions about matters affecting the environment on Category I lands. The federal regime on Category II and III lands operates under the authority of a person or persons appointed by the Governor in Council, also known as the "Administrator." At the time of the events addressed in this paper, the Chair of the Federal Environmental Assessment Review Office was the "Federal Administrator." The provincial regimes are under the authority of an "Administrator" appointed by the Lieutenant Governor in Council. The Québec Minister of the Environment was the "Provincial Administrator" for matters under provincial jurisdiction during the events related to the Great Whale project.

The scope of environmental impact assessments is also addressed in paragraphs 22.2.4 and 23.2.4. According to these sections, responsible governments must give "due consideration" to a series of principles that include:

- protecting the "hunting, fishing and trapping rights of the Native people";
- "minimizing the impact on Native people by developmental activity";
- protecting Native people, societies, communities and economies, and wildlife resources, physical, biotic and ecological systems with respect to developmental activity; and
- minimizing the "negative environmental and social impacts of development of Native people and on Native communities."

The environmental regime, then, had several features through which the negotiators attempted to enhance its effectiveness in protecting hunting economies. The regimes created permanent committees with Cree and Inuit participation, mandated by the Agreement. The regimes had a clear place in government decision-making procedures and were obligatory rather than voluntary. Funding and staffing were provided for in the Agreement (though only in general terms). Expert information was to be made available to these bodies. The assessment of the proposed development was to consider the effect of development on hunting economies, and a series of principles to protect these economies attempted to limit the exercise of powers by federal and provincial governments. In this way, the Native peoples attempted to ensure that the values and priorities of hunting economies would be taken into account in development activities and that these activities would address the potentially negative effects on Native communities and cultures.

However, sub-paragraphs 22.2.2(f) and 23.2.2(f) also indicate that the environmental regime provides for "the right to develop" in the area referred to in the Agreement, and paragraphs 22.2.4 and 23.2.4 include the right to develop in the principles to which governments and agencies "shall give due consideration." Moreover, clauses and arrangements in the Agreement do not tell the whole story. The meaning of particular phrases may be contested, implementation may be slow or partial, federal and provincial governments may choose not to implement particular elements, and Native peoples may not have the resources to fully participate in the bureaucracy that the Agreement created. Evaluating the effectiveness of particular arrangements also requires an examination of instances where particular provisions were put into practice. Reed's (1990) study of the environmental impact assessment process negotiated under the Inuvialuit Agreement showed that practice frequently diverged from the procedures outlined in the Agreement. The following section provides some insights into how parts of the environmental regimes in the James Bay and Northern Québec Agreement were implemented. The context is the proposed hydroelectric development on the Great Whale River, in the territory referred to in the Agreement.

Environmental Impact Assessment of the Great Whale Program

The Cree and Inuit depend on the land for their sustenance, and they consider it their own (Brooke, 1995; Penn, 1995). In addition to hydroelectric development, forestry, contamination from mercury and airborne substances such as PCBs, roads, and mining development represent major threats to the Cree and Inuit subsistence base. These threats are not distributed evenly. In the southern part of Cree territory, for example, forestry is taking more land out of the hunting economy than hydroelectric development. Hydroelectric development has not affected Inuit lands as much as it has Cree lands. This section focuses on hydroelectric development because more published material is available on the relationship between its environmental impacts and the contents of the Agreement.

The Hydroelectric Projects

A description of some aspects of the La Grande project illustrates the scope of the undertaking. La Grande Phase I was started in 1973 and completed in 1986. With eight main dams, 198 dikes, and five major reservoirs, the project diverted seven rivers, nearly doubling the flow of the La Grande River in winter (Day and Quinn, 1992). La Grande Phase I flooded in excess of 10,000 km^2 of land. Construction on La Grande Phase II began in 1991 and is now complete. This phase involved constructing one reservoir, diverting one river, and flooding an area of 765 km^2.

In 1989, Québec announced plans to begin a second hydroelectric development project on the Great Whale River. The proposed Great Whale River project, often called James Bay II, would have diverted water from four major rivers into the Great Whale River. The project entailed 4 reservoirs that together cover 4387 km^2, 5 dams, 133 dikes, and 3 generating stations. The Cree have argued that the ongoing and proposed hydroelectric projects and their short- and long-term social and environmental effects have had and will continue to have serious negative effects on their hunting economies and their society (Hodgins and Cannon, 1995). As a consequence, they took the lead in an ongoing legal and public relations battle to halt construction or, failing that, to ensure that developers minimized the damage to Cree lands.

Environmental Impact Assessment on the Great Whale River Hydroelectric Project

Québec decided to reactivate work on the Great Whale River hydroelectric project (the project had been on the drawing board since 1975, but further work was postponed in 1982) without consulting any of the review bodies set up under the Agreement. Instead, the province indicated that an environmental impact assessment would be conducted under provincial guidelines. According to section 8.1.3, hydroelectric projects in the area referred to by the Agreement are exempt from social impact assessment and are "subject to the environmental regime only in respect to ecological impacts." The exemption, however, does not bypass other aspects of the environmental impact assessment process defined in the Agreement.

Ottawa's initial position was that because the project might affect areas under federal jurisdiction, an assessment was required under the Federal Environmental Assessment Review Process (EARP) and, in addition, that a federal review was required pursuant to sections 22 and 23 of the Agreement. Québec argued that no federal review was required because the project was under provincial jurisdiction (Posluns, 1993).

In October 1990, Québec proposed to split the environmental review of the Great Whale project into two stages. In the first stage, a commission would study the impact of access infrastructure, such as roads, airports, bridges, and docks. Québec contended that Ottawa should not be involved in this phase because these were not matters under federal jurisdiction. The second stage would examine the impact of the dams, dikes, reservoirs, and powerhouses. Neither of these reviews was to be conducted pursuant to the Agreement. While the first phase of the provincial review was held up in the courts by legal challenges, Québec began construction. A troubling implication of construction starts and of the split review was that, once millions of dollars worth of access infrastructure was in place, it would be politically difficult for the public review panel to deny the project, or even recommend substantial changes.

The following month, the federal government abruptly reversed its earlier position, informing the Cree that Ottawa had no mandate for a separate federal review of the Great Whale project. (See Posluns [1993] for a good overview of these events.) Instead, Ottawa and Québec agreed to a single joint review, the federal portion would be submitted to the provincial review body. The federal review would follow the federal EARP guidelines rather than those of the James Bay and Northern Québec Agreement.

While the Inuit entered into negotiations with Hydro-Québec concerning project impacts and mitigating measures, the Cree launched a motion to the Federal Court to try to force the federal government to live up to its responsibilities under the Agreement (Cree Regional Authority v. Robinson, [1991] 2 C.N.L.R. 41). The challenge for Cree lawyers was twofold. They had to demonstrate that the James Bay and Northern Québec Agreement had the status of a federal law, and they had to prove that sections 22 and 23 applied to the Great Whale project. Justice Rouleau of the Federal Court, Trial Division ruled on the first question on 13 March 1991. He indicated that the Agreement did not have the status of a contract, as federal lawyers had argued. Parliament had intended it to operate as a law of Canada. He added:

> I feel a profound sense of duty to respond favourably [to the Cree application]. Any contrary determination would once again provoke, within the native groups, a sense of victimization by white society and its institutions. This agreement was signed in good faith for the protection of the Cree and Inuit peoples, not to deprive them of their rights and territories without due consideration. (Cree Regional Authority v. Robinson [1991] 2 C.N.L.R. 41 at 48)

On 28 March 1991, the Federal Court of Appeal upheld this decision (Hydro-Québec v. A.-G. Canada et al. [1991] 3 C.N.L.R. 41 at 82), and on 14 May 1991 the Supreme Court of Canada refused the Attorney General of Québec and Hydro-Québec's motion for leave to appeal.

In July 1991, Jean Charest, the new federal Minister of the Environment, announced that the federal government was legally bound to a separate review of the Great Whale Project because of its possible impact on areas under federal jurisdiction, for example fisheries, migratory birds, and navigation on harnessed rivers. He indicated, however, that the federal review would be under the federal Environmental Assessment Review Process (EARP). The federal government maintained that it could not prevent Québec from beginning to build $775 million worth of access infrastructure, including roads, marine terminals, and airports, during the estimated two years of hearings. Moreover the federal government argued that in cases of overlapping jurisdiction, it did not have the constitutional authority to block the project (Day and Quinn, 1992).

In September 1991, Justice Rouleau of the Federal Court, Trial Division ruled that the federal government had a public, nondiscretionary duty to carry out an independent federal review, pursuant to the terms of the Agreement. Justice Rouleau spelled out the nature of federal decision-making responsibilities:

> Sections 22.5.15 and 23.4.9 of the JBNQ Agreement impose a mandatory duty on the federal administrator to decide whether or not an environmental and social impact assessment and review is required, and to determine the nature and extent of any such assessment and review.
> Sections 22.6.15 and 23.4.23 assign to the federal administrator the obligation to either advise the developer respecting the alternative submitted or to decide, based on environmental and social impact consideration, whether or not a proposed development should proceed. (Cree Regional Authority v. Robinson [1991] 4 C.N.L.R 84 at 85)

Justice Rouleau also pointed out that the federal government, if it did not already have a fiduciary responsibility toward the Cree, incurred that responsibility when it extinguished their Native rights pursuant to the Agreement.

On 23 January 1992, with a view to carrying out a global and coordinated assessment of the proposal, the federal and Québec governments, the Cree Regional Authority, the Grand Council of the Crees (of Québec), the Kativik Regional Government, and the Makivik Corporation signed a *Memorandum of Understanding* concerning the coordination of environmental assessment and review processes stemming from sections 22 and 23 of the James Bay Agreement with the federal EARP. The Great Whale Public Review Support Office, created by virtue of the *Memorandum*, had as its objective the harmonization and coordination of the work of the review bodies (Evaluating Committee et al., 1992).

Guidelines for the environment impact assessment of the proposed Great Whale Hydroelectric Project were transmitted to Hydro-Québec in September 1992. In August 1993, Hydro-Québec submitted a feasibility study containing the environmental impact statement (EIS) associated with the Great Whale project.

In their evaluation of Hydro-Québec's submission, the review bodies concluded that "As submit-

TABLE 2. Enviromental impact assessment under federal, Quebec and Agreement processes.

	Federal EARP	Quebec Legislation	James Bay Agreement
Mandatory review and assessment	Yes	Yes	Yes
Mandatory assessment of social impacts	Yes	No	Yes
Mandatory consideration of measures to protect hunting economies	No	No	Yes
Cree and Inuit jurisdiction over development decisions	No	No	See note[1]
Incorporation of Cree and Inuit values in impact assessment	No	No	No
Incorporation of Cree and Inuit values in decision making processess			
--participation on review bodies	No	No	Yes
--culturally appropriate decision-making processess	No	No	No
Incorporation of Cree and Inuit values in defining desirable futures	No	No	Yes

[1]Limited - Cat. I land; Advisory - Cat. II and Cat. III land

ted by the Proponent, the EIS is presently neither sufficiently complete nor adequate for the decision-making process" (Provincial Review Committee et al., 1994:3):

In its introduction, the *Joint Report* highlighted seven major inadequacies, including:

- ambiguities related to study area boundaries and project schedule;
- treatment of principal assessment criteria…;
- knowledge of human societies affected;
- approach to the study of the combined and integrated effects of the proposed project;
- justification of the proposed project;
- appreciation of the uncertainty associated with the proposed project's impacts;
- selection of mitigation measures and the short- and long-term management of the proposed project.

(Provincial Review Committee et al., 1994:4)

A general discussion of the major shortcomings of the report was followed by more than 100 pages of detailed instructions for revisions.

The *Joint Report* was completed on 16 November 1994. The report was never officially released, but it was forwarded to Hydro-Québec. On 18 November 1994, Québec announced that the Great Whale project would be put on hold. Clearly this decision was not a response to the *Joint Report*. The Grand Council of the Crees, with other activists, had mounted an intensive and highly successful public relations campaign on the eastern seaboard of the United States. Moreover, the low price of natural gas in the United States at that time allowed for the low-cost production of electricity in local generators. These developments resulted in the cancellation of several contracts to purchase Hydro-Québec electricity (Posluns, 1993; Grand Council, 1995:23). Because the Great Whale project was cancelled, some of the implications of the environmental regime in the Agreement were not fully worked out for this project. Nevertheless, the case demonstrates some of the issues that are important for an evaluation of the environmental regime.

The Agreement and Federal and Provincial Environmental Impact Assessment Regimes

Table 2 compares three environmental regimes—the federal EARP and its successor, the 1992 *Canadian Environmental Assessment Act*; Québec's environmental assessment regulations; and the environmental regime set out in sections 22 and 23 of the Agreement—with respect to processes in place for assessing impacts on and protecting subsistence economies. The Agreement supersedes provincial legislation with respect to environmental impact assessments. However, provincial legislation is described here for purposes of comparison. The assessment criteria used to compare federal, provincial, and Agreement provisions have emerged from a growing literature on Native peoples and environmental assessment. These criteria are discussed individually below.

Mandatory Review and Assessment

All three review processes are mandatory. At the beginning of the attempt to implement the environmental assessment process of the Agreement for the Great Whale River project, the federal EARP was viewed as discretionary. However, rulings from court cases concerning Rafferty-Alameda Dam (Canadian Wildlife Federation Inc. v. Canada, 1989) and the Oldman River Dam (Friends of the Oldman River Society v. Canada 1992) indicated that the federal government had the mandatory duty to perform an environmental impact assessment of any project that involved aspects of federal jurisdiction. In 1992, the *Canadian Environmental Assessment Act* (CEAA) was put in place, in part to make these requirements clear. Québec's *Environmental Quality Act* (1972), amended in 1978, requires an environmental assessment of all projects identified by the regulations of that Act. The environmental regimes under the Agreement place a legal obligation upon all development proponents to comply with its provisions. The legal status of the regimes is based on the James Bay and Northern Québec Final Agreement, 1975, and validated by the federal *James Bay and Northern Québec Native Claims Settlement Act, 1977* and Québec's *Act approving the Agreement concerning James Bay and Northern Québec, 1976.*

It is worth noting, though, that it took a court decision to force the federal government to live up to its responsibilities and to agree to an environmental assessment pursuant to the federal environmental regimes specified in the Agreement.

Mandatory Assessment of Social Impacts

Provincial, federal, and Agreement regimes vary with respect to the requirement for social impact assessment. Social impacts are not a necessary part of the provincial evaluation procedure. In the 1984 Order-in-Council, the scope of the EARP was defined to include assessment of socioeconomic impacts that resulted directly from changes in the biophysical environment. The Guide published by the Federal Environmental Assessment Review Office (FEARO) in 1986 notes:

> Although the initiating department will determine the extent to which socio-economic impacts are to be taken into account in initial assessment, as a minimum, the potential social change associated with the biophysical impacts of a proposal must be considered. (Canada, Federal Environmental Assessment Review Office, 1986:26)

The 1992 *Canadian Environmental Assessment Act* requires the examination of "environmental effects" that include:

> any effect of any such change on health and socio-economic conditions, on physical and cultural heritage, on the current use of lands and resources for traditional purposes by aboriginal persons or on any structure, site or thing that is of historical, archaeological, paleontological or architectural significance. (s. 2[1])

Under the Agreement, while hydroelectric development is exempt from social impact assessment, other economic development is not. Sections 22 and 23 of the Agreement are explicit with respect to the requirement for a social impact assessment. The assessment is described as an environmental and social impact assessment throughout these sections, and Schedule 3 of each section, which describes the contents of the impact statement, includes the specification of social impacts. However, a requirement to evaluate social impacts does not guarantee that the resultant decision making will take them into account. The Cree experience has been that the Québec Ministry of the Environment took the position that it was empowered to make decisions only in areas where it had statutory or regulatory authority to do so. Decisions concerning social, economic, or cultural issues could not be imposed on other departments that did not participate in or have input into the assessment procedure (Penn. 1995). Despite the clear requirement for social impact assessment, then, its implementation has proved elusive.

Mandatory Consideration of Measures to Protect Hunting Economies

The definition and measurement of social impacts is a value-laden exercise, and standard approaches may not provide techniques or concepts that are effective in protecting subsistence economies. Researchers have argued that it is particularly difficult to construct appropriate methodological frameworks and processes for aboriginal peoples (Shapcott, 1989; Ross, 1990; Notzke, 1994).

The assessment of social impacts for northern aboriginal communities requires a different paradigm than that required for non-aboriginal communities (Elias and Weinstein, 1992, Vol. 1:5f.; Weinstein et al., 1992:8f). For example, many of the activities that comprise subsistence economies do not have an easily calculated market value; as a result, benefits from the introduction of a wage economy can easily be overvalued. More importantly, a model that views the replacement of a hunting, fishing, trapping, and gathering society with one based on wage-labour and business as a simple substitute of one kind of economy for another does not provide an adequate conceptualization of the meaning of the land-based activities that make up subsistence economies. Notzke (1994:277) argues that:

> The past two decades have shown that the effects of industrial resource developments on small, northern aboriginal communities are fundamentally different from those on non-aboriginal communities. Studies conducted in various communities. . . show the enduring importance of wildlife, fish and plant resources in the livelihood of native people. The reason for this is that renewable resources do not merely constitute the economic base for aboriginal communities, but their harvesting provides *the* major integrative social force [emphasis in original].

The loss of harvesting opportunities represents more than the loss of an income-generating activity for which a substitute can be found through wage-labour or social assistance.

Because there is no requirement in Québec legislation for a social impact assessment, an evaluation of impacts on hunting economies is not required either.

Social impact assessment was part of the federal environmental assessment review process, and it is part of the current *Canadian Environmental Assessment Act*; however, its presence does not guarantee the appropriate assessment of impacts on subsistence economies, for the reasons outlined above. The Canadian Environmental Assessment Research Council's research prospectus (1985) identifies the types of social change usually investigated as part of a social impact assessment as demographic, economic, resource-related, and cultural. This provides a good summary of other approaches (see also Barrow, 1997:234) and demonstrates some of the limitations of current social impact assessment practices for dealing with subsistence economies. Clearly, changes in hunting economies cut across all of these categories, and there is little provision for an analysis that recognizes their interrelatedness. Moreover, economic changes are conceptualized primarily in terms of income-generating activities, with the result that many subsistence activities are marginalized.

The Agreement potentially provides more scope for directing attention to the particular nature of subsistence economies. As noted above, Sections 22 and 23 introduce a series of guiding principles designed to protect Native economies. In addition, paragraphs 22.2.2 and 23.2.2 describe the environmental protection regime as one that provides for the "protection of the Cree/Inuit people, their economies and the wildlife resources upon which they depend." Schedule 3 of both sections states that "the impact assessment procedure should contribute to a further understanding of the interactions between Native people, the harvesting of wild life resources and the economic development of the Territory" (Canada, Québec, 1976:332, 357). The list of social conditions that form part of the impact assessment specifically includes "harvesting patterns" and the use and importance of various species in harvesting activities.

However, these principles and clauses do not provide detailed methods for assessing the effects on hunting economies. In a 1985 review of the Agreement, Penn (1988:130) noted that:

> [I]t has proved difficult to translate such general principles into operational language. However simple and straightforward the principles may sound, we are still some way from understanding how to translate them into the detailed and subject-specific language of impact assessment or environmental policy. However, the concept of the guiding principles is sufficiently important in the Agreement that operational definitions of these principles will have to be found.

Nevertheless, sections 22 and 23 of the Agreement do insist that hunting economies be taken seriously in the evaluation of impacts of developing. In this way they provide some scope for constructing an appropriate approach to subsistence economies in the assessment of development in the territory referred to in the Agreement.

Cree and Inuit Jurisdiction over Development Decisions

Neither federal nor provincial impact assessment procedures provide Native peoples with jurisdiction over industrial development. Under the federal EARP and the CEAA, the Minister of the Environment of the initiating Minister (the minister responsible for the agency proposing to undertake or authorize the project) makes decisions about whether or not the development should take place and under what terms and conditions. Under provincial legislation, authority for decision making about development projects is given to the Lieutenant-Governor in Council, or delegated to a committee of ministers.

The Agreement allows the Native people limited control over development in the territory referred to in the Agreement by giving Cree and Inuit local government administrators jurisdiction over development decisions on Category I lands. These lands make up only a small proportion (approximately 1%) of the lands in the area, however, and this level of control can have little impact on the protection of subsistence economies. Federal and provincial administrators are responsible for decisions about development in the remainder of the area referred to in the Agreement. While federal and provincial decision makers must consider recommendations from the review bodies established under the Agreement, these bodies (except for the KECQ) are nevertheless advisory in nature.

The KECQ's decision-making authority has been an important element contributing to its perceived effectiveness among northern EIA bodies (Jacobs and Kemp, 1987). The power allocated to the KECQ by its legislative base has meant that, although its evaluation process is applied in conjunction with the provincial Ministry of the Environment, it retains a semiautonomous role. Mulvihill and Keith (1989, 404–405) note that:

> On one occasion, the commission suspended its approval of a regionally significant project (the relocation of a village) in which the Province was the proponent. Despite heavy pressure from the Ministry of Housing and the Ministry of the Environment, the KEQC was able to hold firm and withhold its approval until all of its conditions were met.

However, although most of these bodies lack jurisdiction over decisions about whether or not development should go forward, the requirement for social impact assessment may increase somewhat their level of political influence. Researchers have argued that the process of social impact assessment can facilitate community empowerment despite the communities' lack of formal jurisdiction over development decisions (Corbett, 1986). The process can provide a bargaining tool, allowing communities to negotiate the terms of development and in this way manage to some extent its social and economic impacts. It can also help build local leadership and community decision-making capacity (see also Gagnon et al., 1993). At the same time, it is important to remember the costs, both social and economic, of these kinds of battles. These costs have disproportionately been borne by the Native peoples.

Incorporation of Cree and Inuit Values in Defining Effects of Development

From her study of the Haida's participation in the Joint Canada/British Columbia West Coast Offshore Exploration Panel in 1984 and 1985, Shapcott (1989:68) highlighted the ethnocentric nature of environmental impact assessments in Canada:

> Neither the courts nor the British Columbia Environmental Appeal Board can factor the social values of another culture into their decision-making. The values of the dominant culture are so imbedded in the process of EI (including that administered by the Environmental Assessment Review Process at the federal level), that alternative values cannot even be considered. As noted earlier, the underlying values—both the culture and the process—must be changed to making environmental impact assessment meaningful to Native people.

In part, culturally specific values enter into the assessment process through an identification of the aspects of the environment that need to be considered in the evaluation process. Beanlands and Duinker (1983:92) point out:

> It is impossible for an impact assessment to address all potential environmental effects of a project. Therefore it is necessary that the environmental attributes considered to be important in project decisions be identified at the beginning of an assessment.

Clearly, different cultures might vary widely in what are referred to as Valued Ecosystem Components or VECs.

Neither provincial nor federal impact assessment processes address the issue of VECs. There is nothing in sections 22 and 23 of the Agreement that requires that Cree and Inuit evaluations of aspects of the environment be employed in impact assessment. However the 1992 Guidelines set by the joint scoping and review committees asked Hydro-Québec to describe the environment not only in light of existing written scientific knowledge, but also according to the precepts, values, and knowledge of the aboriginal people:

> No description of the environment can ever be complete and exhaustive. Thus, it is preferable to carry out a systematic description focused primarily on *valued* ecosystem components... [T]he description of the different environments to be carried out by Hydro-Québec shall take into account the knowledge of, and attitudes toward, the environment specific to the Cree and Inuit cultures. (Great Whale Public Review Support Office 1992:2, emphasis in the original)

In their joint evaluation of Hydro-Québec's impact assessment of the project, the scoping and review committees found that the study's incorporation of Cree and Inuit values was inadequate (Provincial Review Committee et al., 1994). The Committees noted that the assessment was based on limited and outdated knowledge of the societies and cultures of the Native peoples in the territory referred to in the Agreement, and that information about the environmental knowledge and values of local communities was entirely absent. The Cree and Inuit have noted the general paucity of data to measure and monitor the effects of development projects (Brooke, 1995; Penn, 1995; Wilkinson and Vincelli, 1995). At present, the lack of data makes it difficult to demonstrate that subsistence harvests or tenure systems are being adversely affected. Wilkinson and Vincelli (1995) argue that Hydro-Québec, as proponent of the major hydroelectric projects in the area, has principal responsibility for ensuring baseline data required for environmental assessment are collected systematically, and that this responsibility has not been exercised adequately.

Québec's cancellation of the Great Whale project means that we do not know exactly how the requirement to incorporate Cree and Inuit values concerning the ecosystem would have been implemented in the final impact assessment. The existence of these provisions in the scoping guidelines for the Great Whale project does not ensure that similar guidelines will be employed for subsequent impact assessments of industrial development in the territory referred to in the Agreement. However, scoping is a difficult exercise. It is likely, therefore, that future scoping committees will refer to these *Guidelines*, and they may provide a standard against which future exercises will be measured. As such, they can serve as an important precedent for incorporating Cree and Inuit values into the assessment process. Without information available about these components, however, the requirement to include them in an impact assessment may be a meaningless gesture.

Incorporation of Cree and Inuit Values in Decision-making Processes

Cree and Inuit values can be incorporated in decision-making processes through representation on various review bodies and through provisions for public participation. While Cree and Inuit representatives are not excluded from review bodies in federal and provincial impact assessment procedures, neither is their presence mandated.

Under the Agreement, Cree and Inuit representatives must be represented in the bodies that set guidelines for and evaluate impact assessments. Discovering the extent to which this participation creates a meaningful conduit for Cree and Inuit interests requires attention to more than the text of the Agreement, however. The dependence of the environmental regime on advisory committees with Native representation reflects the assumption of the Native negotiators that the interests of a small and economically marginal population could influence public policy through the creation of an interface with government representatives (Brooke, 1995; Penn, 1995). Experience with the review committees suggests that the significance of Native participation has been undermined by a number of factors during the tenure of the Agreement. These factors include the appointment to these committees of low-level civil servants, who have little authority for decision making; the frequent turnover of government representatives, which limits their ability to develop expertise on issues or real familiarity with Cree or Inuit cultures; and the unavailability of an independent research staff or research budget (Voinson, 1988; Brooke, 1995; Penn, 1995; Wilkinson and Vincelli, 1995). Wilkinson and Vincelli's (1995) review of the operation of these review bodies points out that Cree and Inuit representatives have been absent from more meetings than government representatives, and also that they have occasionally abstained from voting. The interpretation of these patterns is not entirely clear, and they may have much to do with issues of cultural appropriateness.

There are many indications that the processes of communication and decision making created by the Agreement are less than appropriate for Cree and Inuit cultures. Penn (1995) notes that Cree representatives to review bodies are often empowered to express collective viewpoints with respect to certain issues, but they are not empowered to negotiate or bargain without going back to their constituencies (see also Brooke, 1995). Wilkinson and Vincelli (1995) note that Cree and Inuit cultures are based on the oral transmission of knowledge and concerns and on decision making by consensus of locally or directly affected persons. Yet the environmental impact assessment regimes in the Agreement make fewer provisions for public participation than either existing federal or provincial regimes, and no provision was made in the Agreement for intervenor funding. Many of the reviews take place in southern locations. In addition, reviews rely heavily on written materials; the vast majority of these are in French or English, and many employ highly technical language. Brooke (1995) indicates that Inuit participants' expectation that their intimate knowledge of the land and wildlife would make a valuable contribution to the workings of the review bodies was not borne out. There has been little incorporation of traditional environmental knowledge in the deliberations of the committees to date (Brooke, 1995; Wilkinson and Vincelli, 1995). The result has been the increasing representation of the Cree and Inuit by younger people (who are not necessarily the most knowledgeable about or dependent on the environment) and heavy dependence on non-Native consultants. Wilkinson and Vincelli (1995) conclude that, although the review processes were expected to work differently in the area referred to in the Agreement because of attempts to incorporate Cree and Inuit values and participation, in fact they have not met those expectations.

Incorporation of Cree and Inuit Values in Defining Desirable Futures

Beyond assessing VECs from Cree and Inuit perspectives and ensuring Native participation in decision-making processes, some more general issues must be considered. These issues have to do with how economic development itself is evaluated. Tester (1992) has noted that environmental impact assessment processes in Canada have emphasized mitigative measures, and cases of developments cancelled as a result of social or environmental assessments have been very rare. Underlying these results are the assumptions that industrial development is positive, and that negative spin-offs are secondary results, which can be addressed through compensation or measures designed to minimize their impacts. Impact assessment processes provide virtually no consideration of broader values, for example, questioning the equation of development and progress or the inevitability of modernization, and considering alternative relationships between people and environments. Indigenous peoples' attempts to preserve subsistence economies often reflect an assessment of the desirability of industrial development very different from that of the developers.

Neither federal nor provincial legislation provides scope for evaluating existing practices or frameworks for managing industrial development. However, the James Bay Advisory Committee on the Environment (JBACE) and the Kativik Environmental Advisory Committee (KEAC) created under the Agreement have as part of their mandate the responsibility to evaluate legislation, policies, and regulations in the context of the principles laid out in sections 22 and 23. Unlike environmental impact assessment processes, which can only respond to proposals to change the status quo, these committees have the authority to evaluate current practices, including environmental assessment regimes, and to do so under their own initiative.

At the same time, these Committees are merely advisory to governments. Despite their status as the "preferential and official forum" for environmental protection measures in the territory referred to in the Agreement, they have not been given a clear role in government decision-making structures, which infrequently consult with them (Wilkinson and Vincelli, 1995). Moreover the ability of the Committees to undertake analysis of existing practices depends to a large extent on the degree to which these bodies are supported and consulted by federal and provincial governments. The Cree and Inuit have noted the general paucity of data available for measuring and monitoring the effects of development projects (Brooke, 1995; Penn, 1995; Wilkinson and Vincelli, 1995). The Cree have argued that federal and provincial governments have lacked a commitment to fully implement these Committees. As a result, they have not operated continuously or effectively; technical and scientific information, advice, and assistance have not been forthcoming; and an adequate secretariat has not been provided (Grand Council, 1987, 1991; Mainville, 1991; Penn, 1995; Wilkinson and Vincelli, 1995).

Apart from issues of implementation, there are also issues having to do with the relationship between rights to develop and rights to protect subsistence economies in the Agreement. The right to develop is one of the principles governing the operation of the environmental regime. All of the principles address the effects on Native economies of development. In other words, there seems to be an assumption that development is inevitable and desirable, and that the purpose of the principles enunciated in the Agreement is to minimize the negative effects of this development on Cree and Inuit hunting economies, rather than to halt development altogether.

At the same time, it is clear that the Agreement contemplates the possibility that the various review bodies may recommend that no development take place, and these recommendations could be made on the basis of negative effects on hunting economies. It is also clear that the Agreement gives the Cree and Inuit a number of rights, which include the protection of hunting economies and societies and the "guarantee of levels of harvesting equal to present levels of harvesting of all species in the Territory" (paragraph 24:6.2). It is not easy to ascertain which of these rights should take precedence in any particular situation. For example, paragraph 5.5.1 states that:

> the rights and guarantees given to the Native people by and in accordance with the Section on Hunting, Fishing and Trapping shall be subject to the rights to develop Category III and Category II lands on the part of Québec... However, the developers shall be submitted to the Environmental Regime which takes into account the Hunting, Fishing and Trapping Regime. (Canada, Québec, 1976:70)

The Agreement, then, appears to present mixed opportunities for affecting the direction of development in the territory referred to in the Agreement. While it seems to assume that industrial development will occur, with attempts to mitigate its effects, it also provides some avenues for the consideration of alternative futures. At some degree and scale of industrial development, it is clearly not possible to reconcile the protection of Cree and Inuit subsistence economies and the right to develop written into the agreement. To date, it seems that development initiatives have won out, particularly on Cree lands (Penn, 1995). But it is not clear whether this is so because the Agreement has not been fully or appropriately implemented, or because the rights the Agreement gave the Cree and Inuit to protect subsistence economies are subordinate to the right to develop in the territory referred to in the Agreement. As a result, it is difficult to ascertain the extent to which the Agreement allows for critical consideration of whether economic development is desirable in the territory referred to in the Agreement.

Conclusion

At the signing of the James Bay and Northern Québec Agreement, Cree and Inuit expectations were that the Agreement would give them the tools with which they could exercise meaningful influence over development in the territory referred to in the Agreement, in order to protect their subsistence economies. The Agreement has not met the expectations of the Native parties to the Agreement. Some of the events and situations causing this dissatisfaction result from the wording and contents of the Agreement: it was written before there was much Canadian experience with environmental assessment, and in this sense it represents a pioneering document. This status also means that the experience of other groups could not inform the construction of the Agreement. Many sections of the Agreement appear to be vague and difficult to translate into workplace principles. The advisory committee structure and process has not been well suited to Cree and Inuit cultures. The right to develop is woven through all sections on environmental protection. Finally, committee structures are exceedingly complex, involving federal and provincial governments, with separate areas of jurisdiction, in addition to Native representatives. This complexity is exacerbated in situations such as the Great Whale Project, which overlaps Cree and Inuit territories. When, in addition governments have to be drawn into participating through legal means, it may become difficult to generate the consensus necessary to create less cumbersome procedures (see Brooke, 1995).

However, it is also very clear that there are major issues concerning implementation of the provisions of the Agreement. This makes it very difficult to evaluate the effectiveness of specific provisions. In this context, perhaps the strongest lesson to be learned from the Agreement has to do with the necessity of detailed plans for implementation.

Questions for Discussion

1. The first reading covers several environmental impacts of Phase I. Notice that there are basically two types of impacts: (1) permanent impacts, whose intensities are not expected to decrease with time; and (2) temporary impacts, which are associated only with construction or from which the environment will recover. Make a list of each type and, for temporary impacts, predict their time line. That is, how long will recovery take? Discuss the relative importance of these impacts. Are they tolerable? Could the intensity of impacts be decreased?

2. How is mercury important to natural ecosystems? How does excessive mercury affect living things? Why is the mercury content of fish particularly interesting to natives in the James Bay region? (Hint: You may want to surf the Web to answer some of these questions.)

3. In other publications, Hydro-Québec has argued that, on balance, its projects had little effect on natural ecosystems. Admittedly, some terrestrial ecosystems were flooded, but aquatic ecosystems were created. What is your response to such reasoning?

4. According to the Peter paper, native peoples (Crees and Inuits) are not entirely satisfied with the Environmental Agreement. Why not? Specifically, what are their complaints? How might the Agreement be rewritten to assuage their concerns?

5. At the time the Environmental Agreement was signed, it was widely assumed that if native subsistence interests were provided for, the environment would be sufficiently pristine to satisfy environmentalists. Do you agree with this assumption? Why or why not?

Going Beyond the Readings

Visit our website at <www.harcourtcollege.com/lifesci/envicases2> to investigate alternate energy issuses in additional regions of the United States and Canada.

unit 4

Great Lakes: Water Resources

Introduction
Readings:
Reading 1: A Strategy for Virtual Elimination of Persistent Toxic Substances
Reading 2: Comments on the *Canada–United States Strategy for the Virtual Elimination of Persistent Toxic Substances in the Great Lakes Basin*
Reading 3: Pollution, Thirst for Cheap Water Threaten Great Lakes, Study Says
Reading 4: Protecting Lake Superior: A Community-Based Approach
Questions for Discussion

Great Lakes Cleanup

Introduction

The Great Lakes of North America are collectively the world's largest body of freshwater. They also host one of the most complicated and vexing environmental problems in the world. Everything about the Great Lakes watershed is huge. It contains one-fifth of the world's surface freshwater within an area of 298,500 sq. mi. (773,000 km2). At its deepest point, Lake Superior is 1290 ft. (390 m) deep. The St. Lawrence River and its tributaries connect the lakes with the Atlantic Ocean.

Biologically, the region is a treasure trove providing habitat for countless animals and plants. Fish are particularly important. Significant populations form important links in food chains supporting dozens of species of birds and mammals. Birds nest extensively throughout the region in spring and summer. In addition, the lakes are important to over-wintering waterfowl.

People have long recognized the region's importance to wildlife and have set aside important sites for wildlife and habitat protection. The watershed boasts seven national parks and lake shores, six national forests, seven national wildlife refuges, and hundreds of state and provincial parks, forests, and sanctuaries. The lakes also provide 46 million people with water for drinking, energy, recreation, transportation, agriculture, and industry. The region's human population grows incessantly. Millions of jobs and thousands of businesses are dependent, directly or indirectly, on waters of the Great Lakes.

Politically, the region is complicated. It forms nearly a third of the border between Canada and the conterminous United States. Eight states and two provinces enjoy its shores and banks. No single political entity has jurisdiction of more than a fraction of the region. This vastly complicates potential management decisions.

The Problem

Traditionally, the Great Lakes have provided people with more than recreation and sources of water. To people living along the shores, the lakes seemed logical places to dump sewage, garbage, and industrial wastes. Additional wastes entered inadvertently, leaching from landfills, farms, and yards, and settling from the atmosphere. As long as the region's human population was small, there was no problem. Three factors in recent decades have limited the lakes' ability to render harmless such human wastes.

First, the region's human population has grown dramatically for 300 years and it continues to do so. More people means more wastes.

Second, the lakes' capacity to process wastes has diminished. Initially, forests, wetlands, and lakes formed an ecological partnership to receive and recycle nutrients. They treated many components of human wastes as new sources of nutrients and processed them accordingly, but as human population grew, forests were cut and wetlands were drained and filled. Their contributions to nutrient recycling have been lost to the system.

Third, the region has become one of the world's major industrial centers. Key industries include iron mills, international shipping (via the St. Lawrence Seaway), and an extensive auto industry. Industrialization produces large quantities of special wastes, which were conveniently dumped into the lakes. Indeed, one incentive to build in the region was the presence of an obvious place to conveniently and economically get rid of unwanted materials.

Industrialization contributed to, but was not solely responsible for, a fourth factor. In recent decades, some of the most worrisome wastes dumped in the Great Lakes have been synthetic compounds, largely unknown to the natural world. These compounds, chemically complex and essential to modern life, do not break down naturally. Biological agents pick them up from the environment and incorporate them into tissues, but lack mechanisms to break them down into simpler, less harmful compounds.

Furthermore, some of these compounds bioaccumulate. They may occur in low concentrations in lake water or bottom muds. Plants gather these compounds from the environment, store them in tissues, and intensify concentrations. When herbivores eat contaminated plants, compounds accumulate further. Carnivores eat herbivores and continue the process. Aquatic ecosystems such as lakes may host several levels of carnivores; at each step, concentrations build. Levels that were barely detectable in water and mud become dangerously high—even toxic—further along the food chain.

Chemical pollutants have been implicated in increased numbers of birth defects, decreased birthrates, and shortened lives in Great Lakes' animals. Many of these animals are of special interest to humans, including waterfowl, eagles, mink, domestic pets, and livestock. Humans are not immune. Nursing mothers are advised to limit consumption of Great Lakes fish lest their milk pass to offspring unacceptable concentrations of bioaccumulating compounds.

The buildup of persistent toxic substance is a particularly acute problem in aquatic ecosystems. Today, such substances are widely used throughout the world and are purposely dumped or accidentally leak into waterways. Coastal communities dump directly into oceans, where potentially toxic substances quickly dilute to less than measurable levels. At present, effects of such substances on plants, animals, and people are also less than measurable. Can this situation continue indefinitely? Areas dumping persistent toxic substances into rivers or estuaries—areas where rivers meet oceans, ranging in size from small bays to much larger systems, such as the Chesapeake Bay—have a more acute problem. In these waters, substances do not immediately dilute. However, rivers flow. So do estuaries, albeit slowly. If dumping is stopped, rivers and estuaries will, in time, clean themselves. But in lakes particularly large, deep lakes such as the Great Lakes—flow-through times are measured in centuries. The questions facing those concerned with water quality in the Great Lakes are: How might these systems be cleaned up? Will they ever "self-clean"?

Seeking Solutions

As early as 1909, the U.S. and Canadian governments recognized the special challenges of jointly managed lake areas. In that year they passed the Boundary Waters Treaty. Among other things, the treaty established the International Joint Commission, charged with the responsibility to identify, avoid, and recommend solutions to potential disputes involving the watershed. In 1972, the governments signed the first Great Lakes Water Quality Agreement to control pollution throughout the region. In 1978, a new agreement was signed, which added the commitment to work together to rid the Great Lakes of "persistent toxic substances," that is, those that do not readily degrade. In 1987, the governments signed a protocol to clean up problem areas and promote sustainable development throughout the region.

In 1990, the commission established a special task force to devise a strategy "to virtually eliminate the input of persistent toxic substances into the Great Lakes Basin Ecosystem." The two-volume report, often referred to as the Virtual Elimination Strategy, or VES, was published in 1993. It was not universally welcomed. Some industrialists and politicians worried about how much implementation would cost and who would pay. As usual, citizens were caught in the middle.

The following readings contain a portion of the VES (the entire report can be found at the International Joint Commission's Web site), along with a sampling of responses and discussion of two ongoing projects designed to increase water quality in the Great Lakes.

reading 1

"A Strategy for Virtual Elimination of Persistent Toxic Substances"

International Joint Commission, Windsor, Ontario, 1993, Vol. 1, pp. 2–23 and 61–66.

1. The Issue and The Investigation

1.1 The Agreement and Persistent Toxic Substances

The Parties' stated purpose for the 1978 Great Lakes Water Quality Agreement "is to restore and maintain the chemical, physical, and biological integrity of the waters of the Great Lakes Basin Ecosystem." In particular, the Parties undertook an obligation to virtually eliminate the input of persistent toxic substances. This commitment was strengthened by the 1987 amendments to the Agreement. Article II of the Agreement states that "It is the policy of the Parties that . . . the discharge of toxic substances in toxic amounts be prohibited and the discharge of any or all persistent toxic substances be virtually eliminated."

Specifically with regard to persistent toxic substances, the intent is to undertake actions, programs, and other measures to:

- Protect human health.
- Ensure the continued health and productivity of living aquatic resources, including their use by humans.
- Ensure further ecosystems protection.

To fulfill these requirements, it is necessary to:

- Virtually eliminate present inputs of persistent toxic substances.
- Anticipate and prevent future inputs and problems.
- Remediate problems from past and present inputs.

1.2 The Commission and Persistent Toxic Substances

For more than a decade, the Commission, as it has tracked the Parties' progress, has become increasingly vocal in its concern with regard to persistent toxic substances. In its Fifth Biennial Report, the Commission urged the Parties to

"take every available action to stop the inflow of persistent toxic substances into the Great Lakes environment."

Specifically, the Commission recommended that

"the Parties complete and implement immediately a binational toxic substances management strategy ... for accomplishing, as soon as possible, the Agreement philosophy of zero discharge."

These recommendations were made on the basis of a number of important conclusions that the Commission reached in the course of its research and analysis. It became clear to the Commission that concern for fish and wildlife health was well founded, and that this concern should be extended to humans as well. Thus, it was concluded that

"What our generation has failed to realize is that, what we are doing to the Great Lakes, we are doing to ourselves and to our children."

and

". . . the Commission must conclude that there is a threat to the health of our children emanating from our exposure to persistent toxic substances, even at very low ambient levels."

The Commission based these conclusions and recommendations on mounting evidence which, it concluded, ". . . cannot be denied." It its *Sixth Biennial Report* (3), the Commission concluded that

"because persistent toxic substances remain in the environment for long periods of time and become widely dispersed, and because they bioaccumulate in plants and animals—including humans—that make up the food web, the ecosystem cannot assimilate these substances."

and thus they

"are too dangerous to the biosphere to permit their release in any quantity."

Further,

"the presence and impact of persistent toxic substances on all sectors of the ecosystem . . . defies boundaries and is not easily resolved through traditional technologies and regulations . . . These substances cross jurisdictional, geographic and disciplinary lines that have tended to circumscribe previous efforts to restore and protect the ecosystem.... There are no preordained boundaries in the way the natural system functions and in how humans interact with and within it."

The Commission concluded that, despite the Agreement requirement to virtually eliminate the input of persistent toxic substances to the Great Lakes basin and to protect human and environmental health, *"we have not yet virtually eliminated . . . any persistent toxic substance."*

The Commission observed therefore that, as part of the solution,

"[do] we . . . want to continued attempts to <u>manage</u> persistent toxic substances after they have been produced or used, or [do] . . . we want to . . . <u>eliminate</u> and <u>prevent</u> their existence in the ecosystem in the first place . . . Since it seems impossible to eliminate discharges of these chemicals . . . , a policy of <u>banning</u> or <u>sunsetting</u> their manufacture, distribution, storage, use and disposal appears to be the only alternative."

1.3 The Commission's Charge to the Task Force

In its *Fifth Biennial Report*, the Commission urged Governments to develop and implement

"a comprehensive, binational program to lessen the use of, and exposure to persistent toxic chemicals found in the Great Lakes environment."

The Commission recognized, however,

"that problems associated with persistent toxic substances cannot be simply defined or solutions easily implemented."

To contribute to the definition and resolution of the issue, the Commission charged the Virtual Elimination Task Force to investigate the Agreement requirement to virtually eliminate the input of persistent toxic substances into the Great Lakes Basin Ecosystem. Specifically, the Task Force was charged to provide advice and recommendations to the Commission about what a virtual elimination strategy should contain and how the strategy could be implemented. The Commission will, in turn, provide its advice to Governments.

1.4 The Task Force's Point of Departure

The Commission specifically charged the Task Force to focus on **persistent** toxic substances, rather than toxic substances. Also, the Task Force was not asked to investigate whether persistent toxic substances have caused injury. The virtual elimination commitment incorporated by the Parties into the Agreement in 1978, and the stance taken by the Commission in its *Fifth and Sixth Biennial Reports* support the conclusion that the evidence is more than sufficient to advocate for virtual elimination of the input of persistent toxic substances.

The members of the Task Force, individually and as a whole, accepted that some problems remain with persistent toxic substances; the question is, how to resolve those problems.

The Task Force recognized that the Commission's call for far-reaching action requires clear evidence that damage has occurred and continues to occur, and that persistent toxic substances are among the causes of this injury. Only with strong evidence will there be a stimulus for development of, and the timely commitment to implement a virtual elimination strategy and thereby eliminate or prevent resultant injurious effects to health and the ecosystem.

Since the Commission's *Fifth Biennial Report*, issued in 1990, the evidence has continued to mount. Important scientific and government consensus has emerged to further cement the basis for the Commission's conclusions and position described above. Specifically, the Task Force observed that there is broadened understanding and acceptance that:

- A number of human-made persistent toxic substances have and continue to cause significant adverse effects on, and substantial damage to, fish and wildlife species.
- Persistent toxic substances are a threat to human health, to fish and wildlife health and, indeed, to the entire ecosystem.

In addition, adverse effects have been reported in the children of women who ate contaminated fish from Lake Michigan, and the reported injury occurred mainly prenatally:

Therefore, as a crucial component of this report, the Task Force reviewed evidence and conclusions developed by knowledgeable experts in various scientific disciplines and published in the peer-reviewed literature. Appendix D provides perspective about the injury caused by some persistent toxic substances, and the danger they pose. A brief summary is provided below.

1.5 The Injury

There is general agreement that several contaminants routinely found in the Great Lakes basin already meet the definition of a persistent toxic substance (see Chapter 2). Despite considerable environmental improvement (discussed later in this chapter), long-term exposure to these contaminants presents a continuing threat to the health of the ecosystem and to the life that constitutes it. A focused strategy, together with a concerted effort, are required to virtually eliminate inputs of persistent toxic substances to the ecosystem, so as to virtually eliminate their presence in the ecosystem and to eliminate impairment of ecosystem health. A strategy is also required to protect the ecosystem by

TABLE 1

Critical Pollutants Identified by the Water Quality Board

- Total polychlorinated biphenyls (PCB)
- DDT and metabolites
- Dieldrin
- Toxaphene
- 2, 3, 7, 8-tetrachlorodibenzo-p-dioxin (2, 3, 7, 8-TCDD)
- 2, 3, 7, 8-tetrachlorodibenzofuran (2, 3, 7, 8-TCDF)
- Mirex
- Mercury
- Alkylated lead
- Benzo(a)pyrene
- Hexachlorobenzene

preventing future inputs of persistent toxic substances, prior to their introduction into use.

In 1985, the Commission's Great Lakes Water Quality Board identified 11 Critical Pollutants (Table 1) that are persistent, bioaccumulate in living organisms, cause adverse human and environmental health effects, and have been subject to extensive regulation. However, actions to date are insufficient and incomplete. For example, bans or use restrictions for PCBs and some chlorinated hydrocarbon pesticides are not absolute.

- It is estimated that more than 50% of all PCBs ever produced are still in use. Loadings continue from a variety of known and unknown sources.
- Many bans or restrictions on pesticides (such as DDT, dieldrin, endrin, aldrin, chlordane, toxaphene, heptachlor, and mirex) apply only to domestic uses and may not come into effect until existing stocks are depleted. Thus, commercial products containing many of these pesticides are still for sale in Canada and the United States. This includes DDT, which can still be purchased despite a 1990 ban on its sale. Further, large quantities of banned or restricted pesticides are still produced in the United States and exported.

Continued production, sale, use and/or export provides numerous opportunities for release to the environment and, ultimately, additional inputs to the Great Lakes. Because inputs continue, persistent toxic substances still pervade the ecosystem and its food chain at levels sufficient to cause injury.

Fish, particularly the predators at the top of the food chain, are excellent indicators of ecosystem health because they bioaccumulate and biomagnify many aquatic contaminants. Birds and other wildlife (such as mink and otter) that eat fish display a wide range of contaminant-related problems, including population decrease, effects on reproduction, eggshell thinning, behavioural changes, biochemical change, and increased mortality

The adverse reproductive and developmental effects observed in wildlife may foreshadow human population effects. Wildlife may be the "canary in the coal mine," warning of a potential blight on present and future generations. There are few comprehensive studies of such effects on humans but, given effects in Great Lakes wildlife, some researchers are now focusing on possible human health effects. Generally, for a number of persistent toxic substances, an association has been made between human body burdens and the regular inclusion of fish in the diet.

- One study in Michigan demonstrated that sport anglers who ate Great Lakes fish (especially trout and salmon) had higher blood and tissue levels of PCBs than individuals who seldom or never ate such fish.
- A 1993 report identified an association between blood levels of DDT/DDE and breast cancer, and an elevated (but not statistically significant) risk of breast cancer associated with PCBs.
- There is suggestive evidence from another study that women who ate several meals of Lake Michigan fish a month for at least six years preceding their pregnancies bore children who had lower birth weights, shorter gestational periods, and smaller head circumferences at birth, and who showed discernible cognitive, motor, and behavioral deficits when tested later, compared to infants born to women who had not consumed Lake Michigan fish prior to or during their pregnancies. The discernible cognitive, motor, and behavioural effects persisted in tests at seven months and four years.

Physical growth and short-term memory deficits appear to be specifically related to in utero exposure. This concept of in utero injury to the unborn, due especially to persistent toxic substances that interfere with the extremely subtle and sensitive workings of endocrine systems, including sex steroid metabolism, is of profound consequence. In a recent consensus conclusion, a multidisciplinary group of experts stated that:

"The concentrations of a number of synthetic sex hormone agonists and antagonists measured in the U.S. human population today are well within the range and dosages at which effects are seen in wildlife populations. In fact, experimental results are being seen at the low end of current environmental concentrations."

This is consistent with a 1992 review and a related 1993 study where it is hypothesized that fetal exposure to estrogens or estrogenic chemicals (endocrine disruptors such as DDT, PCBs, dioxins, furans, and hexachlorobenzene, among other organochlorines and metals) may be responsible for declining sperm counts and a rising incidence of abnormalities in the human male reproductive tract.

Persistent toxic substance contamination has also injured the economy and society, through real and suspected human injury and health costs, real environmental costs, and loss of economic value, for example, as a result of the loss of commercial fisheries. Society has accumulated costs in the form of an "environmental deficit"—a debt of problems, cleanup costs, and risks that are shifted to the future, and to society at large.

Taken as a whole, the weight of evidence accumulated over the past three decades indicates that exposures to persistent toxic substances are indeed associated with injury, disease, and death in a wide variety of life forms. In its Sixth Biennial Report, the Commission recommended that such an approach be applied "to the identification and virtual elimination of persistent toxic substances." The weight-of-evidence approach has been endorsed in the United States by the National Academy of Sciences and the Office of Science and Technology Policy, and has been widely adopted by numerous government regulatory agencies for the evaluation of scientific information.

The weight-of-evidence approach assists scientists and others in answering the question: "Is the available information sufficient to conclude that the observed or predicted phenomenon will lead to an adverse effect in humans or aquatic life?" The approach considers the full spectrum of relevant factors, both positive and negative, and gives appropriate weight to the scientific evidence on a case-by-case basis. For example, factors typically considered in evaluating the weight of evidence include the quality of data, the number of positive versus negative studies, species differences, relevance of animal data to humans, strength of association, mechanism of action, and other relevant data.

Although evidence of injury is clear for some persistent toxic substances and there is ample justification to develop and apply a virtual elimination strategy to deal with them, doubt exists for a number of other substances, especially in regard to injury to future generations. In addition, there are different interpretations in regard to observed injury. Because of uncertainty, a precautionary approach is needed.

1.6 Progress To Date

The Commission and the Task Force both recognize that considerable progress has been made to reduce inputs of persistent toxic substances to the Great Lakes Basin Ecosystem. As a result, ecosystem health today is improved from conditions 20 years ago. This is the direct result of several activities such as construction of municipal and industrial waste treatment systems, remedial efforts to mitigate contaminants already in the ecosystem, and restrictions, phaseouts, and bans on the manufacture and/or use of certain persistent toxic substances.

Early treatment methods focused on the control of traditional pollutants, such as phosphorus, biochemical oxygen demand, and suspended solids. This coincidentally reduced other contaminants, especially contaminants (many being persistent) that associated with the particulate phase of an effluent. More recently, releases of some persistent toxic substances have been reduced as a consequence of manufacturing process changes, and as the movement to reduce and phase out persistent toxic substances continues to gain momentum.

Appendix E lists and describes examples of specific technological changes, regulatory programs, and voluntary measures that account for the successes achieved, and that are emerging as possible vehicles for future delivery of virtual elimination. Many of these examples contain elements similar to those recommended for use as part of the virtual elimination strategy presented in Chapter 3 and, in particular, multi-stakeholder consultation and dialogue.

As a result of such measures, ecosystem concentrations of persistent toxic substances dropped markedly, especially during the late 1970s. Collectively, these measures have contributed to increases in bird populations, reductions in bird malformities, and reduction in contaminants in fish tissue. In its *Sixth Biennial Report*, the Commission noted that:

"nesting pairs [of bald eagles] reintroduced to the north and south shores of Lake Erie continue to survive, which can be seen as evidence of improved ecosystem quality. The viability of many of their eggs also attests to improvements."

However, in many cases, contaminant concentrations have leveled off (see, for example, Figure E-1) and, in some cases, appear to be increasing (see Figure E-5). In addition, as discussed above, the actions taken to date, despite leading to significant improvement in ecosystem quality, are insufficient to eliminate biological injury. To illustrate, four bald eagles born in 1993 along the Michigan shoreline of the Great Lakes have life-threatening deformities: twisted beaks or clubbed feet. Two of the eagles were from Lake Erie nests. This is further stimulus to develop and implement a virtual elimination strategy.

1.7 The Task Force's Investigation

In its investigation, the Task Force focused on the overall concept of a virtual elimination strategy and the specific components required to achieve and maintain a healthy ecosystem. In addition, the Task Force evaluated how virtual elimination can be achieved, and applied the strategy to three case examples. The Task Force has endeavoured to develop a strategy and advice that it believes is necessary and right. To accomplish this goal, the Task Force investigated not only the input of persistent toxic substances to the ecosystem, but also their presence in the ecosystem. To ensure the credibility of its work with a wide spectrum of stakeholders and to provide a fair assessment, the Task Force attempted to maintain a fair, open-minded, nonpartisan perspective.

Specifically, the Task Force focused on:

- What injuries have persistent toxic substances caused, and what danger do they pose?
- Definitions of key terms—including persistent toxic substance, virtual elimination, and zero discharge—to ensure a common and clear basis for discussion and understanding.
- Selection criteria and a procedure to develop a framework within which to identify chemicals that would be subject to the virtual elimination strategy.
- Sources and uses of persistent toxic substances. Where are they found in commerce? How do they enter and move within the ecosystem? What is their fate in the ecosystem? What are the quantities associated with the various sources and uses, and what level of concern should be placed on the location and movement of persistent toxic substances in the ecosystem?
- Evaluation of the legislative, regulatory, technological, economic, and educational tools and opportunities to achieve virtual elimination.
- Identification of performance indicators, or measures of success, to conclude that virtual elimination of inputs of persistent toxic substances has been achieved, that the injury has been eliminated, and that the ecosystem has been restored and is protected.
- Other particular issues associated with development and/or implementation of the virtual elimination strategy, such as remediation of contaminated sediment; sources of contaminants to the atmosphere; and waste storage, disposal, and destruction.

In its investigation of the components of the virtual elimination strategy and the application of the strategy to case examples, the Task Force also considered what tools to apply and opportunities to exploit, how and when, and by whom. Further, the Task Force considered and built on a range of relevant initiatives that have been or are being undertaken by others. The material comprising the remainder of this report is organized along these general lines. To the extent possible, the material represents the consensus of the Task Force members.

The initial advice presented in the Task Force's Interim Report (1) about the overall concept of the virtual elimination strategy and the specific components of the strategy served as the point of departure for the discussions in each of the following chapters. The Task Force also concurs with the concept of sustainable development, wherein a healthy economy and a healthy environment are inseparable and mutually achievable. Further, to undertake a virtual elimination initiative carries a degree of risk and uncertainty. However, these are usually accompanied by new opportunities for all stakeholders. These opportunities, in turn, foster cooperation and a sharing of responsibility for environmental protection. In its investigations and advice, the Task Force has endeavoured to recognize and build upon these to achieve the virtual elimination goal. The conclusions and recommendations presented in Chapter 11 represent the Task Force's advice to the Commission about the development and implementation of the virtual elimination strategy.

2. Terminology

In its charge to the Task Force, the Commission requested a definition of key terminology, including persistent toxic substance, zero discharge, and virtual elimination. The time spent defining these terms at the Task Force's public workshops, at the Commission's roundtables, in written comments to the Task Force, and among the Task Force membership is heartening: it indicates that commitments in the Agreement are being taken seriously.

The real challenge, however, is not to reach unanimous agreement on terms, but to achieve the goal of the Agreement: to restore and maintain ecosystem health. To accomplish this goal, the Task Force considered it necessary to include in its investigation the **presence** of persistent toxic substances in the ecosystem along with **inputs** to the ecosystem, as charged by the Commission (see Chapter 1).

For the purposes of this report, the definitions used are based on the language of the Agreement. However, in some cases, the Agreement language is not sufficient to develop a strategy to implement the policy of virtual elimination. Where appropriate, these definitions have been expanded.

2.1 Persistent Toxic Substance, Half-life, and Bioaccumulation

Article I of the Agreement defines toxic substance as one:

"which can cause death, disease, behavioural abnormalities, cancer, genetic mutations, physiological or reproductive malfunctions or physical deformities in any organism or its offspring, or which can become poisonous after concentration in the food chain, or in combination with other substances."

In Annex 12, persistent toxic substance (see sidebar) is defined as:

"any toxic substances with a half-life in water of greater than eight weeks."

Half-life is defined as:

"the time required for the concentration of a substance to diminish to one-half of its original value in a lake or water body."

A more extensive definition of persistent toxic substance is provided in the Commission's *Sixth Biennial Report*. The Commission recommended that:

"The Parties expand the definition of a persistent toxic substance to encompass all toxic substances: with a half-life in any medium—water, air, sediment, soil or biota—of greater than eight weeks, as well as those toxic substances that bioaccumulate in the tissue of living organisms."

The Task Force notes that the concept of half-life, as presented in the Agreement, has no accompanying scientific rationale. Half-life must consider all processes associated with the input to and removal of the substance from the ecosystem. Half-life is difficult or impos-

---

The terms *toxic substance* and *persistent toxic substance* are *not* interchangeable. While a persistent toxic substance always exhibits the characteristics of a toxic substance, the reverse is not the case. The virtual elimination strategy is driven by the characteristic of **persistence.**

---

sible to measure or calculate, and the value determined can vary depending, for instance, on where the substance enters the ecosystem and its propensity to move among media. In general, however, the longer that a substance remains in the environment, the longer and more accessible that substance is to living organisms.

The Task Force believes that half-life should be based on chemical, biochemical, and photochemical degradation processes and should not be based on such considerations as dilution processes.

The **bioaccumulation factor (BAF)** refers to the concentration of a chemical in the biota, received via all routes, divided by the dissolved concentration of that chemical in water. Substances with higher BAFs will accumulate in animals/humans to a higher level creating a greater potential for biological damage. Substances that bioaccumulate (including those that may combine with other chemicals and then bioaccumulate) should receive priority for virtual elimination, but other valid criteria must be considered in deciding on needed substance action (see Chapter 4).

Human activities have augmented the availability of metals, and the potential for them to cause injury to living organisms. Some metals (such as iron), though "persistent" according to the definition in Annex 12, should not be subject to the same stringent regulatory policies as other persistent toxic substances. However, other metals (notably mercury and lead), because of their potential to bioaccumulate after combining in the ecosystem with other substances (methylation), must be included in the definition of persistent toxic substance, as should a number of anthropogenic organometals and other metallic products.

2.2 Zero Discharge

As presented in Annex 12 of the Agreement, zero discharge is a "philosophy adopted for the control of inputs of persistent toxic substances" to guide regulatory strategies and ultimately to achieve virtual elimination. When applied to a chemical, the zero discharge philosophy implies adopting measures to eliminate any use or synthesis, or its existence anywhere in society. The Task Force concurs with this concept. Whereas the general intent of the phrase "zero discharge" is clear, its detailed implementation remains controversial.

In the Task Force's judgement the intent was to express the idea that it is necessary to eliminate inputs of persistent toxic substances, because the capacity of the ecosystem to assimilate these chemicals is small, or non-existent, and thus additional inputs will prolong impairment of ecosystem health.

For new substances that meet the definition of a persistent toxic substance (see Chapter 4), application of the zero discharge concept is straightforward: no synthesis or production—no release. The Task Force also recognizes that minuscule quantities of persistent toxic substances already in the environment may escape capture or interception before entering the Great Lakes, even with the application of prevention, treatment, or control measures. Previous laws, regulations, and courts have also recognized the reality that application of the "zero discharge" philosophy cannot necessarily mean achievement of absolute zero. The Task Force believes this necessary interpretation should not impede progress towards the virtual elimination goal.

2.3 Virtual Elimination

The virtual elimination of inputs of persistent toxic substances is an obligation undertaken by the Parties in the

1978 Agreement and strengthened by the 1987 amendments to the Agreement. This commitment clearly intends that virtual elimination be one of the cornerstones to achieving an absence of injury and the Agreement goal of restoring and maintaining ecosystem health.

The Task Force offers the following observations and conclusions regarding virtual elimination. These are discussed further in Chapter 3.

- Current government programs controlling toxic substances, for the most part, fail to recognize any distinction between **toxic** and **persistent** toxic substances, as called for in Article II of the Agreement.
- Virtual elimination is an overall **strategy** that requires different approaches—some preventive, some remedial—to control or eliminate different inputs and in situ contamination.
- The virtual elimination strategy must apply to **all sources**—point and nonpoint—from **all media.**
- The virtual elimination strategy must apply to new potentially persistent toxic substances that may be created, as well as existing persistent toxic substances.
- The virtual elimination strategy also must apply to persistent toxic substances **already present** in the Great Lakes Basin Ecosystem. Once persistent toxic substances have been released into the ecosystem, it is not practical to completely remove them, especially from the open waters or the bottom sediments of the lakes, or from groundwater contaminated, for example, by leaking landfills. Therefore, the qualifier "virtual" is appropriate as applied to eliminating the presence of persistent toxic substances from the ecosystem.
- The virtual elimination strategy must **prevent** the deliberate input of any additional quantities of persistent toxic substances to the ecosystem. Given our technological capability to measure lower and lower concentrations of contaminants in the ecosystem, virtual elimination of existing persistent toxic substances may never be zero. Rather, the strategy challenges us to continuously strive to reduce the amount entering the environment, en route to fulfilling the Agreement's virtual elimination obligation.
- Because some persistent toxic substances already are present in the ecosystem, and because life in the Great Lakes Basin Ecosystem is vulnerable to contamination from persistent toxic substances, implementation of the virtual elimination strategy requires that the policy of zero discharge be applied to prevent further releases **from all sources** of persistent toxic substances.

3. The Conceptual Approach

The virtual elimination strategy should provide a comprehensive, multi-dimensional approach that addresses all problems associated with persistent toxic substances. It will affect each of us, and must guide industry as well as regulatory agencies by providing a road map to a Great Lakes no longer threatened by persistent toxic substances. If the strategy is to work, it must be fully understood and implemented both in the short and the long term. This chapter summarizes the basic concept of the strategy to virtually eliminate the input of persistent toxic substances to the Great Lakes Basin Ecosystem, specifically:

- A vision for the virtual elimination strategy.
- The need for the strategy.
- Limitations of current approaches toward persistent toxic substances.
- Evolution of approaches to dealing with persistent toxic substances.
- Principles of the virtual elimination strategy.
- Implementation of the strategy: action components and the decisionmaking process.
- Conclusions and recommendations.

Subsequent chapters examine the adequacy of available or required tools or processes to implement the strategy.

3.1 A Vision for the Virtual Elimination Strategy

The virtual elimination strategy for persistent toxic substances must be guided by a vision. The Task Force's vision is ecosystem integrity, characterized by a clean and healthy Great Lakes Basin Ecosystem and by the absence of injury to living organisms and to society. The Task Force believes the virtual elimination strategy to achieve this vision must be compatible with and foster healthy, sustainable, economic activity.

3.2 The Need for the Virtual Elimination Strategy

To understand why a strategy is needed that focuses specifically on persistent toxic substances, it is necessary to examine the limitations of our past approaches to these contaminants. Once we understand why we have not yet virtually eliminated persistent toxic substances, we can design a strategy with principles and components to help society achieve the virtual elimination goal.

A special strategy for the virtual elimination of persistent toxic substances is needed because these substances continue to damage ecosystem health, including subtle effects to the endocrine, immune, reproductive, and other sensitive biological systems. This is discussed more fully in Appendix D. This injury to living organisms continues to occur because of society's failure in the past—and to a large extent even today—to recognize fundamental differences between persistent toxic substances and other con-taminants, especially their ability to resist degradation and, for some, to bioaccumulate in living organisms. A traditional assimilative capacity approach thus is not applicable to persistent toxic substances because even minute, undetectable quantities may build up over time to levels that cause biological injury.

3.3 Limitations of Current Approaches

While current practices to deal with persistent toxic substances have reduced the quantity released to the Great Lakes Basin Ecosystem, *a number of limitations preclude present practices from delivering virtual elimination.*

- *Limitation:* **Proof of harm** *must be established before responsive action is taken. Years could be required to prove a conclusive link, by which time the damage has already occurred.*
- *Limitation: Even after injury has been established, the traditional focus has been on* **management** *and* **control of releases,** *rather than prevention.*

Thus, management practices that allow continued discharge of even the most damaging persistent toxic substances, although based on current regulatory objectives and available technology, may no longer be acceptable. Once a persistent toxic substance has been produced and used, it is impossible to completely control releases, including unintended releases. Recapturing every last molecule is impossible. Even when releases during the manufacturing process are controlled, releases can occur after the final product is discarded. Further, spills or accidental releases can occur during transportation and handling.

- *Limitation: With few exceptions, releases are controlled under current practices by* **single-medium** *laws and regulations designed to protect only air, land, or water. As discussed in Chapter 5, persistent toxic substances enter the Great Lakes via many pathways and, once released, migrate among media, become widely dispersed in the ecosystem, and can end up in Great Lakes biota.*

The traditional way of dealing with contaminants has assumed that the waters in the Great Lakes basin have an **assimilative capacity.** However, as noted above, this concept is inappropriate for persistent toxic substances. The Task Force believes that the current approach must change, because the following precepts do not necessarily hold for persistent toxic substances:

- *An ambient level exists below which residual risk is minimal.* Acceptable ambient levels are generally unknown for most persistent toxic substances. For some, the current scientific evidence indicates ambient levels so low as to be unmeasurable by the most sensitive analytical methods currently available.
- *If a "safe" ambient level exists, then protective water quality standards or numeric criteria can be established for persistent toxic substances.* To set a limit assumes that scientists are able to understand all possible effects of chemicals acting singly or in combination with one another on living organisms. Previous endeavours established limits for some persistent toxic substances which, in light of more recent information, were not protective (i.e. they were too high). In effect, for many persistent toxic substances, existing ambient environmental levels are already above the calculated or observed "no effect" level.
- *If a "safe" ambient level is determined and a water quality standard or criterion can be established, then it is possible to derive and allow for waste load allocations.* Allowing waste loads for persistent toxic substances adds to the exposure of and burden on the biota in the lakes.

However, as one component of the virtual elimination strategy, scientifically valid standards and criteria should be developed to serve as benchmarks to monitor progress in pollution cleanup and prevention.

In addition to limitations posed by current practices, concepts, and assumptions, technical and programmatic limitations also have prevented achievement of the virtual elimination goal. These include the failure to fully implement existing programs, enforce existing laws, and comply with existing policies, as well as a lack of funding and an adequate information base on loadings, sources, and available technologies. In Canada, for example, a National Pollutant Release Inventory is only now under development, while information from the U.S. Toxic Release Inventory underestimates total releases and lacks focus on persistent toxic substances.

Current estimates of the total number of chemicals in use range from 60,000 to 200,000, and the number continues to grow. Current practices cannot adequately screen existing chemicals nor screen all new chemicals (created either intentionally or as byproducts) for possible dangerous effects, especially chronic, sublethal effects on living organisms, and to determine which meet the definition of a persistent toxic substance. There also is no clear mandate to eliminate releases of those confirmed to be persistent toxic substances by any date. Moreover, present mechanisms are not adequate to eliminate the most dangerous substances from use, production, and disposal, even if it is determined that they are too dangerous to be allowed to enter the ecosystem.

TABLE 2

The Evolution of Approaches to Persistent Toxic Substances

	I →	II →	III
	CONTROLLING RELEASES	**PREVENTING USE OR GENERATION**	**TOWARD SUSTAINABLE INDUSTRY AND PRODUCT/MATERIAL USE**
Focus	Release	Chemical use/generation	Materials
Policy	Control abatement technologies (Control technology change)	Use reduction Process/product changes (Process change)	Source/use profile (Use tree and life cycle concepts, industrial sector change)
Goal	Reductions in emissions levels Pollution control (acceptable levels)	Zero discharge/ sunsetting of targeted chemical Pollution prevention (clean production)	Zero production/use of certain elements/compounds Sustainable industry (Materials evaluation)

3.4 Evolution of Approaches

Historically, varied attempts have been made to cope with the problem of persistent toxic substances, usually commensurate with the level of understanding rather than the prevalence of the problem. The three phases in the evolution of attempts to deal with persistent toxic substances are summarized in Table 2 and discussed below.

Phase I: Controlling Releases of Persistent Toxic Substances

Initially and even today, the problems of water, air, and land pollution have been dealt with using treatment and control. The fundamental assumption governing the approach was the assimilative capacity concept, where methodologies were developed to find "acceptable" limits of pollutant releases. The goal is to reduce releases and eliminate any adverse effects. This approach reduced loadings to the environment. However, the levels of many persistent toxic substances remained at lower but still unacceptable levels through the 1980s and into the 1990s.

Phase II: Preventing the Use or Generation of Persistent Toxic Substances

Pollution control reactively addresses the problem once the substances have been used or generated. Prevention attempts to avoid use or generation in the first place through process change, product reformulation, and raw material substitution. In effect, prevention has required the focus to "move up the pipe" to examine the earliest source of the persistent toxic substance itself. The goal is clean production processes, closed loop recycling, and elimination of the use and generation of persistent toxic substances.

To date, neither government nor industry has been able to fully implement a pollution prevention approach. While some progress has been made, most programs tend to be media specific and fragmented compared to the need for comprehensive, integrated approaches (see Chapter 6). By one estimate, only 11% of United States companies filing reports under the Toxic Release Inventory were voluntarily using pollution prevention.

Phase III: Toward Sustainable Industry and Product/Material Use

In addition to implementing a prevention approach, inputs to industrial processes and societal practices need to be examined. This broader and much longer term approach involves an evaluation of the materials used in production processes and questioning the environmental appropriateness of those materials and the products.

This product/materials use notion raises many questions. In the present context the use of certain materials has the potential to result in the generation, use, or release of persistent toxic substances. Product/materials use makes us ask how and why we produce, use, transform, consume, and dispose of materials and products. This approach requires such questions as: Is it possible to eliminate the release of mercury when coal is burned to generate electricity?

The product/materials use approach not only asks what are sustainable and non-polluting production processes (as in Phase II), but also examines the benefits and negatives of entire industrial sectors, the building blocks of production, and various types of social activities. The goal of this approach is to move to sustainable societal activities and industries. This is where the development of a long-term virtual elimination strategy must start. Aids for understanding this framework include the use tree and the life cycle approach, discussed in more detail below.

3.5 Principles of the Virtual Elimination Strategy

The unique properties of persistent toxic substances, coupled with the limitations of present practices and the evolution of strategic thinking, as described above, led the Task Force to articulate a set of principles that must guide a virtual elimination strategy focused on persistent toxic substances. The major principles that underlie the goals, objectives, and implementation of that strategy are **anticipation and prevention** and **remediation, treatment, and control.**

Anticipation and Prevention

Anticipation and prevention of pollution must be adopted for all substances that meet the criteria to be a persistent toxic substance. The virtual elimination strategy applies to all persistent toxic substances. All are presumed to be candidates for phaseout (sunsetting), particularly those with high bioaccumulation potential (see Chapter 4), unless data are available to show that their continued use is safe to human and ecosystem health.

In 1990, the President's Council on Environmental Quality concluded that:

> Thus it appears that the only chemicals to have declined significantly in the Great Lakes ecosystem are those whose production and use have been prohibited outright or severely restricted.

The production and use the most harmful persistent toxic substances must be phased out in the near future following a strict negotiated timetable. The production and use of all other persistent toxic substances must be substantially reduced over the time period required to negotiate and arrange for their virtual elimination. The primary intent is to eliminate formation and/or use of persistent toxic substances, since this is the only way to virtually eliminate such substances from the ecosystem. Once created, it is impossible to recapture or totally eliminate every last molecule of a substance.

Remediation, Treatment, and Control

The virtual elimination strategy recognizes the clear, present need to **treat and control** all persistent toxic substances while they are being virtually eliminated from the Great Lakes Basin Ecosystem, and to **remediate** problems from past and present inputs. These efforts must address the legacy of industrial manufacturing, uses, and disposal over the past 150 years to the present, in concert with prevention and sunsetting mechanisms.

Other Principles

The virtual elimination strategy also adopts eight other principles. The strategy:

- Adopts a **precautionary principle** (see sidebar). Where there are threats of serious, cumulative, and/or irreversible damage, an incomplete understanding of the underlying science and an inability to arrive at a precise risk assessment value should not be used as a reason to postpone measures to prevent environmental degradation and to sustain the ecosystem resource.
- Addresses the **complete life cycle** of persistent toxic substances in society, including beneficial considerations, manufacture (deliberate or inadvertent), import, export, use, transport, disposal, destruction, and remediation.
- Applies to **all sources and pathways.**
- Applies to **all media**—water, sediment, soil, air, and biota—and the movement of contaminants from one to the other. The intent of the strategy is to eliminate a problem, not move it.
- Applies **globally.**
- Requires use of the principle of **reverse onus,** that is, the producer, user, or discharger of a substance bears the responsibility to demonstrate that neither the substance nor its degradation products or any byproducts are likely to pose a threat to the ecosystem.
- Involves **all stakeholders,** including a description of the relationship of business and industry to the people and wildlife that cohabit the region, and assumes maintenance of a robust economy that provides jobs and amenities to its residents).
- Applies the principle of **risk management** to select and evaluate proposed response options, once a substance has been identified as meeting the definition of a persistent toxic substance.

3.6 Action Components for Implementation of the Virtual Elimination Strategy

The key components of the strategy to virtually eliminate persistent toxic substances from the Great Lakes Basin Ecosystem are: **elimination; adoption of a product/materials use policy; use reduction; and control, treatment, and remediation.** This strategy emphasizes the importance of prevention. However, application of the components will depend on the nature of the persistent toxic substance under consideration, as well as other factors such as its sources and uses. The components are described below. Clearly, the virtual elimination strategy will continue to evolve, as additional information and opportunities become available and as it builds on actions taken and successes achieved to date. These components of the virtual elimination strategy are intended to complement and enhance the programs and measures employed for the past two decades.

Elimination—Sunsetting Persistent Toxic Substances

The Commission's Sixth Biennial Report defined sunsetting as a "comprehensive process to restrict, phase out and eventually ban the manufacture, generation, use, transport, storage, discharge and disposal of a persistent toxic substance." Implicit in the concept is that uses of certain chemicals may be phased out using different timetables. For example, it may be possible to eliminate uses of mercury in batteries in the new future. However, eliminating all uses of mercury, including those for medicinal purposes, may occur over a longer time period.

The overriding goal is to eliminate the formation and use and, thus, the releases of all persistent toxic substances. However, that is not possible in the short term for all persistent toxic substances. In the interim, to lead towards the virtual elimination goal, a preventative approach should be applied to all persistent toxic substances.

Immediate bans and phaseouts according to a strict timetable are required for a "short list" of selected substances subject to the virtual elimination strategy. As a matter of urgency and to address the immediate hazards of those persistent toxic substances which have, and continue to cause environmental damage, action should be taken to ban at once, or phase out in the very near future, all production, manufacture, import, export, use, release, transport, and disposal of the 11 Critical Pollutants (Table 1). All are known to cause detrimental effects on living organisms and continue to exist in the ecosystem at unacceptable levels. This is discussed further in Chapter 4.

To determine priorities for phasing out additional persistent toxic substances, government, in consultation with stakeholders, must devise comprehensive criteria and decisionmaking procedures to evaluate all persistent toxic substances not on this short list, following a stringent timeline. These criteria must be applied to all persistent toxic substances, whether they are created intentionally or as byproducts. "Sunrise" criteria are also needed to evaluate new chemicals, including chemicals that may be created as alternatives to those that are slated to be phased out. These criteria are described in Chapter 4, and decisionmaking procedures are suggested later in this chapter.

THE PRECAUTIONARY PRINCIPLE

This (precautionary) principle was agreed to at the World Industry Conference on Environmental Management in 1984 and at the 1989 Paris summit of the seven richest industrial nations (the G7). It was strengthened in the 1990 U.N. Economic Commission for Europe meeting in Bergen: *In order to achieve sustainable development, policies must be based on the precautionary principle. Environmental measures must anticipate, prevent and attack the causes of environmental degradation. Where there are threats of serious or irreversible damage, lack of full scientific certainty should not be used as a reason for postponing measures to prevent environmental degradation.*

Clearly, action is required [to bring about fundamental changes in our economic behaviour and our international relations]. But which actions, and when, given the huge uncertainties involved. This is the sort of issue that business copes with daily.... There are costs involved, but those are costs the rational are willing to bear and costs the responsible do not regret, even if things turn out not to have been as bad as they once seemed. We can hope for the best, but the precautionary principle remains the best practice in business as well as in other aspects of life.

Adoption of a Product/Materials Use Policy—The Use Tree and Life Cycle Approaches

Government and industry, in consultation with stakeholders, must evaluate classes of chemicals and chemical families through use tree analysis, to determine whether and how particular uses should be phased out. As depicted in Table 2, government and industry historically have tried to control releases, but that has not been enough. The importance of "moving up the pipe" is now recognized as a management option, and both are now shifting to prevention and other related measures to avoid the use and generation of persistent toxic substances. The next step is to question some of the raw materials used by industry and society, as well as societal practices.

One means for doing this is the use tree concept. A use tree outlines the end uses and products of chemicals, and then traces those to identify the families of chemicals back to the base element, compound, or mixture. Such a methodology helps to clarify the sources or origins of a persistent toxic substance.

Use tree analysis has been used by industry and engineers for some time; what is different is using it in terms of environmental policy and/or regulations to determine the most appropriate point of intervention. Some substances can be dealt with at the "release" stage, at least in the interim. Some persistent toxic substances, however, can and should be dealt with at the root level on the use tree.

A use tree is only a tool. It does not indicate whether something should be phased out, nor does it prescribe how to do so in terms of regulatory and non-regulatory initiatives. The use tree simply enables one to ask the fundamental question: What is the source of a persistent toxic substance?

In terms of policy and regulations, a use tree analysis allows the following kinds of questions to be asked:

- Where should society intervene to deal with a substance or class of substances: At the release level? At the point of production of precursor chemicals? Or at some intermediate point?
- How do we evaluate where and how to intervene? What is the relationship among environmental, social, and economic considerations? What research is needed to shed light on these questions?
- For new materials, how do we respond to the above questions?

For example, the root of the mercury tree may be mining. One option would be to phase out certain mining sectors, while other options would be to reduce the use of certain applications or uses in certain circumstances. Thus, the use tree provides a framework that enables society to consider the question: Where do we intervene to most appropriately prevent further problems?

Use Reduction

Sunsetting for persistent toxic substances will take time. Therefore, government, in consultation with stakeholders, must implement a preventative strategy for all persistent toxic substances, using the concept of "sustainable manufacturing." This term incorporates pollution prevention, use reduction, and product life cycle analysis. Some key elements of use reduction and elimination programs are:

- Progressively reducing releases of persistent toxic substances, achieved through cooperative, voluntary approaches.
- Making information publicly available on current chemical uses and inventories, as well as facilities' current plans and progress for reducing uses and releases.
- Issuance of permits or approvals for operation and contaminant release, only if use reduction plans for persistent toxic substances have been submitted.
- Provision of technical assistance, especially to smaller facilities, to help determine the most appropriate use reduction method or technology.
- Societal provision for worker retraining and other technical assistance for those whose jobs are lost as a result of the phasing out of a chemical.

Control, Treatment, and Remediation

Elimination, product/materials use policy, and use reduction will not occur overnight for all identified persistent toxic substances. Therefore, treatment and control actions must be applied as intermediate or interim measures, and possibly as long-term measures where necessary, en route to achieving virtual elimination. Treatment and control should focus on intercepting or capturing the persistent toxic substance once it has been produced or used, but before it can enter the ecosystem. Technology can be applied to treat and control point source discharges, air emissions, and nonpoint sources.

Remediation focuses on cleanup of contaminants already in the ecosystem. Technology is an essential tool, but our ability to remove contaminants is limited once they enter the ecosystem.

The full extent of the environmental problems attribute to past releases which now reside in contaminated sediment, leaking landfills, or other uncontrolled sites is unknown. In the United States, through the Resource Conservation and Recovery Act (RCRA) and the Comprehensive Environmental Response, Compensation and Liability Act (CERCLA), an inventory of locations of inputs is underway and has been completed in some states. In Canada, the Contaminated Sites Program, under the auspices of the Canadian Council of Ministers of the Environment, is involved in a similar effort. However, releases of contaminants have not been quantified at all sites; estimates are available only in limited cases.

As discussed in Appendix D, the costs of cleanup are only now beginning to be reckoned. To remove or contain contaminants, to operate and maintain facilities, and to monitor contamination is estimated to cost billions of dollars, with time frames for remediation of 30 years and more. This is not unreasonable, given the experience with RCRA and CERCLA.

A long-range plan is required to systematically focus on sediment, landfills, and other unregulated sources, to monitor and assess the varying degrees of contamination from these sources, and to develop plans to address first the sites that are the most likely sources of inputs to the ecosystem. Included in a site-by-site assessment would be an analysis of whether, given best available technology, the contamination would be better left in place, with no further action taken other than monitoring.

A remedial management program is needed for contaminated sediment, even with a long-range plan. Particular consideration should be given to the environmental effects from dredging. This means employing a multi-media approach, with best available technology for the management, control, and disposal of dredged spoils.

Remedial Action Plans represent a useful mechanism to identify cleanup needed and to move toward the virtual elimination goal; in fact, Annex 2 of the amended Agreement recognizes the connection. Similarly, lakewide management plans, point source impact zones, and watershed management plans—also Agreement requirements—offer opportunities to apply the strategy to virtually eliminate inputs of specific pollutants.

Programs must provide economic incentives to drive improvements and the development of better and less expensive technologies for remediation, cleanup, and control. Persistent toxic substances continue to cause injury to the economy and society in the form of environmental debt, attributable to 150 years of industry and manufacturing in the Great Lakes region. Government and industry, in consultation with stakeholders, must adopt and maintain programs targeted toward remediation or control of inputs of past and present contamination, while continuing to recognize that prevention must be paramount.

Recognizing that it will not be possible to clean up and control all inputs, all at one time, the virtual elimination strategy must include a comprehensive

3.7 The Decisionmaking Process

Important considerations in the implementation of the virtual elimination strategy include:

- Having a decisionmaking framework within which to operate.
- Having the tools, including legislation, technology, economic instruments, and consultation mechanisms.
- Having a strong mandate.

The decisionmaking framework presented here provides a coherent means to examine the nature and dimensions of the problems created by the need to virtually eliminate persistent toxic substances. The framework provides for plausible short- to long-term implementation responses, as well as for input from all sectors of society (government, industry, labour, public). It draws on a variety of disciplines and employs a wide range of tools in order to anticipate the consequences of the decisions that are ultimately reached.

Described below is a decisionmaking process that provides a logical means for implementing the virtual elimination strategy. Subsequent chapters describe and evaluate the various operational components of the strategy in more detail.

As the Task Force discussed the key elements of the virtual elimination strategy, a logical sequence for connecting each element emerged. This sequencing was developed into the decisionmaking process illustrated in Figure 1. By following this process for the virtual elimination of persistent toxic substances, the Task Force believes that the integrity of the ecosystem will be maintained and where necessary restored. A description of the decisionmaking process follows. More detailed discussion on several of the elements of the process can be found in subsequent chapters.

Before applying the decisionmaking process, stakeholders must discuss and agree on definitions of key terms, including persistent toxic substance and virtual elimination, as well as on the principles under which the strategy will be implemented. Clear definitions and agreed-to principles provide a "level playing field" for evaluation of chemicals and application of the strategy. The Task Force's definitions are presented in Chapter 2, and the principles earlier in this chapter.

It is important to recognize that the issue of persistent toxic substances will not be adequately addressed by dealing solely with substances currently in use or in the ecosystem. Provisions are made in this strategy to deal with proposed substances. Figure 1 illustrates the two pathways that a decisionmaker will follow, depending on whether a substance is currently in commercial use or whether there is a new substance under consideration for commercialization. The decisionmaker at that step can be either in government or business.

Screening Proposed New Commercial Substances

The screening process and the criteria should be bilateral and include manufactured (deliberate and inadvertent) and imported chemicals. Both Canada and the United States have set up procedures to establish if a chemical should be approved for manufacture in commercial quantities. In Canada, the requirements are described in the Canadian Environmental Protection Act (CEPA). In the United States, the Toxic Substances Control Act (TSCA) requires the submission of a pre-manufacturing notification (PMN) to the U.S. Environmental Protection Agency (EPA). The contents of a PMN should allow U.S. EPA to assess the appropriateness of the application. Since this strategy deals specifically with persistent toxic substances, the first decision is to assess if the substance under consideration fulfills the criteria to be defined as a persistent toxic substance. See Chapter 4 for more details on selection criteria. Knowledge of the substance's composition or chemical structure, along with the physical and chemical properties, will guide decisionmaking. A substance that is unlikely to meet the definition of a persistent toxic substance can be excluded from further consideration. The adequacy of CEPA and TSCA as screening mechanisms should be evaluated to ensure they allow screening and also are consistent with the principles of the virtual elimination strategy.

Screening Existing Substances

The process outlined below is for substances currently being manufactured (deliberate and inadvertent) and in use, still in use though no longer produced in commercial quantities, or no longer produced or used.

- **Element 1–Apply selection criteria.** As was established for a new substance, the first element in the process to deal with an existing substance is to assess whether it fulfills the criteria to be defined a persistent toxic substance and is therefore to be dealt with under the strategy (see Chapter 4 for more details on selection criteria and the screening process). Knowledge of the substance's composition or chemical structure, along with the physical and chemical properties, will guide decision-making. A substance that does not meet the definition of a persistent toxic substance can be excluded from further consideration from this flow diagram, but is referred to the existing regulatory regime for toxic substances for appropriate consideration and action.
- **Element 2–Prioritize persistent toxic substances of concern.** Since it is unlikely that all issues involving persistent toxic substances can be dealt with at one time, it is necessary to decide which should be dealt with first. At this point, it may be decided to deal specifically with one substance or a set of substances.
- **Element 3–Identify sources and uses of the substance(s).** The intent is to establish where the substance is entering the environment. Since human activities, industrial process, and industrial sectors generate the substance or class of substances, a use tree and life cycle approach is appropriate for this undertaking.
- **Element 4–Evaluate alternative solutions for achieving virtual elimination and select preferred options.**

Figure 1 provides a breakout of Element 4 (see also sidebar on the following page). Persistent toxic substances fall into four broad areas: those currently produced (deliberately or inadvertently) and in use; currently in use but no longer produced; no longer used or produced; and those resident as contaminants in sediment, soil, groundwater, sludge, and sites that have received hazardous waste. As discussed earlier, the actions to be taken fall into the categories of prevention, treatment, control, and remediation. Figure 1 lists some options available (see also sidebar on the following page).

There is a difference of opinion among Task Force members as to the priority for action. For some, only prevention options, which lead directly to elimination of persistent toxic substance formation in the first place, are acceptable. However, preventative solutions may have a relatively long time frame. Other options, such as treatment and control, would therefore be required to achieve a more rapid and positive initial benefit. In reality, all options must be considered and implemented concurrently and as appropriate, to **mutually** contribute to achieving the virtual elimination goal.

A great many factors must be considered in selecting and implementing solutions, for instance, significance of the risks to health and the environment, availability of technology to achieve the desired end point, social and economic impacts, and consensus among stakeholders. All proposed solutions must be subjected to risk assessment and socioeconomic impact assessment, in order to prepare timetables and, where required, mitigation measures.

- **Element 5–Identify endpoints and indicators.** In this element, endpoint goals for the selected solutions are adopted, and indicators of movement toward these goals are specified. See Chapter 10 for more discussion of indicators.
- **Element 6–Implement preferred solution(s).** In this element, the preferred solution(s) that lead to reduction and ultimately elimination of inputs of persistent toxic substances are implemented following an agreed-to timeline. Provisions are needed to mitigate possible negative social and economic impacts resulting from actions taken.

Two monitoring streams follow Element 6:

- **Element 7A–Monitor effectiveness of implementation, as per adherence to schedule and achievement of virtual elimination.** Once the

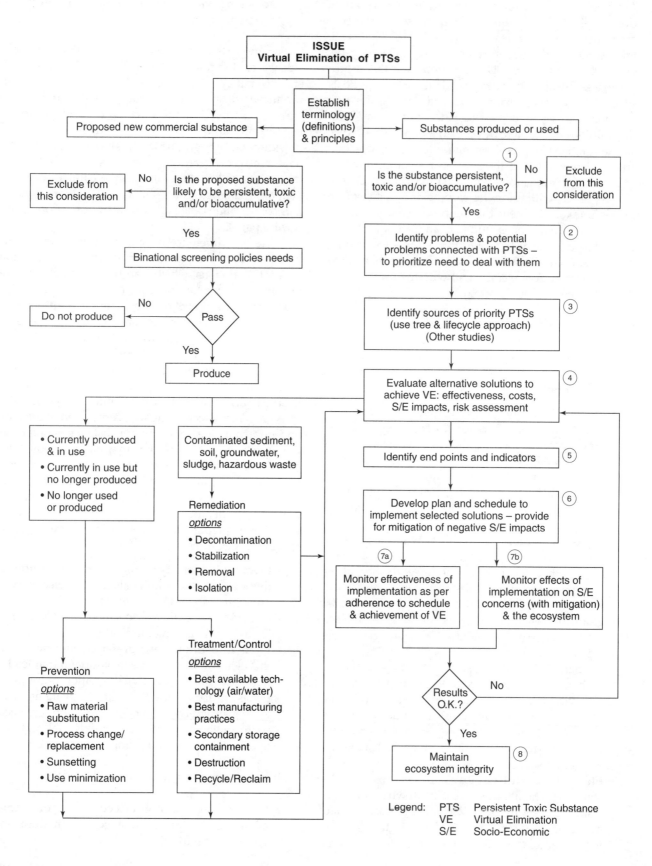

108 Unit Four: Great Lakes: Water Resources

ALTERNATIVE SOLUTIONS TO ACHIEVE VIRTUAL ELIMINATION

(See Figure 1, Element 4)

For persistent toxic substances **currently produced and in use,** the focus is on the manufacturing process used. Available solutions fall broadly into the area of prevention, for example, raw material substitution, process change or replacement, material recycling or reclamation, and use minimization.

For persistent toxic substances **in use but no longer produced**, the focus is on substitute substances or systems.

For persistent toxic substances **no longer used or produced,** treatment and control actions will reduce contaminant release toward the desired virtual elimination endpoint, for example, best available technology and best manufacturing practices. For stored or disposed substances, secondary containment may be advisable. Destruction, reclamation, and recycling may be appropriate for substances in storage, where the result would be a reduces loading to the environment.

Contaminated sediment, soil, groundwater, sludge, and hazardous waste sites involve persistent toxic substances **already in the ecosystem.** The basis for action must consider whether it is preferable to remove or isolate the substance, and which solution produces the minimum overall impact on the environment. Spill containment and control are ways to prevent further ecosystem contamination and thus also contribute to preventive action.

solution has been implemented, monitoring must be maintained to ensure that the solution is being implemented as required and is leading to virtual elimination of the targeted persistent toxic substance or set of substances.

- **Element 7B–Monitor effects of implementation on socio-economic concerns and the ecosystem. The implemented solution may have effects on society, the economy, or the ecosystem that were not predicted.** Monitor- ing is required to ensure that such concerns do not develop. If they do, mitigating actions need to be taken.

Elements 7A and 7B lead to the question: Have the implemented solution(s) led to the desired end points of the indicators? If not, further solutions need to be addressed by returning to Element 4. If yes, the issue is resolved, and one proceeds to Element 8.

- **Element 8—Maintain ecosystem integrity.** Monitor ecosystem integrity, addressing the indicators discussed in Chapter 10.

3.8 Conclusions and Recommendations

The Virtual Elimination Task Force articulates a simple vision regarding persistent toxic substances: ecosystem integrity, characterized by a clean and healthy Great Lakes Basin Ecosystem and by the absence of injury to living organisms and to society. The Task Force has considered what must be done to achieve this vision.

The Task Force concludes that many principles of past pollution-response practices are not appropriate when dealing with persistent toxic substances. The Task Force also observed an evolution in thinking, from control to prevention, toward sustainable industry and product/material use. Consequently, the Task Force has articulated the essential principles and components of a strategy to virtually eliminate the input of persistent toxic substances to the ecosystem, and has also developed a decisionmaking process to implement that strategy. The Task Force firmly believes that implementation of the strategy will achieve the Task Force's vision and the Agreement's virtual elimination goal.

Therefore, the Virtual Elimination Task Force recommends that:

1. **The Commission and the Parties adopt the vision: ecosystem integrity, characterized by a clean and healthy Great Lakes Basin Ecosystem and by the absence of injury to living organisms and to society.**
2. **The Commission and the Parties immediately adopt the Task Force's strategy to virtually eliminate the input of persistent toxic substances to the ecosystem, including its fundamental principles and components and the decisionmaking process to implement the strategy.**

11 Conclusions and Recommendations

11.1 The Strategy

The Virtual Elimination Task Force articulated a simple vision regarding persistent toxic substances: ecosystem integrity, characterized by a clean and healthy Great Lakes Basin Ecosystem and by the absence of injury to living organisms and to society. The challenge is to achieve this vision. As a society, we have not yet virtually eliminated the input of persistent toxic substances to the ecosystem, and injury to living organisms continues to occur. To develop an effective strategy to achieve virtual elimination and an absence of injury, the Task Force asked the question, What must we do to eliminate injury attributable to persistent toxic substances?

The Task Force recognizes that progress has been made to reduce the input to, and the impact of persistent toxic substances on the Great Lakes. This is evidenced, for example, by trends in PCB levels in fish from Lake Ontario and Lake Superior. In addition, a number of recent initiatives such as Canada's ARET (Accelerated Reduction and Elimination of Toxics) process and the binational Lake Superior Program should lead to further progress.

From its deliberations, the Task Force concluded that many principles of past pollution-response practices are not appropriate when dealing with persistent toxic substances. The Task Force also observed an evolution in thinking, from control, to prevention, toward sustainable industry and product/material use. Consequently, the Task Force has articulated the essential principles and components of a strategy to virtually eliminate the input of persistent toxic substances to the ecosystem, and has also developed a decisionmaking process for implementation of that strategy. The Task Force believes that implementation of the strategy will achieve the Task Force's vision and the Agreement's virtual elimination goal.

Therefore, the Virtual Elimination Task Force recommends that:

1. **The Commission and the Parties adopt the vision: ecosystem integrity, characterized by a clean and healthy Great Lakes Basin Ecosystem and by the absence of injury to living organisms and to society.**
2. **The Commission and the Parties immediately adopt the Task Force's strategy to virtually eliminate the input of persistent toxic substances to the ecosystem, including its fundamental principles and components and the decision-making process to implement the strategy.**

The Task Force has also developed a series of supporting recommendations that will facilitate implementation of the strategy, achieve virtual elimination of inputs and the absence of biological injury, while remaining responsive to social and economic realities. These are presented below.

11.2 Chemical Selection

In order to focus a virtual elimination strategy on the correct substances, experts from academia, governments, industry, and other stakeholders collectively must:

- Identify criteria for use in the chemical selection and phaseout processes, and adopt uniform quantitative values for each criterion.
- Develop and recommend a uniform screening procedure to identify chemicals that meet the definition of persistent toxic substance and to schedule their phaseout.
- Develop and recommend a uniform agreed-upon procedure, preferably incorporating the use tree and life cycle approach described in Chapter 3, to select persistent toxic substances for phaseout.

The Task Force concludes that four criteria—bioaccumulation factor (BAF); persistence; chronic toxicity to aquatic organisms; and evidence of specific causality and/or injury to biota—are the most important in the selection and classification process. The Task Force has also proposed numerical values for BAF, persistence, and chronic toxicity, to be applied for initial screening of substances, as well as more stringent values to be applied to identify those chemicals that meet the definition of persistent toxic substance and which should be virtually eliminated.

The Virtual Elimination Task Force recommends that:

3. **The Parties, in consultation with stakeholders, jointly develop, quantify, and apply criteria to screen chemicals, which will lead to development of a list of persistent toxic substances to be evaluated through the decisionmaking process, and to selector persistent toxic substances for phaseout.**

Since considerable work has already been undertaken to identify and develop the basis for selection criteria, the Task Force believes the criteria can be confirmed and quantified within six months after release of this report. As a pint of departure, the Parties should give serious consideration to the criteria and numerical values proposed in Chapter 4. They should also closely examine chemical classes and processes as well as industry sectors related to the generation and use of persistent toxic substances.

11.3 Timing

In some cases, immediate sunsetting is feasible, for example, because alternatives to the particular persistent toxic substance or to a particular production process are available. However, this is not always the case. Therefore, a specific timetable should be established for the phaseout of targeted persistent toxic substances, which would allow industry and the research community an opportunity to develop suitable alternatives. The timetable should also include benchmarks to demonstrate progress toward complete phaseout.

The Virtual Elimination Task Force recommends that:

4. **The Parties set specific timetables for the phaseout of persistent toxic substances not amenable to an immediate ban.**

Particular attention should be focused on those persistent toxic substances which are responsible for injury to the ecosystem.

11.4 Immediate Action

Notwithstanding the development of selection criteria, a screening process, and a list of persistent toxic substances, the Virtual Elimination Task Force concludes that sufficient evidence exists to warrant immediate phaseout of the 11 Critical Pollutants identified by the Water Quality Board in 1985 (Table 1). All 11 substances are persistent and cause such serious injury to living organisms that any entry into, or presence in the ecosystem is unacceptable. The Task Force notes that the 11 Critical Pollutants have, in effect, already been subjected to evaluation, as called for in the virtual elimination strategy. They appear on most, if not all, toxic chemical lists. All are subject to regulation, and actions taken over the past 20 years have significantly reduced ecosystem concentrations. However, levels in the ecosystem continue to be elevated. The Task Force believes that application of the strategy and its decisionmaking process, presented in Chapter 3, will achieve virtual elimination of these persistent toxic substances.

The Virtual Elimination Task Force recommends that:

5. **The Parties, through application of the decisionmaking process, immediately initiate measures to sunset the 11 Critical Pollutants, including all aspects of their manufacture, import, export, use, and disposal.**

The Task Force is aware of the myriad of issues that must be faced and resolved to fully sunset the 11 Critical Pollutants. Among these are continued use and disposal practices, remediation, foreign use, long-range atmospheric transport, and natural occurrence. If we are serious about virtual elimination and fulfilling the requirements of the Agreement, these and other similar questions must be resolved. The use tree and life cycle approach presented in Chapter 3 is an appropriate mechanism within which to consider confounding factors. Appendix A presents further discussion of the problems and factors to consider, and the measures that can be taken when dealing with two of the Critical Pollutants, PCB and mercury.

11.5 Global Considerations

The Task Force notes that large quantities of some persistent toxic substances (such as DDT), although "banned" for domestic use in the United States and Canada, are still produced in the United States for export, as well as produced and used in a number of other countries. While charged to investigate the Agreement requirement to virtually eliminate the input of persistent toxic substances to the Great Lakes Basin Ecosystem, the Task Force concludes that a Great Lakes regional focus is clearly insufficient. The virtual elimination strategy must be applied globally, because of the ability of persistent toxic substances to disperse globally, in large measure through long-range atmospheric transport.

Because of their knowledge and experience with persistent toxic substances, the Great Lakes governments must take a leadership role to implement the strategy globally. To assist in this endeavour, the Virtual Elimination Task Force recommends that:

6. **The Commission, in partnership with Great Lakes governments, convene an international conference to focus on international implementation of the virtual elimination strategy.**

The biological injury caused by the 11 Critical Pollutants can serve as a focus to actively promote the need for the strategy, as well as the means necessary to achieve virtual elimination. The conference can also serve as a forum to obtain quantitative information about the amounts of persistent toxic substances presently in use globally.

11.6 Data and Information: Needs and Management

Information about sources and pathways by which persistent toxic substances enter the ecosystem, and the associated quantities, is required for implementation of the virtual elimination strategy. Information is also required about the life cycles of persistent toxic substances, their precursors, and the quantities used, released, in storage, and disposed of. Accurate and consistent information about the relative and absolute contributions from various sources and pathways provides an accurate baseline against which to measure progress. Such information also helps to prioritize reduction and elimination programs. This information must be available not only for Canada and the United States, but also internationally.

Sources and pathways include municipal and industrial effluents; surface runoff; combined sewer overflows, storm sewers, and treatment plant bypasses; emissions to the atmosphere; contaminated sediment; groundwater; and spills from ships and shore-based facilities. The Task Force concludes that information about sources and pathways and the associated contaminant quantities is inadequate. The Virtual Elimination Task Force recommends that:

7. **The Parties enhance programs to identify sources and pathways and to quantify loadings of persistent toxic substances to the Great Lakes Basin Ecosystem.**
8. **The Parties compile reliable and complete quantitative information for the life cycle of persistent toxic substances and their precursors.**

This information must include, as a minimum, the amounts of persistent toxic substances produced, used, released, and disposed of, as well as their fate in the ecosystem. Further the information must be integrated and must be readily accessible by the public.

Particular emphasis must be placed on the atmosphere, the dominant pathway by which many persistent toxic substances reach the Great Lakes. Since sources within and outside the basin contribute, the virtual elimination strategy cannot be confined to the Great Lakes Basin. The Virtual Elimination Task Force recommends that:

9. **The Parties develop quantitative information about the release of persistent toxic substances to the atmosphere from all sources.**

This information must include the form and speciation of the substances being emitted. Further, the information base must be sufficient to determine and differentiate the impact of local versus distant sources of persistent toxic substances, including identification of major sources and source categories.

The institutional framework to assemble integrated information on persistent toxic substances must be improved. Historic data collection programs generally aimed at conventional pollutants and nutrients. Current programs do not necessarily differentiate between toxic substances and persistent toxic substances. These need to be redirected to focus on persistent toxic substances, and to better address such issues as public access, confidentiality, levels of detection, and timely reporting.

The United States Toxic Release Inventory (TRI) and STORET (Storage and Retrieval Information System) are two complementary programs that can contribute to the assembly and management of persistent toxic substance data, and to the development of integrated information. The concept of an inventory of releases of persistent toxic substances should be expanded to Canada. Canada's National Pollutant Release Inventory, now under development, should be compatible with the U.S. TRI. Both should be comprehensive and apply to all source sectors. Further, reporting thresholds should be lowered and products (including inadvertently produced pollutants) accounted for.

A comprehensive database and integrated information about persistent toxic substances would underpin a decision-support system that could be used, for instance, to identify and substantiate remedial actions, with associated schedules and priorities; project future loadings and concentrations of persistent toxic substances, in the form of agreed-to targets towards virtual elimination; and facilitate assessment of the effectiveness of programs undertaken pursuant to the Agreement.

11.7 Mass Balance EcoSystem Fate Models

Mass-balance ecosystem-fate models are valuable tools for a virtual elimination strategy. Existing models are accurate enough to estimate contaminant fate now, and in the future, as remedial measures are implemented. The models can be used to project future ecosystem concentrations and response times, as a result of reduced inputs of persistent toxic substances to the ecosystem. Models can also be used proactively to identify new pollutants and future problems. The Virtual Elimination Task Force recommends that:

10. **The Parties use mass balance-ecosystem fate models as tools in the virtual elimination strategy not only for contaminants of present concern, but also proactively to identify new pollutants and future problems.**

Improved source information and a better understanding of physical, chemical, and biological processes that control contaminant fate in the lakes would improve projections of the rate and extent of ecosystem response.

11.8 Legislation, Regulations, and Programs

The Virtual Elimination Task Force concludes that governments have the legal authority to implement the virtual elimination strategy, but have not acted on their authority. As described in Chapter 6, the Task Force has identified gaps and impediments that hinder achievement of virtual elimination, but a number of short- and long-term reforms could help to realize the virtual elimination goal.

The Virtual Elimination Task Force recommends that:

11. **The Parties review their legal framework for dealing with persistent toxic substances and, if necessary, promulgate legislation to remove barriers and to promote implementation of the virtual elimination strategy.**

The legislation should promote reduced use of persistent toxic substances, examination of product/material use, sunrise/sunsetting, and pollution prevention to reduce and ultimately eliminate creation of persistent toxic substances. The legislation should also promote establishment and conduct of programs that provide requisite data and information, as described above.

As part of the approval process for allowing releases to the ecosystem, releases must not be allowed to increase from present loadings. Future approvals must include reduced limits as part of the ratcheting down process toward virtual elimination, and approvals will only be given if use-reduction plans have been submitted for persistent toxic substances.

11.9 Technology

The Virtual Elimination Task Force recognizes the need to prevent the creation of persistent toxic substances or, if they have been created, to destroy them. Measures short of destruction, such as storage or disposal, do not close the loop on full life cycle consideration of persistent toxic substances, and could lead to future ecosystem contamination and biological injury. Technology offers valuable tools and opportunities to move toward the virtual elimination goal. The Task Force recognizes that a variety of technologies are or will shortly become available, for instance, to deal with contaminated sediment. The Task Force particularly urges the application of technology:

- To modify production processes so as to prevent the creation of persistent toxic substances in the first place.
- To remediate contaminated sediment, groundwater, and other locations where persistent toxic substances are in the ecosystem.
- To destroy existing stocks of persistent toxic substances, including those in hazardous waste facilities and other storage sites, once and for all.

11.10 Economic Instruments

The Virtual Elimination Task Force investigated the potential usefulness of economic instruments (or "incentives") to help achieve virtual elimination of persistent toxic substances. With the assistance of a contractor, the Task Force endeavoured to design, evaluate, and propose a specific program of economic instruments for application in a virtual elimination context. Although the contractor has provided information, time constraints and other considerations precluded adequate Task Force consideration of the material received, as well as formulation of detailed conclusions and advice for the Commission.

The Task Force can nonetheless conclude that economic instruments are an important component of the virtual elimination strategy. However, the instruments selected, and their method of application to persistent toxic substances in a virtual elimination context, may well be different from the instruments (or incentives) used in the treatment, control, and remediation regimes for nonpersistent contaminants. Further investigation is required to identify the instruments and describe their application.

Even though it was unable to consider the economic material received, the Virtual Elimination Task Force recommends that:

14. **The Commission undertake an investigation to identify appropriate economic instruments for use in a virtual elimination context, and to describe their application to virtually eliminate the input of persistent toxic substances to the ecosystem.**

11.11 Communication, Education, and Consultation

Successful implementation of the virtual elimination strategy requires understanding and support by all stakeholders. Communication and education have key roles to play if understanding of the need for virtual elimination of persistent toxic substances, and support for societal change implicit in this implementation, are to be created. People must become aware of the problems and risks, and become strongly committed to effecting a solution. The Virtual Elimination Task Force recommends that:

15. **The Commission reinforce its commitment to its recommendations concerning awareness and education [expressed in its Fifth and Sixth Biennial Reports and its Special Report on Great Lakes Environmental Education] by again recommending their implementation to the Parties.**
16. **The virtual elimination strategy include provision for education initiatives at the local and regional level, particularly initiatives to encourage communitywide participation in local activities to eliminate the input of persistent toxic substances into the ecosystem.**

Community advisory panels modelled after the public advisory committees associated with the Remedial Action Plans would be an excellent way to obtain involvement in the development of such initiatives.

The Task Force supports multi-stakeholder consultation to identify the existence of problems and to implement solutions. Consultation and dialogue are essential to establish priorities, set goals, and define actions using the decisionmaking process. The Virtual Elimination Task Force recommends that:

17. The Parties highlight and adopt consultation and dialogue as the key components for the validation and implementation of the virtual elimination strategy, and provide sufficient resources for design and implementation of the consultation process.
18. The Parties establish formal and regular opportunities for ongoing stakeholder consultation, as part of the virtual elimination strategy.
12. The Parties promote development of technologies—products and processes—that will eliminate the creation of persistent toxic substances and thereby eliminate their input to the Great Lakes Basin Ecosystem.
13. The Parties inventory existing stocks of destructible persistent toxic substances and apply destruction technology to eliminate these stocks.

Particular emphasis should be placed on PCBs.

11.12 Indicators to Monitor Progress

Appropriate indicators are necessary to track progress toward the virtual elimination goal and to demonstrate ecosystem restoration and long-term protection and integrity. Chemical measurements provide information about the presence of contaminants in water, sediment, air, and biota, and bioindicators are required to assess toxicity. A bioindicator is an organism and/or biological process whose change in structure, function, or activity points to changes in the integrity of the quality of the environment. Bioindicators include indicator species, biochemical markers, and biological end points. The Virtual Elimination Task Force recommends that:

19. The Parties adopt, as part of their regular monitoring efforts:
    - indicators that measure the concentration levels of persistent toxic substances in the ecosystem.
    - coordinated bioindicator monitoring programs that measure toxicity and the occurrence or absence of injury to living organisms.

The indicators chosen should be based on sound science, consider socio-economic factors, and provide an accurate and sensitive "barometer" to indicate the success of the virtual elimination strategy.

11.13 Polychlorinated Biphenyl (PCB) and Mercury

The application of the virtual elimination strategy to PCB and mercury focuses on how to deal with persistent toxic substances that are known to cause injury, and have been the subject of intense action by government, industry, and others. Although ecosystem conditions have improved and biological injury has been reduced, injury is nonetheless still occurring.

Appendix A describes a wide range of actions necessary (but not sufficient) to virtually eliminate PCB and mercury from the ecosystem. Virtual elimination of these and other confirmed persistent toxic substances will not occur through reliance solely on treatment and control activities that are applied at the point of release. Prevention must be adopted and rigorously pursued to ensure that no additional quantities of PCB and mercury (those not already circulating in the ecosystem) are created, used, or released to the ecosystem. Current pollution prevention approaches, as applied by governments, will reduce, but not eliminate releases to the ecosystem. Concurrently, remediation of contaminated sediment, waste disposal sites, and other inplace sources of persistent toxic substances must take place.

To virtually eliminate PCB, a "banned" substance, the Virtual Elimination Task Force recommends that:

20. Governments and industry recover and destroy all existing stocks of PCBs in equipment, cease land disposal, and recover and destroy PCBs in sediment and landfills.

To virtually eliminate mercury, a substance with natural and anthropogenic sources, the Virtual Elimination Task Force recommends that:

21. Governments and industry reduce the use of fossil fuels with high mercury content, concurrently implement conservation measures to reduce electric demand and fuel consumption, phase out mercury use in consumer products, as well as mercury-based industrial processes, reduce mercury emissions from smelter operations, and recover (rather than dispose of) mercury in existing consumer and medical products.

For both PCB and mercury, the decisionmaking process presented in Chapter 3 should be used.

11.14 Basic Feedstock Substances

One debate within the Virtual Elimination Task Force was how to apply the virtual elimination strategy to a basic feedstock chemical and, more fundamentally, whether the strategy should be applied to a basic feedstock chemical. Appendix B provides perspective on this issue. The Virtual Elimination Task Force recommends that:

22. The Parties commission an exhaustive investigation that explores all factors and implications related to the implementation of the

proposed sunsetting of a basic feedstock substance such as chlorine.

Such an investigation should be conducted with input and participation from all stakeholders, including industry, environmental and health experts, consumer and labour groups, special interest groups, and the general public.

"Comments on the Canada–United States Strategy for the Virtual Elimination of Persistent Toxic Substances in the Great Lakes Basin"

Great Lakes Natural Resources Center, National Wildlife Federation, November 8, 1996, © by Canadian Enviromental Law Association, Canadian Institute for Enviromental Law and Policy, Great Lakes United, Greenpeace, and National Wildlife Federation. Reprinted by permission of National Wildlife Federation.

Introduction

We welcome the opportunity to comment on the second draft of the bi-national virtual elimination strategy (VES). A strategy, timetable, process, and the resources necessary to eliminate persistent toxic substances from the Great Lakes ecosystem is essential both for the health of all those who inhabit the Great Lakes region and to fulfill the obligations of the U.S. and Canada under the Great Lakes Water Quality Agreement and domestic law.

Based on our review and collective discussions, we support the adoption, signing and rapid implementation of the VES if the changes we recommend below are incorporated into the strategy.

General Comments

The current draft represents an improvement over the previous draft reviewed at the public workshop of August of 1995. The context for this conclusion and for our review is the rich history of policy evolution of the two nations' commitments to virtual elimination of persistent toxic substances in the Great Lakes ecosystem. Much of that policy evolution and development has been led by the International Joint Commission (IJC). In its *Seventh Biennial Report* (page 1) the IJC called for a " . . . clear and comprehensive action plan to virtually eliminate persistent toxic substances that are threatening human health and the future of the Great Lakes ecosystem." The *Seventh Biennial* went on to the state that "the virtual elimination and zero discharge of all persistent toxic substances are the most critical management targets," recognizing that "there must be interim steps en route to those targets."

In addition to the Great Lakes Water Quality Agreement and the biennial reports of the IJC, our comments reflect policies and recommendations for implementing virtual elimination that have been articulated in documents that include A Prescription for a Healthy Great Lakes (National Wildlife Federation and Canadian Institute for Environmental Law and Policy, 1991) and A Strategy for Virtual Elimination of Persistent Toxic Substances (Virtual Elimination Task Force, International Joint Commission, 1993). It is worth noting that when it was released in conjunction with the Biennial Meeting of the IJC in 1993, the Virtual Elimination Task Force report received near unanimous support from all interested parties, including industrial representatives.

Specific Comments and Recommendations

1. Definition of Virtual Elimination

We are pleased that the absence of injury has been removed as the driving force behind the need for additional action to eliminate persistent toxic pollution. "Absence of injury" is an inappropriate standard for a virtual elimination strategy for a number of reasons. There is already more than ample evidence that injury has occurred to people and wildlife in the Great Lakes region from toxic chemicals currently in the Great Lakes ecosystem. Absence of injury assumes that science can define "safe" levels for toxic chemicals and drives management efforts to define those safe levels. It is not appropriate that implementation of the strategy be subverted into endless debates about "safe" levels or "absence of injury" because of the wealth of knowledge already collected on historic damage from exposure, and the inability of science to enable us to know with certainty what levels of chemicals may or may not be safe. This is especially true because of the cumulative effects of combinations of persistent toxic chemicals on human and ecosystem health.

In the present draft, the term "virtual elimination" is not defined. The preferred route is to define "virtual elimination" as interpreted by the International Joint Commission and supported by environmental groups. For example, Prescription for a Healthy Great Lakes clarifies the difference between "virtual elimination" and "zero discharge." In this context, virtual elimination is defined as follows:

> *The objective of the zero discharge strategy outlined in this report is to virtually eliminate the presence of toxics in the Great Lakes ecosystem. Because of the large amounts of these substances already in the Great Lakes, virtual elimination can only be achieved by preventing any additional discharge of these substances (i.e., by implementing a zero discharge strategy), and by cleaning up to the maximum extent possible those contaminants*

we have already released [National Wildlife Federation and Canadian Institute for Environmental Law and Policy, Prescription for a Healthy Great Lakes (1991) p. 17].

The report recognizes that the goal of "virtual elimination" can only be achieved if a zero discharge strategy is adopted. The Zero Discharge Alliance, a grassroots network of concerned activists, defined the term "zero discharge" as follows:

For us "zero" means zero. Pollution must be prevented before it is generated. Production processes (including agriculture) must be reformulated so that these toxic substances are not used, produced or discharged. "Zero" does not mean reducing discharges beneath some arbitrary level or even beneath the level of detection. Zero means none.

The past four Biennial reports of the IJC reiterate this definition. In the Eighth Biennial Report, released earlier this year, the Commission stated:

There are various interpretations of virtual elimination and zero discharge. Virtual elimination is not a technical measure but a broad policy goal. This goal will not be reached until all releases of persistent toxic chemicals due to human activity are stopped.

Zero discharge does not simply mean less than detectable. It does not mean the use of controls based on best available technology or best management practices that continue to allow some release of persistent toxic substances, even though these may be important steps in reaching the goal. Zero discharge means no discharge or nil input of persistent toxic substances resulting from human activity [IJC, Eighth Biennial report, (1996), p. 9, emphasis in the Original].

In light of the clarity that the IJC has brought to the issue, the VES should be implemented employing the IJC interpretation of "virtual elimination."

The TSMP Definition of Virtual Elimination Should Not Be Used

The absence of an express definition of virtual elimination is of particular concern to Canadian organizations as they fundamentally disagree with the definition provided in the 1995 federal Toxic Substances Management Plan (TSMP). On page 3 of VES, it is implied that the Canadian government intends to rely on the TSMP, and the definition contained within that document (which states that virtual elimination means reduction to non-detectable levels). Hence, we are concerned that the definition of virtual elimination in the TSMP will be used for the purpose of the VES, in the absence of any definition in the VES document.

It should be made clear that Canadian environmental organizations consistently opposed the definition of "virtual elimination" contained in the TSMP. The TSMP, by defining virtual elimination as "non-detectable," contradicts IJC's recommendations. It also ignores the submission from environmental organizations which gave detailed arguments against the definition proposed for TSMP. In particular, reference should be made to the document "Comments on the Proposed Toxic Substances Management Policy" submitted by the Canadian Environmental Law Association et al. in November of 1994. Environmental organizations have also argued strenuously against incorporating the definition in the forthcoming amendments to the Canadian Environmental Protection Act.

Recommendation 1: We recommend that the parties include the definition of virtual elimination and zero discharge as interpreted and articulated by the IJC in the Fifth, Sixth, Seventh, and Eighth Biennial Reports in order avoid a lengthy debate during the implementation of the strategy. We urge the parties not to adopt the definition of virtual elimination in the Canadian Federal Toxics Management Policy either expressly in the VES or when implementing the strategy.

2. Failure to Adopt a Zero Discharge Strategy

To implement the goal of the VES, that is, the virtual elimination of the discharge of persistent toxic substances, it is imperative that the strategy incorporate a zero discharge or sunset chemical component. As noted above, any policy or strategy to achieve virtual elimination must be driven by the goal of reaching zero discharge, or zero additional inputs from human sources, of persistent toxic substances. The recommendations expressed in the Fifth, Sixth, Seventh and Eighth Biennial Reports of the IJC clearly state that the virtual elimination of persistent toxic substances is to be achieved through "zero input or discharge of those substances created as a result of human activities" (IJC Sixth Biennial Report). In the draft VES there is only one reference to zero discharge, and that reference is specifically to the goals of the Binational Program to Restore and Protect the Lake Superior Basin (Binational Program). Nowhere else is such a reference made. There is no other explicit reference to, or recognition of, sunsetting persistent toxic substances as the long-term mechanism to achieving the zero discharge goal.

Moreover, the report of the Virtual Elimination Task Force, which was widely endorsed by interested parties from all sectors of society, included the following two recommendations:

4. The parties set specific timetables for the phaseout of persistent toxic substances not amenable to an immediate ban.

5. The parties, through application of the decision-making process, immediately initiate measures to sunset the 11 Critical Pollutants, including all aspects of their manufacture, import, use, and disposal (Virtual Elimination Task Force, 1993, p. 62).

While we acknowledge that the timetables are a welcome and necessary step, they in themselves are not virtual elimination. The strategy must be amended to include a more rigorous process for phasing out persistent toxic substances.

Recommendation 2: Specifically, we recommend that the strategy be revised to correct this deficiency in the following ways:

1. *The Strategy should be revised to describe the process that will be followed to determine the best mechanisms in the two countries to sunset the Level I Toxic Substances.*
2. *The Strategy should be revised to state more clearly that the timetables for reduction in the current draft are interim targets only, and that the ultimate goal is to achieve zero discharge (or to sunset) these chemicals.*

At a minimum, the process of implementing the VES must be structured so as to require thorough discussion and exploration of mechanisms to sunset dioxin and other persistent toxic substances, as has been done by the Center for the Biology of Natural Systems (CBNS) for dioxins from medical waste incineration. If these changes are not made, it is inappropriate and misleading to title the strategy a "Virtual Elimination Strategy" since the content does not reflect the impression left by that title. Without these changes, the strategy represents an important but incomplete step toward virtual elimination.

Pollution Prevention and Substance Elimination Must Be the Governing Principles

Rather than look to zero discharge as the strategy to achieve the virtual elimination of persistent toxic substances, the draft VES largely adopts a mixture of pollution control and pollution prevention techniques for persistent toxic substances substances. While laudable language regarding the use of phaseouts and bans is contained within the VES, the strategy emphasizes achieving "reductions" within the context of "current regulations, initiatives, and programs" and "cost-effectiveness." We are concerned that the draft VES still does not reflect a real implementation of the IJC recommendations and the goals expressed in the Great Lakes Water Quality Agreement (GLWQA).

The Strategy's emphasis on "reductions" instead of "elimination," and the reliance on existing regulatory initiatives condemns the Great Lakes to continuing ecosystem damage for years to come. Current laws and regulations in Canada and the US emphasize control-oriented strategies, as opposed to preventive approaches. This sends a wrong signal to manufacturers, users, and others whose activities result in the release of persistent toxic substances. A more appropriate signal would be that these chemicals will eventually be eliminated. Businesses and industry might choose to make capital investment decisions differently if they know that ultimately, the end goal is zero, instead of part way.

For example, the VES Technical Support Document Attachment 1, "Actions Under the Binational Strategy," details how specific challenges to the two federal governments will be met. To meet the challenge of reducing releases of dioxins, furans, hexachlorobenzene, and B(a)P, the federal governments intend to promulgate new "control options" (Canada) or "standards" (US). More consistency with IJC recommendations could be achieved through focused discussion on sunsetting and phaseout or ban provisions of existing legislation. In the US, for example, the Toxic Substances Control Act could be used for this purpose.

As the recent Center for the Biology of Natural Systems' (CBNS) reports "Zeroing Out Dioxin in the Great Lakes: Within Our Reach" (June, 1996) and a summary of the reports' findings, "Dioxin Fallout in the Great Lakes: Where it Comes From, How to Prevent It; At What Cost" (June, 1996) have demonstrated, achieving zero discharge of dioxin from incinerators and other sources is achievable by phasing out the processes that create dioxin. Replacing municipal waste incineration with intensive recycling, autoclaving medical waste and intensively recycling non-infectious hospital waste, and substituting totally chlorine-free pulp bleaching technology will result in zero dioxin creation and discharge from these sources. These reports also found that these steps could be implemented at little or no cost, and, in some instances, provide substantial economic benefits to the Great Lakes Basin Economy.

Affirmation of Parties' Commitment

We are also concerned by the language in various parts of the VES and in particular under that heading of "Challenges." We note that the original draft stated that the two Federal governments "will achieve the virtual elimination . . . " whereas the current draft states that the governments "will work . . . toward the goal of virtual elimination." Moreover, the parties use the word "challenge" rather than commitments with respect to what they propose to do under the strategy. This point is underscored by the language in the Technical Support Document stating that the actions being contemplated "are illustrative of the many activities currently taking or expected to take place."

In our view, the parties should use clear, unequivocal language and commit to what they intend to do under the strategy. Such language is needed to ensure that all concerned understand the nature and scope of commitments. Moreover, such language is useful to measuring progress and ensuring public accountability.

Recommendation 3: We recommend that the two national governments clarify their commitments to the goals and targets within the strategy.

3. Limited Emphasis on Remediation

Given the extent of the challenge of dealing with the remediation of contaminated sites in the Great Lakes Basin by 2005, we are concerned that the proposed Actions contemplated in the VES are too limited in scope. The Canadian actions contemplated consist largely of studying innovative technologies, and developing a database of contaminated sediment clean up technologies. Actions for clean up are limited to priority RAPs/AOCs. US actions include continuing what existing clean up efforts exist, as well as continuing to develop approaches for contaminated sites. We would like to see a greater emphasis on the actual clean up of sites in the VES final draft.

Recommendation 4: The two governments should make a commitment to finishing remediation of all sites by a specified date.

4. Lack of Action on Level II Substances

The challenge specified in the VES for Level II substances is to "promote prevention and reduced releases" while monitoring for environmental levels and identifying sources. The issue of eliminating sources and releases about which we already know is not addressed in the draft VES. We are concerned that relying on voluntary reduction efforts by industry, while attempting to determine what damage has already occurred, will result in further damage to the Lakes' ecosystem. See below under the heading of "voluntary actions" for further comments.

In contrast with the approach outlined in the draft VES, the Lake Superior Programme has defined the Level II substances as "prevention" substances. There the goal is to prevent them from entering the ecosystem to begin with, a real reflection of a preventive approach to dealing with these substances.

We are concerned that the current draft VES approach of focusing on voluntary reduction efforts and self-monitoring will result in further environmental insult. Rather than waiting for evidence of environmental damage, and only then taking action to eliminate its source, it would be prudent to begin devising a strategy now to prevent such damage. An approach of this nature would be consistent with IJC recommendations on reverse onus and prevention.

Recommendation 5: The parties should develop action plans based on pollution prevention principles for all Level II substances. Regulatory options should be considered in every situation in the development of these action plans.

5. Reliance on Voluntary and Non-Regulatory Measures

One of the serious concerns of the environmental community is the degree to which many of the mechanisms for furthering the VES rely on voluntary and non-regulatory measures. In particular, reference should be made to the total reliance on voluntary actions for Level II substances (see pages 6 and 8 of VES), mercury reductions (see page 14 of VES), for releases of furans, hexachlorobenzene, B(a)P and dioxins (see page 16 of VES), among other activities in the strategy.

The reliance on voluntary measures is problematic for a number of reasons. Experience, in particular with respect to the Canadian programs, shows that they suffer from a number of serious concerns. For example,

(a) Many of the programs, such as the development of memorandums of understanding (MOUs) between industrial sectors and the government have occurred without public input. Some programs continue, such as the Canadian ARET program, despite the fact that environmental and labour groups withdrew their support on the grounds that important substantive principles were missing, such as the recognition of the need to phase-out substances.
(b) Many of the programs "pre-empt" public debate on important public policy issues, such as whether the focus of efforts should be on "emissions" (which is the focus of most MOUs) or the "use" of toxic substances (which environmentalists view as the appropriate approach).
(c) Voluntary programs lack the accountability and enforcement mechanisms of regulatory programs; and
(d) Voluntary programs often create an unlevel playing field in the sense that they advantage large firms and overlook the certainty and predictability afforded by regulation.

These criticisms have been outlined in greater detail in various publications from Canadian environmental groups and research organizations. [For instance, see: Canadian Environmental Law Association, "Deregulation and Self-Regulation in Administrative Law: A Public Interest Perspective" (1996) and the Canadian Institute for Environmental Law and Policy, "The Use of Voluntary Pollution

Prevention Agreements in Canada: An Analysis and Commentary" (1995).]

Rather than relying solely on the voluntary approach, efforts should be made to improve the regulatory system. A report prepared for the Standing Joint Committee of the Canadian House of Commons and Senate for the Scrutiny of Regulations put the issue this way:

Those critical of the use of regulations as a policy instrument typically characterize regulations as inflexible, difficult to amend, and therefore as being inefficient. Although it seems trite, it must be pointed out in response to such criticisms that none of these attributes are capable of being possessed by regulations themselves. In fact, such criticisms relate not to regulations per se, but rather to the process by which regulations are made and amended. There is no inherent reason why the regulatory process cannot be made more responsive to changing circumstances. In the end any process, including the regulation-making process, can only be as effective as those in charge of it [Report on Bill C-62, Prepared for the Standing Joint Committee for the Scrutiny of Regulations, February 16, 1995, at 15–16].

Making the regulatory system work better, in the end, serves the broader public interest better than devising an alternative system with potentially more pitfalls than the current approach.

The benefits for a strong regulatory system cannot be understated. One of the most succinct articulations of these benefits was recently provided by Michael Porter and Claas van der Linde, in the Harvard Business Review. According to these commentators, regulations:

- create pressure that motivates industry to develop innovative products and processes;
- improve "environmental quality in cases in which innovation and the resulting improvements in resource productivity do not completely offset the cost of compliance;"
- provide an education function for industry by informing it of likely resource inefficiencies and areas for improvement;
- improve the chances that "product innovations and process innovations in general will be environmentally friendly;"
- create "demand for environmental improvement until companies and customers are able to perceive and measure the resource inefficiencies of pollution better;" and
- "level the playing field during the transition period to innovation-based environmental solutions, ensuring that one company cannot gain position by avoiding environmental investments" [Michael E. Porter and Claas van der Linde, "Green and Competitive: Ending the Stalemate"

Harvard Business Review, September-October, 1995, p. 128]. The question should not be whether there should be a regulatory structure, but how to improve the regulatory structure in order to fulfill the function they are designed to do: protect the environment and public goods. Regulations should be used to encourage innovation; they should be cost efficient; they should be fair and achieve measurable results.

Recommendation 6: The VES should expressly recognize the regulatory approach as a legitimate approach to addressing all persistent toxic substances.

6. Differences in Goals Between US and Canada

While we applaud the concepts of numerical goals and timetables for reducing some of the Level I substances, we are concerned with the inconsistent goals delineated in the draft VES. The draft VES offers as explanation for the different goals the differences in regulatory systems, legislative mandates, and baseline data dates. A footnote to the Technical Support Document cites a revision to the US's mercury reduction target between the previous and current draft as the only illustrative example.

Recommendation 7: As the Great Lakes is a common ecosystem, we recommend that the federal governments attempt to standardize the elimination goals down to the common figure of zero. While we recognize the reality that milestones to reaching zero may of necessity be different, it would be helpful if more detailed explanations for all of the variances were included in the VES. Simply put, the overall goal for all persistent toxic substances should be zero discharge, with all reductions understood as interim targets on the path to zero discharge.

7. Progress Reporting

We are encouraged that the federal governments intend to use a number of different indicators to determine progress towards the virtual elimination goals, including actual measurements of ambient air and biota concentrations to assess progress towards goals and the development of baseline data. We encourage making this data broadly available to the public well in advance of stakeholder participation in the biennial SOLEC fora. We encourage the governments to include the development of data that allows for the assessment of how effective measures taken by industry and government have been in making progress towards zero discharge.

Recommendation 8: It is recommended that the VES include a method for evaluating progress in achieving the goals of this strategy. This evaluation should be undertaken through a process that ensures full consultation with and oversight of environmental groups. Efforts should be made immediately to develop this monitoring and report regime, including a baseline system.

8. Implementation and Public Involvement

The signing of the binational VES could prove to be a meaningless symbolic step unless it carries with it a momentum for implementation. Implementation will require broad public support for the strategy and for our shared objectives of protecting and restoring the Great Lakes Basin ecosystem.

We are very interested and concerned about the process by which the strategy will be implemented once it is signed. The success of this strategy will hinge on whether the necessary resources are committed and a transparent process is established that provides access and opportunities for the involvement of all stakeholders. We believe that it is essential that discussion about implementation not be allowed to degenerate into endless debates about absence of injury or safe levels or whether the interim targets are appropriate. Instead, we recommend that the focus of the discussions be on the best means of achieving the interim targets (challenges and timetables) as well as on the ultimate goal for these chemicals of sunsetting and virtual elimination.

Recommendation 9: We strongly recommend that the two governments revise the strategy in the text, or an appendix, to describe how the strategy will be implemented and how interested parties and the public will be involved. This description of the process and the commitment of government resources to implementation of the strategy will be crucial in evaluating the strength of the final strategy.

9. Conclusion

This strategy has the potential to advance the goals of the Great Lakes Water Quality Agreement. However, to achieve this potential, the recommendations we have outlined above must be incorporated into the strategy. Implementation of a binational strategy is crucial to the protection and restoration of the Great Lakes ecosystem. We are pleased that the governments around the Great Lakes basin are developing a virtual elimination strategy. We strongly recommend that you seize the momentum before you, make the revisions recommended above and begin the process of implementing this strategy as soon as possible.

reading 3

"Pollution, Thirst For Cheap Water Threaten Great Lakes, Study Says"
Corydon Ireland

Gannet News Service, February 11, 1997. Reprinted by permission of Gannet News Service.

In a century, global climate changes, water diversions and poor conservation may make the Great Lakes 15 degrees warmer, three feet lower, more polluted and ringed by dying wetlands.

And in the near future, Canada and the United States will be under increasing pressure to divert Great Lakes water to parched regions of North America, where water sources are already being sucked dry by agriculture, ranching and development.

These are among the dramatic scenarios posed by a new report by Great Lakes United and the Canadian Environmental Law Association.

Great Lakes United has monitored lakes' issues for 15 years. The Canadian Environmental Law Association is a nonprofit law firm based in Toronto.

"The Fate of the Great Lakes: Sustaining or Draining the Sweetwater Seas?" is a 96-page look at future threats to the five-lake basin that holds 20 percent of the world's fresh water supply.

Lake Ontario, smallest of the lakes, is at the end of the Great Lakes pipeline. It would suffer the most from changes in water levels, climate and pollution levels, the report says.

Even 40 years from now, "we're talking about a Great Lakes that's undrinkable, for starters," said Margaret Wooster, executive director of Great Lakes United, a coalition of 200 advocacy groups based in Buffalo. "They will be vastly diminished as a human and wildlife resource."

By 2035, she added, the flow of water on the St. Lawrence River will shrink by 25 percent, stranding shoreline wetlands now sustained by relatively stable water levels.

But other experts say the future is unlikely to be that alarming.

"With the normal (water) usage we have now, the lakes are pretty much self-sustaining," said Don Deblasio, a spokesman the Environmental Protection Agency's Great Lakes office in Chicago.

Great Lakes water is used for drinking, irrigation, livestock, industry and electrical power.

EPA's efforts focus on preventing pollution, Deblasio added, not promoting conservation. But periodic schemes for diverting the waters are "a concern." Currently, only three major projects divert water from the Great Lakes: Illinois and Wisconsin for drinking water; New York to increase water flow to the upper Hudson River.

Federal protections against excessive water diversion are strong, said Tony Eberhardt, chief of the water control section in the Buffalo office of the U.S. Corps of Engineers.

And U.S. terrain "does not lend itself to divertging water south," he said. Flat areas would flood, and hilly areas would require vast networks of pipes costing billions.

Even without diversions, the future looks rocky for the Great Lakes, said report coordinator Sarah Miller of the law association.

A move to privatize water distribution systems is under way in Buffalo, Montreal and in suburban Toronto—a trend that will given citizens less control over water use, she said.

"We're our own worst enemies in the Great Lakes—and the largest wardens of water on the globe," said Miller.

According to the report, the stresses on the Great Lakes include:

- Excess development, stretching water and sewer systems in populated areas further from the lakes.
- Threatened diversions of water to the American Southwest and other regions, feeding regional "pipe dreams" of unlimited water.
- Low-priced U.S. and Canadian water, more than three times cheaper than in Germany, that discourages water conservation.
- The North American Free Trade Agreement that regards water as a "good" that can be freely traded and transferred, like wheat, copper or lumber.
- Climate change that by the year 2100 could double the Great Lakes basin concentrations of carbon dioxide emissions, triggering the loss of water quality, wetlands, shoreline forests and fish stock.

The report calls for a 50 percent reduction in Great Lakes water use by 2005.

The United States is the world's largest user of fresh water. Americans consume slightly more than Canada, and four times more than Sweden.

Despite restricted water, "people's lifestyles in Sweden are not suffering," said Miller, who fears more water will soon be taken from the Great Lakes than is renewed by rainfall.

The report also calls for an expanded role for the International Joint Commission—IJC—a U.S.-

Canadian Great Lakes advisory group since 1909. Warnings from the group in the late 1960s triggered U.S.-Canadian efforts to clean up the Great Lakes said Tom Baldini of Marquette, Mich., an IJC commissioner.

A 1985 IJC report studied the effects of climate change, demand for water and changing economies. It concluded that the two governments "know with little precision the present and future uses and values of Great Lakes water."

Ontario, Erie, Michigan, Superior and Huron comprise the Great Lakes.

reading 4

"Protecting Lake Superior: A Community-Based Approach"
Lisa S. Yee

National Wildlife Federation, Great Lakes Natural Resources Center, April 1998

(Note: For the complete text of this article and to find out more about the Great Lakes Natural Resources Center of the National Wildlife Federation, visit their Web site at <www.nwf.org/greatlakes>.

Preface

Rugged and pristine, Lake Superior is the world's largest freshwater lake by surface area. Early explorers have more aptly described the lake as a freshwater sea. Although Lake Superior is treasured by many as one of the world's most spectacular natural resources, the lake is not protected adequately from the barrage of toxic chemicals that threaten to degrade its waters for centuries to come. Local communities are taking a stand by developing hands-on community-based pollution prevention projects. Lake Superior is perhaps the world's last best opportunity to show that we can keep the worst toxic chemicals from poisoning our cleanest water bodies which in turn poison the water that we drink and the fish that we eat.

Protecting Lake Superior: A Community-Based Approach is a guide for Great Lakes communities with real-life examples of pollution prevention projects to keep the Great Lakes clean. It is a critical part of the National Wildlife Federation's ongoing efforts to protect and restore clean waters for the benefit of people and wildlife. You can help to protect your local community from toxic pollution by using the project examples in this guide as a model. Local projects that you can undertake include working with your city council to pass resolutions in support of strong watershed protection measures; encouraging local schools and businesses to buy chlorine-free paper; identifying sources of and eliminating mercury from your community; and encouraging local businesses to opt for pesticide-free lawn and grounds practices. The National Wildlife Federation and its state affiliate organizations are committed to working with groups and individuals everywhere to help them protect the places that they know and love. Though this report was inspired by people throughout the Lake Superior basin who are taking measures to protect their particular watershed, the common sense ideas and strategies it contains can be used to help reduce pollution in any community. This knowledge truly is power and we urge you to put it to use by working with the National Wildlife Federation, its affiliates and other concerned citizens to help make your community a healthier place for people and wildlife.

Thank you for your interest in this report and for your willingness to get involved. Together we can make a difference.

—Mark Van Putten, President

1. Introduction

"Those who have never seen Superior get an inadequate, even inaccurate idea, by hearing it spoken of as a 'lake,' and to those who have sailed over its vast extent the word sounds positively ludicrous. Though its waters are fresh and crystal, Superior is a sea."

—George Grant, an early traveler on the lake

A. Keeping Clean Waters Clean

Travelers on Lake Superior's waters, banks, and forests can still experience much of the same beauty and solitude experienced by Native American inhabitants and the earliest European explorers centuries ago. The lake's 2,700 miles of shoreline runs along Michigan's Upper Peninsula, northwest Wisconsin, northeast Minnesota, and southwest Ontario. Much of that shoreline is as remote and wild as early inhabitants knew it. Lake Superior is the cleanest and clearest of the Great Lakes, largely because the climate is harsh and few people have lived along its shores. Lake Superior is a world treasure. This crown jewel of the Great Lakes contains three quadrillion gallons of water—enough water to flood all of Canada, the United States, Mexico, and South America with one foot of water.

The lake is so large that it dominates weather patterns in the region. U.S. and Canadian policy makers, recognizing the importance of protecting Lake Superior, signed the "Binational Program to Restore and Protect the Lake Superior Basin" (Binational Program) in 1991. The Binational Program commits the governments to the long-term goal of "zero discharge" for nine of the most dangerous pollutants that persist in the environment. Efforts to meet the Binational Program's zero discharge goal will protect Lake Superior residents, recreational and tourism industries, and wildlife, from harmful chemicals such as mercury,

PCBs, dioxin, octachlorostyrene (a chemical produced by some metal processing, incineration, and chemical manufacturing industries), and five pesticides.

Thousands of citizens from local organizations, local governments, and tribal governments are advocating a special protective designation, the "Outstanding National Resource Water" designation under the U.S. Clean Water Act, for Lake Superior. Such a designation would halt new or increased discharges of certain chemicals and thus be an important first step toward the zero discharge goal.

Lake Superior residents have wholeheartedly taken on the challenge and opportunity of keeping their lake clean by advocating strong lake protection measures. Local citizens are also taking direct action in their communities to show how local businesses, hospitals, campuses, wastewater treatment plants, and other places within the community can reduce and eliminate pollution.

B. Threats to Lake Superior

Although the lake retains much of its natural character, it faces tremendous threats. Contamination by mercury, PCBs, chlordane, toxaphene, and dioxins have led to lake-wide fish consumption advisories on certain species of fish, including lake trout, walleye, and siscowit. As a result, certain species of Lake Superior fish are unsafe to eat—especially for women of child-bearing years, children, and members of Lake Superior tribes that are especially susceptible to the effects of consuming contaminated fish or who may be disproportionately affected due to high fish consumption rates. Many of these toxic substances come from the atmosphere, entering the lake through precipitation or dust. Pollutants carried for long distances on wind currents and deposited on Lake Superior account for 98% of mercury and 85% of PCBs entering the lake.

New chlorine-based pulp mills and mining activities could add more dioxins and heavy metals to those already being dumped. Lake Superior is particularly sensitive to these threats because it has an extraordinarily long retention time—an average of 189 years—meaning some persistent toxins that are put into the lake now may remain there for seven generations or more. It is easier and more cost-effective to prevent pollution from entering Lake Superior than it is to attempt to clean up pollution after the fact. This guide provides examples of local community pollution prevention projects to eliminate pollution before it is created.

2. Keeping Lake Superior Clean: Community Projects

A. Background

Since 1991, the National Wildlife Federation (NWF) has been working with Lake Superior communities to protect the lake, using a combination of advocacy, education, and, when necessary, litigation. In 1996 and 1997, NWF focused its efforts on promoting community-based pollution prevention projects around the Lake Superior basin. This effort began with two series of meetings in communities that ring the lake. The community meetings provided a forum for citizens to share ideas for starting local projects to reduce and eliminate some of the chemicals that have been targeted for eventual elimination. Additionally, the meetings helped to educate local communities about the Binational Program and the many opportunities for citizens to help make the zero (toxic) discharge goal a reality. Publicity in local newsletters, television news, radio, and newspapers helped to raise additional community awareness about getting involved in community projects to benefit the lake.

Community members from all walks of life participated in the meetings: local, provincial, and tribal government officials; other First Nations peoples; hunters and anglers; local citizens; campus students, staff, and faculty; Lake Superior policy advisors; teachers; and many others. Participants identified and discussed existing and potential community pollution prevention projects to reduce or eliminate the use or procurement of one or more of the Binational Program's nine chemicals of concern.

B. Lake Superior Community Projects

Editor's note: To see listings of resources and contacts for each of these projects, visit the NWF Web site at <www.nwf.org/greatlakes>.

Throughout the meetings, Lake Superior citizens stressed that although the governments play an important role in advancing the zero (toxic) discharge goal for Lake Superior, the commitment of concerned citizens around the lake will make it a reality. This report highlights examples of pollution prevention projects that ordinary people in local communities have initiated and supported to help meet the zero discharge goal. These project highlights are intended for use by local citizens who want to take actions to clean up their communities. Each community project includes a brief description, notable accomplishments, steps taken to initiate the project, suggestions for getting involved in a similar project in your community, a list of resources, and project contacts.

- Lake Superior Campus Chlorine-free Paper Cooperative Project: A project to encourage Lake Superior basin colleges to join together to purchase chlorine-free paper to reduce dioxin pollution.
- Dioxin Reduction: Eliminating Burn Barrels at Red Cliff Reservation: A project to eliminate (or reduce) dioxin and other toxic chemical emissions by eliminating burn barrels.
- Promoting Zero (Toxic) Discharge Technology Through a Special Lake Designation: A project to encourage local governments and other entities to

pass resolutions supporting a special designation for Lake Superior that would prevent new or increased discharges of chemicals targeted for elimination.
- Mercury Reduction in Lake Superior Hospitals: A project encouraging hospitals, dentists, and other healthcare facilities to become mercury-free through the development and implementation of mercury pollution prevention.
- Community-wide Mercury Reductions in Marquette, Michigan: A project to analyze community-wide mercury use and to recommend ways to reduce and eliminate mercury from the community.
- Mercury Pollution Prevention at a Wastewater Treatment Plant: Western Lake Superior Sanitary District: A project involving a wastewater treatment plant's efforts to identify mercury inputs and to reduce and eliminate them.
- Reducing Pesticides: The Green Thumb Project: A project to provide alternatives to pesticide use on lawns and other grounds.

Lake Superior Campus Chlorine-free Paper Cooperative (CPC)

A project to encourage Lake Superior basin colleges to join together to purchase chlorine-free paper to reduce dioxin pollution.

Project Description: Dioxin and other organochlorines are highly toxic cancer-causing chemicals that are by-products of the chlorine bleaching process used to produce bright white paper. Dioxin is also a hormone disrupting chemical. Hormone disrupting chemicals can interfere with the normal functioning of the hormones which control an organism's reproduction, development, behavior, and homeostasis (the control of temperature, blood sugar level, etc.). Dioxin has been found in Lake Superior fish and wildlife. Organochlorines persist in the fatty tissues of humans and wildlife exposed to these chemicals through the consumption of contaminated fish, meat, and dairy products. Alternative paper-making technologies exist that do not use chlorine to bleach paper. The paper mills on Lake Superior's shores, however, do not use totally chlorine-free papermaking technologies.

The idea for a Lake Superior Campus Chlorine-free Paper Cooperative (CPC) project originated from a Zero Discharge Campus Clinic that was sponsored by NWF in April 1997. The CPC was established when a Northland College student in Ashland, Wisconsin, was funded by NWF to research the availability of chlorine-free papers. Additionally, the Northland student is encouraging Lake Superior basin campuses to join a chlorine-free paper purchasing cooperative. When local communities switch to buying chlorine-free papers, organochlorine pollution (including dioxin) can be reduced because an economic incentive is created for paper producers to switch to cleaner papermaking technologies that do not use chlorine. As major paper purchasers, campuses in the Lake Superior basin have a good potential to create a market for totally chlorine free (TCF) papers (virgin paper that is unbleached or processed without chlorine or chlorine derivatives) and process chlorine free (PCF) papers (recycled paper, with at least 20% post-consumer content, in which the recycled content is unbleached or bleached without chlorine or chlorine derivative—any virgin material used for this kind of paper must be totally chlorine free). The CPC will reward environmentally-responsible paper producers by creating a market for chlorine-free paper, giving them a positive company image. The hope is that Lake Superior campuses and other entities that participate in the CPC will be rewarded with a lower price (than they could obtain as individual purchasers) on high volume sales of chlorine-free papers.

Project Accomplishments: The CPC is using Northland College's chlorine-free paper purchasing policy as a model for other campuses. Members of the Michigan Technological University Student Pugwash group in Houghton, Michigan, are working with the Department of Civil and Environ- mental Engineering to implement a study of non-chlorine bleached papers in their computer labs. If the study provides positive results for the use of non-chlorine bleached paper in these labs, additional labs and departments at Michigan Tech will be encouraged to switch to chlorine-free papers.

The Lake Superior Alliance, a grassroots coalition of approximately 30 U.S. and Canadian groups working to protect the lake, has agreed to arrange group buying of chlorine-free paper after hearing a presentation about the CPC. Jeffrey Huxmann, the Chlorine-free Paper Cooperative coordinator, presented information on the CPC to the Lake Superior Binational Forum, an advisory body to the U.S. and Canadian governments on Lake Superior protection policies. The Forum acknowledged that the CPC was an excellent pilot pollution prevention project that should be used as a model throughout the basin. The idea for campus chlorine-free paper purchasing is spreading to campuses nationwide. NWF's Campus Ecology Program is promoting the concept at workshops across the country.

Steps Taken to Initiate Project: At NWF's Zero Discharge Campus Clinic in April 1997, participants raised the idea to establish a chlorine-free paper purchasing network. NWF provided funding for a Northland College student to begin research for the project. A CPC Web site and listserv were established. A survey was developed and distributed to paper mills in the Lake Superior basin to determine the availability and price of chlorine-free papers. In the end, follow-up phone calls with paper company representatives were critical in obtaining the information.

A second survey was developed and distributed to Lake Superior campuses to determine current paper purchasing policies and willingness to switch to chlorine-free papers. Numerous CPC presentations were conducted at various campuses, community organization meetings, and government meetings to encourage participation in the CPC.

How to Get Involved: Go to the NWF Web site for contact people and other Web sites to get more information on getting involved in the CPC. Encourage your campus or organize campuses in your community to pass a chlorine-free paper purchasing policy. Use the model (see Appendix A) chlorine-free paper purchasing policy as a basis for drafting a similar one. Be sure to send the NWF a copy. Involve local environmental organizations, faculty, staff, administrators and campus purchasers to assist you with your efforts. If a new pulp and paper mill applies for a permit to discharge toxic substances, or if an existing mill applies for a renewal permit, organize comments from community members to your state agency demanding that the mill be required to use the best technology available to ensure compliance with minimum water quality standards.

Dioxin Reduction: Eliminating Burn Barrels at Red Cliff Reservation (Red Cliff, Wisconsin)

A project to eliminate (or reduce) dioxin and other toxic chemical emissions by eliminating burn barrels.

Project Description: The use of metal burn barrels to dispose of trash is a widespread practice in many rural areas where solid waste disposal services may be limited. When plastic refuse or paper is burned in these burn barrels, dioxin, a toxic by-product, is released into the air. According to a 1994 study done for the U.S. EPA, burn barrels emit twice as much furans, 20 times more dioxin, and 40 times more particulates than if that same pound of garbage were burned in an incinerator with air pollution controls. According to U.S. EPA estimates, an example of one state's air pollution emissions from burn barrels is 5,000 tons of air pollution emitted annually from burn barrels in Illinois. Dioxin can be carried on wind currents and deposited into Lake Superior where it can make fish unsafe to eat. The Red Cliff band of Lake Superior Chippewa, near the Apostle Islands in northwest Wisconsin, is working on a project to reduce airborne dioxin and other pollution by eliminating the practice of burning trash in burn barrels. Disposal of solid wastes in this community was originally done by burning trash (paper, plastics, leaves, branches, and other refuse) in metal burn barrels. A waste transfer station which collects solid waste and limited recyclable materials was recently opened at Red Cliff. Residents, however, must pay by the bag to dispose of solid waste and there is no cost to burn wastes, so many residents and businesses still burn wastes in burn barrels. NWF has begun working with environmental staff at Red Cliff to research alternative disposal and recycling methods and to prepare a community plan for adoption of an ordinance to ban burn barrels.

Project Accomplishments: This project is still in the initial stages. NWF and Red Cliff are working on a plan to generate community support for the adoption of a burn barrel ban. Several other communities have expressed interest in pursuing similar ordinances to ban burn barrels. Red Cliff's efforts will be replicable for other communities that use burn barrels.

Steps Taken to Initiate Project: Other area tribes and local governments were contacted to get sample burn barrel ordinances.

Red Cliff's Environmental Programs office is drafting a plan, with assistance from NWF, to generate community support for a ban on burn barrels on the reservation. The main strategy to build community support is to educate local community members about the health effects of dioxin and other pollutants created from burn barrel use and to provide examples of alternative solid waste disposal and recycling methods. A draft burn barrel ordinance will be written based on similar ordinances.

How to Get Involved: Find out if your community has a burn barrel ordinance. If not, start a campaign to get one passed. Be sure to send NWF copies. Educate your community about the health effects of dioxin and other pollutants. If your community has a strong burn barrel ordinance in place, ask your local officials to encourage state-wide legislation to ban burn barrels.

Promoting Zero (Toxic) Discharge Technology Through a Special Lake Designation

A project to encourage local governments and other entities to pass resolutions supporting a special designation for Lake Superior that would prevent new or increased discharges of chemicals targeted for elimination.

Project Description: Although Lake Superior is a world treasure, the lake is not adequately protected from discharges of dangerous chemicals that degrade its waters, making its fish unsafe to eat and its water unsafe to drink. NWF and others have proposed a special designation for the lake: the "Outstanding National Resource Water" (ONRW) designation. Under the Clean Water Act, this designation would require dischargers to prevent new or increased discharges of the most dangerous chemicals that have been targeted for elimination. The ONRW designation would maintain Lake Superior's high quality waters and its exceptional recreational and ecological significance by drawing the line at current pollution levels for particular pollutants.

The effort to designate Lake Superior as an ONRW began in 1994 when NWF, Great Lakes United, and certain individuals petitioned Wisconsin, Minnesota, and Michigan to adopt an ONRW designation for Lake Superior. Canada and Ontario were asked to provide a similar designation under the Canada Water Act. The ONRW designation offers the strongest level of protection under the Clean Water Act and provides the best first step for dischargers to work toward zero discharge of certain toxic chemicals.

Here is how an ONRW designation for Lake Superior would work if enacted: suppose a discharger, such as a pulp and paper mill, an industrial plant, or a mining operation, applies for a permit to discharge one of the fifty-six chemicals listed in the ONRW petition, the state would review the permit to make sure that the discharger does not add a new discharge or increase its discharge of a listed chemical that has been targeted for elimination Existing discharges would not be affected. The ONRW designation encourages dischargers to use new pollution prevention technologies to prevent new or increased discharges of targeted chemicals. An ONRW designation also encourages clean economic growth while saving the community money by reducing the need for expensive clean-up measures later. Community members can generate local support for the ONRW designation by meeting with local government officials and asking them to adopt a supporting resolution. For example, NWF and local Lake Superior community organizations made community presentations to local government bodies, including the St. Louis County Board (Duluth, MN), the Duluth City Council (Duluth, MN), and the Douglas County Board (Superior, WI), to educate local decision makers about the ONRW designation and other local opportunities to protect Lake Superior. These local government bodies were asked to do their part by passing a resolution in support of designating Lake Superior as an ONRW and by supporting local pollution prevention projects. Communities thus have an important opportunity to work with their local government officials to adopt protective water quality control measures. Ultimately, Wisconsin, Minnesota, Michigan, and Ontario have the authority to decide which designation Lake Superior should receive. But with strong local government support for an ONRW designation (and an equivalent designation in Ontario), the state and provincial governments are more likely to support an ONRW designation.

Project Accomplishments: A number of local governments, tribal governments, and organizations have passed resolutions in support of an ONRW designation including Bayfield County (Bayfield, Wisconsin), Red Cliff Tribe of Lake Superior Chippewa (Red Cliff, Wisconsin), the Town of Sanborn, Wisconsin, St. Louis County (Duluth, Minnesota), the Minnesota Conservation Federation, NWF, the Great Lakes Indian Fish and Wildlife Commission, and the Keweenaw Bay Indian Com- munity (Michigan).

Steps Taken to Initiate Project: NWF and local community members gave presentations to local government boards to promote strong water quality protection policies for Lake Superior, such as the ONRW designation. These presentations focused on the economic benefits of keeping the community's watershed clean. Many local citizens came to show support. NWF and local community members provided local governments with model ONRW resolutions passed by other local governments. Follow-up included gathering and sending information to local governments to address questions raised at the meetings. When local governments passed resolutions supporting the ONRW designation for Lake Superior, they were asked to send copies to their Governor, Members of Congress, the International Joint Commission, and other state and federal policy makers.

How to Get Involved Give the sample resolution supporting an ONRW designation for Lake Superior to a local or tribal government official (for example, a member of your local County Board, City Council, Tribal Council, etc.) and a leader of a community organization and ask them to consider adopting a similar resolution. Be sure to mail NWF a copy. Set up a meeting with the editorial board of your local newspaper. Use the sample editorial as an example, ask the board to run a similar editorial supporting an ONRW designation. Be sure to send the NWF a copy of any ONRW editorials. For more information on the ONRW designation for Lake Superior visit NWF's web sit and contact the people listed there. The NWF contact person can also help you organize any of the actions listed above.

Mercury Reduction in Lake Superior Hospitals

A project encouraging hospitals, dentists, and other healthcare facilities to become mercury-free through the development and implementation of mercury pollution prevention plans.

Project Description: Mercury is a chemical that has been targeted for elimination because of the link to human and wildlife health problems such as permanent damage to the nervous system and kidneys and damage to sperm and male reproductive organs. Pregnant women and their unborn fetuses are at the highest risk from the adverse effects of mercury exposure. Most exposure to mercury occurs from eating fish from waters that are contaminated by mercury. Hospitals and other health care facilities are significant contributors of mercury to the environment. Some hospital equipment and other products used at a medical facility contain mercury which is released into the environment upon disposal. Typical sources of mercury in hospitals are thermometers, blood pressure devices, tubes used to keep throat passageways open (esophageal dilators), batteries, laboratory chemicals, cleaning products, and flourescent bulbs. By working with local hospitals to replace products containing mercury, significant reductions in mercury pollution can be made. Local wastewater treatment plants are good community allies since they have a vested interest in reducing toxic waste going into their treatment facilities.

NWF began its mercury pollution prevention work with hospitals and the healthcare industry in late 1995 by establishing cooperative projects with the Michigan Health and Hospital Association, the Detroit Water and Sewerage Department, and the Healthcare subgroup of the Michigan Mercury Pollution Prevention Task Force. Specific examples of NWF's hospital mercury pollution prevention work include: organizing a hospital mercury pollution prevention conference; producing two mercury pollution prevention guides; participating in an advisory role to the Western Lake Superior Sanitary District's Mercury Zero Discharge Project; and, in general, promoting and encouraging the efforts of other hospitals, community groups, and wastewater treatment plants to reduce and eliminate mercury from healthcare facilities. This experience forms the basis for our new focus on hospitals in the Lake Superior region. NWF and local communities have begun working with hospitals in the Lake Superior basin to eliminate mercury, and as a related effort, will assist hospitals with dioxin reductions by asking them to join current basin-wide efforts such as making the switch to chlorine-free paper purchasing.

Project Accomplishments: NWF participated in an advisory capacity to the Western Lake Superior Sanitary District as it developed its Mercury Zero Discharge program which included mercury pollution prevention efforts with St. Mary's Medical Center in Duluth, Minnesota. Key Lake Superior community organizations and individuals, including a large hospital purchasing network in Michigan's Upper Peninsula, the Marquette Waste- water Treatment Plant, and a local Sierra Club chapter, were identified as partners in working with local hospitals on becoming mercury-free. Lake Superior communities are actively using the NWF report Mercury Pollution Prevention in Healthcare to guide mercury pollution prevention efforts at medical facilities. A funding source for local community involvement in mercury pollution prevention in healthcare has been identified.

Steps Taken to Initiate Project: NWF distributed copies of its report Mercury Pollution Prevention in Healthcare to hundreds of hospitals around the Great Lakes. NWF formed a partnership with a local Sierra Club chapter in Marquette, Michigan, and the Marquette Wastewater Treatment Plant to work with the Marquette General Hospital on mercury pollution prevention efforts. The local Sierra Club group recently completed the first phase of a community-based mercury pollution prevention project for the city of Marquette, Michigan (additional information on this project follows in next case study). The next step will be to draft a strategy and action plan with the Marquette General Hospital on mercury and dioxin reduction.

How to Get Involved: Organize local physicians, hospital staff and administrators, dentists, environmental groups, and wastewater treatment plant operators to reduce mercury using the NWF report Mercury Pollution Prevention for City Wastewater Plants. Write letters to your dentist and physician asking them if they have mercury pollution prevention programs in place. Let them know you are concerned. Record and document the measured reductions in mercury used and purchased. Use this information to encourage other medical facilities to participate. Learn more about mercury and educate community members about the adverse health effects of mercury and what they can do to reduce and eliminate this dangerous pollutant.

Community-Wide Mercury Reductions in Marquette, Michigan

A project to analyze community-wide mercury use and to recommend ways to reduce and eliminate mercury from the community.

Project Description: Local residents in Marquette, Michigan, are working on a broad scale to reduce mercury from their community. They have completed an initial community-wide mercury reduction project to identify sources of mercury to their community and to identify ways to eliminate these sources. As a part of the Lake Superior Alliance Sustainable Basin Project, the Central Upper Peninsula Sierra Club Group received a $500 grant to develop a Community Mercury Reduction Project in the City of Marquette, Michigan. Project members also did extensive work to educate community members about mercury and how to prevent mercury pollution with the use of display boards, brochures, public service announcements, and presentations. The next step for the community group is to identify additional sources of funding for continued community mercury reduction efforts and to involve hospitals, local businesses, schools, and other community sectors for an integrated approach.

Accomplishments: Mercury collection programs at local schools and businesses in the community were conducted. As a result of this community project, community policies to reduce or eliminate the use of mercury are being implemented. For example, the Michigan Department of Environ-mental Quality now suggests that mercury-containing floats be eliminated from wastewater pump station designs. The city of Marquette has started to replace mercury floats in its treatment plant. Community members generated a good deal of publicity on the mercury reduction campaign. The mercury public service announcement generated radio station interest in mercury. One station has scheduled an interview with the wastewater treatment assistant superintendent in connection with the Marquette County Pollution Prevention Week.

Steps Taken to Initiate Project: A project task force was formed to direct the actions of the project. The task force consisted of individuals from the Sierra Club Central Upper Peninsula Group, the Upper Peninsula Environmental Coalition, the Michigan Department of Environmental Quality, the Marquette County Health Department, the Marquette Wastewater Treatment Plant, and Northern Michigan University. The Marquette Wastewater Treatment Plant assessed current mercury levels going into the wastewater treatment plant. This information will be used as a reference point to assess future increases or decreases in mercury levels as the project progresses. Potential mercury sources were identified and task force members contacted local and regional businesses, schools, laboratories, utilities, and professionals to assess mercury handling practices. Community outreach and education on the importance of mercury pollution prevention were conducted using display boards, presentations, public service announcements, and brochures. The task force helped to organize a one-time hazardous waste collection, with a special emphasis on mercury.

How to Get Involved: Organize mercury awareness and education activities in your community. Work with local government officials to ensure that good mercury recycling and safe disposal collection methods are in place. Organize a mercury pollution prevention task force in your community to identify sources of mercury and to recommend actions to reduce and eliminate these sources.

Mercury Pollution Prevention at a Wastewater Treatment Plant: Western Lake Superior Sanitary District

A project involving a wastewater treatment plant's efforts to identify mercury inputs and to reduce and eliminate them.

Project Description: The Western Lake Superior Sanitary District (WLSSD) has a wastewater treatment facility in Duluth, Minnesota. WLSSD receives wastewater, which may contain mercury from hospitals, dentists, sewer cleaning practices, septic haulers, residential wastewater, industrial laundries, and laboratories. WLSSD is the largest point source (wastes discharged via a pipe, representing a single "point" of discharge, into surface waters) discharger on the U.S. side of Lake Superior. WLSSD has been involved in many voluntary and innovative pollution prevention programs, including a two-year Mercury Zero Discharge Project. The purpose of the project was to examine the sources of mercury to its wastewater treatment plant and to determine how to reduce or eliminate those sources. This project included cooperative initiatives with industries known to be releasing mercury to the wastewater treatment plant, programs aimed at specific uses of mercury, a monitoring program to identify additional sources, and a public awareness campaign. In addition to these external programs, WLSSD also examined its own facilities and practices. Local community members can do a great deal to help wastewater treatment plants reduce the discharge of toxic chemicals into local waterways by preventing household pollutants from entering the waste stream. Specifically, communities can support collection and disposal programs for household mercury wastes and to research mercury recycling opportunities for commercial wastes.

Project Accomplishments: WLSSD worked with Potlatch Corporation's pulp and paper mill in Cloquet, Minnesota, to reduce mercury resulting from the paper bleaching process. The mercury concentration in the mill's effluent was reduced by 98 percent. WLSSD worked with local dentists to change their mercury-containing amalgam (fill material for cavities in teeth) waste handling practice. Mercury concentrations were reduced between 1993 and 1995. WLSSD produced a report, Blueprint for Mercury Elimination, which provides guidance for mercury reduction efforts at wastewater treatment plants.

Steps Taken to Initiate Project: WLSSD sought funding from the Great Lakes Protection Fund to develop and implement the two-year Mercury Zero Discharge Project to examine sources of mercury to its wastewater treatment plant and to determine how to reduce or eliminate those sources. A Mercury Pollution Prevention Advisory Committee was established and convened. WLSSD established partnerships with local industries, dentists, and hospitals to identify sources of mercury and to reduce or eliminate these sources.

How to Get Involved: Find out if your local wastewater treatment plant has a mercury pollution prevention program in place to work with local businesses and hospitals to reduce or eliminate these sources. If they don't, ask them to start one by getting in touch with the WLSSD (go to the NWF Web site for contact information) for ideas and strategies. Support collection and disposal programs for household mercury wastes and research mercury recycling opportunities for commercial wastes. Volunteer to assist your local wastewater treatment plant with community outreach efforts to educate citizens about reducing household mercury inputs. Support local businesses that utilize mercury pollution prevention and let them know why you are supporting them.

Reducing Pesticides: The Green Thumb Project

A project to provide alternatives to pesticide use on lawns and other grounds.

Project Description: Pesticides may be applied in small amounts but their cumulative use can be staggering. It is estimated that 5.2 million to 8 million pounds of pesticides are used on lawns and golf courses in the Great Lakes basin annually. Lawns and gardens receive heavier doses of pesticides per acre than any other land in the United States, including agricultural land. A National Cancer Institute study indicated that children are six times more likely to get childhood leukemia when pesticides are used in the home and garden. The Green Thumb Project is a pollution prevention program that demonstrates alternative lawn and turf management practices and focuses on community education. A major goal of the project is to increase awareness of the impact of pesticides and fertilizers on the Great Lakes ecosystem. The Green Thumb Project started as a bi-national pilot program in 1995. Four cities were chosen to take part including Sarnia, Ontario; Toronto, Ontario; Milwaukee, Wisconsin; and Duluth, Minnesota/Superior, Wisconsin. During the past three years, the Green Thumb Project has worked with several hundred individuals and organizations including groundskeepers, homeowners, businesses, schools, universities, and churches. Currently, the Green Thumb Project is coordinated by the Environmental Association for Great Lakes Education (EAGLE), with support from the Western Lake Superior Sanitary District (WLSSD) and the Great Lakes Aquatic Habitat Fund, sponsored by the Tip of the Mitt Watershed Council.

Project Accomplishments: In 1997 the Green Thumb Project had 42 individual homeowners participate as demonstration sites in Duluth, Minnesota, and Superior, Wisconsin. The Green Thumb Project has established six demonstration sites (at two campuses, a golf course, a church, a community center, and a landscaping company) in the Duluth, Minnesota, and Superior, Wisconsin area, which have agreed not to use pesticides or synthetic chemical-based fertilizers for one growing season and to implement Green Thumb practices.

Steps Taken to Initiate Project: The Green Thumb Project organizers sought and received funding for the project. Project organizers selected pesticide-free demonstration sites and worked with them to implement pesticide-free plans. Project staff organized a major pesticide education campaign.

How to Get Involved: Go to the NWF Web site for more information and the Green Thumb's Web site address. Purchase a copy of the 25 minute "Great Lakes and Great Lawns" video and guidebook (ordering information is also on the NWF site). Show the "Great Lakes and Great Lawns" video and share information contained in the guidebook in your community to garden clubs, PTA's, church groups, high schools, college campuses, local businesses, and community and neighborhood groups. Encourage these groups to eliminate pesticide use on their grounds. Hold a workshop on how to reduce pesticide use in your community and bring speakers from local extension services, environmental groups, etc. Show the video at the workshop. Learn more about pesticides and pesticide-free lawn care. One useful resource is the Green Thumb Project's "Great Lakes and Great Lawns" video and guidebook. The video and guidebook are available for $15.00 plus $3.00 (U.S.) shipping and handling by sending a check and request to "The Green Thumb Project c/o EAGLE," 394 Lake Avenue South, #308, Duluth, MN 55802.

3. Conclusions and Lessons Learned

Pollution comes from many different sources within a community and also from outside pollution sources that are carried long distances by wind currents. Discharge pipes from local industries release toxic chemicals into local waterways; incinerators release dangerous chemicals into the air when hazardous wastes from households and businesses are burned; and pesticides used on crops and lawns can run off the land into local waterways. Cleaning up, reducing, or eliminating these chemicals may seem like a daunting task but local citizens can make a difference. All it takes is for people to care and to begin with a small idea. In Lake Superior communities, reaching the long-term zero discharge goal for the nine chemicals that have been targeted for elimination is a common goal. The state, federal, and provincial governments around the basin have committed themselves to work together to achieve this goal and they are in a better position (than local community members) to reduce and eliminate pollution sources coming from outside of the basin. At the same time, local communities are leading the way to reduce and eliminate local sources of pollution.

All of the projects in this guide demonstrate the leadership and dedication of Lake Superior citizens who are working to make the zero discharge goal a reality and are merely suggestions of what can be done, if people are willing to try. After working with local communities on many pollution prevention projects, NWF can offer a number of suggestions to make your project more efficient and effective.

Some suggestions for making your pollution prevention project successful:

- Focus on projects that will identify and eliminate all sources of a particular chemical in your community.
- Find out if a similar project has already been done in your community or outside of your community. Use the lessons they learned to make your proj-

ect more efficient. Involve a diverse coalition of community partners to assist you with your community pollution prevention project.
- Look for grant money from local, state, and federal governments and private foundations to assist you with your project.
- Encourage your government officials to provide money for community pollution prevention efforts. Document your efforts. Identify your accomplishments: cost-savings, community health benefits, amount of pollution reduced or eliminated, etc. By document ing your efforts, other communities can organize similar projects.
- Publicize your efforts and provide recognition to project partners. Be sure to use your local media (newspaper, radio, and television) to publicize your community pollution prevention efforts. Be sure to recognize local businesses, community organizations, governments, wastewater treatment facilities or others that participated in the project.

Questions for Discussions

For Reading 1

1. What are the purposes of the VES?

2. What cleanup methods have been tried in the Great Lakes? To what degree have they worked? To what degree have they been less than successful?

3. Summarize into a few sentences the recommendations of the VES.

For Reading 2

1. What are some areas of agreement between the VES and the "Comments" paper?

2. What are some areas of disagreement between the two documents?

3. Assume that you are given the responsibility of mediating between the two points of view. What would be your specific recommendations be to both sides?

For Reading 3

1. The environmental issues facing the Great Lakes go well beyond water pollution. The third reading introduces four of those concerns. This article predicts the future of the Great Lakes but doesn't address directly the buildup of persistent toxic substances. It is more concerned with the effects of global warming on the region. Does the VES address any of the concerns voiced in the article?

For Reading 4

1. Describe the proposed community-based approach to protecting Lake Superior. What are its overall goals? How will it be implemented and funded? Who will carry out the work? How will the Hamilton Harbour project improve the environmental conditions of the Great Lakes?

2. How will the Hamilton Harbour project improve the environmental conditions of the Great Lakes?

3. Describe other projects supported by the Great Lakes Natural Resources Center. How are they contributing to the improvement of the Great Lakes?

Going Beyond the Basics

Visit our Web site at <www.harcourtcollege.com/lifesci/envicases2> to investigate water resource issues in additional regions of the United States and Canada.

Unit 5

Southeast: Biodiversity and Nonindigenous Species

Introduction:
Reading 1: Why We Should Care and What We Should Do
Reading 2: Fido-Munching Toads Take Florida by Leaps and Bounds
Reading 3: Florida: A Wacky Wild Kingdom
Reading 4: Management in National Wildlife Refuges
Reading 5: Plant Management in Everglades National Park
Reading 6: Management on State Lands
Reading 7: Florida's Biological Crapshoot with Carp
Reading 8: Will Beetle Bombs Recapture Everglades—Or Overrun It?
Reading 9: Ecological Effects of an Insect Introduced for the Biological Control of Weeds
Questions for Discussion

Controlling Nonindigenous Species in Florida

Introduction

Aliens are coming. In fact, they already have arrived and are, in many cases, thriving. Not from outer space, these aliens are nonnative (also known as exotic or nonindigenous) species that become established in new territories beyond their natural ranges. They are a worldwide problem. Currently, North America hosts 4500 alien species, and the number grows annually.

Some are useful. Honeybees pollinate plants, and wheat, rice, and most garden crops are nearly essential but nonindigenous species. Often they are not useful. Dandelions, house mice, and starlings are at least nuisances. Mediterranean fruit flies, zebra mussels, and kudzu—a nonindigenous vine smothering whole forests in the southeastern states—are serious pests. A few are not easy to classify. Purple loosestrife is an attractive nursery plant, but in wetlands it is a major pest. As a group, nonindigenous species become problems when they outcompete native species, degrade natural ecosystems, or dominate biological communities.

The Problem

Cost resulting from nonindigenous species are undoubtedly high, but they are difficult to document. Conservative estimates place annual losses at hundreds of millions of dollars to agriculture, forests, rangelands, and fisheries in the United States. In high-impact years, losses exceed several billion dollars. Additional losses are offset by expenditures on pesticides and other control measures. Roughly $7.4 billion are spent annually in the United States on pesticides to control mainly nonindigenous species. Economic values are nearly impossible to assess for habitat losses, species extinctions, and aesthetic losses related to nonindigenous species.

Although some species expand ranges naturally, most travel as a result of human activities. Some are introduced on purpose. Ring-neck pheasants, native in the Orient, were introduced throughout North America by sportsmen. Today, breeding populations are found in 42 states. The international pet trade has resulted in numerous introductions. Some pets escape; others are released when owners tire of them or dealers fail to sell them. In New York City, alligators are sometimes found in sewers, subsisting on rats and garbage. Monk parrots, starting as a few released pets, have established breeding populations in 15 states and expand steadily. Modern transportation and shipping bring goods to market, but also transport stowaway species from continent to continent, country to country, and region to region.

Biologically, nonindigenous species are an enigma. Native species are generally highly adapted to their natural environments, but no two regions have identical environments. The best that a species can hope for are conditions similar to—never identical with—hose they left. Theoretically, native species should be better adapted and therefore resistant to the effects of outsiders. Indeed, most new arrivals fail to survive long enough to breed.

Occasionally, though, human activities aid and abet the alien s cause. Pollution, overcrowding, and habitat modification, especially in combination, lower the native species innate resistance; then aliens become established. Generally, in the new environment they find few natural diseases or natural predators. Under these conditions, nonindigenous species thrive. This problem is particularly acute in Florida. More than 900 nonindigenous plants, and well over half of the U.S. nonindigenous amphibians, and reptiles, are found in Florida. Together, nonindigenous species create economic, ecological, and resource management problems.

Why Florida? Several factors contribute. Most of Florida's natural areas and waterways are disturbed to varying degrees, and that makes them prone to invasion. Intense human activities and frequent hurricanes are major causes. Once established, nonindigenous species spread into other disturbed areas and finally into nondisturbed areas. Without natural control agents, they spread farther, outcompeting native species.

Florida's climate also contributes. South Florida in particular is basically sub-tropical. Florida is surrounded by water on three sides, which moderates temperature changes and extremes. Warm, wet, and consistent climates are desirable not only to humans but to a number of other organisms as well, both native and alien.

Routes of entry in Florida are numerous and are difficult to monitor and control. Major airports and shipping ports link Florida with Europe, South America, and Africa. Smuggling by sea and air surreptitiously brings in illicit pets and stowaways. Major highways link Florida with regions to the north. All these transportation corridors bring nonindigenous species into Florida.

Deliberate species introductions in Florida have frequently gone awry. Ornamental plants leave yards and gardens and move into the countryside. The melaleuca tree is a fast-growing tree brought in to to dry out the wetlands. Now it is overtaking the Everglades. Other problem species that started out as deliberate introductions include the Brazilian pepper tree, especially in south Florida, and hydrilla and showy water hyacinth, which are spreading throughout waterways.

A number of Florida s industries deal with nonindigenous species. Dealers bring in a billion dollars worth of ornamental plants annually. Florida's aquaculture industry is the largest of any state. Tropical fish and aquarium plants shipped from Florida total $170 million annually. Pet owners and dealers have been implicated in the release of dozens of species of nonindigenous birds, fish, mollusks, amphibians, reptiles, and mammals.

Finally, Florida's human population growth exacerbates all of the above. Florida continues to be one of the fastest-growing states in the Union. Its population increased by nearly one-third between 1980 and 1990. Larger numbers of people increase pressure to develop land, which results in habitat loss and greater disturbance—all factors favored by nonindigenous species.

Seeking Solutions

In dealing with its nonindigenous species, Florida faces two immediate problem: (1) How can further introductions be minimized? and (2) How can existing nonindigenous species be controlled and/or eradicated? Behind these immediate problems lurks a more basic problem: In a world beset with problems, invasions by nonindigenous species often appear less critical. How can those concerned with invasive species garner sufficient resources for solutions? The readings that follow explore these questions.

"Why We Should Care and What We Should Do"
Daniel Simberloff, Don C. Schmitz, and Tom C. Brown

Granted with permission from *Strangers in Paradise*, © 1997 by Daniel Simberloff, editor. Published by Island Press, Washington, DC, and Covelo, CA.

Florida harbors a plethora of nonindigenous species. Why should we care about them? Staggering economic costs are associated with some invaders, but, except for control costs, this book has touched on economics largely in passing. Rather we have focused on the ecological impact of these invaders on native species, communities, and ecosystems. Biologists, economists, and philosophers have advanced ethical, psychological, and economic reasons why we should concern ourselves with the conservation of native biodiversity. (See, for example, Norton 1987; Wilson 1992; Callicott 1994.) Cox et al. (1994) briefly discuss the staggering economic value of Florida's native species, communities, and ecosystems in terms of consumptive and nonconsumptive natural resource uses. If ethical considerations matter at all, Florida's burden is particularly great. Of 632 species and subspecies in the United States listed as endangered under the Endangered Species Act (Chadwick 1995), 80 are found in Florida (Anonymous 1995a)—one of the highest concentrations in the nation. Another 26 have the next highest classification: threatened. A recent Nature Conservancy study shows Florida to have the third highest total of highly threatened species and subspecies (522) in the United States, after California and Hawaii (B. Stein, pers. comm.). Many of Florida's species are endemic—about 17 percent of the 668 terrestrial and freshwater vertebrates and 8 percent of the 3500 or so vascular plants (Muller et al. 1989). Moreover, of the 81 native plant communities (Florida Natural Areas Inventory 1990), 13 are endemic (Muller et al. 1989). Rare and endemic communities include mangrove swamp, pine rockland, sandhill, and scrub. All of these communities are diminished and fragmented by activities associated with human population growth, and nonindiginous species threaten to damage these communities further by displacing native species and degrading ecosystem function.

Regional Differences

Many chapters in this book point to an apparent disparity between northern and southern Florida in numbers and impact of nonindigenous species. For some taxa—such as freshwater fishes (Chapter 7) and especially amphibians and reptiles (Chapter 8)—the number of nonindigenous species in the south dramatically exceeds that in the north. For others—such as birds (Chapter 9)—the same pattern is clear but not so pronounced.

Despite numerous hypotheses (Chapter 3), the reason for this pattern is unclear. It is true that, for most taxa, the number of native species declines from north to south, encouraging some version of a "biotic resistance" hypothesis (Chapter 1) to explain the greater number of nonindigenous species in the south, but no real evidence shows that arriving species survive or fail in differential numbers because of differing resistance posed by the resident biota. In fact, there are few data on the numbers of failed introductions. If we wish to understand the reason for the gradient in number of surviving nonindigenous species, we will probably require at least a systematic consideration of failure rates.

Many workers have argued that southern Florida is more "disturbed" than northern Florida and that this disturbance has made it more prone to invasion by interlopers. Yet the many kinds of disturbance, both natural and anthropogenous, cannot all be ranked on the same scale as to severity. In any event, northern Florida has been greatly disturbed, although the results are perhaps not so apparent to the human eye (Chapter 3). Surely an understanding of the role of disturbance in establishment of nonindigenous species will require a careful consideration of the specific kind of disturbance and the specific ways in which both native and nonindigenous species respond to each kind (Chapters 1 and 4). The preliminary but detailed data on the effects of Hurricane Andrew in southern Florida in 1992 show that this disturbance benefited certain native species over certain nonindigenous potential invaders, while other nonindigenous species were favored over other natives. Further, the responses of the various species to anthropogenous disturbance did not adumbrate their responses to the hurricane (Chapter 4). With respect to the relative invasibility of north and south, it would be extremely enlightening to compare ongoing similar studies of the impacts wrought by Hurricane Kate in 1985 in northern Florida, but published research (such as Platt and Rathbun 1993) does not address nonindigenous species.

A simple "null" hypothesis for the greater number of established nonindigenous species in the south has been posed by C. Lippincott (pers. comm.) for plants: more propagules of more species arrive in southern Florida, either by human means or on their own. The data for some taxa—such as plants (Chapter 2), reptiles and amphibians (Chapter 8), and fishes (Chapter 7)—are at least superficially consistent with this hypothesis, and it would be informative to have systematically gathered data on numbers of arriving propagules for many species.

Differences Among Taxa

Ecological effects of Florida's nonindigenous species vary enormously. Two types of invaders are particularly likely to have major ecosystem effects: species that constitute new habitats and species that modify habitats by altering ecosystem processes (Chapter 3). Because plants form the biological matrix for most communities, and several nonindigenous plant species in Florida have formed or affect habitats over substantial areas, plants currently have the greatest impact among all nonindigenous taxa in the state. For terrestrial vertebrates—birds (Chapter 9), mammals (Chapter 10), and reptiles and amphibians (Chapter 8)—the documented effects on native ecosystems are few. Almost all of the contributors to this volume confess that most data on effects are anecdotal and sketchy, and, as noted in Chapter 1, many effects of nonindigenous species are subtle and can easily be overlooked. Nevertheless, relative to invertebrates and plants, terrestrial vertebrates are quite well studied, and it seems unlikely that many of their nonindigenous species are problematic in Florida without our knowing it. Perhaps feral pigs cause the most damage today (Chapters 12 and 18–20). Other terrestrial vertebrates could become great threats in the future. The brown tree snake, for example, should it reach Florida, might devastate many native species (Chapter 8), and an increase in monk parakeet populations could be greatly detrimental to native species, although opinions vary (Chapter 9).

Ecological impacts of nonindigenous freshwater fishes in Florida are probably even more poorly known than are those of the terrestrial vertebrates, but there are compelling arguments (Chapter 7) that, even though no extinctions of native species can be attributed to nonindigenous fishes, many of the latter are substantial threats. In other parts of the United States, nonindigenous fishes have often played a key role in recent cases in which native fishes have been endangered or extinguished (Chapter 7).

Insect species, of course, greatly outnumber all vertebrate species combined. Their impacts, too, are usually poorly understood, so it is not surprising that the impacts of nonindigenous insects, except those that affect agriculture, are barely studied (Chapter 5). It seems possible, however, that nonindigenous insects already affect native systems more strongly than do vertebrates. Part of this effect is beneficial—for example, there are introduced insects that control alligatorweed (Chapters 3, 5, and 15). In contrast, the cactus moth threatens two native cactus species (Chapters 1, 5, and 15), a nonindigenous leaf beetle threatens two native endangered morning glory species (Chapter 5), a Mexican weevil devastates native bromeliad populations (Chapter 5), and the gypsy moth, should it become established in Florida (Chapter 5), could devastate deciduous trees and communities based on them. Although the imported fire ant affects primarily disturbed habitats in Florida (Chapter 5), it potentially threatens many native species as a predator and competitor (Vinson 1994)—and the advent of a polygynous form exacerbates the threat (Porter et al. 1988; Vinson 1994).

Among nonindigenous freshwater invertebrates in Florida, the Asiatic clam may already affect entire native ecosystems, and the zebra mussel, which may arrive soon, would probably have far-reaching impacts (Chapter 6). Several nonindigenous snails already present may affect native species by either competition or herbivory (Chapter 6). If terrestrial and freshwater invertebrates are poorly understood, marine invertebrates are even more so; in many cases their geographic range is unknown and their status as native or nonindigenous is a mystery (Chapter 11). One cannot even hazard a guess as to whether nonindigenous marine invertebrates are affecting native ecosystems, although major demonstrated impacts in well-studied areas (Carlton 1989) suggest that nonindigenous marine invertebrates may already seriously affect some Florida systems. Carlton and Ruckelshaus (Chapter 11), for example, contend that the isopod Sphaeroma terebrans is nonindigenous; a similar congener affects an entire ecosystem in Costa Rica by boring red mangrove roots (Perry 1988). The extent of invasion of Florida's coastal waters by nonindigenous marine plants, and their impact, is virtually unknown, but major effects elsewhere are well documented (Chapter 11).

Predicting Ecological Effects

As noted in Chapter 1, nonindigenous species can have many types of effects (direct, indirect, and synergistic) on native species, communities, and ecosystems. One could reasonably argue that if a potential nonindigenous species can form a new habitat or greatly alter ecological processes, its importation should be forbidden. Among plants, for example, if a potential invader can shade out other species or modify the fire regime, we have reason to expect problems (Chapter 3). The great majority of nonindigenous plant species, however, are not invasive or environmentally harmful. The fraction that have substantial effects has not been estimated in Florida, but other estimates center on about 10 percent (references in Simberloff 1991; Chapter 3).

Most ornamentals and agricultural plants stay where they are planted. Do they then need no regulation? Even an environmental purist would be hard pressed to object to certain nonindigenous species. the Venus's-flytrap (Dionaea muscipula), for example, is native to a small, boggy region in North Carolina. Because of habitat destruction and poaching for the ornamental trade, its range has dwindled from 18 to 11 counties, and it has become a "species of special concern" in North Carolina (Culotta 1994). In 1992, it was added to the Convention on International Trade in Endangered Species of Wild Fauna and Flora list of species requiring a permit for export, but its range continues to contract. It has been planted in northern Florida, survives as small, self-sustaining populations at a few sites, and shows no evidence of even slight invasive tendencies (S. Hermann, pers. comm.). Should this introduction be regulated?

Well, . . . yes! Though most introduced species do not become invasive, scientists have not been very good at predicting which introductions will be scourges and which innocuous (Hobbs and Humphries 1995). Effects are often subtle and surprising (Chapter 1). Further, a previously noninvasive species can become invasive because of subsequent environmental change or because of another introduction: ornamental Asian fig trees in southern Florida were not invasive until their pollinator wasps arrived separately (McKey and Kaufmann 1991). And sometimes a species becomes a major problem only after a long, inexplicable lag (Chapter 3).

Because of the inherent dispersal abilities of living organisms (Chapter 1), there is no guarantee that they will not, on their own, disperse from an area of release where they are unproblematic to sites in which they are pests. Three biological control agents (a wasp, a nematode, and a fly) were introduced into Florida in the 1980s, for example, to control pestiferous nonindigenous mole crickets. Although in Florida native mole crickets are not attacked (J.H. Frank, pers. comm.), no apparent consideration was given to the possibility that the biological control species might spread. An American mole cricket found further north (Gryllotalpa major) is a candidate endangered species. Might a biocontrol agent spread to within its range? Moreover, a nonindigenous species that is not a pest in areas in which it was intended to stay can be carried by humans to areas where it can be troublesome. One hypothesis for the spread of the cactus moth (Chapters 1 and 5) to Florida is that it was carried in the international trade in cut flowers.

The effects of nonindigenous species are inherently unpredictable in another way that is different from, for example, the effects of a chemical pollutant. Species can evolve. In North America, the Dutch elm disease fungus evolved more pathogenic strains (von Broembsen 1989), and numerous initially virulent diseases have evolved benign strains (Ewald 1983). Such evolution is one possible reason for the long time lags occasionally seen before invasiveness is manifested (Ewel 1986). Although we know of no demonstration in Florida of increased ecological damage caused by evolution of a nonindigenous species—no one has even attempted to quantify the probability of such an event—the potential harm is enormous.

In sum, then, the inherent unpredictability of nonindigenous species means that their entry must be much more tightly regulated than it is. They should be assumed guilty until proven innocent (Ruesink et al. 1995). The first and least costly line of defense is keeping them out of Florida. Although political and economic considerations make complete exclusion of all nonindigenous species an unrealistic goal, every single proposed introduction should at least receive thoughtful consideration. The "blacklist" approach—only species determined a priori to be potentially harmful are subject to regulation—has never worked well (Ruesink et al. 1995; Wade 1995). All nonindigenous species are potentially harmful. Thus, a "whitelist" approach is needed: every potential import should be assumed harmful until shown to pose a low risk, after which point it can be placed on a whitelist. No blanket exceptions should be allowed. Further, at the first observation of unexpected invasiveness, a species should be removed from the whitelist and further importation forbidden.

If an invasive species gets into Florida because of inadvertent transport, on its own, or because it was erroneously put on the whitelist, an effort should be made, where feasible, to eradicate the species quickly (Chapter 13)—for once it has spread, the cost of either eradication or continuing control will be much greater. If eradication fails, a comprehensive plan for maintenance control must be established and consistently followed (Chapters 14 and 15).

Controlling Nonindigenous Species

Myers and Ewel (1990b) list two recurrent themes in *Ecosystems of Florida* (Myers and Ewel 1990a): each of Florida's terrestrial ecosystems is but a fraction as large as it was at the beginning of the century and nature can no longer maintain what is left without human management. Similarly, many of Florida's watersheds have experienced increased nutrient loadings because of burgeoning urbanization and agricultural runoff (Canfield et al. 1983b; Schmitz et al. 1993), and intensive management is needed to preserve even a vestige of our native aquatic ecosystems. The present volume has shown that nonindigenous species are a key component of the degradation of Florida ecosystems, and better management of them is crucial, but it has also shown that our understanding of Florida's invasion biology is incomplete—in fact, for same taxa it is rudimentary. Similarly, ecological restoration of sites infested by nonindigenous species often fails because of a lack of understanding of invasion biology (Chapter 12). Improved knowledge will aid management, but management must be improved even as research into the biology of nonindigenous species continues. In light of the facts adduced here, what should be done now?

Preventing Entry

Presently there is virtually no screening of the vast majority of imported plants and animals for their potential invasiveness (Chapter 22). Federal efforts are fragmentary at best and primarily aimed at preventing new agricultural pests (Wade 1995; Chapter 21). Activities by the state of Florida mirror those of the federal government: priority is given mainly to agricultural pests (Chapter 22). Gordon and Thomas (Chapter 2) argue that because the federal response to biological invasions is inadequate, Florida should independently develop a comprehensive Florida Noxious Weeds list that includes species which invade nonagricultural lands. We would go further: no species on the noxious weeds list could be admitted to a whitelist, but neither could any other species be admitted without a detailed consideration of its potential invasiveness. Similarly, Courtenay (Chapter 7) pleads for more stringent regulation of importation of nonindigenous fishes combined with a better effort to educate the public about the dangers of releasing nonindigenous fishes. From the standpoint of state government, Brown (Chapter 22) observes a failure to establish a screening protocol for whether a nonindigenous species of any taxon poses a threat to Florida's ecosystems. Similar failures have beset other blacklist approaches (Wade 1995). Brown also notes a failure to expand blacklists of known invasive species and points to the internal inconsistency of allowing agencies that introduce nonindigenous species to veto the addition of invasive species to blacklists designed to protect public lands and waterways.

As noted earlier, the inherent unpredictability of living entities dictates a philosophy of guilty unless proven innocent—or at least until substantial research buttresses a claim of low probability of harmful effect.

Eradication of Maintenance Control

Management of nonindigenous species that arrive in Florida has usually been poorly funded and fragmentary at best. A success story is the maintenance control of water hyacinth (Chapter 14), which has been reduced to a minor component of the aquatic flora at an approximate annual cost of $2.7 million. Similarly, the total eradication of the snail *Achatina fulica* cost less than $2 million (Chapter 13). These campaigns contrast with many others in which management funding is insufficient and inconsistent. In 1985, for example, $2.5 million was needed to manage hydrilla in Florida's public waterways. Because funding did not keep pace with hydrilla's growth and expansion, the plant's "coverage doubled between 1992 and 1994. Thus $14 million was needed to manage hydrilla adequately in 1995; but only $5 million was appropriated (Chapter 14). Early investment may allow eradication that becomes impossible once a species has spread (Chapter 13), and the cost of implementing maintenance control can quickly exceed that of early eradication. The basic problem is that living organisms do not sit around waiting for us to find funding to deal with them: they grow, move, and evolve.

Certain potentially invasive nonindigenous species are not managed at all—for example, many plants (Chapter 17), most vertebrates (Chapters 7–10), almost all insects that do not affect agriculture (Chapter 5), and freshwater and marine invertebrates (Chapters 6 and 11). Even when there is a management effort, it is usually piecemeal. No lead agency is designated to coordinate efforts of state agencies to manage nonindigenous species in nonagricultural situations. Nor do substantial incentive programs or assistance to private landowners encourage their integration into the control process (Chapter 22).

Prospects

Florida is besieged by nonindigenous species. Virtually every ecosystem is already heavily affected or a potential target. Nevertheless, there is little scientific research on most aspects of invasion biology in Florida; for some taxa, virtually nothing is known. Several factors contribute to this state of affairs: inadequate publicity about the problems, the idiosyncrasies of researchers' taste, and failure of governmental and private sources to provide funding for research. To some extent, these same problems beset all aspects of conservation. But they are particularly acute for nonindigenous species because of their importance to conservation and their continuing spread.

Efforts to exclude nonindigenous species are inadequate, and the dearth of knowledge about many aspects of invasion biology argues for much more rigorous exclusion. Once exclusion has failed, technologies often exist to eradicate or control invasive newcomers, but governmental structure and inadequate funding usually hinder the implementation of these methods.

In sum, problems of invasive nonindigenous species are largely soluble. Exclusionary regulations would help immensely, and methods can be developed to deal with established species. The key is our willingness to try hard enough.

reading 2

"Fido-Munching Toads Take Florida by Leaps and Bounds"
Tom Wells

Associated Press, September 29, 1996. © 1996 by Associated Press. Reprinted by permission.

In one of those examples of man foolishly messing with nature, a huge kind of South American toad was let loose in Florida and other parts of the world earlier this century to try to control sugar cane pests.

Big mistake.

The *Bufo marinus*, or marine toad, can grow to 7 inches or more and weigh more than 3 pounds. They are overrunning part of Australia. They're scaring tourists in the Caribbean. And occasionally they're killing dogs in Florida with a poison so potent that the family pets die in a matter of minutes.

"Someone got this wild idea it would increase sugar cane production. It didn't," said University of Miami biologist Jay Savage. "You'd have to have bumper-to-bumper toads to increase a crop."

"Now they've become a pest themselves," Savage said.

Bufo marinus resembles Jabba the Hutt of "Star Wars," with deeply pitted, swollen glands behind each eye, extending down the back. The glands contain a milky white toxin that the toad secretes when threatened. The animal's call sounds like a tractor in the distance.

Dogs who find all this too tempting to pass up suffer the consequences if they touch the toads with their tongues. Vivian Gil's dalmatian, Jazz, barely survived one such confrontation recently outside her home in western Dade County, on the edge of the Everglades.

"It looked like he was having spasms or something. It felt like his brain was going to explode," she said. "It was very scary."

Some people try to use the poison of the toad and its cousins as an aphrodisiac and a hallucinogen. The substance was sold until last year in grocery stores and tobacco shops in the United States but is now banned.

"Four New York men purchased it, thinking it was an oral aphrodisiac. They died," said Dr. Rossanne Philen, an epidemiologist at the Centers for Disease Control and Prevention in Atlanta. The substance was supposed to be rubbed on the genitals.

"A few years ago there was a fad of licking the toads because people thought they were a hallucinogen," Savage said. "Other people tried to extract and sniff it."

In fact, Maya Indians in Mexico use the toxin as a hallucinogen in their religious ceremonies.

In Florida, "there are a few dogs killed every year. Cats are usually smarter. When they see something that big, they leave it alone," Savage said. "If a dog gets a good shot of the toxin, it can kill it. The first sign is that the dog starts frothing at the mouth."

If a dog spits out the toxin, the toad just ends up ruining the dog's day. But if the dog swallows the poison, death can be quick. In humans, the toxin generally isn't enough to stop the heart, but it can irritate the mucous membranes, Savage said.

Bufo marinus is native to Latin America from Brazil to Mexico. It was introduced into Puerto Rico in the 1920s, into Hawaii, the Philippines and Australia in the 1930s and South Florida in the 1940s, Savage said. It has spread to southern Texas but cold winters keep it out of the rest of the United States.

The toads will eat just about anything, including insects, small birds, snakes, table scraps and vegetation. Sometimes they can be found feeding on cat food or dog food if the pet's dish is left outside at night.

Florida wildlife officials haven't tried to eradicate the toads and can't say for sure how many there are.

In Queensland, Australia, the toads have become so numerous that they are poisoning ranchland water holes when they get into ponds and die, Savage said. Scientists there are testing a virus that seems to kill the toads.

"Florida: A Wacky Wild Kingdom"
Deborah Sharp

USA Today, July 9, 1995. © 1995 by *USA Today*. Reprinted by permission.

People in southern Florida have long known to protect small children and pets from alligators lurking in canals and ponds. These days, they also have to keep a watchful eye out for nasty monkeys in trees and huge lizards under the hoods of their cars.

"It's like Wild Kingdown run amok," says trapper Todd Hardwick, who gets calls at his headquarters south of Miami to capture exotic pets on the loose, from escaped pet cougars to cat-swallowing pythons.

Just last week police in West Palm Beach subdued a carnivorous Asian water monitor along busy Military Trail. Motorists slammed on their brakes as the 5-foot lizard strolled across six lanes of traffic.

A few days later, trappers near Naples caught up with a 65-pound snow leopard. Katu escaped from its owner, a licensed rare-cat exhibitor.

"A snow leopard in the Florida Everglades. Can you imagine that?" Naples trucker Jim McMullen says of the endangered Himalayas native.

Miami's port is a chief entry point for the international animal trade. And there has been explosive growth in the number of Floridians licensed to sell or keep exotic pets: *In 1980, the game commission granted 620 licenses. By 1993, the latest year available, 3,917 licenses had been issued.*

"We're very concerned," says Capt. Jerry Thompson of the Florida Game and Fresh Water Fish Commission. "That's one of the reasons we have an inspection program: To curtail accidental or intentional release of these exotic species into Florida's natural ecosystem."

With similarities between an exotic jungle and Florida's subtropical environment, escaped pets are prospering–often at the expense of less-aggressive native species.

Says Ron Magill of Miami's Metrozoo: "South Florida is like an Ellis Island for exotic animals. . . . They're thriving here."

Though humans are startled by the occasional odd reptile, the most stubborn problem caused by the alien lizards, snakes and birds is the threat posed to indigenous species. The giant poisonous bufo toad is not only trying to outmuscle smaller, less-aggressive Florida toads and frogs for food, the imported toad also endangers household dogs, which often die after biting the creatures.

Brazilian fire ants are devastating the nests of endangered sea turtles on the beaches of island sanctuaries off Key West. The biting creatures attack the eggs.

The large water monitors are gobbling up food that chameleons and other local lizards depend upon. The Florida indigo snake's food supply is similarly threatened by larger imported snakes who feast on rodents and other prey.

Flocks of monk parakeets are pushing out blue jays and mockingbirds. The bright green birds feed on fruit trees, which worries agriculture leaders in Homestead.

Monkeys are frolicking in the mangrove trees along Dania Beach Boulevard.

Hardwick says he still has the 6-foot monitor he removed in 1992 from a stalled engine for a surprised Miami motorist. "I've got my own urban safari every day," says Hardwick, 32, a Florida native.

Hardwick's most famous case was a 250-pound, 22-foot python beneath a suburban Fort Lauderdale home. It took four men to drag it out.

He says he now captures 100 exotic reptiles a month. Boa constrictors. Pythons. Iguanas. All kinds of giant lizards. And that's not to mention the larger animals—from buffalo on the Florida Turnpike to an angry capuchin monkey that bit and pelted Harwick with mangoes: "I keep my rabies shots up to date," he says.

reading 4

"Management in National Wildlife Refuges"
Mark D. Maffei

Granted with permission from *Strangers in Paradise*, © 1997 by Daniel Simberloff, editor. Published by Island Press, Washington, DC, and Covelo, CA.

The U.S. Fish and Wildlife Service (FWS) manages close to 400,000 ha of land in Florida (Table 16.1). From the hardwood hammocks of St. Marks National Wildlife Refuge in North Florida to the tropical islands of Key West National Wildlife Refuge, it manages a wide range of biological communities. Among the management goals for virtually all of these lands is the maintenance or restoration of the function, structure, and species composition of the native ecosystem.

The FWS has a national mandate to protect fish and wildlife and their habitats and has, within the last few years, begun to take an ecosystem approach to its mandate (U.S. Fish and Wildlife Service 1994). Because the invasion of native communities by nonindiginous species significantly threatens the systems managed by the FWS, its land managers devote substantial resources to reducing their impact. Nonindigenous trees are controlled by fire, herbicides, or mechanical removal. Aquatic weeds are controlled with herbicides or water management regimes designed to limit their spread. Controlling nonindigenous animals may require sacrificing native species—as when rotenone is used to remove carp from lakes and ponds—but is necessary because invasive nonindigenous species are one of the greatest threats facing the management of FWS lands.

All the national wildlife refuges (NWRs) in Florida have been invaded by nonindigenous life forms. Some invasions are of little consequence; others pose a serious threat. Some nonindigenous plants and animals found on NWRs in Florida are listed in Tables 16.2 and 16.3. The effort to control any of these species varies from refuge to refuge and depends on the threat it poses.

Nonindigenous Plants

Although most nonindigenous plants pose little threat to NWR lands, some species aggressively invade undisturbed areas or outcompete native species—including threatened and endangered ones (Duever et al. 1979; Grow 1984)—for disturbed areas. They crowd out native plants (Myers 1983) on which wildlife depend, alter ecosystems by increasing evaporation, and poison and irritate wildlife and nearby plants (as do Brazilian pepper and melaleuca; Austin 1978a; Williams 1980). In Florida alone, control and eradication of nonindigenous plants on national wildlife refuges in fiscal year 1993 exceeded $200,000.

Melaleuca quinquenervia (*melaleuca*) is one of the most problematic alien plants on NWRs in Florida. It imperils NWR resources in southern Florida and is capable of becoming a problem in central Florida as well. The species thrives in freshwater marshes and upland areas (Woodall 1980; Myers 1984; Hofstetter 1991; Chapter 3 in this volume)—habitats well represented on NWR lands—and is a severe problem on A.R.M. Loxahatchee, National Key Deer, Crocodile Lake, Hobe Sound, and Merritt Island NWRs, where it is aggressively invading the marsh communities. It has been removed from J.N. "Ding" Darling NWR. Because of its impact on the freshwater marshes of the Everglades, this tree threatens the endangered Everglade kite (*Rostrhamus sociabilis*) and wood stork (*Mycteria americana*); indeed, it is a threat to all organ-

TABLE 16.1. National Wildlife Refuges in Florida, September 1993

Refuge	Size (ha)
Archie Carr NWR (Indian River and Brevard counties)	21
Arthur R. Marshall Loxahatchee NWR (Palm Beach County)	59,500
Caloosahatchee NWR (Lee County)	16
Cedar Keys NWR (Levy County)	337
Chassahowitzka NWR (Citrus County)	12,322
Crocodile Lake NWR (Monroe County)	2656
Crystal River NWR (Citrus County)	27
Egmont Key NWR (Monroe County)	133
Florida Panther NWR (Collier County)	9465
Great White Heron MWR (Monroe County)	77,932
Hobe Sound NWR (Martin County)	397
Island Bay MWR (Charlotte County)	8
J.N. "Ding" Darling MWR (Lee County)	2165
Key West NWR (Monroe County)	84,335
Lake Woodruff MWR (Volusia County)	7913
Lower Suwannee (Levy and Dixie counties)	20,299
Matlacha Pass NWR (Lee County)	207
Merritt Island NWR (Brevard County)	55,977
National Key Deer NWR (Monroe County)	3318
Okeefenokee NWR (Baker County)	1489
Passage Key NWR (Pinellas County)	26
Pelican Island NWR (Indain River County)	1908
Pine Island NWR (Lee County)	222
Pinellas NWR (Pinellas County)	13
St. Johns NWR (Brevard County)	2532
St. Marks NWR (Wakulla and Jefferson counties)	26,547
St. Vincent NWR (Franklin County)	506

Source: U.S. Fish and Wildlife Service (1993).

TABLE 16.2. Partial List of Nonindigenous Plant Species Found on National Wildlife Refuges in Florida

Scientific name	Common name
Acacia auriculiformis	earleaf acacia
Alternanthera philoxeroides	alligator weed
Brachieria mutica	para grass
Casuarina equisetifolia	Australian pine
Casuarina glauca	scaly-bark beefwood
Clerodendrum speciosissinuem	Java glorybower
Colocasia esculenta	wild taro
Colubrina asiatica	lather leaf
Cupaniopsis anacardiodes	carrotwood
Dioscorea bulbifera	air potato
Egeria densa	Brazilian elodea
Eichkornia crassipes	water hyacinth
Hydrilla verticillata	hydrilla
Ipomoea aquatica	water spinach
Leucaena leucocphala	lead tree
Lonicera japonica	Japanese honeysuckle
Lygodium japonicum	Japanese climbing fern
Melaleuca quinquenervla	melaleuca
Myriophyllum spicatum	Eurasian water milfoil
Panicum repens	torpedo grass
Paspalum notatum	bahia grass
Pistia stratiotes	water lettuce
Psidium guajava	guava
Salsola kali	Russian thistle
Salvinia rotundifolia	water spangles
Sapium sebiferum	Chinese tallow tree
Scaevola taccada var. sericea	scaevola
Schinus terebinthifolius	Brazilian pepper
Sesbania punicea	purple sesban
Sesbania Vesicaria	bladderpod
Sonchus asper	prickly sow thistle
Sonchus alenaceus	common sow thistle
Thespesia populnea	seaside mahoe
Vitex trifolia	vitex
Wisteria sinensis	Chinese wisteria
Yucca aloifolia	Spanish bayonet

isms that depend on marshes (Austin 1978a; Mazzotti et al., 1981). Since implementation of a control program (Maffei 1991) in April 1992, more than a million melaleuca trees have been removed from A.R.M. Loxahatchee NWR alone. Control programs for this plant are also under way at Crocodile Lake, Florida Panther, Merritt Island, National Key Deer, and Hobe Sound refuges.

Coastal refuges are seriously affected by Australian pine (*Casuarina equisetifolia*). Beaches and dunes are rapidly invaded by this tree, the canopy of which shades out native vegetation. This tree is a particular problem on refuges, such as Hobe Sound, Merritt Island, and Archie Carr NWRs, with beaches used as nesting sites for sea turtles, all of which are federally listed as endangered or threatened species. When a female sea turtle encounters the roots of the Australian pine while excavating her nest, she abandons her nesting attempt. The large tangles of roots exposed when the trees topple, as they often do, have trapped and killed both adult and hatchling sea turtles. Control of Australian pine is under way at Merritt Island, Hobe Sound, J.N. "Ding" Darling, Egmont Key, Crocodile Lake, and Florida Panther NWRs.

Brazilian pepper (*Schinus terebinthifolius*) has invaded Merritt Island, National Key Deer, Crocodile Lake, Hobe Sound, A.R.M. Loxahatchee, Egmont Key, Key West, Great White Heron, J.N. "Ding" Darling, and Florida Panther NWRs. Control efforts are under way at all these sites. On Florida Panther NWR, shading caused by dense stands of Brazilian pepper and Australian pine kills plants used by white-tailed deer (*Odocoileus virginianus*), which are an important food of the endangered Florida panther (*Felis concolor coryi*).

Other nonindigenous plants for which control efforts are under way on refuges in Florida include the Chinese tallow tree (*Sapium sebiferum*) at St. Vincent NWR; lather leaf (*Colubrina asiatica*) at National Key Deer Refuge; water hyancinth (*Eichhornia crassipes*) at Lake Woodruff and A.R.M. Loxahatchee refuges; lead tree (*Leucaena leucocephala*) at National Key Deer, Crocodile Lake, and Key West refuges; and Japanese climbing fern (*Lygodium japonicum*) at Hobe Sound.

Nonindigenous Animals

The dangers of nonindigenous animals that invade national wildlife refuges in Florida are difficult to assess. Although it is relatively easy to determine the extent to which nonindigenous plants can invade native areas, the impact of nonindigenous animals on native biological communities, and on those native species with which they compete directly, is often less obvious. The Norway rat (*Rattus norvegicus*), for example, found throughout the state, may directly compete with the endangered Key Largo wood rat (*Neotoma floridana smalli*) and cotton mouse (*Peromyscus gossypinus allapatocola*) in the Florida Keys NWRs, but the direct impact is difficult to measure.

The impacts of certain nonindigenous animals are less obscure. Feral pigs (*Sus scrofa*) threaten both plant and animal communities. They destroy the nests of sea turtles, gopher tortoises (*Gopherus polyphemus*), indigo snakes (*Drymarchon corais couperi*), shore and wading birds, and other species, and they prey on native reptiles and amphibians. Their rooting disrupts the soil, causing dune destabilization or damage to wetlands and other plant communities. They compete with native animals for acorns, an important food resource. On some refuges, such as St. Vincent and Florida Panther NWRs, the pig is a reservoir for pseudorabies

TABLE 16.3. Partial List of Nonindigenous Animal Species Found on National Wildlife Refuges in Florida

Scientific name	Common name
Anolis sagrei	brown anole
Astronotus ocellatus	oscar
Belonesax belizanus	pike killifish
Bufo marinus	marine toad
Cervus unicolor	sambar deer
Cichlasoma bimaculatum	black acara
Clarias batrachus	walking catfish
Columba livia	rock dove
Dasypus novemcinctus	armadillo
Eleutherodactylus planirostris planirostris	greenhouse frog
Felis catus	feral cat
Felis yagouaroundi	jaguarundi
Hemidactylus turcicus turcicus	Mediterranean gecko
Leiocephalus carinatus armouri	northern curly-tailed lizard
Marisa cornuarietus	golden-horn marisa
Molothrus bonariensis	shiny cowbird
Oreochromis aureus	blue tilapia
Osteopilus septentrionalis	Cuban tree frog
Passer domesticus	house sparrow
Pomacea domesticus	spike-topped apple snail
Rattus norvegicus	Norway rat
Rattus rattus	black rat
Sphaerodactylus elegans	Indo-Pacific gecko
Streptopelia risoria	ringed turtledove
Sturnus vulgaris	European starling
Sus scrofa	pig
Tilapia mariae	spotted tilapia

virus, which threatens large carnivores such as the endangered Florida panther and is deadly to the endangered red wolf (*Canis rufus*). Feral pigs are present on A.R.M. Loxahatchee, Chassahowitzka, Lower Suwannee, Merritt Island, National Key Deer, St. Marks, and St. Vincent NWRs. Hunting to control populations occurs on Lower Suwannee, St. Marks, and St. Vincent refuges. Pigs are taken on Merritt Island NWR by a contract hunter and by authorized refuge personnel.

The armadillo (*Dasypus novemcinctus*), considered nonindigenous by some (Chapter 10 in this volume), affects native life forms. Armadillos excavate seaturtle nests and feed on the eggs. Like feral pits, they disrupt soils and can kill seedling plants. Armadillos are found on NWRs throughout Florida except in the Florida Keys. No refuge removes armadillos systematically, but on Hobe Sound and St. Vincent NWRs, they are removed when encountered.

Many other nonindigenous species that may affect native species are found on Florida NWRs. The marine toad (*Bufo marinus*) can poison larger animals that attempt to eat it. The brown anole (*Anolis sagrei*) appears to be displacing the native green anole (*A. carolinensis*) in the Florida Panther NWR (J. Krakowski, pers. comm.). Nonindigenous fish species, such as oscars (*Astronotus ocellatus*), may be competing with native sport and nongame fishes for nest sites and food (Chapter 7 in this volume). Oscars and other nonindigenous fishes also prey on juveniles of native fishes and on small nongame species. Fire ants (*Solenopsis invicta*) compete with and displace native ants throughout Florida and may be affecting ground-nesting birds and burrowing animals (Chapter 5 in this volume). On St. Vincent NWR, nonindigenous sambar deer (*Cervus unicolor*) damage or kill trees on which they rub their antlers.

Future Needs

Management and control of nonindigenous species of plants and animals on NWRs in Florida is generally paid for out of a refuge's base operations and maintenance funding. But refuge budgets are simply inadequate for the control of invasive nonindigenous species. Thus control programs for these species on national wildlife refuges can be characterized as "when time allows" programs. Refuge managers facing the choice between spending scarce dollars on controlling invasive plants and animals (which may not yet have a substantial impact on the refuge's resources) and completing other refuge projects and maintenance generally opt to delay work on nonindigenous species. When these species become a problem, however, the cost of removal is much greater than it would have been had there been early control (Chapters 14 and 23 in this volume).

On some NWRs, the impact of nonindigenous species is so great that special funding is provided. A.R.M. Loxahatchee NWR has special funding for control of melaleuca, for example, and Crocodile Lakes NWR for control of Australian pine, Brazilian pepper, and melaleuca. Funding to monitor and control potential problem plants and animals should be available before a crisis develops. Research on the ecosystems of national wildlife refuges is needed so that potential impacts of nonindigenous species can be determined before they occur. Control efforts can then be initiated more efficiently and economically.

Finally, interpretation and education are needed. The support of the public is necessary if efforts to prevent the introduction of nonindigenous species, or to control those already present, are to succeed. Only when the value of protecting and maintaining native ecosystems is explained will the public understand the damage caused by nonindigenous species and support the efforts of the FWS and other land management agencies to control them.

reading 5

"Plant Management in Everglades National Park"
Robert F. Doreen and David T. Jones

Granted with permission from *Strangers in Paradise*, © 1997 by Daniel Simberloff, editor. Published by Island Press, Washington, DC, and Covelo, CA.

Everglades National Park, a World Heritage Site and Biosphere Reserve, encompasses more than 600,000 ha and is the only subtropical wilderness in the continental United States. The park is located in southern Florida at the southern terminus of the vast wetland complex known as the Everglades (Figure 17.1). Established in 1947 to preserve the unique biological resources of the area, the park contains a variety of habitats within its boundaries: shallow-water marine habitat (240,000 ha), saltwater wetland forests and marshes (192,000 ha), freshwater marshes and prairies (162,000 ha), and upland pine and tropical hardwood forests (6000 ha). The distribution of vegetation is largely controlled by the hydrological regime, surface geology, and overlying soil type. Natural disturbances (fires, freezes, hurricanes, sea level changes) and human activity (drainage, development, introduction of nonindigenous plants) also have powerful effects on vegetation patterns. Remarkably, this complex ecological wilderness lies within a hundred miles of 3.5 million people.

Of the 850 plant species reported from the park, 221 (26 percent) are nonindigenous in origin (Whiteaker and Doren 1989). This number includes more than half of all the nonindigenous species known to be established in southern Florida: Indeed, Loope (1992) considers the park among the four U.S. national parks worst affected by nonindigenous species. Many of the nonindigenous species were planted by settlers before the park's establishment. Natural disturbances that are part of the southern Florida environment have allowed weedy species to become established, an effect amplified by human activities. The most successful nonindigenous species are so well adapted to an altered habitat that they outcompete native species (Ewel et al. 1982).

Management of nonindigenous plants in the park has developed in response to laws, general directives, and policies. Under the National Park Service (NPS) Organic Act of 1916, the NPS is charged with management of the parks to "conserve the scenery and the natural and historic objects and the wildlife therein and to provide for the enjoyment of the same in such manner and by such means as will leave them unimpaired for the enjoyment of future generations." NPS policy states that nonindigenous species will be managed "up to and including eradication . . . whenever such species threaten park resources . . . [and] high priority will be given to [nonindigenous species] that have a substantial impact on park resources . . . " (National Park Service 1988, 1991).

Management of nonindigenous plant species is given high priority in the park's Resource Management Plan (Everglades National Park 1982) and is articulated in the *Everglades Exotic Plant Control Handbook* (Doren and Rochefort 1983), which establishes guidelines, priorities, and methods for controlling nonindigenous plants in each of the park's four districts: Pine Island, Flamingo, Florida Bay, and Northwest. Work is shifted among these districts according to other work assignments and funding.

Historical Perspective

The first efforts at plant eradication in the park began before the park's establishment and were directed at a native species—wild cotton (*Gossypium hirsutum*)—in an effort to control the spread of the cotton boll weevil (*Anthonomus grandis*; La Rosa et al. 1992). In the years after the park was established, little attention was given to nonindigenous species or their effects on the environment. Many plants were associated with abandoned homesites or dwellings of both native Indians and Europeans. The earliest control work focused on aquatic weeds in the Royal Palm Pond. After Hurricane Donna in 1960, park staff noticed an increase in populations of Australian pine (*Casuarina* spp.), and the first control was attempted in 1963. Australian pine trees continued to spread, however, after Hurricane Betsy in 1965 (LaRosa et al. 1992). In 1969, the first nonindigenous plant management plan for the park was directed at Australian pine (Klukas 1969).

The first formal, comprehensive, nonindigenous plant management plan for the park, written in 1973 and updated in 1977, included over 100 additional pest plants not previously reported. The highest-priority actions were to control Australian pine and to eradicate melaleuca (*Melaleuca quinquenervia*). The latest update, in 1988 (Whiteaker and Doren 1989), listed 221 species of nonindigenous plants in the park. The update provides information on current distribution, potential to spread and invade native plant communities, and management approaches.

Australian pine, melaleuca, and Brazilian pepper (*Schinus terebinthifolius*), three of the park's most disruptive and widespread nonindigenous plant species, have been the focus of most of the management effort.

Two additional species, however, lather leaf (*Colubrina asiatica*) and shoebutton ardisia (*Ardisia elliptica*), are becoming increasingly widespread in the park and are now the target of some control measures.

Australian Pine

Casuarina equisetifolia and *C. glauca*, two of eight species of Australian pine introduced into Florida, are found in the park (Figure 17.2). The former species invades many vegetation types but is restricted in its distribution by long hydroperiods. It survives best on well-drained sites and tolerates brackish soils and sea spray, colonizing open sand and shell beaches and coastal prairies. Scattered individuals can be found in higher-elevation mangrove stands and in the interior of keys in Florida Bay. The latter species has invaded saw-grass (*Cladium jamaicense*) marshes and southern Florida slash-pine (*Pinus elliottii* var. *densa*) forests but is most common along roadsides and berms and in burned tree islands.

As a result of its prolific root suckering and widespread root system, it often forms compact, dense stands.

Stands of Casuarina are detrimental to wildlife and affect certain threatened and endangered species in the park. Loggerhead turtles (*Caretta caretta*) and green sea turtles (*Chelonia mydas*) require gently sloping beaches with soft sand for successful nesting. Because Australian pines impede nesting, beaches dominated by Casuarina are rarely used by these turtles. Australian pines have displaced native beach-stabilizing vegetation, allowing wave action to erode sand adjacent to roots, resulting in a steeper embankment.

As early as 1956, individual trees were reported at several locations in the park: in the southeastern corner near Card Sound, along the Ingraham Highway to the mangrove zone, on several keys in Florida Bay, and along the Gulf coast beaches (LaRosa et al. 1992). No action was taken until 1963, when post-hurricane surveys revealed rapidly increasing populations on Cape Sable and Highland beaches. From 1963 to 1970, mechanical measures (tree cutting and uprooting) and

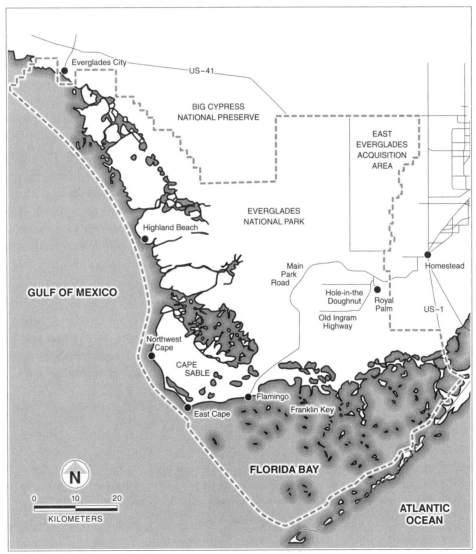

FIGURE 17.1. Map of Everglades National Park and environs.

chemical control were carried out. Prescribed burns killed scattered tress within prairies but were ineffective in dense stands, because of a lack of fuel in the herbaceous understory.

By 1970, Australian pine covered several thousand hectares, prompting a control program that extended from 1971 through 1978. During this period, 86,300 trees were treated in the park interior (including Highland Beach and the northern shore of Florida Bay). Because of budget constraints, only 12,000 trees were treated from 1979 through 1985. A survey in 1983 estimated that dense stands covered 7200 ha in the southeastern corner of the park.

Since 1989, Australian pine has been treated in the East Everglades Acquisition Area (East Everglades) of the park. This 42,400-ha area, located in the northeastern corner of the park (Figure 17.1), is characterized by seasonally inundated saw-grass and muhly (*Muhlenbergia* sp.) prairies with scattered tree islands. Australian pine is widespread throughout the exposed pinnacle rock ("rocky glades") portions of the area and remains a threat to the slightly elevated sites that once supported bayhead or tropical hammock vegetation that was removed by severe wildfires (Schomer and Drew 1982).

Melaleuca

The potential for displacement of native vegetation by melaleuca may be greater than for any other introduced plant species in the park. Its numerous vegetative and reproductive adaptations allow it to compete in native habitats, especially transition zones (ecotones) between different vegetation types and disturbed areas. Cypress-prairie and pine-cypress ecotones seem the most vulnerable, but melaleuca has also invaded saw-grass marshes, muhly prairies, slash-pine forests, tropical hardwood

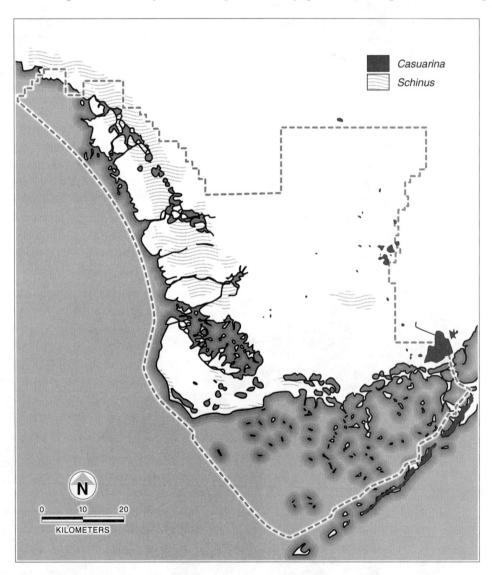

FIGURE 17.2. Map of distribution of Australian pine (Casuarina) and Brazilian pepper (Schinus) in Everglades National Park (based on 1987 data.)

hammocks, and the buttonwood-mangrove association. Desiccation of its seed capsules, brought on by freezing, drought, fire, herbicide treatment, or breakage, can result in seed release. It is the timing of release—under wet or under dry conditions—that determines the success of seed germination and establishment (Bodle et al. 1994).

Large-scale alterations to the hydrological regimes of southern Florida and concomitant changes to the fire regime have also increased the distribution of this plant. In fact, melaleuca is, on the average, able to increase numbers of stems each year by a factor of ten, especially after wildfires (Laroche and Ferriter 1992). In addition to its effects on native vegetation of the park, melaleuca provides poor habitat for native fauna: the expansion and growth of this species essentially eliminates nonavian wildlife habitats, and it appears to be undesirable forage for deer (Schortemeyer et al. 1981).

Melaleuca was first reported in the park in 1967 and occurred as isolated trees near park headquarters and along the eastern and northern boundaries (LaRosa et al. 1992). Treatment before the mid-1970s consisted of felling the trees and applying herbicide to the remnant stump. Between 1979 and 1984, some 8300 individuals were treated, mostly seedlings that were pulled up by hand. Larger individuals were girdled or frilled and herbicides were applied in the cuts (the "hack-and-squirt" method).

The primary concentration of melaleuca in the park is in the East Everglades: approximately 28,000 ha of the area is infested. Melaeuca occurs in the prairies of this area as single plants, even-aged stands, and monotypic forests of several age classes. Aerial reconnaissance in 1993 revealed approximately 200 monotypic stands, ranging from less than 1 ha to more than 20 ha, in the northeastern corner of the area alone. Because its expansion in this area threatens native plant communities, intensive efforts to control its spread have been carried out there since the mid-1980s.

Brazilian Pepper

Brazilian pepper can rapidly colonize disturbed areas and persist through later successional stages (Ewel et al. 1982). As a pioneer species in areas undergoing secondary succession, it grows rapidly, produces many seeds that are consumed and dispersed by animals (especially birds, both native, particularly American robins, *Turdus migratorius*, and nonindigenous), and sprouts readily. Seedlings can establish themselves and survive in open areas, as well as under dense canopies. The fact that few other plant species fruit during the winter, when Brazilian pepper seeds are dispersed, may explain the success of this plant in southern Florida (Ewel 1986).

Brazilian pepper has invaded many habitats in the park, including saw-grass marshes, muhly prairies, tropical hardwood hammocks, coastal hardwood hammocks, slash-pine forests, and the salt-marsh–mangrove ecotone (Figure 17.2). Outliers and small to medium-sized populations on the edges of the sandy soils between East Cape and Northwest Cape threaten nesting habitat for the gopher tortoise (*Gopherus polyphemus*), a threatened species in Florida. It rarely grows on sites flooded longer than three to six months and hence is rarely found in marshes and wet prairies. It thrives on disturbed soils created by natural disruptions—such as hurricanes—and is especially invasive in areas affected by human activity, particularly abandoned farmlands, roadsides, and canal banks. In southern Florida, farming practices, especially rock plowing, alter the substrate and allow Brazilian pepper to outcompete native species on these sites (Meador 1977) and alter successional vegetation patterns (Loope and Dunevitz 1981b). Brazilian pepper does not compete with melaleuca for sites: they differ in their relationship to fire and water regimes and in their degree of dependence on human modification of natural conditions (Ewel 1986).

Although Brazilian pepper was noted to be invading mangrove areas around Everglades City in 1961, control was not attempted there (LaRosa et al. 1992). Most efforts at controlling it have been concentrated in the Hole-in-the-Doughnut—an area of abandoned farmland (4000 ha) in the middle of the park—and adjacent pinelands. Rock plowing in half of the area altered the substrate, resulting in conditions favoring Brazilian pepper establishment, that is, higher nutrient levels and aerobic conditions (Doren et al. 1990). Attempts at restoring the native wetland communities in the Hole-in-the-Doughnut were initiated in 1972 and included mowing, disking, rolling, chopping, and bulldozing (Koepp 1979).

Experiments with use of herbicide and fire on Brazilian pepper have also been conducted in the Hole-in-the-Doughnut. Herbicide applied directly to basal bark proved the most effective means of killing Brazilian pepper had minimal impact on the surrounding vegetation (Ewel et al. 1982; Doren et al. 1991b). Repeated fires may slow down the invasion rate, but they do not exclude Brazilian pepper establishment. Although invasion progresses with or without fire (Doren et al. 1991a), prescribed burning at three-year to seven-year intervals has severely restricted its establishment within the park (Loope and Dunevitz 1981b).

Current Management Practices

The strategy employed by the park in managing its most serious nonindigenous plant pests, primarily Australian pine and melaleuca, is to focus on individual plants and outliers rather than dense stands (Moody and Mack 1988). Concentrated efforts to control the spread of these two species in the East Everglades, an area of significant invasion, began in the mid-1980s with the establishment of the East Everglades Exotic Plant Control Project by the multi-agency East Everglades Resource Planning and Management Committee (DeVries 1995). With the assistance of the Florida Exotic Pest Plant Council (EPPC), the park developed the control project and acquired funding from several cooperating state agencies.

The treatment season for melaleuca and Australian pine in the East Everglades extends from December to May. For melaleuca, the most isolated, distant seed trees are treated first, and treatment progresses toward the center of the infestation. Control is achieved by annual treatment of regrowth and recruits at previously treated sites. Surveillance is conducted by helicopter for outliers, and low to medium densities of melaleuca are treated as found. Locations of high-density, monotypic stands are recorded via the Global Positioning System for subsequent foliar spraying. Field data (location, number of resprouts, number of stems treated, environmental conditions, seed tree heights) are entered into a geographic information system to create project maps. The methods used to control melaleuca include mechanical means (pulling seedlings by hand), chemical or herbicidal treatment (cut stump, basal bark, and foliar), and physical measures (prescribed burn or prolonged inundation after cutting). Larger trees are felled with chain saws and the exposed cambium treated with a herbicide (Arsenal 50%). Monotypic stands of seedlings are either cleared with brush cutters or treated with a herbicide (Arsenal 1.5%) by foliar application. Australian pine is controlled by pulling of seedlings and treatment of larger stems with a herbicide (Garlon 4) by the basal bark method.

Total treatment cost for the control of melaleuca and Australian pine in the East Everglades for the ten years from 1986 through 1995 was about $1.2 million. Table 17.1 provides a detailed breakdown of annual costs. During this period, approximately 4.5 million melaleuca and 100,000 Australian pine stems were treated by hand pulling, by cut stump and basal-bark method, and by foliar application (DeVries 1995).

The removal of Australian pine from the keys in both Florida Bay and Dry Tortugas National Park is carried out when time and manpower permit. These areas include critical habitat for endangered animal species, such as nesting beaches for sea turtles (at Dry Tortugas keys) and American crocodiles (*Crocodylus acutus*) (at Florida Bay keys). These areas have high priority for control of Australian pine, which can eliminate nesting habitat.

Control of Brazilian pepper in the Hole-in-the-Doughnut is part of one of the park's largest wetlands restoration efforts. A study report by Doren et al. (1990) indicates that the complete removal of disturbed substrate from abandoned, rock-plowed lands, and the subsequent increase in hydroperiod, has altered secondary successional patterns in favor of native wetland vegetation and the concomitant exclusion of Brazilian pepper. This technique will be used in the Hole-in-the-Doughnut on all former wetlands that were invaded by

TABLE 17.1. Treatment Costs And Labor Hours for *Melaleuca* and *Casuarina* Treatment in the East Everglades: 1986–1995

Year	Flight cost ($)	Herbicide cost ($)	Cost per year ($)	Labor hours, NPS[a]	Labor hours, non-NPS[a]
1986	2261	3	2624	32	32
1987	12,965	187	13,152	224	336
1988	25,495	556	26,051	708	48
1989	55,198	442	55,640	926	442
1990	70,451	788	71,239	1419	112
1991	111,053	7744	118,797	2755	72
1992	146,295	7400	153,695	4942	64
1993	93,553	11,540	105,093	2252	1160
1994	110,000	11,358	121,358	4509	0
1995	145,000	9000	154,000	2789	0
Total	772,271	49,018	821,649	20,556	2266

[a] "NPS" indicates National Park Service personnel

Brazilian pepper after abandonment of the farmlands, an area of about 2000 ha.

Many areas of the west coast mangrove forests in the park have been damaged by hurricanes (Donna in 1960, Betsy in 1965) and freezes (in 1977 and 1981) and have subsequently been invaded by Brazilian pepper (Armentano et al. 1995). An unpublished study by the park in 1987 reveals that some 42,000 ha of west-coast mangrove and interior coastal prairie habitat within the park contained this plant (Figure 17.2). Removal of Brazilian pepper from mangrove ecosystems is difficult: mechanical removal disturbs mangrove community substrate, which favors Brazilian pepper establishment; the foliar application of herbicides in mangrove forests could kill nontarget species. Removal by hand and direct injection of herbicides have eliminated this species, but only on a small scale. Further research is urgently needed to determine the best means of removing large infestations of Brazilian pepper within mangrove communities.

Other nonindigenous plant species, such as lather leaf and shoebutton ardisia, are problems in the park and becoming increasingly widespread. The control of these plants is carried out sporadically, if at all, because of labor, time, and financial constraints. Lather leaf (*Colubrina asiatica*), native to tropical Asia, has rapidly spread through coastal vegetation around parts of Florida Bay and on certain keys (Rankin Key, for example) and is becoming a highly visible feature of the Everglades landscape. By virtue of its climbing habit, lather leaf forms dense carpets of growth covering buttonwood, mangroves, and associated costal vegetation. The fan palm (*Thrinax*) and mahogany hammocks along the Florida Bay coast in the park are particularly vulnerable to the destructive effects of lather leaf. Its potential to spread onto the keys throughout Florida Bay and other coastal areas is high because its buoyant seeds can be dispersed by water. In the few areas of the park where lather leaf is managed, such as Flamingo area and Rankin Key, herbicides, (Garlon 4) are used to treat it.

Shoebutton ardisia (Ardisia elliptica) originated in Southeast Asia and has invaded disturbed hardwood hammocks and abandoned agricultural lands in the Royal Palm area of the park. It also occurs in the Flamingo area, where it was first reported in 1995 (Seavey and Seavey, unpublished data). It comprises a major part of the understory in the Brazilian pepper-dominated sections of the Hole-in-the-Doughnut. Seavey and Seavey (1994) preduct that habitats with high humidity and an intact overstory, such as tropical hardwood hammocks, adjacent to a seed source are especially vulnerable to invasion by this species; pine forests, inland and coastal prairies, and mangrove forests appear to be the least susceptible. Treatment, which has been concentrated in Paradise Key at Royal Palm, is by mechanical means (hand pulling) and chemical methods (Garlon 4 herbicide).

Prospects

The control of melaleuca in the East Everglades is currently the only funded nonindigenous-species management priority in the park. It has been funded since 1987 through compensatory off-site wetland mitigation and contributions from several state agencies. Specific funding for other projects is still lacking. The eradication of Australian pine, for example, will require significant resources to eliminate the large, dense stands along the eastern boundary and southeastern corner of the park. Removal and control of Brazilian pepper within the Hole-in-the-Doughnut will be funded through a Dade County mitigation bank. Brazilian pepper control in the salt-marsh–mangrove ecotone along the west coast needs to be addressed.

Biological control methods are an important component of an integrated pest-plant management approach. Such methods are expensive and will take a long time, but they could result in eventual elimination or significant reduction of continued manual and chemical control in the region and the park.

All management programs should be supplemented with ongoing monitoring for evaluation. An extensive and permanent series of plots is needed throughout the park in order to monitor the status of nonindigenous species by plant community. These plots would serve as the basis for planning control actions and providing critical information on long-term trends and site vulnerability to invasion.

Regional coordination between affected agency land managers has been excellent, and the park continues to cooperate in all such interagency efforts to manage nonindigenous plants. The EPPC has provided one of the most effective forums for the exchange of ideas and conflict resolution concerning nonindigenous species in the state. Regional control plans continue to be developed by EPPC for high-priority species such as melaleuca (Laroche 1994) and Brazilian pepper. Moreover, some funding and management work have been channeled to the park by the other agencies through the promotion of coordinated efforts by the EPPC.

"Management on State Lands"
Mark W. Glisson

Granted with permission from *Strangers in Paradise*, © 1997 by Daniel Simberloff, editor. Published by Island Press, Washington, DC, and Covelo, CA.

Florida has one of the nation's most intensive land-acquisition programs. Between 1974 and 1994, through various legislative initiatives, the state acquired over 2.1 million hectares of land at a cost of over $1 billion (Office of Environmental Service 1995). Coupled with nonsovereign state landholdings, these purchases have brought significant portions of Florida's undeveloped lands into state ownership and management.

Many lands purchased for conservation or recreation purposes are acquired with established populations of nonindigenous plants and animals, and all are subject to invasion from nearby sources. As a result, state land managers are on the front line of defense against nonindigenous species in Florida and have become acutely aware of the risks to the state's native plants and animals. Moreover, the thousands of miles of rights-of-way in Florida's highways system must be monitored and maintained to reduce the threat of nonindigenous species introductions.

State Agencies

The majority of state lands under active management in Florida are administered by the Florida Park Service, the Game and Fresh Water Fish Commis-sion, and the Division of Forestry. Maintenance and protection of highway rights-of-way are administered by the Florida Department of Transportation.

Florida Park Service

The Florida Department of Environmental Protection's Division of Recreation and Parks manages 143 properties totaling approximately 868,260 upland hectares and 212,506 submerged hectares (Office of Park Planning 1995). Fewer than 90 of the parks, recreation areas, preserves, reserves, special feature sites, museums, ornamental gardens, and trails that comprise the Florida Park Service (FPS) are staffed full-time. All are available for public access of some type.

The mission of the FPS is "to provide resource-based recreation while preserving, interpreting and restoring natural and cultural resources" (Florida Park Service 1994a). FPS policies call for identification and removal of nonindigenous species. Although removals ar tracked annually by park and by district, there is currently no mechanism for tracking the costs associated with combating invasive nonindigenous species. Removal of nonindigenous species is frequently associated with the FPS commitment to restoring altered natural systems and hydrological regimes, but removal efforts have occasionally been slowed or stopped where there has been significant conflict with recreational needs or public sentiment. At John U. Lloyd Beach State Recreation Area (Broward County), for example, the removal of Australian pines (*Casuarina equisetifolia*) has been slowed significantly by complaints from beachgoing visitors who object to the resulting loss of shade.

Game and Fresh Water Fish Commission

The Florida Game and Fresh Water Fish Commission (FGFC) is to some degree responsible for management and public use on approximately 14 million hectares in the state (Boyter 1994). The Types I and II wildlife management areas make up a system that protects resources and provides recreational and educational opportunities for both consumptive and nonconsumptive users. Type I wildlife management areas are generally those on which the FGFC provides both resource management and public use and include most lands acquired by the FGFC through state land purchase programs. Type II areas are owned and managed by forestry companies, water management districts, and the U.S. Air Force. The FGFC maintains a minimal management presence on these lands but does provide some law enforcement (Boyter 1994). The FGFC's mission is "to manage freshwater aquatic life and wild animal life and their habitats to perpetuate a diversity of species with densities and distributions that provide sustained ecological, recreational, scientific, educational, aesthetic and economic benefits" (Boyter 1994).

The FGFC has an especially complex and challenging management relationship with nonindigenous species. Because it regulates nonindigenous animals, including the occasional legal introduction, it cannot simply designate all nonindigenous species for removal but must instead adhere to certain guidelines:

- Approval of the executive director must be obtained before initiating a "hands-on" study outside laboratory conditions of a species not native

to Florida that is not already in public or private lands or waters.
- Species not native to Florida will be evaluated to determine negative and positive impacts on native fish, wildlife, their habitats, and people prior to introduction into the wild.
- Only those species not native to Florida that are clearly shown to have positive benefits to humans and no significant negative impacts to native fish, wildlife, and their habitats will be considered for introduction.
- Where practical and necessary, established populations and isolated occurrences of species not native to Florida will be monitored, controlled, used, or eliminated.
- Importation of species not native to Florida will be as provided by sections 372.26 and 372.265 of the Florida Statutes (Boyter 1994).

Division of Forestry

The Florida Department of Agriculture and Consumer Services' Division of Forestry (FDOF) has some management responsibility on more than 2.3 million hectares of public land. It is the lead managing agency for 32 state forests totaling 1.3 million hectares and manages more than 454,664 additional hectares through special agreements with other state agencies, several counties, and water management districts (Hardin 1994).

The FDOF's mission is "to protect and manage Florida's forest resources through a stewardship ethic to assure these resources will be available for future generations" (Hardin 1994).

FDOF policies call for identifying and removing nonindigenous species, but management and control are emerging issues as species such as cogon grass (*Imperata cylyndrica*) become an exponentially increasing problem. Although FDOF districts are aware that invasive nonindigenous species exist, no districts or state forests have conducted systematic assessments. Most occurrences have been observed during normal management activities (Hardin 1994).

Department of Transportation

The Florida Department of Transportation (FDOT) is responsible for safety, maintenance, and vegetation protection on florida's rights-of-way. The FDOT's goal is a high-quality transportation system. The FDOT's Environmental Policy for State Transportation Facilities states: "the Department will cooperate with the State's efforts to avoid fragmentation of habitat and wildlife corridors by including these considerations in all phases of its operations" (Caster 1994). The FDOT's nonindigenous removal program is driven by three considerations: interference with the safe operation of motor vehicles on state roads; concern for the increased costs associated with nonindigenous pest plant control and removal; and protection and replacement of native trees and other vegetation (Caster 1994).

Impact on Native Species

All three land management agencies and the FDOT report extensive nonindigenous plant species infestations on the lands they manage. Each reports higher concentrations in southern Florida and different primary invasives in different geographic regions of the state. Subtropical coastal areas of Florida have been singled out as particularly threatened.

Few summary data are available on nonindigenous plant removal on state lands. FPS records for 1994 indicate that 58 state parks treated 1.5 million individual nonindigenous plants of various species, as well as 4 metric tons of mostly elodea (*Egeria densa*) and hydrilla (*Hydrilla verticillata*), and cleared 152.46 hectares of such species as kudzu (*Pueraria montana*), cogon grass, and air potato (*Dioscorea bulbifera*). Statewide, 94 nonindigenous plant species were individually reported as being removed from parks, together with an additional 25 species grouped as "other" (Florida Park Service 1994).

In 1994, the FDOF estimated that over 5436 hectares in six state forests are infested with nonindigenous plant species and that infestations are acknowledged but unestimated in three additional forests (Hardin 1994). Moreover, a myriad of grasses can invade rights-of-way and challenge the FDOT's commitment to stopping the spread of nonindigenous plants on transportation corridors. Torpedo grass (*Panicum repens*), Johnson grass (*Sorghum halepense*), smut grass (*Sporobolus jacquemontii*), crowfoot grass (*Dactyloctenium aegyptium*), and vassey grass (*Paspalum urvillei*) are examples (Caster 1994).

Australian pine (*Casuarina equisetifolia*) is of particular concern in southern Florida and along the southern coasts of the state: in 1990, Australian pines were found in 30 state parks (Stevenson 1990); in 1994, some 84,628 Australian pines were removed from state parks (Florida Park Service (1994 b).

Recent surveys of upland coastal vegetation in southern Florida (Johnson and Muller 1993) indicate that Australian pine has colonized large sections of the remaining undeveloped portions of barrier islands along both the Atlantic and Gulf coasts. On the Gulf coast from Pasco to Collier counties, Johnson and Muller (1993) identified 60.2 miles of coastline as undeveloped (out of a total coastline length of 178 miles)—of which 20.1 miles, or 33 percent, were heavily invaded by Australian pine. Similarly, on the Atlantic coast from Indian River County to northern Dade County, 46 percent of the undeveloped barrier island coast is heavily invaded (Johnson 1994). Australian pines have also been found to interfere with sea-turtle nesting activities in the Florida Keys, where beaches are typically narrow and pine growth can occur close to the water line (Duquesnel 1994).

Another serious threat to southern Florida and the southern coasts is Brazilian pepper (Schinus terebinthifolius). It differs from Australian pine in that it does not require bare soil to invade (Duquesnel 1994). Brazilian pepper can become established in native vegetation, or it can form monospecific stands on disturbed sites (Johnson 1994). It is consistently present in the coastal forest understory on outer barrier islands on both the Atlantic and Gulf coasts, from Brevard and Pasco counties southward (Johnson 1994a). The plant was present in 42 state parks in 1990 (Stevenson 1990). In 1994, some 161,431 individual Brazilian pepper plants were removed from state parks (Florida Park Service 1994b). It is a threat on wildlife management areas (Boyter 1994) and is widespread, if not dense, on DuPuis Reserve (Martin and Palm Beach counties) and Picayune Strand (Collier County) state forests (Hardin 1994). It causes maintenance and safety problems for the FDOT, as it grows through right-of-way fences and eventually lifts them clear off the ground, allowing wildlife and humans onto high-speed, limited-access highways (Caster 1994).

Cogon grass is widespread and occurs in extremely thick stands along rights-of-way in central and northern Florida. The FDOT reports 900 large monospecific stands accounting for more than 741 hectares of right-of-way (Caster 1994). In 1994 FPS staff treated approximately 59 hectares of monospecific stands and many unreported smaller stands (Florida Park Service 1994). The FDOF initiated a five-year work plan in 1993 to eliminate cogon grass in state forests (Hardin 1994). It is presently considered a threat to nine state forests. The effects of cogon grass on native species have been observed and reported by the FDOF and FGFC (Hardin 1994; Weimer 1994) and are of concern to the FGFC (Boyter 1994). Anecdotal information suggests that gopher tortoises (Gopherus polyphemus) abandon areas infested with cogon grass. Loss of the gopher tortoise results in the disappearance of the numerous species that depend on gopher-tortoise burrows, including gopher frogs (Rana aesopus), eastern indigo snakes (Drymarchon corais), and scarab beetles (Scarabaei-dae) (Myers 1990). Where cogon grass inhibits longleaf-pine regeneration, endangered red-cockaded woodpecker (Picoides borealis) nesting and foraging may be affected. Moreover, the woodpecker's nest trees are vulnerable to the intense fires that can occur in cogon grass ground cover (Hardin 1994). The FDOF reports that most of the Withlacoochee State Forest (Citrus, Hernando, Sumter, and Pasco counties) may be infested with cogon grass within a generation (Hardin 1994).

Melaleuca (Melaleuca quinquenervia) poses a serious threat to Florida's highway rights-of-way. Along the Homestead Extension of Florida's Turnpike (Dade County), for example, melaleuca is the most serious nonindigenous invader, continually interfering with the operation and maintenance of drainage conveyances. Because of limited staff, equipment, and funding, the FDOT removes melaleuca trees only from drainage conveyances, public areas such as rest stops, structures, and mitigation areas (Caster 1994). Melaleuca is one of the most invasive plants in the state parks of southern Florida (Florida Park Service 1994b) and has infested approximately 247 hectares of Golden Gate State Forest, where the highly fragmented nature of state holdings and lack of an ownership map compound the problem (Hardin 1994).

Skunk vine (Paederia foetida) and air potato threaten hammocks on public lands (Hardin 1994), and ardisia (Ardisia crenulata) is becoming the dominant understory plant in hammocks at both the Lake Overstreet addition to Maclay State Gardens (Leon County) (E. Johnson, pers. comm.) and Withlacoochee State Forest, where, along with skunk vine, it threatens the federally listed Cooley's water willow (Justica cooleyi) (Hardin 1994).

Other species reported as being particularly invasive include Chinese tallow tree (Sapium sebiferum), described as one of the most invasive of northern Florida's nonindigenous plants (Weimer 1994); tropical soda apple (Solanum viarum), spreading through pastures and open areas of central Florida (K. Alvarez, pers. comm.); and beach naupaka (Scaevola taccada) on southern Florida coasts (Duquesnel 1994). Table 18.1 lists endangered and potentially endangered plants from southern Florida state parks that are threatened by invasive nonindigenous species.

State land managers have few summary removal data and little understanding of the impacts of nonindigenous animals on the lands they manage. Removal records either are not maintained at all or exclude important aquatic invaders such as blue tilapia (Oreochromis aureus). FPS records indicate that some 1778 nonindigenous animals were removed from 25 state parks in 1994, but the costs are unrecorded (Florida Park Service 1994).

Feral pigs are the major nonindigenous animal threat to state lands. The FPS reports 1205 individuals removed from state parks in 1994, including 886 from Myakka River State Park (Sarasota County) alone (Florida Park Service 1994b). They are major pests in Seminole, Withlacoochee, and Goethe state forests (Levy County) and are present in others (Hardin 1994). On wildlife management areas, populations are controlled to some degree through liberal hunting regulations (Boyter 1994).

Feral pigs impose an obvious and extensive toll on native ground cover through their rooting habits (Hardin 1994). In the Panhandle, they may threaten Applalachian relict species such as columbine (Aquilegia candensis var. australis), bellflower (Campanula americana), and mayapple (Podophyllum peltatum) (Ludlow 1994). They pose a threat to humans through the poten-

TABLE 18.1. Endangered and Potentially Endangered Plants Threatened by Nondigenous Plant Species in Southern Florida State Parks.

Scientific name	Common name
Acacia choriophylla	tamarindillo
Cordia sesbestena	geiger tree
Jacquemontia reclinata	beach clustervine
Jacquinia keyensis	joeweed
Argusia gnaphalodes	sea lavender
Okenia hypogaea	burrowing four-o'clock
Pseudophoenix sargentii	buccaneer palm
Remirea maritima	beach star
Scaevola plumieri	inkberry
Suriana maritima	bay cedar
Encyclia tampensis	butterfly orchid

Source: Duquesnel (1994).

tial transmission of brucellosis, trichinosis, and pseudorabies, which the FGFC has also determined to have been transmitted to endangered Florida panthers (*Felis concolor coryi*), (Boyter 1994).

Armadillos (*Dasypus novemcinctus*) were the second most removed nonindigenous animals from state parks in 1994; 409 were removed statewide (Florida Park Service 1994b), and they are well established on many lands in public ownership (Boyter 1994; Hardin 1994). The rooting habits of armadillos affect ground-cover species in much the same ways as do those of pigs. Domestic and feral dogs (*Canis familiaris*) and cats (*Felis catus*) are present on most public lands, particularly near campgrounds and other high-use areas (Florida Park Service 1994b; Hardin 1994). Cats are voracious predators; a single cat may kill hundreds of birds in one year (D. Bryan, pers. comm.). They may also take a toll of threatened rodents such as the Anastasia beach mouse (*Peromyscus polionotus phasma*) (Bard 1994) and the Key Largo wood rat (*Neotoma floridana smalli*) (Duquesnel 1994). Dogs disrupt resting and feeding patterns of various species and are known predators of gopher tortoises and wild turkeys (*Meleagris gallopavo*).

Conflicts and Challenges

For public land managers, the removal and control of nonindigenous species is only one element of a diverse set of resource management policies. None of the agencies is able to focus primarily on this objective, and each attempts to control invasive species as part of its daily management activities. Some major removal projects have been carried out through mitigation and grants and where natural forces have destroyed nonindigenous species and offered restoration opportunities. At Cape Florida State Recreation Area (Dade County), for example, an extensive stand of Australian pine and other upland nonindigenous species was completely destroyed by Hurricane Andrew in August 1992. Earlier attempts by the FPS to remove the Australian pines had been largely thwarted by local opposition to removal of any shade tree. But given the opportunity afforded by Andrew, a complete restoration of natives is now under way.

In another example of conflict between nonindigenous species control and public policy or sentiment, the FGFC deals with feral pigs by imposing liberal hunting regulations on them in wildlife management areas. The agency believes it has effected a fair measure of control of pig populations through regulated hunting (Boyter 1994); other public land managers, however, advocate pig removal rather than management (Florida Park Service 1994b; Hardin 1994). At Silver River State Park, local residents and the Florida legislature combined to prevent the FPS and the FGFC from removing nonindigenous rhesus monkeys, even though they are infected with a disease potentially lethal to humans. On all managed lands, the pervasive cats and dogs that threaten native wildlife can frequently belong to neighbors. In areas of high visitation, such as campgrounds and picnic areas, removal of cats and dogs can be an extremely touchy proposition.

Most public land managers currently have no formal mechanism for tracking nonindigenous plant removal costs. According to a detailed survey conducted by the FPS in 1989, state park costs are between $250,000 and $350,000 annually. In fiscal 1993, the FGFC spent $200,000 on vegetation control, a large percentage of which was spent controlling nonindigenous species (Boyter 1994). The FDOF reported specific removal projects totaling $11,905 in 1993 but estimated that $400,000 would be needed over the next several years to control cogon grass in Withlacoochee State Forest alone (Hardin 1994).

Including mowing, manual elimination, and chemical removal, the FDOT estimates that over $7 million was spent on nonindigenous plant control on rights-of-way in fiscal 1993–1994. In addition, $4.5 million was allocated for fiscal year 1996 for the removal of Australian pine and melaleuca and their replacement with more desirable species (Caster 1994).

The actual costs and management impacts of removing nonindigenous plants on public lands are extremely difficult to measure. Much of the removal occurs as part of daily work duties and is not reported. In areas where nonindigenous plants have been removed, surveys and hand pulling of invading seedlings are necessary to prevent reinvasion, and these removals are rarely reported. Nonindigenous plant removal is often part of natural-systems restoration projects but is rarely listed as a separate budget item.

As Florida's lawmakers begin to shift their emphasis from land acquisition to land management over the next decade, the need for a systematic approach to the issue of nonindigenous invaders will be critical. Land managers, including the FDOT, agree that separate accounting and budgeting processes are important to future planning and management of nonindigenous plant control (Boyter 1994; Caster 1994; Hardin 1994).

What future steps can be taken by the legislature? The key steps are those that would specifically prohibit the introduction or planting of invasive nonindigenous plants. Clear policies governing the introduction of new species and prohibiting the use of state funds to plant nonindigenous species are needed, as are tough laws governing sale and use. The short-term economic impact of such legislation would be more than offset by the savings associated with reducing control costs.

Actual removal costs associated with nonindigenous animals on state lands are relatively low. State parks frequently employ contractors to remove pigs, which has helped to reduce staff time devoted to their control, but even so there is a significant drain on the workforce.

Additional research is needed on the environmental impacts associated with nonindigenous animals. Moreover, the incidence of nonindigenous microogranisms and their impact is poorly understood, although the Florida Department of Agriculture and Consumer Services' Division of Plant Industry has expressed interest in using state lands as research sites for compiling lists of insect and microorganism species.

Public education and understanding of the threats posed by nonindigenous animals, particularly pets, would go a long way toward solving some of the most sensitive problems associated with nonindigenous animals on public lands. On remote public lands, where no local animal control service exists, removal of dogs or cats can be difficult. Land managers must work to gain the cooperation of neighbors in order minimize removal needs and control domestic animals.

"Florida's Biological Crapshoot with Carp"
Mike Clary

Los Angeles Times, June 10, 1997. © 1997 by *Los Angeles Times*. Reprinted by permission.

Soon after the Overturfs pulled the truck alongside the canal to end their 24-hour drive from Arkansas, the fish were flying. Beverly hooked an 8-inch hose up to a tank, Chuck hoisted the hose onto his shoulder like a bazooka, told Beverly "go" and suddenly it was raining carp. This wasn't shooting fish in a barrel; it was more like fish barreling through a chute. The silver-sided 10-inchers splattered into the warm waters of the C-100 canal, righted themselves, formed schools, then swam off to explore their new home—a vast system of waterways that provide flood control and make dryland living possible in South Florida.

"They probably won't start eating right away," says biologist Gordon Baker, overseeing the operation for the South Florida Water Management District. "But they will."

They'd better. The 5,000 grass carp on the Overturf's truck were the first of 21,000 dumped into the canals here last month in an effort to eradicate exotic weeds. Grass carp are known as voracious consumers of aquatic vegetation, especially hydrilla and other unwanted species that are impeding the flow of water and fouling floodgates.

The idea of spending about $80,000 to introduce yet another exotic animal into the fertile South Florida ecosystem—already alive with a Noah's Ark of foreign critters, most of them unwelcome—may seem imprudent. After all, accidental introductions have resulted in thriving populations of pests, such as the African giant snail, the cane toad, monk parakeet, various lizards and the fire ant, among many others.

At least 19 nonnative fresh-water fish already swim in Florida's canals, rivers and lakes, according to the federal Office of Technology Assessment.

Indeed, the fish branded Florida's most troublesome, the blue tilapia, was introduced by the state's Game and Fresh Water Fish Commission as a sport fish and weed-eater. But it proved to be unpopular with anglers, to have an anemic appetite and has multiplied so mightily that it now competes with native species.

"It is ironic to chase one exotic with another, but that's what we're faced with," says Don C. Schmitz, a biologist with Florida's Department of Environmental Protection. "We want to make sure we don't unleash a new ecological disaster, but if we can find a good biological control agent, we have to use it."

Baker says the grass carp is a good agent, in part because the fish are sterile. As raised by J.M. Malone & Son, a Lonoke, Ark., fish farm, these white amur carp, native to the rivers of Russia and China, are genetically engineered to have an extra set of chromosomes through a process in which fertilized eggs are shocked with a jet of water.

These carp are created, says Baker, to eat weeds and die.

Since the hybrid was produced in 1983, the grass carp has been introduced into the water districts in Southern California's Imperial and Coachella valleys, the Panama Canal and the Nile River in Egypt, said Doretta Malone, the fish farm proprietor. "Ten to 12 fish per acre will clean up and control the worst aquatic vegetation," Malone says. "And a fish will eat for five to seven years, and grow to 80 pounds or more, before it gets old and lazy."

If the carp proves successful in South Florida, the Water Management District plans to restock the fish at a rate of about 10% a year, to replace those that become alligator food or die in other ways. Life expectancy is about 10 years.

Dan Thayer, director of vegetation management for the water district, admits the carp come with some risks. Once they eat all the exotic plants, they could begin to dine on eel grass and other natives favored by manatees. And they could get into the Everglades ecosystem and further rock the balance of nature there.

Malone predicts the grass carp will win friends in South Florida. "They are fast swimmers, and they will play and roll over like a porpoise," she says. "Working with this fish all these years, I can see that they have an intelligence that other species don't have. I see it in their eyes. They will respond to you. And they are also weather forecasters. They are very active when a front is coming through."

Baker is more concerned about the carp's taste for hydrilla and East Indian hygrophila. If the fish does not attack the weeds, Baker says, water managers would have to use herbicides, "and that's a last resort."

Another concern is human taste for the carp. Fishing magazines, Baker reports, have written about ways to catch the fish, and recently a Chinese-language newspaper in South Florida printed a recipe for steamed carp in a wine, soy and ginger root sauce.

Baker reminds anglers that taking the carp from the canals is illegal. And besides, "those fish are designed to eat vegetation and keep the water flowing."

"So people have to decide: They can either eat dinner or have a clean canal. Please, put them back."

"Will Beetle Bombs Recapture Everglades—Or Overrun It?"
Mike Clary

Los Angeles Times, June 10, 1997. © 1997 by *Los Angeles Times*. Reprinted by permission.

A voracious predator from a distant antipodean world has been roaming the marshlands of South Florida for about six weeks now.

Before its release, many residents were fearful. But eight years of laboratory study convinced scientists that the confetti-sized Australian weevil would eat only the pesky melaleuca trees, and eventually public apprehension subsided. So in April, about 700 of the tiny snout beetles—each marked on its underside with a dollop of paint—were turned loose in the Everglades.

Early reports are promising. "They are doing well," said entomologist Ted Center of the U.S. Department of Agriculture on Monday. "We set some in a clump of about 100 trees, and every single tip on every branch in that clump has been destroyed by bugs.

"We have also found two unmarked weevils, indicating that they are reproducing."

This past weekend, however, House Speaker Newt Gingrich (R-Ga.) took a spin through the sawgrass on an airboat and, while vowing his support for Everglades restoration, made an off-the-cuff remark comparing the introduction of the bettle to what he called the "Jurassic Park" effect. Wondering aloud about the danger of adding another exotic animal to the fecund subtropical stew of life, Gingrich noted that the mongoose imported to the Caribbean years ago for control of rats is now killing native birds.

Well, Center said, "that Jurassic Park comment hasn't helped at all. Paralleling the insect with a dinosaur really bothers me. Newt's comment will cause us to have to explain again and again and again.

"This is a very safe approach. Worldwide, more than 200 species of insects have been introduced for weed control, 14 in Florida, and there have been virtually no unintended effects. Insects don't eat everything."

Still, the notion that an exotic species set loose to solve one problem turns into an even bigger problem is not just the stuff of thriller fiction. The melaleuca tree, for example, was widely planted in Florida to dry up unwanted wetlands, just as its Australian cousin, the blue-gum eucalyptus, was planted throughout California as a potential source of lumber. Both proved to be expensive mistakes.

But scientists, the public and Gingrich—a former environmental studies professor—are more mindful of unwanted consequences now. After years of studies and careful quarantine, Center said, "I have perfect faith in this insect's safety. We would not have proceeded with releasing it unless we were certain."

Center's confidence comes from a wealth of evidence that *Oxyops vitiosa* eats melaleuca buds and leaves, and only that.

Moreover, Center said, the gray-brown bug is unlikely to skew the delicate balance of nature by becoming another creature's favorite food because it secretes "an oil which forms a slimy layer over its body.

"It probably tastes bitter. Even fire ants won't touch it."

Although the snout beetle is science's best hope against the melaleuca, which has crowded out native species and interfered with alligator nesting while taking over about 600,000 acres of South Florida wetlands, it is not the only biological weapon waiting to be unleashed. Researchers also have several other insects in quarantine, including a melaleuca-munching saw fly that could be set free within two years, Center said.

Restoring the imperiled Everglades ecosystem—compromised by exotic plants, pollution from farm runoff and canals that have altered the natural water flow—is now a bipartisan concern in Washington. The Clinton administration budget for the next fiscal year makes rescuing the Everglades the largest environmental project in U.S. history. And Gingrich backed a $200-million addition to the farm bill for buying up adjacent wetlands.

Yet when it comes to using bugs to atone for human error, "there is so much entomophobia out there," Center said. "Already I have heard rumors that this beetle is attacking citrus. Not true."

"Ecological Effects of an Insect Introduced for the Biological Control of Weeds"

S. M. Louda, D. Kendall, J. Connor, D. Simberloff

Science, August 22, 1997 277(5329):1088-1090.

Few data exist on the environmental risks of biological control. The weevil Rhinocyllus conicus Froeh., introduced to control exotic thistles, has exhibited an increase in host range as well as continuing geographic expansion. Between 1992 and 1996, the frequency of weevil damage to native thistle consistently increased, reaching 16 to 77 percent of flowerheads per plant. Weevils significantly reduced the seed production of native thistle flowerheads. The density of native tephritid flies was significantly lower at high weevil density. Such ecological effects need to be better addressed in the future evaluation and regulation of potential biological control agents.

The perception of high economic, health and environmental costs of chemical pest control has stimulated interest in biological control, specially the importation of specialized natural enemies to limit invasive coevolved pest species. When bio-control is successful, pest populations are suppressed below the economic threshold by a self-sustaining interaction between the pest prey species and its introduced antagonist. Successes in the United States include biological control of insect pests, such as cottony cushion scale and red scale on citrus in southern California, and weeds, such as Kalmath weed (*Hypericum perforatum* L.) in northwestern rangelands and alligatorweed (*Alternanthera philoxeroides*) in Florida waterways. However, not all biological control efforts work. Estimates of success for herbivorous insects introduced to control weeds in the United States vary, from 41% of projects with evidence of some control to 20% that have exerted significant control. All successful programs, and many unsuccessful ones, leave nonindigenous species in the environment.

Biological control of invasive weeds is seen as an especially attractive option for large natural areas, such as parks, reserves, national forests, and open rangelands. However, the use of biological control has generated controversy over the environmental risks associated with deliberate introductions on nonindigenous species. Many advocates of biological control argue that there is no evidence of significant adverse ecological effects by carefully screened insects on post-release use of non-target host plants leave the issues unresolved. Intensive study is required to identify the role of insects herbivores in the limitation of plant growth, abundance, and distribution, so the lack of evidence for ecological costs may simply reflect the paucity of quantitative studies after deliberate introductions.

The flowerhead weevil *Rhinocyllus conicus* Froeh, was the first of four insects reported as released in North America for the biological control of Eurasian thistle of the genus *Cardiuus L.*, including musk thistle. After extensive prerelease screenings of host preference, oviposition, growth, and fitness of this species in Italy and Canada, weevils from France and Italy were released in Ontario and Saskatchewan in 1968 and were immediately redistributed to Manitoba, Quebec, and British Columbia. Weevils from Canada were released in the United States—in Virginia (1969), Montana (1969), California (1971), and Nebraska (1972)—and then redistributed from these sites. Currently, *R. conicus* is also reported from Arizona, Colorado, Idaho, Iowa, Illinois, Kansas, Kentucky, Maryland, Minnesota, Missouri, New Jersey, North Dakota, Oregon, Pennsylvania, South Dakota, Tennessee, Texas, Utah, Washington, and Wyoming. Redistribution continues. The original releases were made even after initial feeding trails indicated that the weevils hosts range included the native North American genera *Cirsium*, *Silybum*, and *Onopordum*. Stronger oviposition preference for *Carduus*, plus more successful larval development on *Carduus*, were expected to limit use of native North American plants by *R. conicus*.

We documented the continuing expansion of the host range by this weevil; three new host associations—with *Cirsium canescens* Nutt, *C centaureae* (Rydb.) K. Schum., and *C. pulchellum* (Greene, Woot & Standi)—were found. Infestation rates are given in Table 1. Three of the six native thistle species in Rocky Mountain National Park—namely *C. centaureer*, *C. tweedyi* (Rydb.) Petrak, and *C. undulatum* (Nutt. Spreng.)—had *R. conicus* developing within their flowerheads. The two lower elevation species had 43 to 70% of their flowerheads attacked (Table 1). Extensive *C. undulatum* infestation was also found in Mesa Verde National Park (38.7%), Wind Caves National Park (77.5%), and two Sandhills prairie preserves (Table 1). We also found *R. conicus* developing within flowerheads of Platte thistle, *C. canescens*, a characteristic species of Sandhills prairie. Studies before 1993 detected no R. conicus weevils on Platte thistle.

The frequency of damage by *R. conicus* to flowerheads of native plants increased sharply for all study species at all sites for which we had observations in both 1992 and 1996 (binomial probability $P <$ 0.008,

N=7). The infestation levels observed were as high as or higher than those previously reported. This is true for infestation of native thistles by both native insects and *R. concius* as well as for infestation of exotic thistle by exotic insects.

The direct effect of *R. conicus* on seed production was severe wherever it was quantified. For example, in 1996 the average number of viable seeds produced by flowerheads of Platte thistle infested with weevils was 14.1% of that produced by similar heads with no insects or only native insects; 35.4 viable seeds per head without the weevil (SE 6.00, N=40 heads) versus 4.8 viable seeds per head with the weevil (SE 1.31, N=181 heads) (t test, t_1= 7.385, P<0.001). Likewise, in Mesa Verde National Park in 1996, viable seed produced by wavyleaf thistle (*C. undulatum*) flowerheads infested with weevils averaged 28% of that produced by similar heads with no insects or only native insects.

A reduction in viable seed of the magnitude observed will reduce regeneration from seed by these native plants. Thistle are fugitive species with large seeds that generally depend on current seed production for establishment and persistence. For Platte thistle, field experiments have demonstrated that seed availability limited both local population density and lifetime maternal plant fitness, even before *R. conicus*

TABLE 1. Flowerheads of native and exotic thistles with **R. conicus** damage in 1996. Results are expressed as pecentage of flowerheads per plant (X) with evidence of **R. conicus**. Range for all sites, except Rocky Mountain National park, is 0 to 100%.

Site	Species	Location (elevation, in meters)	X (%)	Range (%) or SE	Plants (No.)
		Native species			
Rocky Mountain National Park, CO	*Cirsium centaureae* (elk thistle)	Beaver Meadows (2960)	45	6–63	24
	Cirsium tweedy	Trail Ridge Road at Ute Trailhead (4150)	<1.0	0–1	35
	Cirsium undulatum (wavyleaf thistle)	Park Utility Area (2815)	70	63–75	11
		Trail Ridge Road at Beaver Meadows (2960)	43	17–63	18
Mesa Verde National Park, CO	*Cirsium pulchelium* (shade thistle)	Knife Edge Trail (2406)	24.3	7.71	21
	Cirsium undulatum	Sagebrush Valley (2119)	38.7	10.88	17
Wind Cave National Park, SD	*Cirsium undulatum*	Bison Flats Prairie Restoration (1250)	77.5	7.15	17
Sandhills Prairie, Nature Conservancy Preserves, NE	*Cirsium canescens* (Platte thistle)	Arapaho Prairie Preserve (1120)	58.1	6.31	32
		Niobrara Valley Preserve (795)	63.6	5.48	42
	Cirsium undulatum	Arapaho Prairie Preserve (1120)	55.3	11.29	10
		Niobrara Valley Preserve (795)	16.6	7.44	15
		Exotic Species			
Rocky Mountain National Park, CO	*Carduus nutans* (musk thistle)	Horseshoe Park (2990)	29	13–49	30
		Hwy. 36, near Harvest House (2805)	25	14–35	30
Mesa Verde National Park, CO	*Carduus nutans*	Sagebrush Valley (2119)	100	–	14
Wind Cave National Park, SD	*Cirsium arvense* (Canada thistle)	Norbeck (1311)	77.3	7.15	22
Sandhills Prairie, Nature Conservancy Preserves, NE	No exotic thistles				

established. Because Platte thistle is sparsely distributed and geographically restricted to Sandhills prairie, further decreases in seed, leading to decreases in local densities within the Sandhills, could threaten its global persistence. Platte thistle is also the putative progenitor for Pitcher's thistle [*C. pitcher* (Torr.) Torrey & Gray], a federally listed threatened species in the Great Lakes dunes. The species are ecologically similar, including their susceptibility to insects. Thus, the impact of *R. conicus* on Platte thistle suggests that there may be comparable effects on Pitcher's thistle if the weevil establishes in the Great Lakes dune ecosystem.

Native picture-winged flies (*Tephritiae*) often exploit the same stage and size heads as *R. conicus*, suggesting the potential, indirect effects. The recent data are consistent with this hypothesis. From 1994 to 1995-96, as the number of R.conicus increased significantly, from 0.1 per head (SE 0.04) to 3.1 (SE 0.61) (t test, $t^1 = 3.83$, $P<0.001$), the average number of *Paracantha culta* (Weid) per Platte thistle flowerheads per plant (N = 27 plants in 1994, 46 plants in 1995) decreased significantly, from 4.1 per head per plant (SE 0.55) to 0.7 per head (SE 0.13) (t test, $t^1 =7.553$, $P<0.001$). Similarly, in Mesa Verde National Park, *Orellia occidentalis* (Snow) disappeared from sampled flowerheads of wavyleaf thistle in 1994, at the peak of *R. conicus* density.

Some of these results are not surprising. Prerelease testing demonstrated that *R. conicus* is oligophagous. Cirsuim species were included in its diet in Europe. Thus, host range expansion to North American species is not completely unexpected. However, the frequency and magnitude of nontarget plant seed destruction, the time delay from introduction to host range expansion where documented (1972 to 1993 in Nebraska), the geographic extent of spread to native species, and the continuing increase in weevil feeding on native species were not predicted. The results strongly reinforce the recommendation that diet specialization is one of the crucial criteria in the selection of a biological control agent. Our study supports suggestions that further evaluation of ecological interactions be required before the deliberate release of an exotic organism. However, the outcome also reinforces suggestions that ecological consequences may be difficult to predict in advance.

The acceptable potential hosts for *R. Conicus*, namely most thistles, often co-occur in disturbed areas and naturally dynamic habitats. Theory suggests that the carrying capacity of an herbivore, such as *R. conicus*, in the presence of co-occurring prey species will be set by the joint availability of its resources, in this case flowerheads. Thus, by using flowerheads of both exotic and native species, *R. conicus* should be able to drive the native thistle population down without declining in abundance itself. A simple equilibrium that allows persistence of both a predator and a prey species, as predicted by basic biological control theory, is not expected when the predator has multiple prey species.

Biological control may be a solution to some weed problems. However, our results challenge the general expectation of little environmental risk with the release of biological agents for weed control. The breadth of diet, potential host range, and ecological effects need to be investigated and then carefully weighed against the environmental costs of the pest and of alternative management options. Intensified follow-up monitoring of species that have already been released is a key step in assessing environmental costs and improving the predictability of biological control. The eradication of a nonindigenous species after establishment is extremely difficult at best so the responsibility for demonstrating that a release will have no unacceptable ecological consequences must reside with the advocates of the introduction. The potential risks to both biodiversity and ecological stability are high when a mistake occurs. This provides strong justification for the intensive study of species already released and for an increased emphasis on rigorous, ecologically focused research on potential agents before they are released.

Questions for Discussion

For Reading 1

1. What are some of the reasons we should be concerned with nonindigenous species in North America? Is it possible to rank these concerns or are they of nearly equal importance?

2. What are some specific ways in which new nonindigenous species can be prevented from entering North America?

3. Broadly speaking, how can we control those nonindigenous species that have become established?

For Readings 2 and 3

1. How are nonindigenous species affecting the quality of life in Florida?

For Readings 4, 5 and 6

1. What is the biggest problem for the Fish and Wildlife Service involving nonindigenous species in Florida?

2. Discuss methods the National Park Service is using to control Australian pine, melaleuca, and Brazilian pepper tree in the Everglades National Park. What are some of the factors that will determine the success of these methods?

3. What are some of the problems facing state agencies in controlling nonindigenous species?

For Readings 7, 8 and 9

1. What are the promises and dangers of two management control programs discussed in these articles?

2. What are the dangers of biological control programs discussed in these articles?

3. Weighing the costs and benefits of biological control, should this method ever be used to control alien species? If so, when? What precautions would you recommend before biological control plans implemented?

Going Beyond the Readings

Visit our Website at <www.harcourtcollege.com/lifesci/envicases2> to investigate issues of biodiversity and nonindigenous species in additional regions of the United States and Canada.

Midwest: Soil Conservation

Introduction

Readings:

Reading 1: Conserving the Plains: The Soil Conservation Service in the Great Plains

Reading 2: The Great Plains Conservation Program, 1956-1981: A Short Administrative and Legislative History

Reading 3: Summary Report: 1997 National Resources Inventory

Reading 4: The Buffalo Commons, Then and Now

Reading 5: Comment on the Future of the Great Plains: Not a Buffalo Commons

Reading 6: The Bison Are Coming

Questions for Discussion

Avoiding Another Dustbowl

Introduction

It is said that those who don't understand history are doomed to repeat it. Could a disregard for history in the U. S. Midwest lead to a disastrous repetition of the American dust bowl?

In 1933, the term "dust bowl" became synonymous with a significant national crisis. A year of drought turned wheat fields in Oklahoma, the panhandle of Texas, and portions of New Mexico, Colorado, and Kansas into fields of dust. Wind storms—not uncommon in the region—whirled powdery topsoil into blizzards of dust: dust that turned day into night; dust that buried houses and roadways under 9-foot (3-m) drifts; dust that settled into lungs, spawning a new respiratory disease—"dust pneumonia." Seemingly overnight, what had been America's breadbasket became its dust bowl.

The drought lasted nine years. Many farms and businesses went bankrupt. Thousands of "Okies"—displaced refugees—drifted west, north, and east into regions that could not easily absorb them. The nation was already suffering its deepest depression. It was a dark chapter in U.S. history.

How could this happen in so rich a country and so rich a farming region? Seeds of the dust bowl were planted decades earlier and grew slowly into crises. When significant numbers of settlers arrived in the region in the 1800s, they replaced native grasses with wheat—a domesticated grass. Native grasses have deep roots that hold even drought-ridden soil in place. Wheat has shallower roots and demands more water than native grasses. At the end of the growing seasons, farmers plowed stubble under, leaving soils bare and exposed to wind until the next growing season.

This was not a problem as long as the number of farms was low. But in the 1920s, prices of wheat soared. The region's farms grew in number and expanded in size. By the early 1930s nearly every plot of the region's potentially arable land was under cultivation. Then the drought hit.

Today we understand that what happened in the dust bowl is part of a much larger problem. Worldwide, the application of questionable farm practices to marginal, semiarid farmland results in similar problems. Farmers are lured into such regions by rich soils, promising markets, and extensive, cheap, available land. As long as rains come, even occasionally, farmers profit, but in semiarid areas, droughts are inevitable. When they arrive, conditions change rapidly for farmers and the region. At best, farms fail. At worst, environmental conditions change permanently. Droughts that would have been only periodic and short lived become permanent regional features. What was once semiarid becomes permanent desert. Roughly 60 percent of arable farmland worldwide is prone to desertification.

Fortunately for the dust bowl region, permanent desertification did not occur. The United States has vast resources and its people have a strong spirit that will, when forced, embrace change. These efforts saved the region. In 1935, the federal government established the Soil Conservation Service, dedicated to study, monitor, and remediate soil erosion. That agency, a part of the U. S. Department of Agriculture and now called the Natural Resources Conservation Services, provides similar functions today.

Good luck also helped. A series of wet years ended the drought in 1940. Throughout the 1940s and early 1950s, profitable farming returned to the region. Cattle, cotton, and wheat became important regional commodities. Then, in the 1950s, drought returned. But the nation was ready. Congress quickly approved programs designed to protect soil under adverse conditions. Farmers were paid to leave part of their land fallow for a year so that moisture could collect in subsoils. Plowing practices, less destructive to soil, were encouraged. Trees were planted to serve as windbreaks. The drought passed with little effect.

Today's Problem

In the 1960s, the region's occasional shortage of water was seemingly solved forever. Beneath the dust bowl region and the high, dry plains north into Colorado, Nebraska, and South Dakota, and south and west into Texas and New Mexico, is the largest groundwater deposit in North America, the Ogallala aquifer. For millions of years this vast underground lake, a sandwich of freshwater between two layers of shale and rock, waited for farmers and their thirsty crops. A new irrigation technology quickly developed.

Now pumps bring water to the surface and feed in into huge sprinkler systems consisting of long pipes mounted on wheels. As water pulses, wheels turn and the whole rig pivots around the well head. The largest rigs may water roughly 2.5 acres (1 hectare) of thirsty fields. Trees in the rig's way were eliminated, and literally millions were removed.

Wind erosion, of course, continued to be a problem, but seemed amenable to solution. First, plant stubble was left in the fields after cultivation to hold soil in place between growing seasons, but stubble still has roots capable of siphoning available water lingering in subsoils. New techniques were developed that cut roots below the surface but left stubble in place. The feeling was "Let winds blow. Let droughts come." High-tech farming throughout the Great Plains seemed capable of being long-term and profitable in spite of innate environmental challenges. Finally, economic stability seemed to have arrived to a region that had known its share of booms and busts.

Today, roughly 200,000 wells tap into the Ogallala aquifer to provide water to more than two million people. Water from the aquifer is being seriously overdrawn. In New Mexico, some farmers extract 5 feet (1.5 m) of water a year, whereas natural replacement is only about one-quarter of an inch (0.6 cm) per year. In 1930, the average thickness of the saturated aquifer was 58 feet (18 m). Today, it is less than 8 feet (2.4 m). Land above the aquifer has begun to sink, compressing the aquifer and permanently diminishing its capacity to receive and store water. Some wells in Nebraska have already gone dry. Experts predict that at current rates of withdrawal, the rest of the Ogallala aquifer will be exhausted by the year 2020.

What then for the region? If a drought hits (and it seemingly always does), an expanded dust bowl could once again darken skies and dampen hopes. Is it time to develop a new land-use pattern for the region? Or can the region's farmers revert back to techniques developed in the 1950s, less dependent on groundwater and less prone to wind erosion? These are some of the questions facing the U.S. midsection in coming years. The readings that follow explore these questions further.

reading 1

"Conserving the Plains: The Soil Conservation Science in the Great Plains"
Douglas Helms

Agricultural History, Spring 1990, 64(2):58–73. Reprinted by permission of University of California Press.

Hugh Hammond Bennett, in early April of 1935, found himself on the verge of achieving an ambition that had dominated his professional life for years, the establishment of a permanent agency dedicated to soil conservation. True, his temporary Soil Erosion Service in the Department of the Interior had received some of the money Congress appropriated to put people back to work during the Depression providing him an opportunity to put some of his ideas about soil conservation to work in demonstration projects across the country. But this had never been the ultimate objective; he had from the beginning yearned for something that would survive the Depression and attack soil erosion until it was eliminated as a national problem. Friends of the soil conservation movement had introduced bills into Congress to create a specific agency for that purpose. Now, as Bennett sat before the Senate Public Lands Committee, he needed to make a convincing case. The sky darkened as dust from the plains arrived. The dust cloud's arrival was propitious, but not totally unexpected—at least not to the main witness. The Senators suspended the hearing for a moment and moved to the windows of the Senate Office Building. Better than words or statistics or photographs, the waning daylight demonstrated Bennett's assertion that soil conservation was a public responsibility worthy of support and continuing commitment to solve one of rural America's persistent problems. Bennett recalled that, "Everything went nicely thereafter."

In the beginning, as so often would be the case in the future, the Great Plains seemed to be at the center of developments in soil conservation policies. Probably the soil conservation bill would have passed in any event. Bennett's crusading zeal converged with the opportunity offered by the Depression to get the work started, but the situation in the Great Plains provided the final impetus for legislation. The Depression awoke the nation to the interrelated problems of poverty and poor land use. The public glimpsed some of this suffering in the South in the photographs of the Farm Security Administration and those in Walker Evans and James Agees, *Let Us Now Praise Famous Men*, that told a tale of poor land, poor people, complicated by tenancy and racism. But it was the Great Plains that captured the national attention. Newspaper accounts of dust storms, the government-sponsored documentary classic, *The Plow That Broke the Plains*, and John Steinbeck's novel, *Grapes of Wrath,* evoked powerful images. For Americans, the Dust Bowl set the image of the human condition complicated by the problem of soil erosion. It remains a powerful historical touchstone for the public's ideas about soil erosion. We may collect data, analyze, and argue, as we do about the relative seriousness of soil erosion in our most productive agricultural regions like the Corn Belt or the wheat region in the Palouse. Occasionally stories appear in newspapers on salinity on irrigated land. But none of these situations compares with the inevitable question that accompanies each prolonged drought in the Great Plains: Is the "Dust Bowl" returning?

The Dust Bowl also proved to be the most popular area in the United States for historians studying soil erosion. Within the past decade historians have produced three books on the Dust Bowl—that section of the plains encompassing western Kansas, southeastern Colorado, northeastern New Mexico, and the panhandles of Oklahoma and Texas. If the wheat and grass sometimes wither in the plains, historical interpretation seems to flourish where the fates of man and land are so intertwined and subjected to the vagaries of climate. To summarize the themes briefly, Donald Worster in *Dust Bowl: The Southern Plains in the 1930s* found the Dust Bowl to be the result of a social system and an economic order, capitalism, that disrupts the environment and will continue to do so until the system is changed. For Paul Bonnifield in *The Dust Bowl: Men, Dirt, and Depression*, plains farmers struggled successfully not only against drought and depression, but also against too much government idealism, whose most threatening manifestation was the soil conservation district with its potential to make plainsmen "tenant farmers for an obscure and distant absentee landlord." R. Douglas Hurt in *The Dust Bowl: An Agricultural and Social History* believed that "farmers in general learned from the Dust Bowl and adjusted their farming practices, so that when drought returned in the 1950s so did wind erosion, but not the black blizzards." These volumes detailed many of the specific farming practices that the Soil Conservation Service advocated in the Great Plains. In this article, I will concentrate on some of the later developments since the Dust Bowl. Finally, on pain of being labeled a geographical determinist, I want to make a few points as to how the Great Plains influenced national soil conservation programs and policies.

The establishment of the Soil Conservation Service created a locus for pulling together all the information on the best methods of farming, but farming safely within the capabilities of the land. The Soil Conservation Service at first worked through demonstration projects and the Civilian Conservation Corps camps. President Franklin Roosevelt in 1937 encouraged the states to pass a standard soil conservation districts act. Afterward, the U.S. Department of Agriculture could sign a cooperative agreement with the district. Much of the SCS's contribution to the districts has been providing personnel to the district. In this manner an agency concentrating on conservation established a presence in the countryside working directly with farmers and ranchers in a relationship that had two fortunate results. First, it made all the disciplines work together on common problems. Thus on the demonstration projects, it drew together the engineers, agronomists, and range management specialists. They were to work together on common problems rather than concentrating solely on their own discipline. Second, the Soil Conservation Service provided a means to work on what we now call technology transfer from both ends of the spectrum. This seemed particularly appropriate in the plains where farmers had struggled with wind erosion and devised a number of methods to combat it. State agricultural experiment stations and later USDA stations specializing in soil erosion provided answers. When SCS began operations, there were already some ideas on answers. To provide vegetative cover SCS advocated water conservation through detention, diversion and water spreading structures and by contour cultivation of fields and contour furrows on rangeland. The vegetative strips in stripcropping and borders of grass, crops, shrubs, or trees served as wind barriers. The young soil conservationists also encouraged the adaptation of crops and cultural practices to fit the varying topographic, soil, moisture, and seasonal conditions. Organic residues should be used to increase organic content and they should also be kept on the surface, as in the case of stubble-mulching, to prevent wind erosion. Critically erodible land should be returned to permanent vegetative cover. Rangelands could be improved by good range management through distribution, rotation, and deferment of grazing. Probably the most far-reaching recommendation was that farmers shift from extensive cash crop farming, wheat in particular, to a balanced livestock and farming operation, or that they shift to a livestock operation and the growing of livestock feeds only. While technology has changed through the years, these essential elements still guide the soil conservation program.

In retrospect, progress in using rangeland more within its capabilities seems one of the more obvious achievements since the 1930s. By most measures, the condition of rangeland in the Great Plains and elsewhere has improved since the 1930s. Henry Wallace's preface to the Western Range report in 1936 predicted it would take fifty years to restore the range to a condition that would support 17.3 million livestock units. That goal was reached in the mid-1970s. Other assessments by the Soil Conservation Service over the last twenty years reveal improvements in rangeland conditions.

It would be difficult to attribute responsibility for this to particular agencies, be they federal or state. Even today, SCS works with approximately half of the ranchers in the Great Plains, though many of those not participating are part-time farmer-ranchers, with other sources of income. What is clear is a growing appreciation for the principles of range management in livestock raising. That is a definite shift from the attitude of the early-twentieth century when the concept that rangeland could be grazed too intensively was anathema to many cattlemen. The controversy about grazing intensity was such that Secretary of Agriculture James Wilson in 1901 wrote on the manuscript of a USDA bulletin on the subject: "all too true, but not best for us to take a position now." Shortly after the dust storms in 1935, SCS Associate Chief Walter C. Lowdermilk was addressing a group of plains cattlemen only to have them terminate the meeting when he mentioned the baleful term "overgrazing."

It has been quite a journey from that attitude to general acceptance of range management as being in the interest of the land and the rancher. Several elements seemed crucial to the development. SCS people working with local soil conservation districts and ranchers had to convince them that range management was in their best interests. The field people work for the most part with owner-operators and consequently in a less adversarial climate than the Forest Service and Department of the Interior range specialists, who had to try to improve range conditions by imposition of stocking rates and grazing fees on federal lands. Also, knowing that an educational job lay ahead, the range specialists had to develop a system to promote range management that was understandable to the SCS field technicians and ranchers alike. That necessity took what had generally been regarded as a research activity into the farm and ranch setting. The key for ranchers in wisely using rangeland was to know the condition of the range, so as to know when and how much it might be grazed without further deterioration. Thus, SCS needed to develop a system of range condition classification, based on scientific principles, that field staff of SCS and ranchers could understand and use.

Early range management pioneers recognized that the composition of the range changed with heavy grazing as cattle selected the taller, more palatable grasses leaving the shorter, less palatable ones. Following thirteen years of research on National Forest rangelands in the West, Arthur W. Sampson elaborated on this concept and observed that the surest way to detect overgrazing was by observing succession, or the "replacement of one type of plant by another." Furthermore, the grazing value of rangelands was highest where "the cover represents a stage in close proximity to the herbaceous climax and lowest in the type most remote from the climax." Sampson's research prefaced the application of Frederic Clement's ideas about plant communities to practical range problems. A pioneer in prairie ecology, Clement theorized that grasslands were a community of plants in various stages of plant succession progressing toward a climax stage.

Range management experts in the Soil Conservation Service needed a classification system that could be used in the field in working with ranchers. Most range management systems in the 1930s and 1940s recognized the validity of ecological concepts for range management. The distinctiveness of the SCS system was that it would be a quantitative system that applied ecological concepts to range classification and management. Other systems were judged to be too qualitative for practical application in the field. The idea was to develop floristic guides of plant population for the various range condition classes. For instance, as rangeland is grazed by animals certain plants will show an increase in the percentage of cover under heavy grazing; others will decrease, and in other cases heavy grazing leads to an invasion of plants onto the site. Thus, SCS field staff learned to inventory rangeland for particular "decreasers, increasers, and invaders" in determining whether the range condition fell into one of four categories—poor, fair, good, or excellent.

So as not to make too general a recommendation that would be of limited value, SCS added the concept of "range site" to the study of range management and improved range management practices. Foresters had originally developed the concept of site as an ecological or management entity based on plant communities. Soil type, landscape position, and climate factors would be involved in determining the climax vegetation and should be taken into account when making recommendations for using rangeland following general instructions the local SCS soil conservationists had to delineate range sites in their soil conservation district. Field staff could then work with ranchers to develop a conservation plan that included advice on how best to use the land for grazing and at the same time maintain or improve range condition. In working with farmers SCS tried to ensure that ranchers understood the key plants and their response to light or heavy grazing and deferment. Overall the system was not supposed to focus solely on those plants that benefited cattle most. In concept it adhered to the suggestion of Clement that "There can be no doubt that the community is a more reliable indicator than any single species of it." Advice to farmers might also include information on fencing, development of water supplies, and rotation grazing as range management theories changed over the years. But the reliance on range site and condition as the foundation has persisted to the present.

The range management experience illustrated two important points about the desirability of an interdisciplinary approach to problems and the need to link scientific theory to practical application. Because of its large field staff, SCS was able to test its ideas about using ecological quantification for range classification at numerous sites in the Great Plains. Isolated researchers have no such means for testing theory and classification in practice. The other point involves the emphasis on soil in range classification. Certainly the early ecologists emphasized soil as a part of the biotic environment. Nonetheless, it is quite likely that having both soil scientists and range managers in the same agency led to greater recognition of the importance of soil in site identification than might have been the case otherwise. Range management was but one of the cases in which the so-called action agencies such as SCS had to translate the scientific into the practical. In so doing it removed the prejudice often held toward what was considered strictly research or theoretical musings. The ecological emphasis and the recognition of the other values of rangeland for wildlife and water, not just the forage produced, seem to have increased the popularity of range management with ranchers.

Cultural practices, especially tillage methods, that reduced wind erosion found favor with farmers. Subsurface tillage, or stubble-mulch farming, eliminated weeds that depleted moisture during the summer fallow period while at the same time leaving wheat stubble on the surface to control wind erosion. Farmers employed the rotary rod weeder, or the large V-shaped Noble blade, or smaller sweeps in this work. Developments in planting and tillage equipment and in herbicides have added a whole array of planting and cultural methods that leave crop residues on the surface as well as increasing the organic content of the topsoil. These practices, such as no-till, ridge-till, strip-till, mulch-till, and reduced tillage fall under the general rubric "conservation tillage." The Conservation Technology Information Center, which promotes conservation tillage, estimated in 1988 that 23 percent of the acreage in the southern plains and 32 percent of acreage in the northern plains was planted with conservation tillage. Larger farm equipment can have some adverse effects on conservation, but the powerful tractors make for timely emergency tillage operations to bring moist soil to the surface to control wind erosion.

SCS's work in the Great Plains always emphasized retiring the most erodible soils to grass. Thus they worked on introducing grass and devising planting methods for the range. The land utilization projects provided a means to test some of these methods. But some plains farmers and absentee owners have continued to use erodible soils for cropland that would be better suited to rangeland or pasture. Nonetheless, as farmers have learned about their land through the hazards or erosion or poor crop production potential, or perhaps through the teachings of the Soil Conservation Service, there have been some adjustments from the homesteading days or the World War I era of wheat expansion. The system of land capability classification developed by the Soil Conservation Service in the late 1930s and recent surveys of land use provided some clues to this shift. In making recommendations to farmers, SCS learned to classify land. In class I are soils with few limitations that restrict use, class II soils require moderate conservation practices, class III soils require special conservation practices, and class IV soils have very severe limitations that require very careful management. Soils in class V and VI are not suited to common cultivated crops. The system takes into account several limitations on use. Where the major limitation is susceptibility to erosion, the subclass designation "e" is used. Generally less than 20 percent of the land in the worst classes, VIIIe and Vie is currently use for cropland, and less than half of the IVe land is used for cropland. So there have been some adjustments.

Wind erosion is still a problem on the plains. While dust storms are not common generally, several years of drought, such as occurred recently can still set the stage for dust storms such as the one that occurred in Kansas on March 14, 1989. The 1988-1989 wind erosion season was the worst since 1954-1955 when SCS started keeping records. Nonetheless, one can perceive the cumulative effects of conservation practices that break up the flat, pulverized landscape and thus prevent dust storms from gathering force uninterrupted. Chief among them seem to be leaving crop residues on the surface, higher organic content of the soil, wind stripcropping, field windbreaks, and interspersed grasslands. The Conservation Reserve Program, authorized in the 1985 farm bill, that pays farmers to keep highly erodible land in grass has proven most popular in the Great Plains. This is not surprising, because the plains influenced it as they did so many other conservation programs.

The drought that struck the Great Plains in the 1950s led once again to emergency drought measures, but also eventually to new soil conservation programs and policies. The Colorado legislature made $1,000,000 available to plains farmers in March 1954. The U. S. Department of Agriculture spent $13.3 million on emergency tillage in 1954, and another $9,275,000 in 1955. The Agricultural Conservation Program spent $70,011,000 on drought emergency conservation measures in twenty-one states during 1954-1956. Colorado, Kansas, Oklahoma, New Mexico, and Texas used $37,848,000 of the funds. Additional funds went to other drought relief measures.

As it turned out, the 1950s drought provided an opportunity for SCS to promote a new program for dealing with conservation and drought in the Great Plains. They suggested to USDA's drought committee that any financial assistance be used to assist farmers to convert cropland back to grassland by paying 50 percent of the cost with the proviso that it remain in grass at least five years. The full committee's report seized on the idea of long-term contracts for restoring grass. It went even further in saying that to discourage a subsequent plow-up it might be necessary to use "restrictive covenants and surrender of eligibility for allotments, loans and crop insurance." Meanwhile, USDA representative met with members of the rejuvenated Great Plains Agriculture Council to work on a program. It called for measures it was hoped would prove more lasting than the cyclical assistance in emergency tillage and emergency feed and seed programs. The report called for "installing and establishing those practice which are most enduring and most needed but which are not now part of their normal farm and ranch operations." President Eisenhower introduced the bill that was to become the Great Plains Conservation Program into Congress on June 19, 1956. Under the bill, the Secretary of Agriculture could enter into contracts, not to exceed ten years, with producers. No contract could be signed after December 31, 1971. The Secretary was to designate the counties in the ten Great Plains states that had serious wind erosion problems. The contracts were to stipulate the "schedule of proposed changes in cropping systems and land use and of conservation measures." The House committee reported favorably on the bill with a few reservations. Only one major farm group showed up to testify in favor of the bill. John A. Baker of the National Farmers Union favored the bill, but even he reported that plains' farmers and ranchers had "some qualms and some apprehensions about these master plans."

After the President signed the bill on August 7, 1956, (Public Law 84-1021) Assistant Secretary Ervin L. Peterson designated the Soil Conservation Service to implement the program. Cyril Luker, a native Texan who had worked in Amarillo in charge of erosion control practices, chaired an inter-agency group that would write the basic guidelines and program structure. Jefferson C. Dykes, Assistant Administrator and a student of the history of the Great Plains, chaired the work group on farm and ranch planning. Donald Williams, administrator of the Soil Conservation Service, ordered the state conservationist of the ten Great Plains states to make proposals to the inter-agency group. The government officials also held meetings with cattle- and sheep-raising groups as well as farm groups.

In working with the inter-agency committee, SCS wrapped nearly two decades of experience into the program guidelines. Essentially, they wanted the individual contracts with farmers to bring about soil conservation while at the same time assisting in the development of economically stable farm and ranch units. Though he did not work on the Great Plains program,

H. H. Finnell, former head of SCS's regional office at Amarillo, wrote in *Soil Conservation*, the official magazine of the Soil Conservation Service:

> A more logical and permanent remedy would be the development of an intermediate type of agriculture to use marginal land. This land is just as capable of being efficiently operated as any other lands, provided the demands made upon it are kept within its natural moisture and fertility capabilities. Ranching is not intensive enough to resist economic pressures; while grain farming is too intensive for the physical limitations of the land. A special type of agriculture for marginal land is needed. It must use the land more intensively than ranching and at the same time more safely than grain farming. Men of stable character and more patience than those who ride on waves of speculation will be needed to work this out.

The contracts with farmers certainly did not dictate what was to be done; there would be mutual agreement. But it would nonetheless be a contract, and the contract would promote the idea of soil conservation and stability. The idea of risk reduction through diversification was certainly not new in the plains, or to other agricultural areas of the United States. Diversification helped farmer-ranchers withstand fluctuations in weather and prices. Surveys during the 1930s showed that failure in the plains came primarily among two groups, strict dry farmers who had no cattle, and cattlemen who grew no feed. Those who combined ranching and farming most often succeeded. SCS people such as Luker and Dykes recognized that stability was good for soil conservation. The Great Plains Conservation Program was to aim for both. The debate in the work group about farm and ranch planning over sharing the cost of irrigation illustrated the emphasis on the stability of operating units. Many members of the work group believed irrigation should be ineligible for cost-sharing, since it could not be considered a soil conserving practice. Dykes, however, argued that irrigation would be needed on some of the small ranches to achieve the goal of economic stability by providing supplemental feed.

Irrigation was of course only one of the farming and ranching practices that contracts with the Great Plains Conservation Program would include. USDA would share the cost of some of these practices with the farmer. Assistant Secretary Patterson also decided that SCS should be responsible for making the cost-sharing payments for soil conservation practices to farmers and ranchers. It was a decision to which SCS attached the utmost importance. USDA began paying part of the cost of soil conservation practices under the Agricultural Conservation Program which was provided for in the Soil Conservation and Domestic Allotment Act of 1936. USDA seized on the soil conservation rationale to reenact production controls after the Supreme Court invalidated portions of the Agricultural Adjustment Act of 1933. Farming practices that were eligible for conservation payments became a point of contention between SCS and the agencies responsible for administering the Agricultural Conservation Program. Currently it is the Agricultural Stabilization and Conservation Service. SCS regarded some practices, such as liming, as annual production practices. SCS preferred sharing the cost of "enduring" soil conservation practices, such as terracing, that brought long-term benefits. Another long-held preference SCS people brought to their task was the matter of the whole farm conservation plan. Since the 1930s they taught that farmers should regard all their needs and concerns in planning for soil conservation while at the same time taking the need for cash crops, pasture, forage, and other needs into account. Of course, farmers could start using this plan at the rate they preferred. But the Great Plains program would involve a contract that provided for rather generous cost-sharing. Thus, it was required that the farmers and ranchers have a plan for the whole farm and that they install all the conservation measures, though the government might not be sharing the cost of all of them.

The three- to ten-year contracts called for a number of conservation practices—field and wind stripcropping, windbreaks, waterways, terraces, diversions, erosion control dams and grade stabilization structures, waterspreading systems, reorganizing irrigation systems, wells and water storage facilities, fences to distribute grazing, and control of shrubs. But by far the greatest emphasis was on converting cropland on the erodible sandy and thin soils back to grassland and improving rangeland and pastures to further diversified farming-ranching in the plains. A recent program appraisal revealed that 53 percent of the GPCP contracts had been with combination livestock-crop farms, 30 percent with principally livestock farms or ranches, and just over 10 percent with crop and cash grain farms. About 85 percent of the units were under the same management when the contracts expired.

The Great Plains, its climate, geography, and history, influenced another national program, the small watershed program as it is generally called. The Watershed Protection and Flood Prevention Act of 1954 made USDA one of the federal participants in flood control work. SCS took the leadership in working in upstream tributary watersheds of less than 250,000 acres. The flood control side of the project provided federal funding for floodwater retarding structures, channel modifications, and other engineering works to reduce flooding along streams. Watershed protection involved soil conservation practices on farms and ranches in the watershed to reduce the sediment moving to the streams and reservoirs. For much of its history, SCS has generally added soil conservationists to these watershed project areas to assist farmers with the soil conservation practices. USDA has been involved in 1,387 projects covering more than 87 million acres.

The Flood Control Act of 1936 gave USDA authority to work on flood control in the upstream areas. Some SCS people certainly favored retarding structures as part of the program to be submitted to Congress for approval, but they were stymied at the department level. The Flood Control Act of 1944 authorized eleven projects for work by the Department of Agriculture. SCS did build a few retarding structures, but the USDA General Counsel ruled against building any additional ones. In the late 1940s and early 1950s SCS was having difficulty getting additional programs approved. There the matter rested until floods hit the Missouri River in the early 1950s. Kansas City, Topeka, and Omaha demanded completion of the Pick-Sloan plans for flood control on the tributaries of the Missouri. Farmers and residents who would lose their farms and homes stridently resisted. They offered soil conservation and small dams in the headwaters as an alternative. The most vocal were the residents of the Big Blue Valley, north of Manhattan, Kansas. They were joined by residents of Lincoln, Nebraska, who had formed a Salt-Wahoo group to promote a small watershed program. Elmer Peterson, a journalist from Oklahoma, promoted small dams as an alternative in Big Dam Foolishness.

That this debate should emanate from Oklahoma, Kansas and Nebraska was in part related to the climate and geography of the plains where farmers could raise corn in the moist bottomland to supplement the hilly grasslands that were too dry to support crops. A small watershed program would provide flood protection to land already used for agriculture, while large dams would inundate the best agricultural land and leave the land suited to grazing or wheat. Because of soil type and moisture the flood plains of the Missouri River tributaries were prized by farmers. Consider the case of N. A. Brubaker, who had 283 acres of land on the Vermillion River in Kansas. The 83 acres of bottom land that supplied feed for his livestock were about to be lost to the Tuttle Creek Dam. His 200 acres of hill land was nontillable. He posed this dilemma to Senator Arthur Capper, "Now if my bottom land will be effected by the water from the Dam, and taken away from me, what use would I have for the 200-acre pasture, as I would not have any land to raise feed for the live stock, and as there would be so much pasture land left in the same way, there would not be much chance of leasing it." A chemistry professor at nearby Kansas State College believed similarly, that the bottomland was the only productive cropland in the Blue River watershed. "The Flint Hills upland provides grazing for cattle but is useless for cropping. There farmers must raise corn on bottomland to finish their cattle. This combination of bottom land for corn and truck farming, and upland for grazing has made the Blue Valley a productive, prosperous region. Without bottom land the entire region will be impoverished and depopulated." The Tuttle Creek Dam and others of the Pick-Sloan plan were built, but the small watershed forces persisted. They met with President Eisenhower and secured his blessing. The small watershed program, authorized in the Watershed Protection and Flood Prevention Act of 1954, spread to the rest of the country. In addition to flood control on agricultural land, it has been used for protection of rural communities, small towns, recreation, water supply, irrigation, and drainage.

The Great Plains also influenced the conservation provisions in the recent Food Security Act of 1985. The plains have been central to questions of landowners' responsibilities to neighbors in not letting erosion impact on their farms. This, of course, can happen with water erosion, with one farmer in the upper part of the watershed influencing the runoff and sedimentation taking place on a farm in the lower part of the watershed. But the most dramatic examples are usually wind erosion from cropland affecting a neighbor's fields. Generally the cases cited have laid the blame on outside investors looking for a quick profit in wheat. Whether this is an accurate portrayal in all cases, the breaking of rangeland for cropland did in part speed passage of some drastic changes in soil conservation laws and policies. It was undoubtedly one of the factors influencing the conservation provisions of the Food Security Act of 1985.

Probably the opening wedge in events that would change the conservation programs took place with the rise in grain prices following the large Soviet grain deals in the early 1970s. Grain exports for 1973 were double those of 1972, and the price quadrupled from 1970 to 1974. At the time Secretary of Agriculture Earl L. Butz released production controls, including the annual set-aside acres. He declared, "For the first time in many years the American farmer is free to produce as much as he can." Farmers in many sections of the country responded, but the plains received the most publicity, mostly for the removal of wide windbreaks for center pivot irrigation system. A Soil Conservation Service survey later found that new, narrower windbreak plantings between 1970 and 1975 offset the losses.

As stories of increased soil erosion spread, groups that had played a large role in the environmental movement increasingly turned attention to soil erosion. They—along with allies in Congress—questioned the effectiveness of existing soil conservation programs. The Soil and Water Resources Conservation Act of 1977 mandated studies of the soil and water conservation programs and the development of new policies to attack the problem. The lobbying and studies resulted in some changes in policies, but the drastic changes came with the 1985 farm bill. Events in the plains played a key role in the new conservation authorities that would appear in the bill. Between 1977 and 1982 wheat farmers planted large tracts of grassland in Montana (1.8 million acres), South Dakota (750,000 acres), and Colorado (572,000 acres). In some places the resulting wind erosion proved a nuisance to neighbors. Some vocal and effective local landowners such as Edith Steiger Phillips of Keota, Colorado, wanted action. The Coloradans persuaded Senator Williams Armstrong in 1981 to introduce a bill that would deprive those who plowed fragile lands of price support payments. Such payments have long been seen as inducing speculation and reducing normal caution in planting very erodible land to wheat. Mainline groups like the Colorado Cattlemen's Association and the American Farm Bureau Federation supported the legislative effort. Several counties in Colorado, including Weld County where Edith Phillips lived, and Petroleum County in Montana passed ordinances to try to prevent plowing on grasslands.

The Armstrong bill, finally dubbed the "sodbuster bill" did not become law. USDA wanted to wait for the next reauthorization of the general farm bill to consider any new provisions, but the pressure from the Great Plains gave some grass roots support for changes in the conservation provisions. The Food Security Act linked soil conservation to eligibility for other USDA programs. The act included sodbuster as well as other conservation provisions. The framers of this act especially wanted to eliminate the possibility that commodity price support programs encouraged poor soil conservation practices. Under the conservation compliance section farmers have until 1990 to begin applying a conservation plan on highly erodible land, and until 1995 to fully implement the conservation plan in order to stay eligible for other USDA programs.

The sodbuster provision applies to any highly erodible field that was neither planted to an annual crop nor used as set-aside or diverted acres under a USDA commodity program for at least one year between December 31, 1980 and December 23, 1985. If farmers wish to bring such land into production, they would lose eligibility for USDA programs unless they applied an approved conservation system to control erosion on the fields. The swampbuster or wetland conservation stipulated that farmers would lose eligibility for USDA programs if they drained wetlands after December 23, 1985, the date of the passage of the act. A conservation coalition that lobbied for this provision included old-line soil conservation organizations like the Soil and Water Conservation Society of America and the National Association of Conservation Districts as well as environmental groups. Prominent officials in USDA such as John Block and Peter Myers favored many of the provisions. But the grass roots examples of support from the plains influenced the Congress even more. This is a prime example but not the only one of the way commodity programs instigated the use of land for cropland that would be better suited to rangeland. Emotionally, the conversion of rangeland to cropland has an appeal that catches the public attention more than erosion from cropland in the humid east. The 1985 provisions are some of the most far-reaching we have seen in agriculture. They are premised on the idea that some USDA programs induced the use of erodible land that would not have occurred otherwise. The Great Plains, as they so often did, served as the prime example for changes in soil conservation policies.

reading 2

"The Great Plains Conservation System, 1956–1981: A Short Administrative and Legislative History"

Douglas Helms

SCS National Bulletin, November 24, 1981, No. 300-2–7. Reprinted by permission of the author.

Enthusiastic supporters of the Great Plains Conservation Program recently gathered to celebrate the 25th anniversary of the authorizing legislation, signed August 7, 1956. The program was the latest of the nearly three-quarters of a century of local, state, and federal efforts to deal with drought, dust storms, and the resulting agricultural instability on the Great Plains. The novel feature of the program was that it provided for the government's sharing the cost of conservation measures with farmers and ranchers under a contract.

Settlement and Early Droughts

The proponents of this new concept had reason to believe that something new was needed to adjust man's agricultural endeavors to the climatic and geographic realities of the plains. Most had witnessed the drought of the 1930s and had heard tales of the ones in 1887-97 and 1910-13. The emphasis in the new program on developing enduring conservation practices rested on an understanding that drought would return to the Great Plains. A review of earlier periods of climatic stress is important because the understanding of recurring drought shaped the thinking of the people who devised and administered the Great Plains Conservation Program.

Reports from 19th century military expeditions led Americans to regard the area between the 100th meridian and the Rocky Mountains as the "Great American Desert." Major Stephen H. Long, after crossing the area, declared it "almost wholly unfit for cultivation, and of course uninhabitable by a people depending upon agriculture for their subsistence." Soldiers returning from the Civil War had plenty of the fertile tall grass prairie left to settle. Eventually settlement pushed westward to the plains as promoters tried to dislodge the notion that the region was not fit for agricultural settlement. The few who had pushed out onto the plains in the mid-1870s had to withstand both drought and grasshoppers.

With the return of favorable weather in the 1870s, movement into western Kansas and Nebraska intensified. In Ellis County, Kansas, it was observed that "incessant breaking for wheat can be seen in all directions." The boom in settlement peaked in the mid-1880s. There were 3,547 homestead entries in Kansas in 1884. New entries in 1885 and 1886 numbered 9,954 and 20,688, respectively. As the boom receded in Kansas it continued in Colorado. There had been only 1,808 homestead entries in 1886; the number increased to 5,081 in 1887 and peaked at 6,411 the following year. During the latter two years, 4,217,045 acres, predominantly in the plains, were filed under the Homestead Act and the Timber Culture Act. The lack of capital and insufficient knowledge about farming in semiarid conditions took its toll when the drought resumed in the late 1880s. That many settlers had departed and that many never took up residence on their claims was evident in the 1890 census. There were only 3,535 farms reported in fifteen eastern Colorado counties. Quite a number of these farms were along the Arkansas and Platte rivers.

The western movement was turned back with the drought that began in the late 1880s and lasted ten years with a few good years interspersed. Population statistics revealed the impact but not the suffering involved. Western Nebraska had a decline of 15,284 residents during the decade of the 1890s. During the same period the western Kansas population dropped from 68,328 to 50,118, and a considerable number had left before the census was taken in 1890. According to one estimate, half the population of western Kansas departed between 1888 and 1892. Twenty vacant towns stood witness to the effects of drought on the entire economy.

Farther south in Texas, farming had not supplanted ranching to any great extent. Generally, the farms were larger than those of the other plains states which had been limited in size by the homestead laws. Having larger farms, Texans were better able to persevere through the drought. Drought also struck the northern plains, and population declined in some areas. As would be the case in the future, drought was not as devastating as it had been in Nebraska, Kansas, and Colorado. Emergency relief measures did not begin with federal assistance in the 1930s. Already in the 19th century state governments were being called upon for assistance. A Mendota, Kansas, housewife wrote the Governor Lewelling in 1894, "I take my pen in hand to let you know that we are starving to death. It is pretty hard to do without anything to eat here in this God forsaken country…. My husband went away to find work and came home last night and told me that he would have to starve…. If I was in Iowa I would be all right." With such conditions widespread, several state and private organizations undertook relief measures. The Nebraska legislature appropriated $200,250 in 1891, mainly for food and grain. Colorado provided $21,250 to supply farmers in eight counties with seed for the 1891 planting season. Kansas spent $60,000 for the same purpose in 1891. In response to the 1886 drought in Texas, the state gave $100,000 in aid to 28,000 individuals.

The drought dislodged the belief among farmers as well as the scientific community that rain followed the plow; that growing crops and plowed fields induced greater rainfall. With that faith destroyed, farmers and agriculturalists were ready to make concessions to the climate and turned their attention to adjustments in farm management, cultivation methods, and drought resistant crops.

The hardy qualities of the "Turkey Red" wheat brought to the plains by Russian-German immigrants around 1873 became obvious during the dry years. Mark Carleton and others now set out to discover other crops suitable to the area.

Farmers began to adapt their cultural practices to the climate. Hardy Webster Campbell became the chief promoter of dry farming, although some of the measures predated his involvement in the campaign. Campbell's Soil Culture Manual (1902) recommend deep fall plowing, thorough cultivation before and after seeding, light seeding, alternating summer fallow, tillage during fallow and crop years, sub-surface packing, and inter-row cultivation.

With the return of favorable weather in the first decade of the 20th century, dry farming spread across the plains. Cattle raising was also prospering. Both ventures received a shock with the return of drought in 1910. The dry farming method had some sound elements, but it was no panacea for withstanding drought. The dry farming movement was practically destroyed in South Dakota, leading one critic of its more exaggerated claims to surmise that it was time to "to cut out the cheap talk about dry farming and talk cows." Actually the cows were not fairing all that well either. Selling during the drought, 1910-11, and losses during the winter of 1911-12 reduced Great Plains herds seventy percent. The reduction drove many ranchers out of the business. The turnover of ownership benefited the land. Newcomers had a better idea of the value of good range management, both to their pocketbooks and to the conservation of the range.

The 1910-13 drought in the southern Great Plains brought another problem. A small "dust bowl" developed in Thomas County, Kansas. Although dust storms were not confined to Thomas County, the storms that swept over 65,000 acres from 1912-14 were probably as severe as any since. Responding to the need to reduce dust storms, Kansas State College issued its first bulletin on wind erosion control in 1912.

The return of rain in 1914, high prices, and government exhortations to produce for the war effort led to an expansion of wheat growing in the Great Plains. The wheat acreage in the plains areas of Montana, North Dakota, and South Dakota increased from 2,563,000 acres in 1909 to 4,903,000 acres in 1919. Nationwide profits on wheat rose from $56,713,000 in 1913 to $642,837,000 in 1917. Between 1909 and 1924 plains farmers increased the wheat acreage by 17,000,000 acres. Even the drought in 1917-1921 did not measurably slow the change. Many settlers gave up in the northern plains but acreage figures for wheat held steady. Nor did the drop in wheat prices in the early 1920s have much effect. Farmers responded to declining prices by planting more to recoup dwindling profits. Another 15,000,000 acres went from grass to wheat between 1924 and 1929. Much of the expansion in the late 1920s took place in the southern plains where wheat acreage increased 200 percent between 1925 and 1931. With only a few interruptions the years 1914-1931 had been good in terms of weather.

The Dust Bowl

The 1930s ushered in another prolonged drought. Scant use of structural, cultural, and vegetative water conservation measures further complicated the problem. The lack of rainfall prevented good stands of wheat and left the ground barren for wind erosion. By August 10, 1933 there had been thirty dust storms in the vicinity of Goodwell, Oklahoma. Another year of drought in 1934 left 97,000,000 acres in eastern Colorado, western Kansas, eastern New Mexico, and the panhandles of Texas and Oklahoma susceptible to wind erosion. Newspaper reports brought the storms national attention. A reporter for the Washington (D.C.) Evening Star supplied the term "dust bowl" to describe the area. The dust bowl, or the worst of the general blow area, was in Baca County, Colorado; the six most southwestern counties in Kansas; Cimarron and Texas counties, Oklahoma; Dallam and Sherman counties, Texas; and a portion of Union County, New Mexico.

The Soil Conservation Service and its predecessor, the Soil Erosion Service, had increasingly turned their attention to the area. By the end of 1936, SCS had established fifty-five demonstration projects in the Great Plains with a heavy concentration in the worst wind erosion areas. When the projects began in 1934, only 10,454 acres in the project areas were being farmed using soil and water conservation measures. With its large force of Work Projects Administration and Civilian Conservation Corps labor, plus the work of farmers, the Service made progress. The results at the conclusion of 1936 were impressive—conservation measures in place on 600,000 acres—including 155,000 stripcropped acres, 200,000 contour tilled acres, contour furrows on 85,000 acres of grasslands, and 3,600 miles of terraces on 65,000 acres. Additionally, 200,000 acres of grassland were under management to prevent overgrazing. The acreage of erosion retarding crops had been increased twenty-eight percent. With the adoption of conservation district laws by the states, beginning in 1937, the Service extended its technical assistance to areas outside the demonstration projects. The Service assisted in contour listing (an emergency wind erosion control practice) 2,500,000 acres in 1936. The federal government spent $793,000 for emergency wind control measures under its Agriculture Conservation Program in 1938. The total drought emergency expenditures for cattle and sheep purchases, feed and forage, seed, loans, and erosion were $212,916,000 in 1936, $2,735,000 in 1936, $515,000 in 1937, and $1,000,000 in 1938.

Other government programs involved planting windbreaks in the shelterbelt project supervised by the Forest Service. The Farm Security Administration and the Bureau of Agricultural Economics purchased what were termed "submarginal lands" under the land utilization program. After revegetating the land, the government proposed to lease it for grazing. SCS eventually assumed leadership of both programs.

The Plains in the 1940s

Again the rain and war seemed to arrive at about the same time. Weather in the Great Plains improved in 1940. The government called on farmers to produce food for the military forces and the allies when World War II began. As SCS employees entered the armed forces, the reduced staff was instructed that "Emphasis should be given to the widespread application of conservation practices that contribute the most to maintaining or increasing yields and that can be (1) applied with little or no additional use of farm labor, equipment, power and production supplies and (2) furthered with the minimum of technical assistance." Nationwide, World War II had varying effects on soil conservation. The situation in the Southeast and Mississippi Delta improved in 1943-44 when compared to 1935-39, due partially to the reduction of row crops. The Corn Belt had significant losses compared to 1935-39. The Great Plains showed little change after the recovery from the dust bowl but there was cause for concern about the future.

H. H. Finnell, regional conservationist at SCS's Amarillo (Texas) office and an authority on wind erosion control, was concerned. He conceded that the World War II plow-up had not been as extensive as that of World War I. Nonetheless, he saw future problems. Farmers had planted pinto beans on loose, sandy soils in New Mexico, cotton on sandy land in Texas, and wheat on thin soils in Colorado. Finnell particularly directed his ire at absentee land speculators in Colorado, who had tried to get Colorado's soil conservation law nullified in the state supreme court and who were lobbying to have the lands reclaimed under the land utilization program put up for sale.

Not only was the use of submarginal land for crops detrimental to the soil, according to Finnell, but also it could not be justified economically. The profits from wheat for a few years would not compensate for revenue lost on grazing while the range was being reestablished. Finnell called for a special type of agriculture for the area:

A more logical and permanent remedy would be the development of an intermediate type of agriculture to use marginal land. This land is just as capable of being efficiently operated as any other lands, provided the demands made upon it are kept within its natural moisture and fertility capabilities. Ranching is not intensive enough to resist temporary economic pressures; while grain farming is too intensive for the physical limitations of the land. A special type of agriculture for marginal land is needed. It must use the land more intensively than ranching and at the same time more safely than grain farming. Men of stable character and more patience than those who ride on waves of speculation will be needed to work this out.

The trend continued as prices held up after the war because of demand from countries where war had disrupted the agricultural economy. Between 1941 and 1950 farmers broke out about 5,000,000 acres. The estimate was that 3,000,000 acres of this land was not suitable for cultivation. In fact, some of it had not previously been in crops.

Drought of the 1950s

An extended drought and dust storms returned in the 1950s. Western Nebraska ranchers traveling to their annual convention on June 8, 1950 had hazardous driving conditions and saw roadside ditches filled with soil. Most of the 100,000 windswept acres in Scotts-bluff, Box Butte, Morrill, and Sioux counties were summer fallow fields with no conservation practices or irrigated sandy land for beets and beans. The worst blowing of the 1950s was yet to come. SCS surveyed the plains and located the most susceptible areas. The survey cited the bean growing area of Colorado—Pueblo, Crowley, El Paso, and Lincoln counties. The wheat had died over large parts of the Oklahoma panhandle. Chase and Perkins counties, Nebraska, were listed as critical as was central Kansas. There were problems in the cotton growing areas of Lamesa-Lubbock, Texas. Eastward across the plains, the western cross timbers of Oklahoma and Texas planted in cotton, wheat, peanuts, and watermelons had also experienced blowing.

The Department of Agriculture set up a Great Plains Committee in April 1950 to study the problem and make recommendations. The drought continued, leaving acre after acre without any vegetation to protect it from erosion. The dust storm that signaled the national awakening to the "filthy fifties" occurred on February 19, 1954. H. H. Finnell observed the storm from Goodwell, Oklahoma. He wrote to Tom Dale of SCS:

> ... conditions in the marginal zone are worse than in the 1930s because poorer lands under more arid conditions have been exposed to wind erosion in a wider territory than in the 1930s...it will be more difficult to subdue than the wild lands of the 1930s. Catastrophe to the land has already exceeded that of the 1930s, but due to the absence of financial straits and hysteria which existed in the 1930s, farm abandonment has been much slower to gain headway.... I had hoped the lessons of the 1930s would be more widely grasped and acted upon than they have been. I don't know how many times this thing will have to happen to the Southern High Plains before the idea of safe land use soaks in. The agricultural potential of the area was measurably lessened by the experience of the 1930s and will be again. Too much Class IV land is being physically transformed into Class VI and VII.

Newspapers treated the nation to stories that depicted little difference between the drought of the 1950s and that of the 1930s, except for the absence of outmigration. The *Washington* (D.C.) *Daily News* proclaimed that the "new dust bowl" was "in roughly the same place on the map as the old one." Actually there had been some significant changes. The area subject to wind erosion was larger and encompassed all of the area of the 1930s. More significantly the centers of the worst areas had shifted and expanded. The area in New Mexico stretched from Quay down to Lea County. Adjoining it in Texas, the blow area was bounded by Palmer County on the north and Ector County in the south. The Colorado blow area extended from the eastern border to El Paso and Pueblo counties. The points of the triangular area in Kansas were Wallace, Finney and Morton counties. With the exception of Baca County, Colorado, and Morton County, Kansas, most of the earlier dust bowl was not included. The conservation measures of the 1930s had obviously helped. After another three years of drought, some of the older dust bowl had been included, but the problems were not as persistent as those of the newer areas that Finnell had pointed to in his 1946 article.

The Colorado legislature made $1,000,000 available to dust bowl farmers in March 1954. The U.S. Department of Agriculture spent $13.3 million on emergency tillage in 1954 and another $9,275,000 in 1955. The Agriculture Conservation Program funds spent on drought emergency conservation measures in twenty-one states, 1954-56, totaled $70,011,000. Colorado, Kansas, Oklahoma, New Mexico, and Texas used $37,848,000 of the funds. Additional funds went to other drought relief measures.

USDA and the Great Plains Agricultural Council

While the relief measures were being extended to the plains states, the USDA continued working through its committee on land use problems in the Great Plains to develop a program to reduce the need to respond periodically with emergency measures. The Soil Conservation Service suggested to the committee that the government use "financial assistance to encourage farmers to convert cropland to grass with the federal government paying at least 50 percent of the cost and making an agreement to continue the program over a 5-year or longer period." The full committee elaborated on the proposal. The report recognized that "diverting the 6 to 8 million acres of cropland that are unsuited for cultivation to grassland is largely a problem of voluntary action or land use regulation, hence it must be handled mainly by State and local governments and individual owners." But "cost-sharing payments...might be increased and spread over a period of 3 to 5 years while grass is being established." To discourage a subsequent plow-up it might be necessary to use "restrictive covenants and surrender of eligibility for allotments, loans and crop insurance."

Meanwhile, the Great Plains Agricultural Council, born during the drought of the 1930s, had begun to develop a long-range program. Representatives of the USDA met with council members on May 31-June 2, 1955, to develop a program. A later meeting, July 25-27, refined the proposals. President Dwight D. Eisenhower transmitted the council's "Program for the Great Plains" to Congress on January 11, 1956. The program did not specify that cost-sharing for conservation practices would be offered through contracts with farmers and ranchers. It did, however, call for sharing the cost of "installing and establishing those practices which are most enduring and most needed but which are not now a part of their normal farm and ranch operations The ACP cost-sharing program on those practices that are intended to bring about those land use adjustments required for a long-range program will be accelerated and rates of payments made more flexible."

The Department of Agriculture was already considering the specifics of how the program might be implemented, including long-term contracting. Donald A. Williams, Administrator of the Soil Conservation Service, wrote to Assistant Secretary of Agriculture Ervin L. Peterson that the soil conservation districts would be a perfect device for implementing whatever plan Congress adopted. Williams made it clear that the districts could incorporate these new activities into their existing programs so as "to insure a permanent, sound coordinated land use and management program in the Great Plains area." To emphasize SCS's interest in the new program Williams made it clear that he was "prepared to ask SCS personnel to aggressively work with the district governing bodies to the fullest extent possible in this effort."

Public Law 84-1021

Congressman Clifford Hope of Kansas introduced a bill (H.R. 11833) on June 19, 1956, that was to become the Great Plains Conservation Program. The bill provided that the Secretary of Agriculture could enter into contracts, not to exceed ten years, with producers. No contract was to be signed after December 31, 1971. The Secretary was to designate the counties in the ten Great Plains states that had serious wind erosion problems. The contracts would outline the "schedule of proposed changes in cropping systems and land use and of conservation measures" to be carried out. The bill further stipulated the obligations of the grower and made the provision that any acreage diverted to grass would not affect commodity acreage allotments for the time of the contract. Not more than $25,000,000 was to be spent in any year, and the total could not exceed $150,000,000. Assistant Secretary Peterson testified before the House Committee on Agriculture on June 28, 1956. Peterson responded mainly to questions concerning how the program differed from the new Soil Bank. Representatives from beef producing states expressed concern over the effects of putting more land to grazing purposes when cattle prices were already depressed.

Karl C. King, a Pennsylvania congressman, but a native of Reno County, Kansas, thought that buying the land would be cheaper than applying conservation measures. Congressman Hope interceded to explain what the program planned to accomplish in terms of farm management. One of the problems of the plains had been the pattern of outmigration during drought followed by a wave of new settlers when the weather improved. Each new group had to learn the tough lessons that came with the drought. The proposed program, as Hope explained it, would assist farmers and ranchers through the drought, improve farming and ranching techniques, and lessen the impact of future droughts.

The hearings concluded after John A. Baker of the National Farmers Union testified in favor of the legislation. Baker, who would later oversee the Great Plains Conservation Program as Assistant Secretary of Agriculture, had some reservations. He wanted it known explicitly that the new program would be a "partial supplement, not a substitute for existing programs." The possibility that the Farmers Home Administration could deny credit to farmers who did not follow a conservation plan was also of concern. Baker stated that plains farmers and ranchers had "some qualms and some apprehensions about these master plans." Nonetheless, the Union supported the bill.

In reporting out the bill on July 7, the committee emphasized that the program was voluntary and that participation would not be a necessary condition for making acreage allotments, FHA loans, agricultural credit, or eligibility for other Department of Agriculture programs. One proposal to speed up the conversion of land not suited for cropping back to rangeland had been to make crops on that land ineligible for federal crop insurance. Although the committee did not specifically mention the insurance program, the report gave their view on possible linkage of USDA programs.

The House of Representatives passed the bill on July 23, and the Senate concurred without changing the bill on July 26. President Eisenhower signed Public Law 84-1021 on August 7, 1956, with the statement that the act authorized the "Secretary of Agriculture to enter into long-term contracts with farmers and ranchers in the Great Plains states to assist them in making orderly changes in their cropping systems and land uses which will conserve soil and water resources and preserve and enhance the agricultural stability of that area."

SCS Selected to Administer Program

It then fell to the Department of Agriculture to develop a plan for administering the program. Actually, the agencies within the Department were at work on plans before the President signed the legislation. Donald Williams of SCS and Paul Koger of the Agricultural Conservation Program Service had discussed implementation. They agreed on a number of points but could not agree on which agency should administer the program. Both wrote to Assistant Secretary Peterson in early August. Williams presented a detailed proposal for administering the program with SCS as the lead agency. Koger pointed out that ACPS had traditionally dealt with the cost-sharing aspects of conservation programs. Both agencies continued to work on plans and awaited the decision. The Commodity Stabilization Service supported the ACPS. The Great Plains Agricultural Council suggested that the county Agricultural Stabilization and Conservation committees handle the cost-sharing aspects of the services.

Peterson resolved the issue in Secretary's Memorandum No. 1408 on December 10, when he assigned responsibility to SCS. He also announced the creation of the Great Plains Inter-agency Group, composed of all the cooperating USDA agencies, to develop the policies and procedures. The same day Williams appointed Cyril Luker to chair the group and called a meeting of the state conservationists of the ten Great Plains states to work on the new program. Assistant Secretary Peterson attended the first meeting of the Inter-agency Group on December 17 and reiterated what he expected from it. He emphasized that "short term activities must be consistent with the long-range objectives." Whatever the group developed had to have the understanding and support of the Great Plains Agricultural Council.

Luker appointed task forces on information, cost-sharing and contracts, farm and ranch planning, and meshing the legislative authorities of the various agencies. The group sought and received advice from outside. Federal, state, and local officials and representatives from cattle and sheep raising groups and farm organizations held a January meeting in Denver to draw up suggestions. During the next weeks the task forces met and reported back to the full group with their majority and minority findings. Again Peterson met with the group and stated that the matters on which there was no unanimity had left the group on "dead center." The differing views should be documented and presented to him for resolution. Peterson resolved several issues at the meeting. The scheduling of practices was a technical matter and should be included in the farm plan, because the single practice concept conflicted with the long-range good of the program. Certification of installment of measures would be the responsibility of SCS. As the work of the group progressed the Assistant Secretary was called on for additional decisions, the main one being whether SCS would serve as the contracting agency because it had responsibility for helping he owner develop the farm and ranch plan for the entire unit. Therefore, SCS should have responsibility for insuring that the practices were installed as scheduled and that they be maintained throughout the life of the contract.

The SCS people participating in drawing up the list of cost-share practices could draw upon over two decades of experience of working with farmers and ranchers. Also, managing the lands acquired under the land utilization program gave SCS technicians an opportunity to test various conservation measures. The conservation practices in GPCP accordingly reflected this field experience.

Great Plains Inter-agency Group

Not surprisingly, the question of cost-sharing for irrigation came up for discussion. The majority of the Farm and Ranch Planning Task Force wanted to exclude irrigation, but J. B. Slack of the Farmers Home Administration and Jefferson C. Dykes of the SCS disagreed. They pointed out that irrigation was needed on some small ranches to achieve the goal of economic stability by providing supplemental feed. It would help bring about the desired land use change on the rest of the farm. The fear that it could encourage carrying more animals than the ranch could support would be corrected in the contract. The minority view prevailed, and irrigation was included.

The matter of establishing the exterior boundaries for the program did not occasion much controversy. The criteria developed by the group included physical and climactic conditions that made crops undependable, erosive and deteriorated soils, and the need for land use change and conservation measures. The group solicited the states' suggestions on counties to be included under the criteria. Under this criteria, the boundary generally corresponded with the one proposed in the Great Plains Agricultural Council's program for the plains. As to which counties would initially be designated, the group added the element of local interest and initiative. It would be better to get the program off to a good start in counties where farmers were asking for assistance and then expand to the rest of the area.

With many of the details worked out, those who worked on the program anxiously awaited the appropriations hearings. Peterson and Williams testified before the House Committee on Appropriations and requested $20 million per year. Again they were called upon to explain how the new program differed from the Agricultural Conservation Program. Peterson emphasized the hope that the money spent on GPCP would reduce the amount needed for emergency drought programs. The committee appropriated $10 million for the year.

In the months following the hearing, the group firmed up the policies and procedures, refined the list of practices, established the percentage of cost-shares for each practice, developed a handbook, and trained the SCS staff in drawing up contracts. The work unit conservationist was well acquainted with developing conservation farm plans, but the element of contracting was new.

Beginning of GPCP

Berthold Sackman of Stutsman County, North Dakota, signed the first contract on December 19, 1957. The same day, Walter L. Wood and Robert H. Hunt of Gaines County, Texas, signed contracts. These three and the subsequent contracts were to provide from 50 percent up to 80 percent of the average cost of conservation measures and included a schedule for the coordinated implementation of measures. The plans called for an assortment of complimentary conservation measures to stabilize the farm or ranch in accordance with the owners' objectives.

There were cost-sharing items for establishing vegetation on lands previously cropped and for reseeding range. Irrigation for pasture and forage, fencing, and development of water supplies supported the shift to rangeland and were designed to prevent overgrazing. Conservation measures for cropland included contour stripcropping, terracing, grassed waterways, land levelling, reorganizing irrigation systems, and windbreaks. The terms "permanent" and "enduring" were used to describe the conservation measures. GPCP architects hoped that farmers and ranchers would maintain the measures after the expiration of the contract. The fact that they were willing to pay part of the cost of installation boded well for long-range retention.

Such reluctance as there was on the part of owners centered on the contractual aspects of the program. Farmers had over twenty-five years of experience in dealing with government supervised acreage allotments and commodity price support programs. The notion of entering into a contract with obligations on both sides was a novelty. The work unit conservationists, as they were called in the 1950s, explained the new approach and pointed out the benefits.

Any reluctance to enter into a contract soon withered as farmers and ranchers saw the benefits neighbors derived from signing up. It was not long before the applications exceeded the amount of money available—a condition that has continued throughout the history of GPCP. By September 1959, twenty months after the first contract was signed, there were 3,142 contracts covering 8,597,385 acres with a federal obligation of $16,794,0441. There were 2,579 applications for assistance in SCS offices throughout the Great Plains states.

Limitation on Irrigation and Contract Size

Despite the impressive start, Williams and Luker found reason to reevaluate some aspects of the guidelines. Some of the early contracts had been larger than anticipated, with a substantial part of the funds going to irrigation. Actually, accelerated land treatment could be carried forward more rapidly under large contracts, but the trend held some dangers for the continuation of the program. With limited funds going into the large contracts, many applications would go unserviced. Eventually, there would be criticism that GPCP was only for large farmers and ranchers. Expensive irrigation construction could easily absorb most of the money provided in individual contracts. There was a fear that the package of interrelated conservation measures for the whole land unit would be neglected and that critics would regard GPCP as a production, not a conservation program.

Williams and Luker proposed to the state conservationists in the Great Plains states that the amount spent on irrigation in individual contracts be limited to one-fourth of the contract with a $2,500 maximum. They developed a set of priorities to be used in selecting contracts to fund. Units having difficulty converting from cropland to permanent vegetation; units having wind and water erosion problems on rangeland or cropland suited to continuous cropping; and units having erosion problems requiring cooperative action by several owners would have priority. They further advised that the size of the farm or ranch should not determine the priority of assistance but that "a sufficient number of medium and small farms and ranches should be scheduled to provide a representative balance in the use of resources."

State conservationists Lyness Lloyd of North Dakota and H. N. "Red" Smith of Texas objected to the percentage limitation on irrigation practices. Lloyd stated that the change would hinder the stabilization of ranches while the conversion to ranching was being made. Irrigation was needed to provide cattle feed and pasture while former cropland was being returned to range. Smith said the alternation in the program would reduce support for GPCP and eliminate a large part of the state from participation. He wrote, "The principal leadership in the Great Plains portion of this state have a strong interest in irrigation farming….The proposed fund limitation for irrigation practices would particularly eliminate irrigated cropland in this state from participation." Objections notwithstanding the limitation of cost-sharing on irrigation practices went into effect. A year later on May 29, 1959, SCS placed a $25,000 limit on individual contracts.

Protecting the Cropland History

The supporters of GPCP managed in 1960 to correct an aspect of the legislation which was viewed as an impediment. Some farmers who were willing to convert cropland to grass or to crops better suited to the land nonetheless wanted to retain the option of keeping the crop allotments and any payments due them. Public Law 1021 had protected the cropland history of the farm for the period of the contract. President Eisenhower signed Public Law 86–793 on September 14, 1960, to protect the cropland history for twice the length of the contract.

Diversity of GPCP Contracts

While the Washington office and state staffs wrestled with administrative and legislative details, significant progress in implementing conservation measures was taking place. GPCP contracts reflected the geographical diversity within the plains, the various types and sizes of agricultural units, and the objectives of individual farmers and ranchers.

D. H. and Charlene Dean of Claunch, New Mexico, made a total conversion from cropland to ranching. To convert 2,000 acres to grazing land, the Deans installed three ponds and three miles of water lines for livestock, six miles of cross fences, and controlled brush on 845 acres.

Rancher-farmers had more of a mixture of conservation measures for cropland and range. Walter Markel of Gray County, Kansas, had an 804 acre farm. He added 1,800 feet of diversions, installed 21,000 feet of terraces, and contour farmed and stubble mulched 231 acres. Thirty-nine acres were furrow seeded. For better grazing distribution he added 330 rods of fences. Markel had belonged to the local soil conservation district since 1949. He was in some ways typical of many who used GPCP to make progress on a farm conservation plan that they had envisioned for years.

GPCP contracts were used near Dumas, Texas, to solve flooding in the town. Ten farmers constructed 22,120 feet of waterways. In the process, 2,560 acres of irrigated cropland were also protected.

In addition to individuals, it was also possible for groups to sign contracts. A dozen FmHA-financed grazing districts in Montana held GPCP contracts in 1968. The contracts called for over 10,000 acres to be seeded and reseeded and for putting up 39,000 rods of fences. The reseeded range provided twenty-five percent more forage by 1968, with other acres remaining to be reseeded under the contracts.

The use of a GPCP contract on the Dee Hankins from in Wichita County, Texas, demonstrated the rehabilitation, both physically and economically, of worn-out land. The 815 acres (665 cropland, 140 acres rangeland, 10 acres farmstead) had been sold six times in four years. Much of the farm was waterlogged and denuded because of salt deposits. The plan called for 65 irrigated acres, 267 dryland crop acres, 161 acres of irrigated pasture and 312 acres of rangeland. Concrete irrigation ditches were used for water conservation on the irrigated part. Two hundred acres of waterlogged and salt denuded land was seeded to sideoats grama and native grasses. The acres planted in coastal Bermuda grass were hayed, grazed and provided strips of sod to sprig other farms. The farm became economically viable and remained so until Hankins sold it for suburban development.

State Trends in GPCP Contracts

Although there was much diversity of conservation practices established on individual farms and ranches, there were some state and regional trends in the 1960s. Based on the percentage of total expenditures for each practice (1957-1972), North Dakota, South Dakota, Montana, and Nebraska led in establishing permanent vegetation on former cropland. Oklahoma and Texas were by far the leaders in reseeding rangeland. Only in North Dakota was stripcropping significant. That state also led in establishing windbreaks, followed by South Dakota. Leading in percentage expenditures on terracing were Kansas (30%), Nebraska (20%), and Texas (17.5%). New Mexico and Wyoming had the most activity in dam construction for erosion control, and Montana easily spent the most on waterspreading. Land leveling was most prevalent in Colorado and Kansas. Only Montana spent over 10 percent of its money on fences. Controlling invading mesquite and other undesirable shrubs was understandably highest in the two southwestern states, New Mexico and Texas.

Congress Extends GPCP

The program had become so popular that each year's allocations to states were usually obligated early in the year for contracts that had already been written. As the expiration date of P.L. 1021 approached, farmers, ranchers, conservation district supervisors, and state officials hoped and worked for the extension of the program. All groups had some idea how the program might be improved, but the main objective was to have it extended. Most senators and representatives from the Great Plains states cosponsored the legislation. At the hearing before the House Committee on Agriculture, Congressmen George H. Mahon and Richard C. White of Texas and Thomas Kleppe and Mark Andrews of North Dakota testified for the extension. Several other congressmen inserted statements into the record. Norman A. Berg, Associate Administrator of SCS, testified for the Department of Agriculture.

Berg could point to 56,601,700 acres covered by 31,122 contracts. Thirty-seven percent of the funds had been spent to establish vegetation or for reseeding. The average contract had been about $3,500, covering 1,822 acres. Earlier Congressman Richard Crawford had inserted even more impressive information from "Red" Smith of Texas concerning the long-range objective of the program. A survey of the 4,0550 expired contracts in Texas determined that 93.3 percent of the conservation measures had been maintained. Many of the 271 owners who had not maintained conservation practices did so in order to participate in commodity allotment and diversion programs.

Along with requesting the extension, Berg supported changes that would confirm the contribution the soil and water conservation districts had been making to GPCP. Farm conservation plans, developed with district assistance, had been used as the basis for contracts. The change in legislation acknowledged this arrangement. Another provision would allow contracts on nonagricultural land that had erosion. Enhancement of fish, wildlife, and recreation in the plains would be eligible for cost-sharing.

At the 1956 hearings, only the National Farmers Union had supported the GPCP. Now the Farm Bureau and National Grange added their support to that of the Union. The National Association of Conservation Districts enthusiastically supported the extension. Lyle Bauer, Area Vice President, spoke for the extension and the provision to define the role of soil and water conservation districts. The House reported out the bill. After a conference to work out some changes suggested by the Senate committee, the legislation was signed on November 18, 1969. Public Law 91–118 extended the program ten years with a ceiling of $300 million and an annual budget not to exceed $25 million.

Boundary Extended

The House of Representatives hearings in 1969 created a new "legislative history" that allowed expansion of the exterior boundary. Most of the counties within the original boundary had finally been included. In fact, SCS had already added five outside the boundary. Within a month of the signing of the first contracts, SCS recommended adding an additional 22 counties. Donald Williams explained the situation to Assistant Secretary Peterson. "The interest of local people had not developed sufficiently to include this list of counties at the time the initial list was submitted for consideration July 3, 1957." By the end of 1958, the Secretary had approved another 78 counties. Thereafter, there was steady growth until there were 417 designated counties on January 1, 1968. State conservationist "Red" Smith proposed in 1963 that the boundary be extended to include the western cross timbers where there had been wind erosion in the 1950s. He made a good case for the needs of the area. Williams responded that the legislative history would not permit such an extension and that, before any extension, the whole boundary should be studied. Furthermore there was already a backlog of applications, and the lower than authorized appropriations created a "need to concentrate the program in the 422 counties within the original approved boundary." F. A. Mark summed up the feeling of the state conservationists. Unless additional funds could be had, any extension would "play havoc with needs in the existing authorized area."

The National Association of Conservation Districts favored extending the principles of GPCP but favored keeping the original boundary. The Great Plains News informed district members that the original boundary should probably have been drawn farther west in the northern plains and farther east in the southern plains. They asked rhetorically, "once the boundary is changed where can the stopping point be?" With the new authority provided in the GPCP extension, the number expanded from 424 in January 1970 to 469 counties in 1972. The number remained there until Public Law 92–263, signed on June 6, 1980, extended GPCP for another ten years. Another 49 counties then entered the program, bringing the total to 518.

Contract Size Increased

The matter of the limitations on contract size and irrigation costs have continually been discussed throughout the life of GPCP. On one side have been state and local people who favored an increase. But the administrators of the program have had to be attentive to criticism during the 1960s of large payments to individual farmers. The differences in the conservation program and its long-term goal and in commodity programs has not always been obvious to those unfamiliar with the specifics of the programs. The fact that plains farms and ranches were, of necessity, larger than those in humid areas has also led to misunderstanding. A group of state officials and other GPCP leaders suggested in 1975 that the contract limitation be raised to $40,000 and irrigation practices to $7,500. There was little consensus among the state conservationists responding to the proposal. Some wanted the increase; some did not.

Some said that the change would neither hinder nor help GPCP. Interestingly, the attitude in Texas had changed. Edward Thomas, state conservationist, wrote that "some restraint is needed to keep the use of irrigation practices compatible with the legislative intent of the program." The limitation remained in effect until Norman Berg, Chief of SCS, raised the limits to $35,000 total and $10,000 for irrigation in November 1980. By then, inflation had more than negated any effect the change would have had on the uniqueness of the program.

Special Practices

Some of the toughest administrative decisions have concerned approving "special practices." These are designed to allow flexibility for state and regional problems for which the standard GPCP cost-share measures are not adequate. Usually the requests are for sound conservation initiatives, but, nonetheless, are recurring, annual practices which do not meet the criteria of being "enduring." Requests to cost-share for stubble mulching and planned grazing systems have been denied. Approval has been given to the construction of stock trails for livestock distribution, initial planting of tall wheatgrass for wind erosion control, and drip irrigation to get windbreaks established. Recently Norman Berg, Chief of SCS, approved conservation tillage as a special practice. Considering the durability of farm machinery and the initial investment required, it would seem to fit into the "enduring" category.

Special Areas

The success and popularity of GPCP have been such that it inspired suggestions that other sections of the United States could benefit from similar programs. Programs for other specifically designated areas have not succeeded in Congress. The problem of wind erosion may actually have been a benefit in getting legislation enacted for the Great Plains. The dust storms that blew over cities in the 1930s and 1950s awakened urban residents to the problem in the plains and created a feeling of empathy. The deterioration of resources in other areas has not been as visible to persons outside the immediate area. Thus, these problems have not received similar national attention. But there has been one significant development. The Agriculture and Food Act of 1981, as reported out by the committees, included a special areas conservation program to "identify and correct erosion-related or irrigation water management" problems. If the law is enacted, the Secretary of Agriculture can provide technical assistance and share the cost of conservation measures. Under this program, the areas would not be designated in the legislation. The Secretary would have the discretion of selecting areas to participate. It need hardly be noted that the record of GPCP convinced senators and congressmen of the value of a similar program for their states.

Other USDA Programs

Throughout the life of GPCP, there have been suggestions and attempts to merge GPCP with other cost-sharing programs. The argument that has spared GPCP from merger or elimination has been SCS's ability to demonstrate the necessity of linking cost-sharing, technical assistance, and good farm and ranch management to attack a special problem in a special area.

Various cost-sharing and loan programs administered by different agencies need not overlap or create rivalries to the detriment of the conservation effort. During the GPCP Inter-agency Group meetings, the Farmers Home Administration offered to adjust its loan procedures to fit GPCP. This adjustment made it possible to advance FmHA loans in consecutive years to owners and, thereby to assist in carrying out the conservation plan under GPCP. The eligibility of GPCP participants for conservation reserve payments under the now expired soil bank, the long-term agreements, and ACP payments administered by the Agricultural Stabilization and Conservation Service has varied through the past twenty-five years. Cost-sharing funds under ACP could contribute to achieving conservation farming and ranching. However, the Agricultural Stabilization and Conservation Service (ASCS) ruled that after January 1, 1979, participants in GPCP would not be eligible for the ACP cost-sharing program. Prior to that time the ability and willingness of the SCS district conservationist and the FmHA and ASCS representatives to develop a working relationship has been crucial to coordinating programs for the best effect.

The matter of meshing acreage allotments and the commodity price supports that go with them has been of greater concern to those who framed or directed GPCP. Generally, these programs were regarded as being incompatible with the objectives of GPCP because these programs encouraged farmers to plant land to crops that were better suited by capability to grassland or less erosion inducing crops.

In assessing the impact of acreage allotments, one must consider the total effect of farm prices on conservation. The experience of the late 1920s and early 1930s is illustrative. When farmers who have mortgage payments to meet are faced with declining commodity prices or prices that do not keep pace with inflation, the tendency is to expand production to reap an ever diminishing profit on each acre—regardless of the capability of the land. Without endorsing a particular commodity price system, it should be recognized that a healthy and stable agricultural economy is conducive, even necessary, to good conservation farming and ranching.

The Part of GPCP in SCS History

The Great Plains Conservation Program has been significant in the development of SCS and can be regarded as a third era in its history. The agency began operations through demonstration projects and provided WPA and CCC labor, seed, plants, equipment, and other supplies. The Service then shifted to working through conservation districts. The labor, equipment, and supplies ceased being available with the onset of World War II. The conservation effort then rested on the ability of conservation district supervisors and SCS conservationists to convince land owners of the benefits of conservation. The Small Watershed Act (1954) and GPCP provided SCS with the inducement of cost-sharing to accelerate the conservation work with local governing bodies and individuals. The lessons learned on contracting and cost-sharing in GPCP have been the model used for land treatment in Small Watershed Projects, the Resource, Conservation and Development Program, the Rural Abandoned Mine Program, and the Rural Clean Water Program.

GPCP also changed the role of the individual SCS conservationist to a limited extent. The GPCP contract was much like a good conservation farm plan, only more detailed. Under the contractual arrangement, he had to certify that both parties, government and individual, met their obligations. Insuring compliance with some aspects of a contract, such as preventing newly seeded range from being grazed too soon, was a new task for the conservationist. These new management roles brought a closer working relationship between the conservationist and the farmer that eventually benefited the land. Not only did farmers and ranchers learn better farm and ranch management techniques, but also the expertise of the conservationist increased. Improved stewardship of land has resulted.

The contract between the individual and the government has been the aspect of GPCP that made it unique. SCS technicians annually reviewed contracts to insure that cost-sharing monies were spent and practices maintained as specified in the contract. Although breaches of contracts were the exception, SCS in some cases cancelled contracts and collected payments made to violators. Such vigilance, combined with a willingness to make changes in contracts when justified, early established the reputation of GPCP as a unique conservation program.

A Unique Conservation Program

The burden of keeping GPCP attuned to its objective also fell on the administrators in the Washington office. During the last twenty-five years, national agricultural policy has fluctuated between using various programs to promote production of commodities and de-emphasizing production programs to reduce surplus commodities. It is usually expected that all agricultural programs be adjusted to the goal. GPCP has had to operate in the varying climate of national agricultural policy and yet retain its objective. As SCS and the National Association of Conservation Districts were preparing in 1968 to ask for an extension of the program, William Vaught, supervisor of GPCP operations, spoke to the Great Plains conservation district leaders about retaining the uniqueness of GPCP.

> Don Williams, in maintaining a personal interest in the program, has held steadfast over the years in his efforts to keep faith with Congress. And I might add that it has not been an easy thing to do. He has been under constant pressure to relax some of the restrictions…as we move into the process of attempting once again to solicit the support of Congress…we can be thankful for his determination. I think we **have** kept the faith with Congress and its intent to provide a unique program—regional in nature—to help us solve those tough wind erosion problems.

The succeeding administrators, Kenneth Grant and R. M. Davis, kept the program on course. The present Chief, Norman Berg, "grew up with the program" and knows the elements that have to be retained to keep it unique. The administrators and chief have relied on specialists to advise and carry out the daily operations of GPCP. Cyril Luker started the program as head of the Inter-agency Group and was followed by Norman A. Berg, William L. Vaught, John W. Arnn, Julius H. Mai, John J. Eckes, and Guy D. McClaskey.

Impact of GPCP

Of necessity, the success of the program must be judged in terms of the land and its condition, compared to the 1950s. What happened to the land? SCS estimated in 1956 that between 11 and 14 million acres were in cultivation in the plains that should be in grass. SCS had to estimate the figure because soil surveys and land capability studies had not been completed. Before the enactment of P.L. 1021, the Service increased the hiring of soil scientists for surveying the plains states. Further-more, the state conservation district associations concurred in plans to shift experienced soil scientists from the prairie and mountain sections to the plains to accelerate the soil surveys. By September 30, 1980, 2,869,062 acres of former cropland had been converted to grassland. An undetermined percentage of this has reverted to crops since the expiration of contracts. Developments in conservation tillage and drought resistant crops have reduced the hazards of cropping marginal lands. With the need to spread the use of conservation tillage, it is desirable not to present it as the new "panacea" that makes complementary conservation measures unnecessary. Drought resistant crops have been of great benefit in controlling wind erosion. However, if the drought is so prolonged on some sandy land that spring germination is impossible, it will make little difference whether the seeds are of drought resistant varieties or not.

Other questions surround the success of GPCP. Did irrigation for pastures and for forage make cattle raising possible for ranchers who did not own enough land for dryland ranching? Have we seen the last of the wild fluctuations in the number of cattle on the range during droughts and good years? Has the program halted the cycles of migration out of the plains during droughts and land speculation in good years that resulted in each succeeding generation repeating the mistakes of the past? Were farmers and ranchers better able to withstand droughts? Studies in North Dakota and South Dakota indicated that this was the case. In short, did GPCP bring about the agricultural and resource stability promised in 1956? A study of these questions and others would be of interest on the county, state, and regional level. All of them may not be answerable by quantification, or by the numbers. Many who participated in GPCP as farmers, ranchers, district conservationists, or conservation district supervisors believe that the judgment is in the affirmative, or partially so, on many questions.

Donald Williams recently summed up his dual feeling of success and frustration over the conservation movement in general. "It seemed like we would get to a certain point and then something would happen. The war would break out. The price of wheat would go up, and the farmers would go out and plow up the land again. So there you are; you had to back up and start over again in a way. But we never went clear back to where we were before. We had a better starting point so that we were able to get ahead." No doubt many regard GPCP as a significant development in the push to "get ahead" with conservation work.

Summary Report: 1997 National Resources Inventory

United States Department of Agriculture, 1999, selected pages.

(Editor's Note: The land-use designation CRP refers to the Conservation Reserve Program, in which farmers are paid to remove farm land from farming and allow it to lay fallow. "Sheet and rill erosion" refers to soil erosion caused by water.)

Introduction

This bulletin presents summary results from the 1997 National Resources Inventory (NRI), conducted by the U.S. Department of Agriculture's Natural Resources Conservation Service, in cooperation with the Iowa State University's Statistical Laboratory. The NRI is a scientifically-based longitudinal panel survey of the Nation's soil, water, and related resources designed to assess conditions and trends every five years. The 1997 NRI provides results that are nationally consistent for all non-Federal lands for four points in time—1982, 1987, 1992, and 1997.

This bulletin includes state and national level estimates for changes in broad land cover/use, cropland use by irrigated and non-irrigated acres, broad land cover/use by land capability class and subclass, prime farmland, erosion and erodibility, and wildlife habitat diversity. These basic summary statistics are presented on the Internet and in hard copy to provide base-line natural resource information to a variety of groups and individuals interested in obtaining insight into the condition of our Nation's non-federal rural lands. Subsequent sections of this bulletin discuss the broader suite of information available from the 1997 NRI and methods for obtaining access to other results.

Background

The Natural Resources Conservation Service, formerly the Soil Conservation Service, was established in response to the Dust Bowl catastrophe of the mid-1930s. Hugh Hammond Bennett, the agency founder and first administrator, convinced the U.S. Congress that soil erosion was a national menace and that a permanent agency within the Department of Agriculture was needed to call landowners' attention to their land stewardship opportunities and responsibilities. The results of the 1934 National Erosion Reconnaissance Survey, which was the first formal study of erosion conducted in the United States, were instrumental in the passage of the Soil Conservation Act of 1935. Through the Act, the Soil Conservation Service was established, and a nationwide partnership of federal agencies, local conservation districts, and communities was developed to provide assistance to the rural and urban sectors in the conservation of natural resources. Today, more than 60 years later, NRCS champions the vitality of the land as USDA's lead conservation agency. No other federal agency speaks for the health of America's private land.

Throughout its history, NRCS has conducted periodic inventories of the Nation's natural resources. The 1945 Soil and Water Conservation Needs Inventory (CNI), a reconnaissance study, was the foundation for the 1958 and 1967 CNI's, the agency's first efforts to collect data nationally for scientifically selected field sites. The 1975 Potential Cropland Study examined the conversion of the Nation's best farmland to urban development. National Resources Inventories were conducted in 1977, 1982, 1987, 1992, and 1997. Several less-intensive, special issue inventories have been performed during the 1990s to investigate topical matters of concern and to supplement the major NRI's.

In addition to these recurrent NRI inventories, NRCS also collects large quantities of field level natural resources data in support of conservation planning activities and the Soil Survey Program. Thousands of NRCS technical specialists, including soil scientists, soil conservationists, range conservationists, foresters, wildlife biologists, and agronomists collect data at the field and farm level in order to provide conservation assistance to farmers and ranchers in the development of conservation systems uniquely tailored to the land and their individual way of doing business. Assistance is also provided to rural and urban communities to help reduce erosion, conserve and protect water resources, and solve other resource-related problems. The information that NRCS collects about natural resources in the United States is critical for sustaining agriculture, promoting the conservation and stewardship ethic, and for preserving the Nation's well-being.

Legislation also has mandated that NRCS collect natural resources data. The Rural Development Act of 1972 was a key statute in authorizing resource inventory activities within NRCS. It directs the Secretary of Agriculture to implement a land inventory and monitoring program and to issue a report on the conditions and trends of soil, water, and related resources at intervals not exceeding 5 years. The Soil and Water Resources Conservation Act of 1977 and other supporting legislation augmented the statutory mandate for periodic assessment of the Nation's natural resources. To fulfill this requirement, the NRI was developed to provide critical information regarding natural resources and to supplement the NRCS Soil Survey Program.

Inventory Procedure

The objectives of NRCS resource inventories have expanded over time, as the focus of agricultural policy has moved toward a balance between short-term production goals, long-term capabilities, and environmental quality. Statistical techniques, data collection protocols, and data handling and dissemination technologies have evolved as inventory goals have become broader and more sophisticated.

The primary objective of the 1997 NRI was to provide natural resource managers, policy makers, and the public with scientifically valid, timely, and relevant information on natural resources and the environment. This information can provide the scientific basis for effective public policies, sound agricultural and natural resource legislation, sensible state and national conservation programs, and targeted USDA financial and technical assistance in addressing natural resource concerns. NRI data are designed to be part of the core components of the agency's strategic planning and accountability efforts, and to help assess consequences of existing legislative mandates, such as the 1996 Farm Bill.

To accomplish these objectives in a cost-effective manner, it was necessary to conduct the 1997 NRI in much the same manner as the 1992 NRI. Careful consideration was given to assure that 1997 NRI data elements were consistent with definitions, categories, and concepts from previous inventories. The sample used for the 1992 NRI was used for 1997 data collection. This enables analysis of trends extending over 15 years (1982, 1987, 1992, 1997).

NRI data are collected at scientifically selected sample sites. The sample constitutes a two-stage stratified area sample of the entire country. Samples are located in all counties and parishes of the 50 states and in Puerto Rico, the Virginia Islands, the District of Columbia, and selected portions of the Pacific Basin. The first-stage sampling unit, or primary sample unit (PSU), is an area/segment of land; the second-stage sampling units are points located within the PSU's. Detailed NRI data are collected for the specific sample points, but some items are also collected for the entire PSU/segment. Some data, such as total surface area, federally owned land, and area in large water bodies, are collected on a census basis external to the sample survey. The NRI database accounts for and represents the total area of the United States, but very little information is given for points on federal lands.

Data for the 1997 NRI were collected for about 300,000 PSU's and 800,000 sample points, using photo-interpretation and other remote sensing methods and standards. Data gatherers utilized a variety of ancillary materials; extensive use was made of USDA field office records, information provided by local NRCS field personnel, soil survey and wetland inventory maps and reports, and tables and technical guides developed by local field office staffs. The NRI is unique because it is based upon NRCS expertise in identifying soil occurrences and patterns, and then utilizing this knowledge of soils (and extensive databases of properties and characteristics) in providing technical assistance and developing conservation plans for land owners. The NRI data gathering process relies heavily upon information provided by the NRCS Soil Survey Program. Knowledge about the specific soil occurring at the sample site and the many properties and characteristics of that soil and surrounding landscape are utilized in the NRI data development process.

Inventory procedures were developed to ensure:

- that data reflect 1997 growing season conditions,
- that inventory results are nationally consistent, and
- that data recorded for the years 1982, 1987, and 1992 are consistent with the 1997 determination.

Intricate quality assurance procedures were developed to make sure that year-to-year differences reflect actual changes in resource conditions, rather than differences in the perspectives of two different data collection specialists or changes in technologies and protocols.

Data gathering for the 1997 NRI occurred from July 1997 through October 1998. This time frame took into account that some aerial photography needed to be flown during a time period that highlighted late growing season conditions. Consequently, delivery of imagery to some data collection sites did not occur until later in the data collection cycle.

Field visits were not required for the 1997 NRI unless available imagery and ancillary materials were not suitable for making determinations for one or more data elements. Field visits were also made for training purposes and other facets of the quality assurance process. All NRI sample sites were visited on-site for the 1982 NRI. Subsequent on-site visits of selected PSU's also occurred in 1987, 1991, 1992, or 1995.

The computer-assisted survey information collection methods developed for the 1997 NRI provided substantial efficiencies in data gathering and data processing and were important facets of the quality assurance process. The system featured direct entry of data into hand-held computers called personal digital assistants (PDA's), modern telecommunication strategies, a centralized database server at the Iowa State University Statistical Laboratory, and elaborate data checking protocols that featured review and edit of data recorded during previous inventories.

Standards and protocols for the NRI were developed nationally by NRCS, in collaboration with the Statistical Laboratory. Oversight and management of data gathering activities were assigned to 21 units established during 1996 and 1997. These units, called Inventory Coordination and Collection Sites (ICCS's), were established according to regional land use patterns and according to state allocations of resources. Geographic boundaries of ICCS organizations ranged from one state to all or portions of several states. Some ICCS's distributed data collection staff among multiple office locations, while other assembled staff at one central location.

Inventory methodology is evolving as part of an ongoing effort to better assess soil conservation, natural resource health, and other agri-environmental issues. The NRI has been conducted as a longitudinal survey designed to assess condition and trends for nonfederal lands every five years. Current initiatives include transitioning into a continuous resource inventory process, developing a multi-agency integrated inventory approach, incorporating a wider variety of assessment tools for resource health, and further developing geospatial analysis and modeling capabilities to support policy analysis and program implementation.

Utilization and Interpretation of NRI Data

Uses of the Data

The NRI database contains millions of pieces of information. It can serve as the foundation for inspection and analysis of the condition of our Nation's natural resources. It indicates:

- how our Nation's nonfederal lands are being used
- the condition of our natural resources
- how land use patterns have changed over time

The NRI database has been constructed in a manner that facilitates the inspection and analysis of these data. Sophisticated statistical procedures developed collaboratively with Iowa State University have been used to provide a database that scientifically incorporates a broad array of data into a format that is easy to use and manipulate.

The 1997 NRI database contains data for four points in time (1982, 1987, 1992, 1997) that are comparable and consistent, and that reflect true trends. Reliable and accurate temporal analysis is available from this data set. Analytical capabilities are greatly enhanced because NRCS's extensive soil interpretations database is an integral and easy-to-use part of the NRI database.

The NRI is conducted to obtain scientifically valid, timely, and relevant data on natural resources and environmental conditions, with the specific goal of supporting agricultural and environmental policy development and program implementation. Historically, NRI information has been used to formulate effective public policies, to fashion agricultural and natural resources legislation, to develop State and National conservation programs, to allocate USDA financial and technical assistance in addressing natural resource concerns, and to enhance the public's understanding of natural resources and environmental issues. Information derived from the NRI is used by natural resource managers; policy makers and analysts; consultants; the media; other federal agencies; state governments; universities; environmental, commodity, and farm groups; and the public.

Interpretation of the Data

Statistics derived from the NRI database are estimates and not absolutes. This means that there is some amount of uncertainty in any result obtained using NRI data.

The NRI database contains linkages to other databases, in particular to the agency's extensive soil interpretations database. Linkages to other databases can be made by using other themes, such as cover/use, forest cover type, and spatial features. Analysis of NRI data in conjunction with other data sources is encouraged but differences in definitions, concepts, and data collection protocols should be carefully examined. Additionally, it is worth repeating that the NRI includes very little data for federal lands.

The 1997 NRI database has been designed for use in detecting significant changes in resource conditions relative to the years 1982, 1987, 1992, and 1997. All comparisons for two points in time should be made using the new 1997 NRI database. Comparisons made using data published for the 1982, 1987, or 1992 NRI may produce erroneous results, because of changes in statistical estimation protocols, and because all data collected prior to 1997 were simultaneously reviewed (edited) as 1997 NRI data were collected. Note, for example, that federal land area for 1992 has been adjusted from 408 to 402 million acres, and that the estimate of 1992 nonfederal rangeland has changed from 399 to 405 million acres.

The NRI provides not only overall estimates of change in resource conditions but also the dynamics of the changes. For example, it is typically more informative to examine gross losses and gains in cropland, rather than just the net change from one year to another and further to determine why cropland was lost (i.e., to urban development), how much had been prime farmland, "where" these losses are occurring. If new cropland is gained, will this cause additional conservation and environmental concerns, because the land is more erodible, the soils are less productive and require higher levels of fertilization, or the site is located in some other sensitive location.

The erosion data cannot be used to compute the actual erosion occurring during a particular year. Erosion rates are estimated average annual (or expected) rates based upon the cropping practices, management practices, and inherent resource conditions that occur at each NRI sample site. Climatic factors used in the erosion prediction equations (models) are based upon long-term average conditions and not upon one year's actual events. Note also that NRI estimates of sheet and rill erosion are based upon the standard Universal Soil Loss Equation (USLE) and not the revised USLE (RUSLE), and that erosion estimates are made only for cropland, CRP land, and pastureland.

The NRI category of "developed land" varies from that used by some other data collection entities. From the NRI, the intent is to identify which lands have been permanently removed from the rural land base. Therefore, the developed land category includes: (a) large tracts of urban and built-up land; (b) small tracts of built-up land, less than 10 acres in size; and (c) land outside of these built-up areas that is in roads, railroads, and associated rights-of-way.

The 1997 NRI shows only minor changes in land under Conservation Reserve Program (CRP) contracts for the time period 1992 to 1997, even though most original CRP contracts expired in the mid-1990's and there were extensive sign-ups during that period. This is because the 1997 NRI reflects conditions as of the 1997 growing season, and most actual on-the-ground changes in CRP land did not occur until later in 1997 or until the 1998 growing season.

For the NRI, land is considered irrigated if irrigation occurs during the year of inventory, or for two or more of the last four years. Other entities typically consider land to be irrigated only if irrigation water is applied for the year of interest.

The NRI has been designed to facilitate geospatial analysis. This not only enhances the analysis process, but also provides the ability to use a map to present analytical findings. This can be quite a powerful medium for displaying geospatial trends. Maps produced from NRI data depict only patterns or trends within an area and do not provide an estimate of conditions for any specific location on the map.

Availability

This report presents selected NRI summary data at the national level. Further information regarding the NRI and additional data summaries can be obtained from the national NRI Web site at: http://www.nhq.nrcs.usda.gov/NRI. Additional data summaries from the NRI will be released periodically as more comprehensive analyses are performed.

Of particular interest are detailed compilations of data at the state level, which can be accessed via this Web site. Active links to individual state Web sites are available for obtaining specific state-level NRI information.

Explanation of the Tables

On the following pages selected national summary data are displayed in tables. National totals include results for the 48 contiguous states, Hawaii, and the Caribbean area. Results for Alaska and the Pacific Basin islands of Guam, Rota, Tinian, and Saipan will be released at a later date.

The category "other rural land," which occurs in many of the tables, includes farmsteads and other farm structures, field windbreaks, barren land such as salt flats or exposed rock, and marshland.

The figures used in the tables are estimates, not absolutes. They are based on data collected at sample sites, not data taken from a complete census. Therefore, sampling variation is present but generally small for state and national totals. However, sampling variation may be significant when using these totals to calculate 5- and 10-year changes. Small changes may not be statistically significant.

Table 1 presents NRI findings on surface area, federal land, nonfederal rural land, developed land, and water area. Since 1982, federal land increased by 4.6 million acres, nonfederal rural land decreased by 36.7 million acres, and developed land increased by nearly 30 million acres.

Tables 2 and 3 present estimates of acreage of land cover/use for 6 components of nonfederal rural land (cropland, pastureland, rangeland, forest land, other rural land, and CRP land). Cropland is classified as irrigated, nonirrigated, cultivated, or noncultivated acreage. Cropland acreage nationally decreased by 45.9 million acres between 1982 and 1997. Rangeland decreased by 12.4 million acres and pastureland decreased by almost 14 million acres (Fig. 1). Table 3 further depicts a shift in irrigated agriculture from west to east across the country.

Table 4 presents acres of land cover/use by land capability class and subclass. The land capability classification system was originally developed by the Natural Resources Conservation Service and provides a quick, uniform, and useful way to evaluate the potential of land for crop production. Each capability class has several subclasses to identify specific limitations on use:

- e = erosion risk
- w = wetness
- s = shallowness or root zone problems, and
- c = climatic limitations.

Class I soils have few limitations that restrict their use. Class II soils have moderate limitations that reduce the choice of plants or that require careful management. Land identified as Class IIe, for example, would be suitable for growing crops if adequate measures were installed to reduce or prevent soil erosion.

Class III soils have severe limitations that reduce the choice of plants, require special conservation practices, or both. Class IV soils have very severe limitations that reduce the choice of plants, require very careful management, or both.

Class V soils are not likely to erode but have other limitations, impractical to remove, that limit their use largely to pasture or range, woodland or wildlife. Class VI soils have severe limitations that make them generally unsuitable for cultivation and limit their use largely to pasture, range, woodland, or wildlife. Class VII soils have very severe limitations that make them generally unsuitable for cultivation and limit their use largely to pasture, range, woodland, or wildlife. Class VIII soils and miscellaneous land types have limitations that preclude their use for commercial crop production and restrict their use for recreation, wildlife, water supply, or esthetic purposes.

Tables 5, 6, 7, and 8 provide an overview of land use changes from 1982-97, 1982-87, 1987-92, and 1992-97 (Fig. 2). These tables show all land conversions, whereas previous Tables presented net land use change. For example Table 5 shows that a total of 72.2 million acres of 1982 cropland was converted to other uses by 1997, which was offset by 26.3 million acres converted to cropland from non-cropland uses since 1982. The net change was therefore a reduction of 45.9 million acres, as shown in Table 2. Table 5 further shows that, of the 72.2 million acres of cropland converted to other uses, 30.5 million acres went to CRP, 19.1 million acres went to pastureland, 3.5 million acres went to rangeland, 5.4 million acres went to forest land, 3.3 million acres went to other rural land, 8.8 million acres went to developed land, and 1.6 million acres went to water areas and federal land. Of the 26.3 million acres converted to cropland from other uses, 15.6 million acres came from pastureland, 6.9 million acres came from rangeland, 1.9 million acres came from forest land, 1.0 million acres came from other rural land, 0.2 million acres came from developed land, and 0.6 million acres came from water areas and federal land.

Table 9 presents the distribution of prime farmland by land cover/use. Prime farmland is land that has the best combination of physical and chemical characteristics for producing food, feed, forage, fiber, and oilseed crops and is also available for these uses. There were 330.6 million acres of prime farmland in 1997, which was down 11.7 million acres from 1982. Most (64%) of the prime farmland is in cropland, but large amounts are in pastureland (35.5 million acres) and forest land (47.7 million acres).

Tables 10 and 11 present estimates from the NRI for soil erosion rates. Table 10 shows rates of sheet and rill erosion, which is erosion caused by water; Table 11 represents estimates of wind erosion. Average erosion rates for 1997 are substantially lower than erosion rates for 1982. The average rate of sheet and rill erosion fell from 4.1 tons per acre per year in 1982 to 2.8 tons per acre per year in 1997. The average rate of wind erosion on the same land base fell from 3.3 tons per acre per year in 1982 to 2.2 tons per acre per year in 1997. The combined wind and water erosion reduction translates to a savings of more than 1.2 billion tons of soil per year on cropland (Fig. 3).

Tables 12 and 13 present acres of cropland (cultivated and noncultivated), pastureland, and CRP land with erosion rates greater than T, the soil loss tolerance or rate at which soil productivity is maintained.

Table 14 shows acreage according to six classes of erodibility index scores. The erodibility index (EI) provides a numerical expression of the potential for a soil to erode, considering the physical and chemical properties of the soil and climatic conditions where it is located. The higher the index, the greater the investment needed to maintain the sustainability of the soil resource base if intensively cropped. EI scores above 8 are equated to highly erodible land.

Table 15 presents statistical information dealing with wildlife habitat composition and configuration. Median diameter of wildlife patch size is an indicator of habitat diversity. For the 1997 NRI, general cover data were collected along X-shaped transects [the length of each diagonal line of the transect was 1,000 feet]. Patches of cover were classified to one of nine general cover types (see glossary). Entries in Table 15 denoted as "1,000" indicate that at least 50% of the transects were classified as having a 1,000 foot length of the same cover type.

Metric Conversion

To convert acres to hectares, multiply the number of acres by 0.405.

To convert tons to metric tons, multiply the number of tons by 0.907.

To convert tons/acre to metric tons/hectare, multiply the number of tons/acre by 2.24.

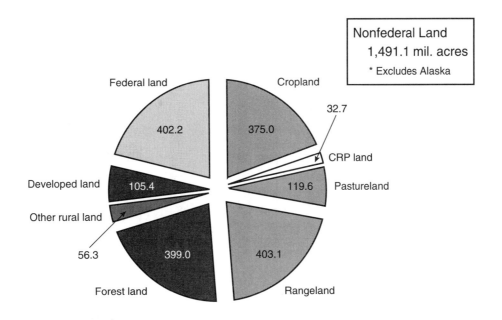

*Figure 1—How Our Land is Used. 1992 Data in Million Acres**

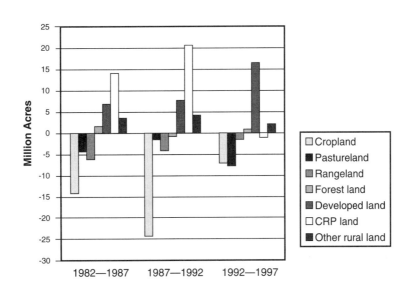

Figure 2—Land Use Changes, 1982–1992

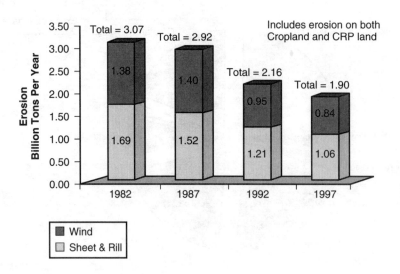

Figure 3—Cropland Erosion

Table 1—Surface area of nonfederal and federal land and water areas, by state and year—page 1 of 7

State		Federal land	Water areas	Nonfederal land			Total surfaced area
				Developed	Rural	Total	
		---------------- 1,000 acres ----------------					
Alabama	1982	945.0	1,162.6	1,643.6	29,672.4	31,316.0	33,423.8
	1987	951.7	1,178.6	1,839.4	29,454.1	31.293.5	33,423.8
	1992	970.0	1,198.4	1,964.5	29,290.9	31,255.4	33,423.8
	1997	997.8	1,241.7	2,409.8	28,774.5	31,184.3	33,423.8
Arizona	1982	30,257.4	184.3	1,101.2	41,421.5	42,522.7	72,964.4
	1987	30,448.3	186.6	1,326.4	41,003.1	42,329.5	72,964.4
	1992	30,426.2	189.4	1,475.8	40,873.0	42,348.8	72,964.4
	1997	30,426.2	208.7	1,675.2	40,654.3	42,329.5	72,964.4
Arkansas	1982	3,030.6	798.5	1,167.8	29,040.0	30,207.8	34,036.9
	1987	3,041.2	837.8	1,203.8	28,954.1	30,157.9	34,036.9
	1992	3,102.5	844.5	1,263.6	28.826.3	30,089.9	34,036.9
	1997	3,102.8	894.6	1,500.8	28,538.7	30,039.5	34,036.9
California	1982	45,620.8	1,913.3	4,192.1	49,784.0	53,976.1	101,510.2
	1987	45.874.0	1,920.9	4,464.4	49,250.9	53,715.3	101,510.2
	1992	46,633.4	1,911.4	4,992.3	47,973.1	52,965.4	101,510.2
	1997	46,633.4	1,951.3	5,687.1	47,238.4	52,925.5	101,510.2
Colorado	1982	23,574.2	338.2	1,277.9	41,434.2	42,712.1	66,624.5
	1987	23,732.2	340.2	1,430.2	41,121.9	42,552.1	66,624.5
	1992	23,802.9	343.7	1,585.3	40,892.6	42,477.9	66,624.5
	1997	23,793.8	350.5	1,705.6	40,774.6	42,480.2	66,624.5
Connecticut	1982	9.7	126.4	749.4	2,309.2	3,058.6	3,194.7
	1987	14.3	126.7	795.4	2,258.3	3,053.7	3,194.7
	1992	14.5	127.2	833.6	2,219.4	3,053.0	3,194.7
	1997	14.5	128.7	897.0	2,154.4	3,051.5	3,194.7
Delaware	1982	28.3	288.2	167.2	1,049.8	1,217.0	1,533.5
	1987	31.0	288.8	184.9	1,028.8	1,213.7	1,533.5
	1992	31.0	289.0	202.5	1,011.0	1,213.5	1,533.5
	1997	31.0	289.5	237.6	975.4	1,213.0	1,533.5
Florida	1982	3,656.3	3,058.3	3,340.3	27,478.8	30,819.1	37,533.7
	1987	3,679.4	3,060.2	3,752.7	27,041.4	30,794.1	37,533.7
	1992	3,784.2	3,088.8	4,503.4	26,157.3	30,660.7	37,533.7
	1997	3,784.2	3,153.4	5,448.7	25,147.4	30,596.1	37,533.7

Table 1—Surface area of nonfederal and federal land and water areas, by state and year—page 2 of 7

State		Federal land	Water areas	Nonfederal land			Total surfaced area
				Developed	Rural	Total	
		---------------- 1,000 acres ----------------					
Georgia	1982	2,093.9	951.1	2,418.6	32,276.9	34,695.5	37,740.5
	1987	2,107.3	969.3	2,698.2	31,965.7	34,663.9	37,740.5
	1992	2,125.6	1,004.8	3,184.9	31,425.2	34,610.1	37,740.5
	1997	2,124.0	1,052.9	4,238.1	30,325.5	34,563.6	37,740.5
Hawaii	1982	319.8	58.5	153.2	3,631.7	3,784.9	4,163.2
	1987	386.3	58.5	156.6	3,561.8	3,718.4	4,163.2
	1992	388.0	58.5	176.8	3,539.9	3,716.7	4,163.2
	1997	388.0	58.5	185.5	3,531.2	3,716.7	4,163.2
Idaho	1982	33,558.6	547.5	604.1	18,777.3	19,381.4	53,487.5
	1987	33,318.2	550.4	678.1	18,940.8	19,618.9	53,487.5
	1992	33,480.9	552.2	690.0	18,764.4	19,454.4	53,487.5
	1997	35,563.1	556.3	810.8	18,557.3	19,368.1	53,487.5
Illinois	1982	464.9	735.5	2,723.3	32,135.0	34,858.3	36,058.7
	1987	464.9	737.6	2,874.9	31,981.3	34,856.2	36,058.7
	1992	492.0	739.2	2,969.3	31,858.2	34,827.5	36,058.7
	1997	490.2	761.2	3,261.5	31,545.8	34,807.3	36,058.7
Indiana	1982	470.3	344.9	1,853.0	20,490.2	22,343.2	23,158.4
	1987	470.9	359.8	1,973.3	20,354.4	22,327.7	23,158.4
	1992	473.5	365.0	2,081.3	20,238.6	22,319.9	23,158.4
	1997	472.4	386.1	2,355.7	19,944.2	22,299.9	23,158.4
Iowa	1982	153.0	439.7	1,647.6	33,776.2	35,423.8	36,016.5
	1987	154.1	445.4	1,675.5	33,741.5	35,417.0	36,016.5
	1992	164.2	449.0	1,699.9	33,703.4	35,403.3	36,016.5
	1997	186.2	476.6	1,802.8	33,550.9	35,353.7	36,016.5
Kansas	1982	494.6	538.7	2,572.3	49,055.2	51,627.5	52,660.8
	1987	495.0	539.2	2,605.5	49,021.1	51,626.6	52,660.8
	1992	504.0	524.2	2,689.3	48,943.3	51,632.6	52,660.8
	1997	504.0	559.9	2,881.8	48,715.1	51,596.9	52,660.8
Kentucky	1982	1,091.5	591.8	1,238.4	22,941.7	24,180.1	25,863.4
	1987	1,140.7	600.1	1,435.7	22,686.9	24,122.6	25,863.4
	1992	1,176.9	615.2	1,601.2	22,470.1	24,071.3	25,863.4
	1997	1,187.1	628.8	1,955.3	22,092.2	24,047.5	25,863.4

Table 1—Surface area of nonfederal and federal land and water areas, by state and year—page 3 of 7

State		Federal land	Water areas	Nonfederal land			Total surfaced area
				Developed	Rural	Total	
		------------------ 1,000 acres ------------------					
Louisiana	1982	1,163.6	3,577.0	1,257.2	25,379.0	26.636.2	31,376.8
	1987	1,242.8	3,620.4	1,412.0	25,101.6	26,513.6	31,376.8
	1992	1,308.1	3,677.2	1,520.4	24,871.1	26,391.5	31,376.8
	1997	1,308.1	3,754.3	1,692.5	24,621.9	26,314.4	31,376.8
Maine	1982	165.9	1,247.5	486.3	19,066.5	19,552.8	20,966.2
	1987	193.2	1,249.5	532.8	18,990.7	19,523.5	20,966.2
	1992	197.7	1,249.2	578.8	18,940.5	19,519.3	20,966.2
	1997	207.2	1,250.1	746.6	18,762.3	19,508.9	20,966.2
Maryland	1982	162.2	1,651.0	921.4	5,135.5	6,056.7	7,869.9
	1987	162.3	1,653.8	1,002.9	5,050.9	6,053.8	7,869.9
	1992	168.9	1,658.5	1,068.3	4,974.2	6,042.5	7,869.9
	1997	168.9	1,663.3	1,290.6	4,747.1	6,037.7	7,869.9
Massachusetts	1982	97.2	366.9	1,034.4	3,840.5	4,874.9	5,339.0
	1987	97.7	369.1	1,140.2	3,732.0	4,872.2	5,339.0
	1992	97.7	376.5	1,267.5	3,597.3	4,864.8	5,339.0
	1997	97.7	379.4	1,549.0	3,312.9	4,861.9	5,339.0
Michigan	1982	3,141.0	1,097.0	2,750.6	30,360.6	33,111.2	37,349.2
	1987	3,213.2	1,091.6	2,953.7	30,090.7	33,044.4	37,349.2
	1992	3,274.7	1,095.9	3,212.9	29,765.7	32,978.6	37,349.2
	1997	3,274.7	1,110.6	3,763.7	29,200.2	32,963.9	37,349.2
Minnesota	1982	3,233.9	3,118.7	1,814.0	45,843.3	47,657.3	54,009.9
	1987	3,299.6	3,117.8	1,941.2	45,651.3	47,592.5	54,009.9
	1992	3,336.3	3,121.8	2,049.6	45,502.2	47,551.8	54,009.9
	1997	3,336.3	3,147.3	2,360.9	45,165.4	47,526.3	54,009.9
Mississippi	1982	1,602.9	727.8	1,198.9	26,997.7	28,196.6	30,527.3
	1987	1,676.7	769.1	1,279.6	26,801.9	28,081.5	30,527.3
	1992	1,751.9	797.1	1,343.2	26.635.1	27,978.3	30,527.3
	1997	1,770.0	862.1	1,655.8	26,239.4	27,895.2	30,527.3
Missouri	1982	1,878.8	752.3	2,137.5	39,845.3	41,892.8	44,613.9
	1987	1,872.2	779.6	2,239.4	39,722.2	41,961.6	44,613.9
	1992	1,902.8	807.9	2,342.0	39,561.2	41,903.2	44,613.9
	1997	1,917.0	849.4	2,652.5	39,195.0	41,847.5	44,613.9

Table 1—Surface area of nonfederal and federal land and water areas, by state and year—page 4 of 7

State		Federal land	Water areas	Nonfederal land			Total surfaced area
				Developed	Rural	Total	
		---------------- 1,000 acres ----------------					
Montana	1982	27,207.5	1,073.7	679.0	65,149.8	65,828.8	94,110.0
	1987	27,166.3	1,069.8	692.5	65,181.4	65,873.9	94,110.0
	1992	27,089.7	1,052.5	758.0	65,209.2	65,967.8	94,110.0
	1997	27,089.7	1,060.7	881.3	65,078.3	65,959.6	94,110.0
Nebraska	1982	575.6	461.6	1,147.5	47,324.9	48,472.4	49,509.6
	1987	583.3	478.8	1,166.4	47,281.1	48,447.5	49,509.6
	1992	649.4	478.9	1,186.7	47,194.6	48,381.3	49,509.6
	1997	649.4	489.7	1,267.9	47,102.6	48,370.5	49,509.6
Nevada	1982	59,704.4	424.7	291.6	10,342.4	10,634.0	70,763.1
	1987	59,722.8	430.4	339.9	10,270.0	10,609.9	70,763.1
	1992	59,870.7	431.5	374.3	10,086.6	10,460.9	70,763.1
	1997	59,870.7	444.6	415.8	10,032.0	10,447.8	70,763.1
New Hampshire	1982	726.7	233.1	385.2	4,596.0	4,981.2	5,941.0
	1987	730.7	233.8	476.7	4,499.8	4,976.5	5,941.0
	1992	756.3	234.9	534.4	4,415.4	4,949.8	5,941.0
	1997	763.2	236.5	641.7	4,299.6	4,941.3	5,941.0
New Jersey	1982	133.1	510.3	1,267.1	3,305.1	4,572.2	5,215.6
	1987	135.9	516.9	1,490.2	3,072.6	4,562.8	5,215.6
	1992	148.3	520.3	1,565.7	2,981.3	4,547.0	5,215.6
	1997	148.3	530.2	1,848.9	2,688.2	4,537.1	5,215.6
New Mexico	1982	25,646.1	147.3	809.8	51,220.1	52,029.9	77,823.3
	1987	26,115.1	148.1	895.5	50,664.6	51,560.1	77,823.3
	1992	26,448.5	146.9	976.1	50,251.8	51,227.9	77,823.3
	1997	26,448.5	154.5	1,324.6	49,895.7	51,220.3	77,823.3
New York	1982	219.3	1,251.6	2,655.7	27,234.2	29,889.9	31,360.8
	1987	218.8	1,261.2	2,756.3	27,124.5	29,880.8	31,360.8
	1992	208.9	1,273.2	2,880.8	26,997.9	29,878.7	31,360.8
	1997	208.9	1,286.0	3,373.2	26,492.7	29,865.9	31,360.8
North Carolina	1982	2,191.1	2,737.7	2,463.3	26,317.2	28,780.5	33,709.3
	1987	2,361.9	2,750.8	2,905.9	25,690.7	28,596.6	33,709.3
	1992	2,506.6	2,758.5	3,399.1	25,045.1	28,444.2	33,709.3
	1997	2,507.5	2,776.9	4,180.6	24,244.3	28,424.9	33,709.3

Table 1—Surface area of nonfederal and federal land and water areas, by state and year—page 5 of 7

State		Federal land	Water areas	Nonfederal land			Total surfaced area
				Developed	Rural	Total	
		------------------ 1,000 acres ------------------					
North Dakota	1982	1,727.8	962.5	1,016.5	41,543.9	42,560.4	45,250.7
	1987	1,744.1	962.7	1,034.4	41,509.5	42,543.9	45,250.7
	1992	1,785.1	965.2	1,102.5	41,397.9	42,500.4	45,250.7
	1997	1,785.1	1,049.1	1,152.2	41,264.3	42,416.5	45,250.7
Ohio	1982	349.4	386.9	2,806.7	422,901.8	25,708.5	26,444.8
	1987	351.7	387.4	3,009.9	22,695.8	25,705.7	26,444.8
	1992	373.2	394.5	3,275.3	22,401.8	25,677.1	26,444.8
	1997	373.2	408.1	3,796.5	21,867.0	25,663.5	26,444.8
Oklahoma	1982	1,192.1	923.1	1,615.4	41,007.5	42,622.9	44,738.1
	1987	1,149.1	1,004.0	1,700.9	40,884.1	42,585.0	44,738.1
	1992	1,148.3	1,037.4	1,772.2	40,780.2	42,552.4	44,738.1
	1997	1,148.3	1,082.2	1,996.7	40,510.9	42,507.6	44,738.1
Oregon	1982	30,961.3	801.5	980.6	29,417.6	30,398.2	62,161.0
	1987	31,069.3	923.7	1,071.3	29,096.7	30,168.0	62,161.0
	1992	31,275.2	661.2	1,145.1	29,079.5	30,224.6	62,161.0
	1997	31,260.0	828.4	1,295.5	28,777.1	30,072.6	62,161.0
Pennsylvania	1982	720.7	468.7	2,781.2	25,024.6	27,805.8	28,995.2
	1987	721.2	469.3	2,965.9	24,838.8	27,804.7	28,995.2
	1992	723.9	473.7	3,212.3	24,585.3	27,797.6	28,995.2
	1997	723.9	480.0	4,335.5	24,455.8	27,791.3	28,995.2
Rhode Island	1982	6.0	151.2	168.1	488.0	656.1	813.3
	1987	6.2	151.2	177.5	478.4	655.9	813.3
	1992	6.8	151.2	194.6	460.7	655.3	813.3
	1997	3.5	151.3	204.8	453.7	658.5	813.3
South Carolina	1982	1,031.7	796.8	1,385.5	16,752.3	18,137.8	19,939.3
	1987	1,029.2	780.6	1,565.5	16,564.0	18,129.5	19,939.3
	1992	1,036.2	790.6	1,785.6	16,326.9	18,112.5	19,939.3
	1997	1,036.2	820.8	2,325.3	15,757.0	18,082.3	19,939.3
South Dakota	1982	2,941.7	867.0	897.3	44,652.0	45,549.3	49,358.0
	1987	3,045.3	873.4	901.2	44,538.1	45,439.3	49,358.0
	1992	3,107.9	874.4	957.9	44,417.8	45,375.7	49,358.0
	1997	3,107.9	883.1	1,034.6	44,332.4	45,367.0	49,358.0

Table 1—Surface area of nonfederal and federal land and water areas, by state and year—page 6 of 7

State		Federal land	Water areas	Nonfederal land			Total surfaced area
				Developed	Rural	Total	
		---------------- 1,000 acres ----------------					
Tennessee	1982	1,195.2	759.1	1,565.2	23,454.1	25,019.3	26,973.6
	1987	1,231.8	762.1	1,775.7	23,204.0	24,979.7	26,973.6
	1992	1,232.2	773.8	2,006.3	22,961.3	24,967.6	26,973.6
	1997	1,232.2	787.2	2,617.9	22,336.3	24,954.2	26,973.6
Texas	1982	2,770.8	3,684.8	6,372.1	158,224.2	164,596.3	171,051.9
	1987	2,849.0	3,840.8	7,051.0	157,311.1	164,362.1	171,051.9
	1992	2,909.9	3,992.8	7,764.6	156,384.6	164,149.2	171,051.9
	1997	2,910.6	4,151.7	8,984.1	155,005.5	163,989.6	171,051.9
Utah	1982	34,533.7	1,757.5	548.4	17,499.3	18,047.7	54,338.9
	1987	34,185.6	2,345.2	594.3	17,213.8	17,808.1	54,338.9
	1992	34,278.2	1,784.0	655.3	17,621.4	18,276.7	54,338.9
	1997	34,278.2	1,800.9	760.4	17,499.4	18,259.8	54,338.9
Vermont	1982	304.6	264.3	255.2	5,329.5	5,584.7	6,153.6
	1987	335.5	264.9	292.1	5,261.1	5,553.2	6,153.6
	1992	370.9	265.0	320.1	5,197.6	5,517.7	6,153.6
	1997	392.8	266.2	346.1	5,148.5	5,494.4	6,153.6
Virginia	1982	2,615.6	1,922.6	1,884.4	20,664.5	22,548.9	27,087.1
	1987	2,630.5	1,929.0	2,127.9	20,399.7	22,527.6	27,087.1
	1992	2,645.7	1,941.9	2,338.0	20,161.5	22,499.5	27,087.1
	1997	2,645.7	1,958.1	2,805.2	19,678.1	22,483.3	27,087.1
Washington	1982	11,878.9	1,533.8	1,575.3	29,047.3	30,622.6	44,035.3
	1987	11,906.7	1,542.5	1,653.4	28,932.7	30,586.1	44,035.3
	1992	11,921.9	1,541.5	1,863.6	28,708.3	30,571.9	44,035.3
	1997	11,923.5	1,554.5	2,213.6	28,343.7	30,557.3	44,035.3
West Virginia	1982	1,092.2	162.6	596.4	13,657.0	14,253.4	15,508.2
	1987	1,128.3	163.3	633.2	13,583.4	14,216.6	15,508.2
	1992	1,214.3	165.3	710.5	13,418.1	14,128.6	15,508.2
	1997	1,215.2	169.9	986.1	13,137.0	14,123.1	15,508.2
Wisconsin	1982	1,813.0	1,277.2	2,012.7	30,817.1	32,829.8	35,920.0
	1987	1,822.8	1,281.9	2,129.6	30,685.7	32,815.3	35,920.0
	1992	1,845.3	1,286.7	2,260.3	30,527.7	32,788.0	35,920.0
	1997	1,845.3	1,297.2	2,543.1	30,234.4	32,777.5	35,920.0

Table 1—Surface area of nonfederal and federal land and water areas, by state and year—page 7 of 7

State		Federal land	Water areas	Nonfederal land			Total surfaced area
				Developed	Rural	Total	
		- 1,000 acres - - - - - - - - - - - - - - - - -					
Wyoming	1982	28,689.9	429.5	629.1	32,854.3	33,483.4	62,602.8
	1987	28,693.5	430.4	681.0	32,797.9	33,478.9	62,602.8
	1992	28,748.0	430.9	662.8	32,761.1	33,423.9	62,602.8
	1997	28,748.0	435.6	715.5	32,703.7	33,419.2	62,602.8
Caribbean	1982	94.3	46.6	279.4	1,886.8	2,166.2	2,307.1
	1987	88.5	47.6	325.4	1,845.6	2,171.0	2,307.1
	1992	91.2	49.2	404.0	1,762.7	2,166.7	2,307.1
	1997	91.2	50.1	557.1	1,608.7	2,165.8	2,307.1
Total	1982	397,537.1	48,624.6	75,519.0	1,422,453.8	1,497,972.8	1,944,134.5
	1987	399,090.8	49,900.8	82,010.4	1,413,432.5	1,495,142.9	1,944,134.5
	1992	402,000.5	49,560.6	89,403.1	1,403,170.3	1,492,573.4	1,944,134.5
	1997	402,185.6	50.868.7	105,369.1	1,385,711.1	1,491.080.2	1,944,134.5

Table 2—Land cover/use of nonfederal rural land, by state and year—page 1 of 7

State		Cropland	CRP land	Pastureland	Rangeland	Forest land	Other rural land	Total rural land
		------------------------------------- 1,000 acres -------------------------------						
Alabama	1982	4,510.8	0.0	3,747.8	84.4	20,770.4	559.0	29,672.4
	1987	3,996.2	207.8	3,577.4	74.9	21,063.5	534.3	29,454.1
	1992	3,147.0	534.6	3,770.5	67.8	21,102.1	668.9	29,290.9
	1997	2,919.4	521.9	3,526.6	67.9	21,072.7	666.0	28,774.5
Arizona	1982	1,218.7	0.0	84.2	32,712.4	4,607.5	2,798.7	41,421.5
	1987	1,228.3	0.0	73.6	32,371.5	4,494.8	2,834.9	41,003.1
	1992	1,197.7	0.0	75.3	32,401.8	4,312.9	2,885.3	40,873.0
	1997	1,230.7	0.0	67.3	32,114.2	4,261.9	3,007.2	40,654.3
Arkansas	1982	8,100.6	0.0	5,863.9	172.6	14,569.6	333.3	29,040.0
	1987	7,973.2	96.5	5,824.8	172.6	14,542.4	344.6	28,954.1
	1992	7,729.4	234.5	5,791.9	167.6	14,530.6	372.3	28,826.3
	1997	7,581.8	230.4	5,452.9	72.7	14,764.8	436.1	28,538.7
California	1982	10,520.3	0.0	1,416.8	18,335.8	15,132.2	4,378.9	49,784.0
	1987	10,223.5	117.2	1,517.6	17,980.7	15,008.7	4,403.2	49,250.9
	1992	10,051.6	180.7	1,170.6	17,320.0	14,680.4	4,569.8	47,973.1
	1997	9,560.5	173.0	1,064.8	17,457.3	14,295.4	4,687.4	47,238.4
Colorado	1982	10,604.1	0.0	1,280.0	24,457.6	4,017.5	1,075.0	41,434.2
	1987	9,750.5	1,113.1	1,277.8	23,901.7	3,962.5	1,116.3	41,121.9
	1992	8,940.3	1,913.2	1,276.4	23,843.2	3,754.3	1,165.2	40,892.6
	1997	8,860.1	1,890.2	1,268.9	23,854.6	3,728.8	1,172.0	40,774.6
Connecticut	1982	244.6	0.0	113.6	0.0	1,810.8	140.2	2,309.2
	1987	232.7	0.0	108.8	0.0	1,774.4	142.4	2,258.3
	1992	228.4	0.0	109.7	0.0	1,742.6	138.7	2,219.4
	1997	198.9	0.0	107.1	0.0	1,728.6	119.9	2,154.5
Delaware	1982	518.6	0.0	37.2	0.0	367.5	126.5	1,049.8
	1987	510.8	0.0	31.4	0.0	360.6	126.0	1,028.8
	1992	499.2	0.8	25.9	0.0	356.3	128.8	1,011.0
	1997	471.5	0.8	22.6	0.0	347.0	133.5	975.4
Florida	1982	3,556.3	0.0	4,290.7	4,557.2	12,698.6	2,376.0	27,478.8
	1987	3,184.2	93.9	4,550.7	4,198.5	12,618.6	2,395.5	27,041.4
	1992	2,997.5	123.4	4,478.3	3,635.7	12,461.0	2,461.4	26,157.3
	1997	2,719.2	119.7	4,176.8	3,192.5	12,255.2	2,684.0	25,147.4

Table 2—Land cover/use of nonfederal rural land, by state and year—page 2 of 7

State		Cropland	CRP land	Pastureland	Rangeland	Forest land	Other rural land	Total rural land
		---------------- 1,000 acres ----------------						
Georgia	1982	6,568.6	0.0	2,961.3	0.0	21,792.1	954.9	32,276.9
	1987	5,908.7	302.0	2,939.4	0.0	21,872.2	943.4	31,965.7
	1992	5,173.0	601.5	3,057.4	0.0	21,636.0	957.3	31,425.2
	1997	4,661.2	595.6	2,853.0	0.0	21,216.3	999.4	30,325.5
Hawaii	1982	302.0	0.0	87.5	946.7	1,619.2	676.3	3,631.7
	1987	294.2	0.0	79.5	936.8	1,588.1	663.2	3,561.8
	1992	274.4	0.0	90.7	945.9	1,530.2	698.7	3,539.9
	1997	243.9	0.0	89.3	945.7	1,514.3	738.0	3,531.2
Idaho	1982	6,390.5	0.0	1,292.6	6,583.7	3,984.9	525.6	18,777.3
	1987	6,052.3	448.4	1,288.8	6,533.1	4,067.8	550.4	18,940.8
	1992	5,600.2	823.4	1,259.9	6,495.1	4,019.9	565.9	18,764.4
	1997	5,499.9	784.4	1,253.2	6,477.7	3,941.9	600.2	18,557.3
Illinois	1982	24,731.0	0.0	3,240.2	0.0	3,526.2	637.6	32,135.0
	1987	24,697.0	120.2	2,977.8	0.0	3,537.3	649.0	31,981.3
	1992	24,099.5	711.7	2,839.3	0.0	3,512.0	695.7	31,858.2
	1997	23,953.7	726.0	2,525.1	0.0	3,631.4	709.6	31,545.8
Indiana	1982	13,781.4	0.0	2,223.0	0.0	3,671.9	813.9	20,490.2
	1987	13,842.1	143.2	1,935.8	0.0	3,677.7	755.6	20,354.4
	1992	13,512.8	413.5	1,873.1	0.0	3,661.3	777.9	20,238.6
	1997	13,358.1	377.6	1,817.6	0.0	3,637.8	753.1	19,944.2
Iowa	1982	26,439.3	0.0	4,609.6	0.0	1,791.5	935.8	33,776.2
	1987	25,714.6	1,243.9	4,041.6	0.0	1,850.7	890.7	33,741.5
	1992	24,988.2	2,093.0	3,775.7	0.0	1,972.9	873.6	33,703.4
	1997	25,262.0	1,739.3	3,553.8	0.0	2,083.5	912.3	33,550.9
Kansas	1982	29,124.7	0.0	2,112.6	15,913.4	1,225.2	679.3	49,055.2
	1987	28,507.0	649.6	2,135.8	15,829.8	1,227.5	671.4	49,021.1
	1992	26,565.5	2,863.8	2,245.4	15,285.5	1,300.6	682.5	48,943.3
	1997	26,459.9	2,849.4	2,212.9	15,179.1	1,289.9	723.9	48,715.1
Kentucky	1982	5,934.4	0.0	5,958.8	0.0	10,250.8	797.7	22,941.7
	1987	5,466.7	203.9	5,906.8	0.0	10,310.6	798.9	22,686.9
	1992	5,091.9	422.9	5,927.6	0.0	10,373.1	654.6	22,470.1
	1997	5,151.2	331.4	5,612.6	0.0	10,440.4	556.6	22,092.2

Table 2—Land cover/use of nonfederal rural land, by state and year—page 3 of 7

State	Year	Cropland	CRP land	Pastureland	Rangeland	Forest land	Other rural land	Total rural land
		---------- 1,000 acres ----------						
Louisiana	1982	6,410.2	0.0	2,253.5	279.0	13,292.1	3,144.2	25,379.0
	1987	6,291.1	42.2	2,272.1	277.9	13,103.0	3,115.3	25,101.6
	1992	5,971.0	141.7	2,294.0	281.0	13,078.9	3,104.5	24,871.1
	1997	5,568.4	140.5	2,376.2	279.9	13,114.3	3,142.6	24,621.9
Maine	1982	521.3	0.0	180.9	0.0	17,688.7	675.6	19,066.5
	1987	506.7	0.0	144.8	0.0	17,679.1	660.1	18,990.7
	1992	447.6	35.6	106.2	0.0	17,679.4	671.7	18,940.5
	1997	418.8	29.7	82.0	0.0	17,633.1	598.7	18,762.3
Maryland	1982	1,795.1	0.0	540.3	0.0	2,447.1	352.8	5,135.3
	1987	1,740.2	1.3	557.0	0.0	2,417.0	335.4	5,050.9
	1992	1,673.0	18.6	552.4	0.0	2,391.6	338.6	4,974.2
	1997	1,597.5	18.9	454.3	0.0	2,330.7	345.7	4,747.1
Massachusetts	1982	296.9	0.0	205.4	0.0	3,028.0	310.2	3,840.5
	1987	288.0	0.0	181.0	0.0	2,944.7	318.3	3,732.0
	1992	272.5	0.0	173.2	0.0	2,835.0	316.6	3,597.3
	1997	270.7	0.0	114.3	0.0	2,657.3	270.6	3,312.9
Michigan	1982	9,443.5	0.0	2,975.9	0.0	15,716.4	2,224.8	30,360.6
	1987	9,308.3	54.6	2,620.2	0.0	15,899.6	2,208.0	30,090.7
	1992	8,985.1	254.8	2,399.4	0.0	15,920.8	2,205.6	29,765.7
	1997	8,439.1	322.0	2,003.5	0.0	16,237.7	2,197.9	29,200.2
Minnesota	1982	23,024.4	0.0	3,923.1	0.0	14,558.1	4,337.7	45,843.3
	1987	22,399.0	780.3	3,609.1	0.0	14,414.9	4,448.0	45,651.3
	1992	21,355.1	1,810.8	3,385.4	0.0	14,424.1	4,526.8	45,502.2
	1997	21,327.9	1,543.3	3,422.6	0.0	14,829.7	4,041.9	45,165.4
Mississippi	1982	7,416.5	0.0	3,981.2	0.0	15,244.2	355.8	26,997.7
	1987	6,662.6	292.8	3,902.9	0.0	15,596.9	346.7	26,801.9
	1992	5,726.0	777.8	4,032.1	0.0	15,758.0	341.2	26,635.1
	1997	5,296.3	797.9	3,698.7	0.0	16,018.7	427.8	26,239.4
Missouri	1982	14,999.2	0.0	12,784.0	167.6	11,189.3	705.2	39,845.3
	1987	14,384.8	570.7	12,380.0	130.7	11,547.5	708.5	39,722.2
	1992	13,347.5	1,602.3	12,005.7	125.6	11,753.6	726.5	39,561.2
	1997	13,709.8	1,606.7	10,946.6	97.7	12,118.3	715.9	39,195.0

Table 2—Land cover/use of nonfederal rural land, by state and year—page 4 of 7

State		Cropland	CRP land	Pastureland	Rangeland	Forest land	Other rural land	Total rural land
		---------------- 1,000 acres ----------------						
Montana	1982	17,197.0	0.0	3,110.2	38,136.2	5,330.8	1,375.6	65,149.8
	1987	16,233.1	1,485.6	3,245.1	37,485.1	5,353.0	1,397.5	65,181.4
	1992	15,034.2	2,781.4	3,452.7	37,239.7	5,284.2	1,417.0	65,209.2
	1997	15,086.0	2,721.1	3,495.4	37,015.6	5,279.0	1,481.2	65,078.3
Nebraska	1982	20,276.7	0.0	2,145.8	23,409.8	747.1	745.5	47,324.9
	1987	19,935.0	589.2	2,091.4	23,122.2	782.2	761.1	47,281.1
	1992	19,239.5	1,362.9	2,083.2	22,936.5	790.3	782.2	47,194.6
	1997	19,420.5	1,244.9	1,975.9	22,863.6	799.1	798.6	47,102.6
Nevada	1982	859.8	0.0	328.5	8,365.0	375.7	413.4	10,342.4
	1987	843.1	0.0	318.3	8,314.7	375.7	418.2	10,270.0
	1992	760.9	1.4	313.5	8,223.8	373.8	413.2	10,086.6
	1997	710.7	2.3	270.6	8,299.9	296.9	451.6	10,032.0
New Hampshire	1982	157.6	0.0	125.0	0.0	4,117.8	195.6	4,596.0
	1987	146.4	0.0	108.8	0.0	4,036.2	208.4	4,499.8
	1992	141.4	0.0	98.4	0.0	3,956.1	219.5	4,415.4
	1997	131.5	0.0	91.6	0.0	3,874.6	201.9	4,299.6
New Jersey	1982	809.5	0.0	239.0	0.0	1,866.7	389.9	3,305.1
	1987	688.0	0.0	179.2	0.0	1,814.0	391.4	3,072.6
	1992	649.6	0.5	158.9	0.0	1,784.7	387.6	2,981.3
	1997	574.0	0.5	108.5	0.0	1,624.7	380.5	2,688.2
New Mexico	1982	2,412.7	0.0	161.2	41,749.8	4,672.6	2,223.8	51,220.1
	1987	1,960.9	425.8	198.9	41,217.7	4,495.8	2,365.5	50,664.6
	1992	1,891.5	481.9	207.4	40,739.1	4,532.0	2,399.9	50,251.8
	1997	1,842.0	467.4	207.0	40,275.9	4,914.5	2,188.9	49,895.7
New York	1982	5,911.6	0.0	3,916.8	0.0	16,672.5	733.3	27,234.2
	1987	5,747.1	18.4	3,436.4	0.0	17,004.1	918.5	27,124.5
	1992	5,616.3	56.6	3,035.3	0.0	17,345.9	943.8	26,997.9
	1997	5,375.0	53.8	2,626.9	0.0	17,532.8	904.2	26,492.7
North Carolina	1982	6,694.8	0.0	1,983.6	0.0	16,835.4	803.4	26,317.2
	1987	6,363.3	30.3	1,996.7	0.0	16,462.1	838.3	25,690.7
	1992	5,959.7	138.2	2,025.1	0.0	16,077.7	844.4	25,045.1
	1997	5,539.1	131.1	1,979.8	0.0	15,677.7	916.6	24,244.3

Table 2—Land cover/use of nonfederal rural land, by state and year—page 5 of 7

State		Cropland	CRP land	Pastureland	Rangeland	Forest land	Other rural land	Total rural land
		------------------------------------- 1,000 acres -------------------------------						
North Dakota	1982	27,043.3	0.0	1,327.8	11,373.7	454.7	1,344.4	41,543.9
	1987	27,099.8	526.6	1,253.2	10,840.0	453.6	1,336.3	41,509.5
	1992	24,743.4	2,901.3	1,223.2	10,760.2	443.6	1,326.2	41,397.9
	1997	24,990.8	2,801.5	1,105.4	10,551.0	442.6	1,373.0	41,264.3
Ohio	1982	12,447.9	0.0	2,790.8	0.0	6,570.6	1,092.5	22,901.8
	1987	12,343.0	57.6	2,460.0	0.0	6,808.9	1,026.3	22,695.8
	1992	11,929.1	315.6	2,334.7	0.0	6,829.0	993.4	22,401.8
	1997	11,503.7	323.6	1,979.6	0.0	6,983.5	1,076.6	21,867.0
Oklahoma	1982	11,567.9	0.0	7,230.1	15,146.7	6,638.9	423.9	41,007.5
	1987	10,902.4	590.3	7,592.5	14,475.5	6,857.8	465.6	40,884.1
	1992	10,080.4	1,162.6	7,812.7	14,124.8	7,087.3	512.4	40,780.2
	1997	9,708.7	1,137.8	7,932.8	13,973.8	7,253.9	503.9	40,510.9
Oregon	1982	4,358.1	0.0	2,115.0	9,811.7	12,450.8	682.0	29,417.6
	1987	3,970.8	393.0	2,024.6	9,674.7	12,356.5	677.1	29,096.7
	1992	3,776.0	522.3	2,012.9	9,764.1	12,333.2	671.0	29,079.5
	1997	3,799.6	482.8	1,904.9	9,556.0	12,294.5	739.3	28,777.1
Pennsylvania	1982	5,896.0	0.0	2,625.1	0.0	15,440.4	1,063.1	25,024.6
	1987	5,747.4	15.9	2,485.0	0.0	15,519.1	1,077.4	24,838.8
	1992	5,595.7	92.6	2,359.7	0.0	15,478.5	1,058.8	24,585.3
	1997	5,245.0	90.2	1,811.6	0.0	15,306.1	1,002.9	23,455.8
Rhode Island	1982	27.3	0.0	34.7	0.0	398.0	28.0	488.0
	1987	25.6	0.0	33.6	0.0	393.1	26.1	478.4
	1992	24.9	0.0	23.6	0.0	383.8	28.7	460.7
	1997	20.0	0.0	23.9	0.0	381.2	28.6	453.7
South Carolina	1982	3,578.8	0.0	1,233.2	0.0	11,207.2	733.1	16,752.3
	1987	3,320.1	95.1	1,218.5	0.0	11,192.8	737.5	16,564.0
	1992	2,982.7	262.1	1,215.8	0.0	11,110.4	755.9	16,326.9
	1997	2,541.7	262.4	1,182.4	0.0	10,957.7	812.8	15,757.0
South Dakota	1982	16,945.7	0.0	2,753.4	22,895.0	564.2	1,493.7	44,652.0
	1987	17,511.5	359.7	2,280.4	22,336.0	570.5	1,480.0	44,538.1
	1992	16,436.3	1,756.6	2,196.0	21,996.2	541.9	1,490.8	44,417.8
	1997	16,738.3	1,685.8	2,078.2	21,763.9	531.7	1,534.5	44,332.4

Table 2—Land cover/use of nonfederal rural land, by state and year—page 6 of 7

State		Cropland	CRP land	Pastureland	Rangeland	Forest land	Other rural land	Total rural land
		- 1,000 acres - - - - - - - - - - - - - - - - - -						
Tennessee	1982	5,592.0	0.0	5,466.5	0.0	11,800.8	594.8	23,454.1
	1987	5,375.0	174.4	5,241.1	0.0	11,858.7	554.8	23,204.0
	1992	4,865.8	440.4	5,272.6	0.0	11,842.1	549.4	22,961.3
	1997	4,565.9	374.5	4,985.2	0.0	11,736.4	674.3	22,336.3
Texas	1982	33,321.0	0.0	17,240.0	96,052.2	9,433.9	2,177.1	158,224.2
	1987	31,198.1	1,582.1	16,923.7	95,434.0	9,912.7	2,260.5	157,311.1
	1992	28,261.7	3,974.0	16,904.0	94,763.5	10,072.5	2,408.9	156,384.6
	1997	26,762.0	3,905.5	15,807.4	95,322.9	10,626.8	2,580.9	155,005.5
Utah	1982	2,038.6	0.0	558.9	10,773.4	1,792.6	2,335.8	17,499.3
	1987	1,889.2	150.6	646.4	10,576.7	1,783.1	2,167.8	17,213.8
	1992	1,815.2	226.0	673.7	10,820.6	1,746.0	2,339.9	17,621.4
	1997	1,676.3	216.5	694.9	10,719.9	1,829.6	2,362.2	17,499.4
Vermont	1982	648.2	0.0	502.9	0.0	4,098.9	79.5	5,329.5
	1987	643.1	0.0	387.3	0.0	4,152.5	78.2	5,261.1
	1992	634.5	0.0	349.3	0.0	4,138.8	75.0	5,197.6
	1997	601.3	0.0	342.4	0.0	4,118.0	86.8	5,148.5
Virginia	1982	3,397.2	0.0	3,292.0	0.0	13,311.0	664.3	20,664.5
	1987	3,109.2	23.1	3,299.2	0.0	13,350.0	618.2	20,399.7
	1992	2,900.9	73.4	3,342.8	0.0	13,226.6	617.8	20,161.5
	1997	2,879.0	70.0	3,071.3	0.0	13,030.2	627.6	19,678.1
Washington	1982	7,796.1	0.0	1,383.9	5,922.1	13,007.7	937.5	29,047.3
	1987	7,296.4	455.3	1,424.7	5,876.6	12,935.8	943.9	28,932.7
	1992	6,744.6	1,015.8	1,393.3	5,744.9	12,842.9	966.8	28,708.3
	1997	6,688.6	1,017.2	1,199.5	5,744.2	12,666.3	1,027.9	28,343.7
West Virginia	1982	1,092.6	0.0	1,865.1	0.0	10,419.0	280.3	13,657.0
	1987	998.0	0.6	1,730.6	0.0	10,556.6	297.6	13,583.4
	1992	915.2	0.6	1,600.3	0.0	10,515.4	386.6	13,418.1
	1997	847.8	0.0	1,502.9	0.0	10,472.1	314.2	13,137.0
Wisconsin	1982	11,465.5	0.0	3,418.3	0.0	13,498.4	2,443.9	30,817.1
	1987	11,315.8	218.1	3,089.2	0.0	13,548.1	2,514.2	30,685.7
	1992	10,812.8	664.9	2,976.4	0.0	13,519.4	2,554.2	30,527.7
	1997	10,537.4	662.4	2,881.7	0.0	13,634.1	2,518.8	30,234.4

Table 2—Land cover/use of nonfederal rural land, by state and year—page 7 of 7

State		Cropland	CRP land	Pastureland	Rangeland	Forest land	Other rural land	Total rural land
		------------------------------- 1,000 acres -------------------------------						
Wyoming	1982	2,587.5	0.0	777.5	27,505.1	1,007.9	976.3	32,854.3
	1987	2,444.1	129.1	873.7	27,366.4	1,008.2	976.4	32,797.9
	1992	2,271.6	251.7	960.5	27,169.5	1,025.4	1,082.4	32,761.1
	1997	2,171.1	247.0	1,181.2	27,149.6	994.8	960.0	32,703.7
Caribbean	1982	408.2	0.0	770.0	153.7	506.6	48.3	1,886.8
	1987	389.4	0.0	749.7	148.3	507.3	50.9	1,845.6
	1992	366.9	0.0	692.2	139.5	508.4	55.7	1,762.7
	1997	354.5	0.0	429.2	138.2	622.3	64.5	1,608.7
Total	1982	420,975.6	0.0	133,559.4	415,504.8	398,220.0	54,194.0	1,422,453.8
	1987	406,652.5	13,802.4	129,222.9	409,276.1	399,630.5	54,548.1	1,413,132.5
	1992	382,316.2	34,041.4	127,434.0	404,991.6	398,577.5	55,809.6	1,403,170.3
	1997	375,044.0	32,697.0	119,572.9	403,113.8	399,030.8	56,252.6	1,385,711.1

Table 3—Cropland use, by state and year—page 1 of 7

State		Cultivated Cropland			Noncultivated Cropland			Total Cropland
		Irrigated	Nonirrigated	Total	Irrigated	Nonirrigated	Total	
		- - - - - - - - - - - - - - - - - - 1,000 acres - - - - - - - - - - - - - - - - - -						
Alabama	1982	76.7	4,151.5	4,228.2	0.6	282.0	282.6	4,510.8
	1987	70.2	3,608.7	3,678.9	0.6	316.7	317.3	3,996.2
	1992	48.8	2,755.1	2,803.9	7.5	335.6	343.1	3,147.0
	1997	51.1	2,531.5	2,582.6	12.0	324.8	336.8	2,919.4
Arizona	1982	1,013.8	23.7	1,037.5	180.8	0.4	181.2	1,218.7
	1987	964.8	85.7	1,050.5	171.2	6.6	177.8	1,228.3
	1992	943.4	73.5	1,016.9	180.5	0.3	180.8	1,197.7
	1997	909.0	82.0	991.0	212.4	0.3	212.7	1,203.7
Arkansas	1982	4,048.8	3,764.2	7,813.0	7.9	279.7	287.6	8,100.6
	1987	4,297.8	3,394.4	7,692.2	4.0	277.0	281.0	7,973.2
	1992	4,718.0	2,718.3	7,436.3	17.9	275.2	293.1	7,729.4
	1997	4,971.2	2,356.5	7,327.7	15.5	238.6	254.1	7,581.8
California	1982	5,818.8	1,287.9	7,106.7	3,150.8	262.8	3,413.6	10,520.3
	1987	5,829.8	1,028.0	6,857.8	3,096.4	269.3	3,365.7	10,223.5
	1992	5,321.1	1,235.8	6,556.9	3,224.5	270.2	3,494.7	10,051.6
	1997	5,064.8	1,132.0	6,196.8	3,140.7	223.0	3,363.7	9,560.5
Colorado	1982	2,012.2	7,404.1	9,416.3	1,042.7	145.1	1,187.8	10,604.1
	1987	2,020.3	6,573.8	8,594.1	1,009.5	146.9	1,156.4	9,750.5
	1992	1,868.8	5,806.3	7,675.1	1,091.2	174.0	1,265.2	8,940.3
	1997	1,920.6	5,721.5	7,642.1	1,048.6	169.4	1,218.0	8,860.1
Connecticut	1982	9.1	107.9	117.0	7.2	120.4	127.6	244.6
	1987	11.8	88.5	100.3	7.7	124.7	132.4	232.7
	1992	14.5	79.5	94.0	12.1	122.3	134.4	228.4
	1997	16.4	63.2	79.6	8.2	111.1	119.3	198.9
Delaware	1982	48.7	456.2	504.9	0.9	12.8	13.7	518.6
	1987	58.0	439.7	497.7	0.0	13.1	13.1	510.8
	1992	65.5	427.8	493.3	0.0	5.9	5.9	499.2
	1997	78.2	387.0	465.2	0.0	6.3	6.3	471.5
Florida	1982	963.0	1,373.9	2,336.9	896.3	323.1	1,219.4	3,556.3
	1987	965.6	990.2	1,955.8	926.3	302.1	1,228.4	3,184.2
	1992	920.8	772.4	1,693.2	1,096.7	207.6	1,304.3	2,997.5
	1997	894.4	555.7	1,450.1	1,112.3	156.8	1,269.1	2,719.2

Table 3—Cropland use, by state and year—page 2 of 7

State		Cultivated Cropland			Noncultivated Cropland			Total Cropland
		Irrigated	Nonirrigated	Total	Irrigated	Nonirrigated	Total	
		---------------- 1,000 acres ----------------						
Georgia	1982	1,095.8	5,066.4	6,162.2	43.0	363.4	406.4	6,568.6
	1987	1,064.8	4,406.1	5,470.9	67.2	370.6	437.8	5,908.7
	1992	1,059.7	3,655.4	4,715.1	68.5	389.4	457.9	5,173.0
	1997	1,056.7	3,032.0	4,088.7	68.5	504.0	572.5	4,661.2
Hawaii	1982	169.4	97.7	267.1	8.2	26.7	34.9	302.0
	1987	163.6	88.5	252.1	10.4	31.7	42.1	294.2
	1992	138.8	89.8	228.6	24.2	21.6	45.8	274.4
	1997	103.0	94.2	197.2	22.4	24.3	46.7	243.9
Idaho	1982	2,843.1	2,559.5	5,402.6	690.9	297.0	987.9	6,390.5
	1987	2,939.1	2,134.3	5,073.4	625.8	353.1	987.9	6,052.3
	1992	2,859.4	1,795.9	4,655.3	639.3	305.6	944.9	5,600.2
	1997	2,806.7	1,723.2	4,529.9	624.1	345.9	970.0	5,499.9
Illinois	1982	165.6	24,032.4	24,198.0	11.5	521.5	533.0	24,731.0
	1987	178.1	24,124.2	24,302.3	9.6	385.1	394.7	24,697.0
	1992	199.7	23,302.9	23,502.6	5.9	591.0	596.9	24,099.5
	1997	172.8	23,334.1	23,506.9	5.5	441.3	446.8	23,953.7
Indiana	1982	163.5	13,163.8	13,327.3	2.3	451.8	454.1	13,781.4
	1987	169.4	13,162.8	13,332.2	4.0	505.9	509.9	13,842.1
	1992	180.0	12,744.2	12,924.2	4.0	584.6	588.6	13,512.8
	1997	215.0	12,507.7	12,722.7	10.5	624.9	635.4	13,358.1
Iowa	1982	134.2	25,486.9	25,621.1	0.0	818.2	818.2	26,439.3
	1987	116.9	24,772.2	24,889.1	0.0	825.5	825.5	25,714.6
	1992	131.7	23,845.4	23,977.1	0.8	1,010.3	1,011.1	24,988.2
	1997	151.5	24,005.5	24,157.0	0.0	1,105.0	1,105.0	25,262.0
Kansas	1982	3,336.3	24,153.9	27,490.2	178.3	1,456.2	1,634.5	29,124.7
	1987	3,389.5	23,828.8	27,218.3	115.5	1,173.2	1,288.7	28,507.0
	1992	3,367.7	21,719.9	25,087.6	207.2	1,270.7	1,477.9	26,565.5
	1997	3,293.1	21,442.0	24,735.1	223.0	1,501.8	1,724.8	26,459.9
Kentucky	1982	12.3	4,727.0	4,739.3	1.1	1,194.0	1,195.1	5,934.4
	1987	5.0	4,207.6	4,212.6	1.1	1,253.0	1,254.1	5,466.7
	1992	5.5	3,617.9	3,623.4	0.0	1,468.5	1,468.5	5,091.9
	1997	57.4	3,426.4	3,483.8	0.0	1,667.4	1,667.4	5,151.2

Table 3—Cropland use, by state and year—page 3 of 7

State		Cultivated Cropland			Noncultivated Cropland			Total Cropland
		Irrigated	Nonirrigated	Total	Irrigated	Nonirrigated	Total	
		- 1,000 acres -						
Louisiana	1982	1,647.5	4,557.6	6,205.1	1.2	203.9	205.1	6,410.2
	1987	1,304.6	4,790.7	6,095.3	1.7	194.1	195.8	6,291.1
	1992	1,733.7	4,508.7	5,792.4	4.1	174.5	178.6	5,971.0
	1997	1,521.9	3,862.9	5,384.8	1.0	182.6	183.6	5,568.4
Maine	1982	5.4	227.5	232.9	3.4	285.0	288.4	521.3
	1987	4.9	199.5	204.4	3.4	298.9	302.3	506.7
	1992	5.0	155.1	160.1	3.4	284.1	287.5	447.6
	1997	19.8	139.7	159.5	3.4	255.9	259.3	418.8
Maryland	1982	45.2	1,592.5	1,637.7	2.8	154.6	157.4	1,795.1
	1987	48.4	1,561.8	1,610.2	5.8	124.2	130.0	1,740.2
	1992	49.8	1,448.5	1,498.3	7.8	166.9	174.7	1,673.0
	1997	55.6	1,341.5	1,397.1	4.4	196.0	200.4	1,597.5
Massachusetts	1982	3.7	77.5	81.2	17.7	198.0	215.7	296.9
	1987	3.9	74.1	78.0	16.6	193.4	210.0	288.0
	1992	4.7	72.1	76.8	22.8	172.9	195.7	272.5
	1997	9.4	55.9	65.3	22.0	183.4	205.4	270.7
Michigan	1982	297.7	7,479.6	7,777.3	66.9	1,599.3	1,666.2	9,443.5
	1987	341.7	7,196.9	7,538.6	65.0	1,704.7	1,769.7	9,308.3
	1992	394.1	6,722.4	7,116.5	71.0	1,797.6	1,868.6	8,985.1
	1997	449.1	6,038.6	6,487.7	58.3	1,893.1	1,951.4	8,439.1
Minnesota	1982	436.6	20,792.1	21,228.7	4.0	1,791.7	1,795.7	23,024.4
	1987	424.4	20,607.5	21,031.9	13.9	1,353.2	1,367.1	22,399.0
	1992	451.4	19,167.6	19,619.0	8.4	1,727.7	1,736.1	21,355.1
	1997	340.1	19,272.2	19,612.3	4.5	1,711.1	1,715.6	21,327.9
Mississippi	1982	644.6	6,551.5	7,196.1	0.0	220.4	220.4	7,416.5
	1987	989.9	5,516.5	6,506.4	1.2	155.0	156.2	6,662.6
	1992	1,150.1	4,332.0	5,482.1	1.2	242.7	243.9	5,726.0
	1997	1,222.7	3,658.5	4,881.2	0.0	415.1	415.1	5,296.3
Missouri	1982	766.0	12,359.2	13,125.2	10.0	1,864.0	1,874.0	14,999.2
	1987	752.2	11,899.5	12,651.7	6.6	1,726.5	1,733.1	14,384.8
	1992	978.9	10,015.8	10,994.7	10.0	2,342.8	2,352.8	13,347.5
	1997	999.2	9,453.7	10,452.9	14.0	3,242.9	3,256.9	13,709.8

Table 3—Cropland use, by state and year—page 4 of 7

State		Cultivated Cropland			Noncultivated Cropland			Total Cropland
		Irrigated	Nonirrigated	Total	Irrigated	Nonirrigated	Total	
		---------------- 1,000 acres ----------------						
Montana	1982	861.6	13,739.8	14,601.4	1,282.3	1,313.3	2,595.6	17,197.0
	1987	923.7	12,897.2	13,820.9	1,114.1	1,298.1	2,412.2	16,233.1
	1992	893.0	11,588.6	12,481.6	1,196.0	1,356.6	2,552.6	15,034.2
	1997	928.5	11,540.8	12,469.3	1,215.6	1,401.1	2,616.7	15,086.0
Nebraska	1982	6,555.8	12,153.6	18,709.4	331.4	1,235.9	1,567.3	20,276.7
	1987	6,903.8	11,748.6	18,652.4	289.7	992.9	1,282.6	19,935.0
	1992	7,101.6	10,709.0	17,810.6	392.3	1,036.6	1,428.9	19,239.5
	1997	7,413.0	10,522.5	17,935.5	354.4	1,130.6	1,485.0	19,420.5
Nevada	1982	265.5	76.7	342.3	516.9	0.6	517.5	859.8
	1987	98.2	56.8	155.0	687.7	0.4	688.1	843.1
	1992	128.1	83.4	211.5	547.2	2.2	549.4	760.9
	1997	67.8	51.5	119.3	589.5	1.9	591.4	710.7
New Hampshire	1982	0.0	40.9	40.9	2.5	114.2	116.7	157.6
	1987	0.0	34.9	34.9	2.5	109.0	111.5	146.4
	1992	0.0	20.7	20.7	2.5	118.2	120.7	141.4
	1997	0.0	20.4	20.4	2.5	108.6	111.1	131.5
New Jersey	1982	89.1	555.0	644.1	36.1	129.3	165.4	809.5
	1987	69.0	491.2	560.2	36.4	91.4	127.8	688.0
	1992	75.4	412.5	487.9	48.0	113.7	161.7	649.6
	1997	78.8	335.7	414.5	51.6	107.9	159.5	574.0
New Mexico	1982	992.7	1,013.4	2,006.1	378.2	28.4	406.6	2,412.7
	1987	756.5	762.5	1,519.0	404.0	37.9	441.9	1,960.9
	1992	715.8	745.0	1,460.8	403.4	27.3	430.7	1,891.5
	1997	612.6	749.1	1,361.7	447.3	33.0	480.3	1,842.0
New York	1982	46.8	3,387.4	3,434.2	16.0	2,461.4	2,477.4	5,911.6
	1987	52.1	3,116.4	3,168.5	20.4	2,558.2	2,578.6	5,747.1
	1992	52.6	2,822.8	2,875.4	22.5	2,718.4	2,740.9	5,616.3
	1997	58.7	2,665.5	2,724.2	21.7	2,629.1	2,650.8	5,375.0
North Carolina	1982	260.3	6,090.7	6,351.0	5.2	338.6	343.8	6,694.8
	1987	310.7	5,719.3	6,030.0	6.6	326.7	333.3	6,363.3
	1992	322.5	5,226.8	5,549.3	9.4	401.0	410.4	5,959.7
	1997	317.0	4,669.1	4,986.1	16.9	536.1	553.0	5,539.1

Table 3—Cropland use, by state and year—page 5 of 7

State		Cultivated Cropland			Noncultivated Cropland			Total Cropland
		Irrigated	Nonirrigated	Total	Irrigated	Nonirrigated	Total	
		---------------- 1,000 acres ----------------						
North Dakota	1982	211.7	24,874.7	25,086.4	27.7	1,929.2	1,956.9	27,043.3
	1987	211.2	24,869.9	25,081.1	21.9	1,996.8	2,018.7	27,099.8
	1992	228.6	22,628.1	22,856.7	17.0	1,869.7	1,886.7	24,743.4
	1997	221.7	22,592.1	22,813.8	19.9	2,157.1	2,177.0	24,990.8
Ohio	1982	26.6	11,454.9	11,481.5	13.1	953.3	966.4	12,447.9
	1987	23.4	11,262.1	11,285.5	15.0	1,042.5	1,057.5	12,343.0
	1992	28.0	10,824.9	10,852.9	14.9	1,061.3	1,076.2	11,929.1
	1997	20.4	10,138.1	10,158.5	10.2	1,335.0	1,345.2	11,503.7
Oklahoma	1982	678.6	10,445.9	11,124.5	56.2	387.2	443.4	11,567.9
	1987	648.7	9,913.3	10,562.0	46.2	294.2	340.4	10,902.4
	1992	594.0	9,041.8	9,635.8	71.8	372.8	444.6	10,080.4
	1997	593.5	8,723.6	9,317.1	47.2	344.4	391.6	9,708.7
Oregon	1982	1,000.3	2,457.0	3,457.3	695.2	205.6	900.8	4,358.1
	1987	941.0	2,024.5	2,965.5	787.6	217.7	1,005.3	3,970.8
	1992	916.4	1,865.1	2,781.5	795.2	199.3	994.5	3,776.0
	1997	854.8	1,846.2	2,701.0	869.8	228.8	1,098.6	3,799.6
Pennsylvania	1982	3.2	3,913.1	3,916.3	9.0	1,970.7	1,979.7	5,896.0
	1987	4.2	3,865.8	3,870.0	10.6	1,860.8	1,871.4	5,741.4
	1992	3.6	3,730.8	3,734.4	12.2	1,849.1	1,861.3	5,595.7
	1997	7.3	3,474.1	3,481.4	11.8	1,751.8	1,763.6	5,245.0
Rhode Island	1982	0.8	8.1	8.9	5.5	12.9	18.4	27.3
	1987	0.0	6.0	6.0	6.3	13.3	19.6	25.6
	1992	0.6	6.5	7.1	5.0	12.8	17.8	24.9
	1997	0.0	4.3	4.3	5.2	10.5	15.7	20.0
South Carolina	1982	72.3	3,321.0	3,393.3	25.2	160.3	185.5	3,578.8
	1987	81.3	3,036.5	3,117.8	27.7	174.6	202.3	3,320.1
	1992	103.6	2,651.2	2,754.8	20.1	207.8	227.9	2,982.7
	1997	129.0	2,162.8	2,291.8	13.0	236.9	249.9	2,541.7
South Dakota	1982	394.3	14,602.5	14,996.8	77.0	1,871.9	1,948.9	16,945.7
	1987	453.4	15,035.6	15,489.0	38.5	1,984.0	2,022.5	17,511.5
	1992	420.1	13,984.6	14,404.7	61.5	1,970.1	2,031.6	16,436.3
	1997	389.9	13,932.4	14,322.3	108.1	2,307.9	2,416.0	16,738.3

Table 3—Cropland use, by state and year—page 6 of 7

State		Cultivated Cropland			Noncultivated Cropland			Total Cropland
		Irrigated	Nonirrigated	Total	Irrigated	Nonirrigated	Total	
		---------------- 1,000 acres ----------------						
Tennessee	1982	14.6	4,706.9	4,721.5	5.7	864.8	870.5	5,592.0
	1987	16.3	4,399.5	4,415.8	6.7	952.5	959.2	5,375.0
	1992	10.7	3,632.4	3,643.1	6.0	1,207.7	1,213.7	4,856.8
	1997	15.1	3,193.5	3,208.6	2.6	1,354.7	1,357.3	4,565.9
Texas	1982	9,682.5	22,822.4	32,504.9	233.6	582.5	816.1	33,321.0
	1987	9,083.6	21,489.2	30,572.8	165.5	459.8	625.3	31,198.1
	1992	8,622.4	18,931.7	27,554.1	176.5	531.1	707.6	28,261.7
	1997	8,216.2	17,938.1	26,154.3	185.8	421.9	607.7	26,762.0
Utah	1982	620.3	671.2	1,291.5	682.2	64.9	747.1	2,038.6
	1987	390.1	515.6	905.7	910.0	73.5	983.5	1,889.2
	1992	469.5	431.2	900.7	842.7	71.8	914.5	1,815.2
	1997	322.6	373.9	696.5	928.9	50.9	979.8	1,676.3
Vermont	1982	0.4	166.1	166.5	1.8	479.9	481.7	648.2
	1987	0.4	185.5	185.9	1.8	455.4	457.2	643.1
	1992	0.4	145.0	145.4	1.8	487.3	489.1	634.5
	1997	0.4	135.2	135.6	1.8	463.9	465.7	601.3
Virginia	1982	108.1	2,441.7	2,549.8	2.3	845.1	847.4	3,397.2
	1987	114.6	2,051.2	2,165.8	5.6	937.8	943.4	3,109.2
	1992	115.5	1,690.9	1,806.4	3.1	1,091.4	1,094.5	2,900.9
	1997	118.8	1,509.2	1,628.0	9.0	1,242.0	1,251.0	2,879.0
Washington	1982	1,119.2	5,791.2	6,910.4	596.8	315.9	885.7	7,796.1
	1987	1,119.7	5,231.0	6,350.7	601.6	344.1	945.7	7,296.4
	1992	1,115.4	4,644.4	5,759.8	682.7	302.1	984.8	6,744.6
	1997	1,035.0	4,573.4	5,608.4	760.8	319.4	1,080.2	6,688.6
West Virginia	1982	0.0	299.8	299.8	0.0	792.8	792.8	1,092.6
	1987	0.0	227.1	227.1	0.0	770.9	770.9	998.0
	1992	0.0	214.1	214.1	0.0	701.1	701.1	915.2
	1997	1.8	159.8	161.6	1.2	685.0	686.2	847.8
Wisconsin	1982	319.6	9,092.9	9,412.5	18.4	2,025.6	2,044.0	11,456.5
	1987	331.8	9,057.1	9,388.9	9.5	1,917.4	1,926.9	11,315.8
	1992	345.1	8,464.0	8,809.1	35.0	1,968.7	2,003.7	10,812.8
	1997	363.2	8,329.8	8,693.0	24.6	1,819.8	1,844.4	10,537.4

Table 3—Cropland use, by state and year—page 7 of 7

State		Cultivated Cropland			Noncultivated Cropland			Total Cropland
		Irrigated	Nonirrigated	Total	Irrigated	Nonirrigated	Total	
		---------------- 1,000 acres ----------------						
Wyoming	1982	451.3	961.9	1,413.2	976.9	197.4	1,174.3	2,587.5
	1987	489.5	731.4	1,220.9	933.5	289.7	1,223.2	2,444.1
	1992	454.3	520.4	974.7	964.3	332.6	1,296.9	2,271.6
	1997	455.2	532.6	987.8	877.5	305.8	1,183.3	2,171.1
Caribbean	1982	79.2	271.5	350.7	2.8	54.7	57.5	408.2
	1987	75.5	252.5	328.0	2.8	58.6	61.4	389.4
	1992	68.1	214.5	282.6	2.2	82.1	84.3	366.9
	1997	43.5	122.3	165.8	7.7	181.0	188.7	354.5
Total	1982	49,612.9	326,854.8	376,467.7	12,299.5	32,208.4	44,507.9	420,975.6
	1987	49,213.4	313,760.7	362,974.1	12,315.7	31,362.7	43,678.4	406,652.5
	1992	49,396.4	285,842.7	335,239.1	13,042.3	34,034.8	47,077.1	382,316.2
	1997	48,644.5	276,513.5	325,158.0	13,195.9	36,690.1	49,886.0	375,044.0

Table 10—Estimated average annual sheet and rill erosion on nonfederal land, by state and year—page 1 of 6

State		Cropland			CRP Land	Pastureland
		Cultivated	Noncultivated	Total		
		---------------- tons/acre/year ----------------				
Alabama	1982	7.6	0.8	7.2	----	0.6
	1987	6.4	0.4	5.9	3.0	0.5
	1992	6.9	0.5	6.2	0.6	0.5
	1997	6.7	0.5	6.0	1.2	0.5
Arizona	1982	0.6	0.3	0.5	----	0.2
	1987	0.6	0.2	0.6	----	0.1
	1992	0.6	0.2	0.6	----	0.1
	1997	0.7	0.2	0.6	----	0.1
Arkansas	1982	3.8	0.7	3.7	----	1.1
	1987	3.8	0.6	3.7	0.7	1.1
	1992	3.5	0.7	3.4	0.7	1.2
	1997	3.5	0.6	3.4	0.6	1.1
California	1982	1.1	0.7	1.0	----	0.2
	1987	1.0	0.8	1.0	2.1	0.2
	1992	1.0	0.5	0.8	1.2	0.1
	1997	0.7	0.5	0.6	0.2	0.1
Colorado	1982	2.1	0.2	1.9	----	0.3
	1987	2.1	0.2	1.9	2.3	0.3
	1992	2.0	0.2	1.7	0.8	0.3
	1997	1.7	0.2	1.5	0.4	0.3
Connecticut	1982	4.8	0.6	2.6	----	0.2
	1987	5.6	1.4	3.2	----	0.2
	1992	6.1	1.5	3.4	----	0.1
	1997	5.6	0.7	2.6	----	0.1
Delaware	1982	2.1	0.2	2.0	----	0.4
	1987	2.0	0.5	2.0	----	0.4
	1992	2.1	0.8	2.1	0.1	0.5
	1997	2.0	0.4	2.0	0.1	0.6
Florida	1982	2.4	0.5	1.8	----	0.1
	1987	2.1	0.4	1.4	0.4	0.1
	1992	1.8	0.4	1.2	0.7	0.1
	1997	1.8	0.5	1.2	0.6	0.1
Georgia	1982	6.3	0.4	6.0	----	0.5
	1987	6.1	1.0	5.7	2.8	0.4
	1992	5.5	0.6	5.1	0.5	0.4
	1997	5.9	0.3	5.2	0.2	0.4
Hawaii	1982	5.1	3.0	4.9	----	0.8
	1987	4.9	2.8	4.6	----	0.8
	1992	4.4	2.8	4.2	----	0.7
	1997	2.5	3.3	2.7	----	0.9

Table 10—Estimated average annual sheet and rill erosion on nonfederal land, by state and year—page 2 of 6

State		Cropland			CRP Land	Pastureland
		Cultivated	Noncultivated	Total		
		------------- tons/acre/year -------------				
Idaho	1982	5.0	0.5	4.3	----	0.4
	1987	4.3	0.3	3.7	2.8	0.4
	1992	3.4	0.4	2.9	1.4	0.4
	1997	3.3	0.4	2.8	1.2	0.5
Illinois	1982	6.3	1.2	6.2	----	1.0
	1987	5.3	1.4	5.2	4.3	1.3
	1992	4.4	1.7	4.3	1.2	1.0
	1997	4.1	0.6	4.0	0.5	1.0
Indiana	1982	4.9	1.1	4.7	----	1.0
	1987	4.4	0.8	4.2	0.8	0.8
	1992	3.4	1.1	3.3	0.3	0.8
	1997	3.0	0.9	2.9	0.3	0.7
Iowa	1982	7.7	1.8	7.5	----	1.3
	1987	6.5	1.5	6.3	0.8	1.3
	1992	5.6	1.1	5.4	0.5	1.2
	1997	4.9	0.8	4.7	0.4	1.1
Kansas	1982	2.7	0.4	2.5	----	0.9
	1987	2.6	0.5	2.5	2.3	0.8
	1992	2.3	0.4	2.2	0.4	0.7
	1997	2.2	0.4	2.1	0.3	0.6
Kentucky	1982	8.4	1.0	6.9	----	2.4
	1987	8.2	1.1	6.6	3.5	2.4
	1992	5.8	1.2	4.5	1.0	2.5
	1997	4.4	1.3	3.4	1.1	2.0
Louisiana	1982	4.8	0.7	4.6	----	0.2
	1987	4.1	0.5	4.0	0.4	0.2
	1992	3.5	0.8	3.4	0.2	0.2
	1997	3.3	0.6	3.2	0.5	0.2
Maine	1982	3.7	0.2	1.8	----	0.2
	1987	4.0	0.3	1.8	----	0.1
	1992	3.0	0.3	1.3	0.1	0.2
	1997	3.7	0.3	1.6	0.1	0.2
Maryland	1982	5.5	1.2	5.2	----	1.1
	1987	5.3	1.9	5.0	1.7	1.1
	1992	4.9	1.8	4.6	1.5	1.0
	1997	4.4	1.2	4.0	1.3	0.7
Massachusetts	1982	5.5	0.1	1.6	----	0.2
	1987	5.7	0.1	1.6	----	0.1
	1992	4.1	0.2	1.3	----	0.2
	1997	4.6	0.1	1.2	----	0.1

Table 10—Estimated average annual sheet and rill erosion on nonfederal land, by state and year—page 3 of 6

State		Cropland			CRP Land	Pastureland
		Cultivated	Noncultivated	Total		
		\- tons/acre/year \-				
Michigan	1982	2.5	0.6	2.2	----	0.3
	1987	2.5	0.7	2.2	3.9	0.2
	1992	2.3	0.6	1.9	0.4	0.2
	1997	2.0	0.5	1.6	0.2	0.2
Minnesota	1982	2.6	0.6	2.4	----	0.3
	1987	2.6	0.4	2.4	1.3	0.3
	1992	2.3	0.3	2.2	0.3	0.3
	1997	2.1	0.3	2.0	0.2	0.3
Mississippi	1982	7.8	2.4	7.6	----	1.3
	1987	6.7	2.2	6.6	4.4	1.2
	1992	5.7	1.2	5.5	2.6	1.2
	1997	5.3	1.1	5.0	1.1	1.2
Missouri	1982	10.9	0.9	9.6	----	2.0
	1987	8.4	0.7	7.5	6.4	1.7
	1992	6.6	0.7	5.6	1.0	1.6
	1997	5.6	0.6	4.4	0.7	1.3
Montana	1982	2.1	0.2	1.8	----	0.2
	1987	2.3	0.1	2.0	0.8	0.2
	1992	2.0	0.2	1.7	0.2	0.2
	1997	1.9	0.3	1.6	0.2	0.2
Nebraska	1982	4.8	0.7	4.5	----	0.9
	1987	4.2	0.5	4.0	1.5	0.8
	1992	3.5	0.5	3.3	0.7	0.7
	1997	2.9	0.5	2.7	0.5	0.7
Nevada	1982	0.2	0.0	0.1	----	0.0
	1987	0.2	0.0	0.1	----	0.0
	1992	0.2	0.0	0.1	0.0	0.0
	1997	0.2	0.0	0.1	0.0	0.1
New Hampshire	1982	4.3	0.4	1.5	----	0.5
	1987	4.7	0.4	1.4	----	0.5
	1992	3.7	0.3	0.8	----	0.4
	1997	3.6	0.4	0.9	----	0.5
New Jersey	1982	6.7	1.0	5.5	----	0.5
	1987	6.8	1.2	5.7	----	0.6
	1992	5.5	0.9	4.3	0.3	0.5
	1997	5.7	0.6	4.3	0.3	0.4
New Mexico	1982	1.2	0.1	1.0	----	0.1
	1987	0.9	0.1	0.7	1.0	0.1
	1992	0.9	0.2	0.8	0.4	0.1
	1997	0.9	0.1	0.7	0.2	0.1

Table 10—Estimated average annual sheet and rill erosion on nonfederal land, by state and year—page 4 of 6

State		Cropland			CRP Land	Pastureland
		Cultivated	Noncultivated	Total		
		---------------- tons/acre/year ------------------				
New York	1982	4.0	0.7	2.6	----	0.4
	1987	4.1	0.9	2.7	8.6	0.4
	1992	4.0	0.8	2.4	0.5	0.3
	1997	3.8	0.7	2.3	0.4	0.3
North Carolina	1982	6.4	1.4	6.1	----	1.1
	1987	6.2	1.0	6.0	6.7	1.0
	1992	5.6	1.4	5.3	3.6	1.0
	1997	5.0	1.0	4.6	1.3	1.7
North Dakota	1982	1.9	0.4	1.8	----	0.4
	1987	2.0	0.4	1.8	1.0	0.5
	1992	1.5	0.3	1.4	0.3	0.5
	1997	1.4	0.3	1.3	0.2	0.4
Ohio	1982	3.8	1.1	3.6	----	2.2
	1987	3.7	1.1	3.4	3.7	1.7
	1992	3.3	1.1	3.1	0.6	1.7
	1997	2.6	1.4	2.5	0.3	1.7
Oklahoma	1982	2.7	0.6	2.6	----	0.9
	1987	3.0	0.6	2.9	1.1	0.7
	1992	2.9	0.5	2.8	0.4	0.7
	1997	2.8	0.5	2.8	0.3	0.6
Oregon	1982	4.7	0.7	3.8	----	0.6
	1987	3.4	0.5	2.7	2.9	0.5
	1992	3.2	0.4	2.5	0.4	0.5
	1997	3.1	0.4	2.3	0.4	0.5
Pennsylvania	1982	6.9	0.7	4.8	----	1.1
	1987	6.9	1.2	5.0	1.2	1.0
	1992	5.8	1.2	4.3	0.8	1.0
	1997	5.1	1.2	3.8	0.2	0.8
Rhode Island	1982	6.9	1.1	3.0	----	0.1
	1987	6.0	2.0	3.0	----	0.1
	1992	5.6	1.6	2.7	----	0.1
	1997	3.3	1.9	2.2	----	0.1
South Carolina	1982	4.0	1.8	3.9	----	0.5
	1987	3.9	1.3	3.8	3.7	0.4
	1992	3.3	1.0	3.1	1.7	0.4
	1997	3.2	0.7	3.0	1.0	0.4
South Dakota	1982	2.8	0.3	2.5	----	0.3
	1987	2.5	0.3	2.3	2.8	0.3
	1992	2.2	0.3	2.0	0.4	0.2
	1997	2.0	0.2	1.7	0.1	0.2

Table 10—Estimated average annual sheet and rill erosion on nonfederal land, by state and year—page 5 of 6

State		Cropland			CRP Land	Pastureland
		Cultivated	Noncultivated	Total		
		---------------- tons/acre/year ------------------				
Tennessee	1982	11.1	0.9	9.5	----	0.8
	1987	10.9	0.9	9.1	9.5	0.7
	1992	9.1	0.9	7.1	0.7	0.7
	1997	7.7	0.6	5.6	0.7	0.7
Texas	1982	2.6	0.9	2.6	----	0.6
	1987	2.5	1.1	2.5	0.6	0.6
	1992	2.6	0.7	2.6	0.3	0.5
	1997	2.6	0.8	2.6	0.2	0.5
Utah	1982	1.4	0.2	1.0	----	0.1
	1987	1.6	0.2	0.8	3.4	0.1
	1992	1.4	0.2	0.8	1.3	0.1
	1997	1.6	0.2	0.8	0.9	0.2
Vermont	1982	4.7	0.2	1.4	----	0.3
	1987	4.3	0.2	1.4	----	0.2
	1992	3.5	0.5	1.2	----	0.1
	1997	3.2	0.7	1.3	----	0.1
Virginia	1982	6.8	1.5	5.5	----	3.4
	1987	6.5	1.5	5.0	0.7	3.4
	1992	6.4	1.4	4.5	0.6	3.4
	1997	6.0	1.5	4.0	0.5	3.3
Washington	1982	6.1	0.4	5.5	----	0.2
	1987	7.0	0.4	6.1	2.3	0.6
	1992	5.0	0.5	4.3	0.5	0.5
	1997	4.6	0.6	4.0	0.6	0.3
West Virginia	1982	7.0	0.7	2.5	----	4.2
	1987	9.2	0.9	2.8	0.7	5.4
	1992	4.7	0.8	1.7	0.3	6.1
	1997	4.3	0.8	1.4	0.0	6.1
Wisconsin	1982	4.7	1.5	4.1	----	0.6
	1987	4.1	2.0	3.7	4.6	0.6
	1992	3.8	0.7	3.2	0.8	0.5
	1997	3.7	1.2	3.3	0.6	0.6
Wyoming	1982	1.4	0.2	0.9	----	0.3
	1987	1.4	0.1	0.8	1.5	0.2
	1992	1.3	0.2	0.7	0.5	0.3
	1997	1.1	0.1	0.6	0.2	0.3
Caribbean	1982	11.1	12.1	11.3	----	6.9
	1987	11.2	13.5	11.5	----	7.1
	1992	12.2	15.7	13.0	----	7.9
	1997	12.5	13.5	13.0	----	6.4

Table 10—Estimated average annual sheet and rill erosion on nonfederal land, by state and year—page 6 of 6

State		Cropland			CRP Land	Pastureland
		Cultivated	Noncultivated	Total		
		---------------- tons/acre/year ----------------				
Total	1982	4.4	0.7	4.0	----	1.1
	1987	4.0	0.7	3.7	2.0	1.0
	1992	3.5	0.6	3.1	0.6	1.0
	1997	3.1	0.7	2.8	0.4	0.9

Table 11—Estimated average wind erosion on nonfederal rural land, by state and year—page 1 of 6

State		Cropland			CRP Land	Pastureland
		Cultivated	Noncultivated	Total		
		---------------- tons/acre/year -------------------				
Alabama	1982	0.0	0.0	0.0	----	0.0
	1987	0.0	0.0	0.0	0.0	0.0
	1992	0.0	0.0	0.0	0.0	0.0
	1997	0.0	0.0	0.0	0.0	0.0
Arizona	1982	8.1	1.1	7.1	----	0.8
	1987	9.3	2.5	8.3	----	0.4
	1992	11.6	1.1	10.0	----	0.2
	1997	9.6	2.1	8.3	----	0.6
Arkansas	1982	0.0	0.0	0.0	----	0.0
	1987	0.0	0.0	0.0	0.0	0.0
	1992	0.0	0.0	0.0	0.0	0.0
	1997	0.0	0.0	0.0	0.0	0.0
California	1982	1.0	0.5	0.8	----	0.6
	1987	0.9	0.5	0.8	0.0	0.5
	1992	0.8	0.3	0.6	0.0	0.4
	1997	0.7	0.2	0.5	0.0	0.4
Colorado	1982	12.9	1.8	11.6	----	2.8
	1987	12.4	1.1	11.1	12.7	2.3
	1992	10.6	0.9	9.2	3.6	1.7
	1997	10.4	1.3	9.2	1.1	1.7
Connecticut	1982	0.0	0.0	0.0	----	0.0
	1987	0.0	0.0	0.0	----	0.0
	1992	0.0	0.0	0.0	----	0.0
	1997	0.0	0.0	0.0	----	0.0
Delaware	1982	0.0	0.0	0.0	----	0.0
	1987	0.0	0.0	0.0	----	0.0
	1992	0.0	0.0	0.0	0.0	0.0
	1997	0.0	0.0	0.0	0.0	0.0
Florida	1982	0.0	0.0	0.0	----	0.0
	1987	0.0	0.0	0.0	0.0	0.0
	1992	0.0	0.0	0.0	0.0	0.0
	1997	0.0	0.0	0.0	0.0	0.0
Georgia	1982	0.0	0.0	0.0	----	0.0
	1987	0.0	0.0	0.0	0.0	0.0
	1992	0.0	0.0	0.0	0.0	0.0
	1997	0.0	0.0	0.0	0.0	0.0
Hawaii	1982	0.0	0.0	0.0	----	0.0
	1987	0.0	0.0	0.0	----	0.0
	1992	0.0	0.0	0.0	----	0.0
	1997	0.0	0.0	0.0	----	0.0

Table 11—Estimated average wind erosion on nonfederal rural land, by state and year—page 2 of 6

State		Cropland			CRP Land	Pastureland
		Cultivated	Noncultivated	Total		
		\-\-\-\-\-\-\-\-\-\-\-\-\-\-\-\-\- tons/acre/year \-\-\-\-\-\-\-\-\-\-\-\-\-\-\-\-\-\-\-				
Idaho	1982	4.2	0.1	3.5	----	0.2
	1987	4.7	0.1	4.0	4.1	0.3
	1992	4.8	0.2	4.0	1.4	0.2
	1997	3.9	0.2	3.3	1.1	0.2
Illinois	1982	0.0	0.0	0.0	----	0.0
	1987	0.0	0.0	0.0	0.0	0.0
	1992	0.0	0.0	0.0	0.0	0.0
	1997	0.0	0.0	0.0	0.0	0.0
Indiana	1982	0.5	0.0	0.5	----	0.0
	1987	0.5	0.1	0.5	0.4	0.0
	1992	0.4	0.0	0.4	0.0	0.0
	1997	0.5	0.1	0.5	0.0	0.0
Iowa	1982	3.0	0.0	2.9	----	0.0
	1987	2.4	0.1	2.3	1.3	0.0
	1992	1.4	0.0	1.3	0.0	0.0
	1997	0.7	0.0	0.6	0.0	0.0
Kansas	1982	2.7	0.5	2.6	----	0.0
	1987	3.0	0.4	2.9	11.7	0.0
	1992	2.1	0.5	2.0	0.8	0.0
	1997	1.5	0.2	1.4	0.3	0.0
Kentucky	1982	0.0	0.0	0.0	----	0.0
	1987	0.0	0.0	0.0	0.0	0.0
	1992	0.0	0.0	0.0	0.0	0.0
	1997	0.0	0.0	0.0	0.0	0.0
Louisiana	1982	0.0	0.0	0.0	----	0.0
	1987	0.0	0.0	0.0	0.0	0.0
	1992	0.0	0.0	0.0	0.0	0.0
	1997	0.0	0.0	0.0	0.0	0.0
Maine	1982	0.0	0.0	0.0	----	0.0
	1987	0.0	0.0	0.0	----	0.0
	1992	0.0	0.0	0.0	0.0	0.0
	1997	0.0	0.0	0.0	0.0	0.0
Maryland	1982	0.1	0.0	0.1	----	0.0
	1987	0.1	0.0	0.1	0.0	0.0
	1992	0.1	0.0	0.1	0.0	0.0
	1997	0.1	0.0	0.1	0.0	0.0
Massachusetts	1982	0.0	0.0	0.0	----	0.0
	1987	0.0	0.0	0.0	----	0.0
	1992	0.0	0.0	0.0	----	0.0
	1997	0.0	0.0	0.0	----	0.0

Table 11—Estimated average wind erosion on nonfederal rural land, by state and year—page 3 of 6

State		Cropland			CRP Land	Pastureland
		Cultivated	Noncultivated	Total		
		---------------- tons/acre/year -------------------				
Michigan	1982	2.5	0.3	2.1	----	0.2
	1987	2.7	0.4	2.3	2.8	0.1
	1992	2.6	0.2	2.1	0.2	0.1
	1997	2.4	0.2	1.9	0.1	0.1
Minnesota	1982	5.9	0.1	5.4	----	0.1
	1987	6.7	0.7	6.3	6.7	0.2
	1992	6.4	0.1	5.9	0.4	0.2
	1997	5.8	0.1	5.3	0.1	0.1
Mississippi	1982	0.0	0.0	0.0	----	0.0
	1987	0.0	0.0	0.0	0.0	0.0
	1992	0.0	0.0	0.0	0.0	0.0
	1997	0.0	0.0	0.0	0.0	0.0
Missouri	1982	0.0	0.0	0.0	----	0.0
	1987	0.0	0.0	0.0	0.0	0.0
	1992	0.0	0.0	0.0	0.0	0.0
	1997	0.0	0.0	0.0	0.0	0.0
Montana	1982	7.9	0.3	6.8	----	0.2
	1987	8.8	0.4	7.6	12.7	0.1
	1992	7.2	0.1	6.0	0.2	0.1
	1997	3.8	0.2	3.1	0.2	0.1
Nebraska	1982	1.6	0.1	1.5	----	0.1
	1987	1.7	0.2	1.6	3.3	0.1
	1992	1.7	0.3	1.6	0.4	0.1
	1997	1.6	0.2	1.5	0.0	0.1
Nevada	1982	39.4	1.0	16.3	----	1.3
	1987	30.0	0.8	6.2	----	1.1
	1992	23.1	1.0	7.1	0.0	1.2
	1997	27.0	1.0	5.3	0.7	1.5
New Hampshire	1982	0.0	0.0	0.0	----	0.0
	1987	0.0	0.0	0.0	----	0.0
	1992	0.0	0.0	0.0	----	0.0
	1997	0.0	0.0	0.0	----	0.0
New Jersey	1982	0.1	0.0	0.1	----	0.0
	1987	0.1	0.0	0.1	----	0.0
	1992	0.1	0.0	0.1	0.0	0.0
	1997	0.1	0.0	0.1	0.0	0.0
New Mexico	1982	15.0	3.7	13.1	----	3.6
	1987	16.0	3.8	13.3	17.4	3.8
	1992	16.8	2.9	13.6	6.7	4.0
	1997	12.3	3.3	9.9	2.6	4.2

Table 11—Estimated average wind erosion on nonfederal rural land, by state and year—page 4 of 6

State		Cropland			CRP Land	Pastureland
		Cultivated	Noncultivated	Total		
		----------------- tons/acre/year -----------------				
New York	1982	0.0	0.0	0.0	----	0.0
	1987	0.0	0.0	0.0	0.0	0.0
	1992	0.0	0.0	0.0	0.0	0.0
	1997	0.0	0.0	0.0	0.0	0.0
North Carolina	1982	0.0	0.0	0.0	----	0.0
	1987	0.0	0.0	0.0	0.0	0.0
	1992	0.0	0.0	0.0	0.0	0.0
	1997	0.0	0.0	0.0	0.0	0.0
North Dakota	1982	6.4	0.3	5.9	----	0.1
	1987	6.4	0.3	6.0	8.4	0.1
	1992	2.1	0.1	1.9	0.2	0.0
	1997	4.0	0.2	3.6	0.2	0.1
Ohio	1982	0.3	0.0	0.3	----	0.0
	1987	0.3	0.0	0.2	0.1	0.0
	1992	0.1	0.0	0.1	0.0	0.0
	1997	0.1	0.0	0.1	0.0	0.0
Oklahoma	1982	2.5	0.6	2.4	----	0.1
	1987	2.6	0.3	2.5	10.1	0.0
	1992	1.9	0.2	1.8	0.4	0.0
	1997	1.5	0.1	1.5	0.3	0.0
Oregon	1982	2.2	0.2	1.8	----	0.1
	1987	2.4	0.3	1.8	1.3	0.1
	1992	1.9	0.1	1.5	0.1	0.1
	1997	2.0	0.1	1.5	0.0	0.1
Pennsylvania	1982	0.0	0.0	0.0	----	0.0
	1987	0.0	0.0	0.0	0.0	0.0
	1992	0.0	0.0	0.0	0.0	0.0
	1997	0.0	0.0	0.0	0.0	0.0
Rhode Island	1982	0.0	0.0	0.0	----	0.0
	1987	0.0	0.0	0.0	----	0.0
	1992	0.0	0.0	0.0	----	0.0
	1997	0.0	0.0	0.0	----	0.0
South Carolina	1982	0.0	0.0	0.0	----	0.0
	1987	0.0	0.0	0.0	0.0	0.0
	1992	0.0	0.0	0.0	0.0	0.0
	1997	0.0	0.0	0.0	0.0	0.0
South Dakota	1982	4.0	0.1	3.5	----	0.1
	1987	3.6	0.4	3.3	3.2	0.2
	1992	2.6	0.2	2.3	0.3	0.1
	1997	2.0	0.1	1.7	0.0	0.1

Table 11—Estimated average wind erosion on nonfederal rural land, by state and year—page 5 of 6

State		Cropland			CRP Land	Pastureland
		Cultivated	Noncultivated	Total		
		---------------- tons/acre/year ----------------				
Tennessee	1982	0.0	0.0	0.0	----	0.0
	1987	0.0	0.0	0.0	0.0	0.0
	1992	0.0	0.0	0.0	0.0	0.0
	1997	0.0	0.0	0.0	0.0	0.0
Texas	1982	12.7	2.6	12.5	----	0.1
	1987	11.5	2.7	11.3	13.4	0.0
	1992	9.4	1.4	9.2	1.0	0.0
	1997	9.4	0.2	9.2	0.8	0.0
Utah	1982	5.9	1.7	4.3	----	1.5
	1987	6.4	2.1	4.2	6.4	1.9
	1992	6.5	1.5	4.0	1.9	1.6
	1997	4.5	0.7	2.2	0.8	1.3
Vermont	1982	0.0	0.0	0.0	----	0.0
	1987	0.0	0.0	0.0	----	0.0
	1992	0.0	0.0	0.0	----	0.0
	1997	0.0	0.0	0.0	----	0.0
Virginia	1982	0.2	0.0	0.1	----	0.0
	1987	0.3	0.0	0.2	0.0	0.0
	1992	0.2	0.0	0.1	0.0	0.0
	1997	0.2	0.0	0.1	0.0	0.0
Washington	1982	3.8	0.6	3.5	----	0.2
	1987	3.9	0.9	3.5	1.8	0.3
	1992	5.8	0.4	5.0	0.2	0.1
	1997	5.1	0.9	4.4	0.0	0.0
West Virginia	1982	0.0	0.0	0.0	----	0.0
	1987	0.0	0.0	0.0	0.0	0.0
	1992	0.0	0.0	0.0	0.0	0.0
	1997	0.0	0.0	0.0	0.0	0.0
Wisconsin	1982	0.2	0.0	0.1	----	0.0
	1987	0.2	0.0	0.2	0.0	0.0
	1992	0.2	0.0	0.2	0.0	0.0
	1997	0.2	0.0	0.2	0.0	0.0
Wyoming	1982	6.4	1.5	4.1	----	2.5
	1987	7.9	1.5	4.7	2.3	2.0
	1992	7.7	0.9	3.8	1.0	1.7
	1997	6.1	0.4	3.0	0.2	0.7
Caribbean	1982	0.0	0.0	0.0	----	0.0
	1987	0.0	0.0	0.0	----	0.0
	1992	0.0	0.0	0.0	----	0.0
	1997	0.0	0.0	0.0	----	0.0

Table 11—Estimated average wind erosion on nonfederal rural land, by state and year—page 6 of 6

State		Cropland			CRP Land	Pastureland
		Cultivated	Noncultivated	Total		
		---------------- tons/acre/year -------------------				
Total	1982	3.6	0.4	3.3	----	0.1
	1987	3.5	0.4	3.2	6.8	0.1
	1992	2.7	0.2	2.4	0.6	0.1
	1997	2.5	0.2	2.2	0.3	0.1

"The Buffalo Commons, Then and Now"
Deborah Popper and Frank Popper

Focus, January 1, 1993, Vol. 3, p. 17. Reprinted by permission of the American Geographical Society.

In 1987 we wrote in Planning Magazine about the development of the Great Plains and were surprised that anyone cared. We described Plains history in quite conventional terms as a series of boom-and-bust cycles, each largely the result of government and private incentives that stimulated profuse development, landscape manipulation, and then repercussions. Plains settlement had taken off after the Civil War. Encouraged by the 1862 Homestead Act, as well as by promoters and railroad companies, settlers made their way out to the region for free or nearly free land. Once there, the high rainfall of the 1870s gave them hope of prosperity. But by the 1890s, the Great Plains entered its first bust as the agricultural region hit drought and the country hit a depression. Fully loaded wagon trains headed east out of the Plains. Population rebounded in the early 1900s with new, more generous homesteading laws. Plains agriculture proved especially profitable during World War I, for it filled the gap left when much European agricultural land became battlegrounds.

The optimism of this period gave way to the despair of the 1930s *Dust Bowl*, which blew the exposed topsoil, no longer held in place by native grasses, hundreds or even thousands of miles away. This era, captured so stunningly by the WPA photographs of grim **dust** clouds and bleak faces, challenged the assumption that the Plains could withstand the settlement practices imposed upon it.

After the 1930s, the Plains once again moved into an optimistic mode. An assortment of government and private incentives—new irrigation projects, energy-led development, foreign wheat deals—resulted in slow economic rebound: The earlier ambitions for a heavily settled region had evaporated, so decline proved less dramatic when it hit in the 1980s. Population continued to drop, especially away from the Interstates, irrigation projects, large rivers and big cities. Soil erosion approached **Dust Bowl** rates, but less noticeably because now it was water-borne. The Ogallala Aquifer, the source of agricultural and urban water for much of the southern two-thirds of the Plains, dropped. In many places, the Plains economy suffered as Americans ate less beef, federal deficits provoked unhappiness with farm surpluses, and missile silos aimed at Russians became outmoded.

We saw little to stem these trends. The look of the Plains provoked nostalgia, but this palimpsest of small-town America had often lost its pastor, its doctor, its banker, its farm implements dealer. The land remained evocative with its oceans-of-grass vistas, enormous horizons, billowy clouds, and somber-serene beauty. The land, so fitfully conquered, appeared to be trying to reassert itself as grasses overtook deserted homesteads and abandoned storefronts.

The Buffalo Commons grew out of our late-1980s expectation that ongoing trends would continue unabated for another generation—more depopulation combined with overgrazing, overplowing, and overirrigating. Only after another generation of desettling did we expect the onset of a new set on incentives for the region, largely coming from the previous incentive provider, the federal government. To us, the Buffalo Commons in the 1980s represented such an alternative. It was neither a prescription nor a plan; instead it served as a metaphor for a fresh way of thinking about the region's land use. The term buffalo evoked the need to remember and value the particular characteristics of place, the value of biodiversity, and the wisdom of less intensive use of the land. The term commons evoked incentives that maintain land as a renewable resource while balancing community and individual rights.

We had set down our musings about the Great Plains without much expectation of response. Instead it set off a torrent of discussion about the region's future. To this day we receive mail about our work, some admiring, some vituperative. Respondents range in age from schoolchildren to retirees. They are farmers and ranchers, bankers, lawyers, dentists, doctors, academics, bureaucrats, writers. We have been invited to the Plains to address an equally diverse set of groups about the Buffalo Commons; we have spoken in lecture halls, ballrooms, cafes, and public squares. The talk served to cross-pollinate ideas—ours and theirs.

In our audiences we found disparate groups agreeing on many issues. A recurring theme was that they were getting too many subsidies that should be cut; they varied, but the sentiment remained constant. There was much concern with loss—of small-town life, animal life, or opportunity. Hope was equally present: for environmentally benign and economically profitable agricultural adaptations, for telecommunication innovations that would make the region's remoteness a nonissue. Many people felt a possessiveness about their Plains—positively, exuding pride and love, and negatively, resenting outsiders.

With the discussion has come change, including a wide variety of new efforts that we never anticipated. There have been several initiatives to increase bison and to preserve grasslands across the Great Plains. The groups most involved are ranchers, Native American tribes, land preservation organizations, and public land managers. These ventures depend less on federal intervention than we had expected in the 1980s, and some yield direct income.

Increasing numbers of cattle ranchers, for example, have found buffalo economically viable and easier to raise. As a native species, bison are better adapted to the physical environment of the Plains, so management leans toward leaving them to fend for themselves. Buffalo graze more widely than cattle and survive harsh weather better. They bring good prices, both for the meat which fits into a growing consumer demand for organic and low-fat foods (bison are a red meat with a cholesterol level as low as chicken) and for other parts of the animal like horns and hides.

Growing numbers of Plains farmers and ranchers have found their land more valuable as habitat than as farmland. They make their living from fees paid by individuals or groups for hunting or backpacking. Land preservation organizations like the Nature Conservancy paid little attention to the Plains until a few years ago, but now they are undertaking large buy-ups and trades and experimenting with non- intensive uses on their holdings; sometimes they exclusively promote biodiversity, and in other cases they encourage ecotourism or lower-impact agriculture.

Native Americans, whose populations are increasing in the rural Great Plains as the white populations diminish, are actively restoring buffalo on their reservations. The Intertribal Bison Cooperative, a consortium of about thirty tribes, represents one of the most active efforts. While each tribe makes its own decisions, the group shares expertise in buffalo acquisition, management, marketing, and slaughtering. The tribes see the return of bison as vital to their cultural, spiritual, and economic renewal.

These examples of Buffalo Commons land uses reflect a shift from the assumption that the value of land lies in the intensity of its use—that it is more valuable when it is plowed, sold, or developed. The new attitude sees some lands as better treated more gingerly. The limitations placed on their development are in fact the wisest way to heighten and perpetuate their value.

The Buffalo Commons began as a metaphor for reenvisioning practices in a region clearly experiencing difficulties—loss of water, soil, people, and community. In the eight years since we first wrote, the metaphor has seen a series of real-world improvisations bring it to life. Where it will go in the future is still unknown. We expect large parts of the Plains will remain much as they are today. The cities are likely to grow, continuing to pick up outmigrants from the region's most rural areas as well as from the country's more urban regions. Those parts of the Plains where farming, ranching, and mining support viable economies will probably continue much as they are.

But the less sustainable rural areas are likely to continue to shift towards Buffalo Commons land uses. Such uses open up exciting new possibilities. The innovations may continue on a piecemeal, individual basis, remaining separate while reflecting a broader change in the ways we value land. Or they may eventually become part of a larger-scale enterprise as these improvisations begin to grow, overlap, and find organic coherence. They may generate conflicts—say, between private and public buffalo herds or between buffalo and cattle ranchers—that call forth a more systematic, less improvisational approach, a legislated Buffalo Commons. For us the core of the Buffalo Commons remains in the metaphor. Over the next generation, as we travel through the Plains, we hope to see all the two-legged and four-legged species roaming the grasslands, thriving in their native habitat.

"Comment on the Future of the Great Plains: Not a Buffalo Commons"

Karen J. DeBres, David E. Kromm, and Steven E. White

Focus, January 1, 1993, Vol. 43:16. Adapted by permission of the American Geographical Society.

The Buffalo Commons Proposal, whether offered in this issue of *Focus* or in its original 1987 version, has served a useful function in alerting the nation to the environmental, economic and social problems long acknowledged in the Great Plains. But having seen and heard the Poppers' ideas in several places, we feel obliged to ask the following question. Are Frank and Deborah Popper deliberately discussing this topic in such a way, so as to create an emotionally charged atmosphere? Are they at times greatly exaggerating certain negative conditions in the Plains while glossing over or ignoring altogether other, more positive, attributes? Again, while this may be a good tactic for a public debate, we wonder about its usefulness when applied in print. We raise this point for several reasons.

We question the Poppers' tactic of creating an environmental context which overstates the negative aspects of the Plains environment. Sometimes their descriptions are plain wrong. For example, neither the hottest summers nor the coldest winters occur in the region. Nor do the shortest growing seasons occur in the Great Plains. They also say that the greenhouse effect "has growing credibility and seems likely to hit the southern Plains particularly hard." We don't think there is any debate about the "greenhouse effect," which we assume that the Poppers are equating with global warming. However, there is still a good deal of uncertainty about global warming as evidenced by a book by geographer Robert Bailing (1992) titled The Heated Debate: Greenhouse Predictions Versus Climate Reality. The Poppers tend to write in vivid, broad, sweeping statements with little or no factual verification, a style which weakens even their valid arguments.

In several recent articles, the Great Plains are described by Frank and Deborah Popper as spare and weather whipped, endlessly windswept, the land of the Dust Bowl (with "dusty" towns, of course) and as having the nation's hottest summers and coldest winters, the worst hail, locusts and range fires, fiercest droughts and blizzards (that throw black grit no less), and its shortest growing season. According to the Poppers, the Great Plains is a place where "privacy is almost too easily obtained" and where "visitor learns to recognize the Plains person's 20-yard stare;" a place running out of water and where towns are withering and dying. The future holds desertion with "safaris across Kansas, Indians in gatherings of the lost tribes of Texas, giant abandoned North Dakota power plants that become environmentalist shrines." What a place! Let the buffalo loose now!

The Poppers' negative description of the Great Plains does not match the inhabitants' vision of their personal reality; the natives' sense of place is much more humane, warm, and positive. We can vouch for the fact the Kansas rarely spit black grit, worry about the dust storms, flee range fires, swat locusts, let the wind beat them to death, feel lonely or stare down those nasty outsiders. Our vision is that the Plains are inhabited by friendly people who perceive their environment in much more endearing terms. We know this because we live nearby, teach the Plains children, and frequently venture into this "frontier" to conduct research.

The Poppers' history also shows a tendency toward exaggeration. For example, they state that Kansas has "hundreds of ghost towns." This is true. What the Poppers don't mention, however, is that most Kansas ghost towns were "speculative" towns which never existed anywhere but on paper or perhaps as a small hamlet next to an ephemeral railroad stop. Why do the Poppers base the quality of life or the economic value of an area on population growth?

In a recent paper, the Poppers are still using their system of county level data from the Census and the County and City Data Book to devise their list of counties in land use distress. The major methodological flaw of the Buffalo Commons proposal is that the variables used to identify counties in land use distress are not land use indicators. Also, county level data misses the variability within counties, as well as rural-urban differences. Another methodological problem which we find with the Poppers' proposal is its lack of citations, either from primary sources other than the Census or from many academic sources. They tend to support their arguments with local newspaper articles or their own ideas.

Finally, we believe the Poppers' model is wrong in portraying the Great Plains as an essentially homogenous region. It is presented as similar throughout with respect to climate, landforms, resource, endowment, vegetation, economic viability, and demographic change. In other words, the Great Plains is a uniform region undergoing transformation into a Buffalo Commons. Nothing could be further from the truth. So vast an area expectedly varies significantly from place to place in terms of winter temperatures, annual precipitation, flatness of land, reserves of metals and fossil fuels, transportation linkages, types of economic activity, and degree of urbanization.

One important variable is the existence of the Ogallala aquifer. Wherever the High Plains aquifer system supports irrigation, there are prosperous farms, growing cities, and general economic well-being. The buffalo would need to crowd out the vibrant farms, towns, cattle feedlots, and meat packing plants. We think not.

We realize that the Poppers have performed a valuable service in alerting the nation to land use problems in the Great Plains. We think, however, that it is time for the Poppers to rework more thoroughly their ideas about this region and we also look to others for new approaches to the problems of the Great Plains.

reading 6

"The Bison Are Coming"
Deborah Popper and Frank Popper

High Country News, February 2, 1998. Reprinted by permission of *High Country News* and the authors.

In the December 1987 issue of *Planning,* we wrote what we thought was an innocuous article on land use in the Great Plains. The piece explored the state of the short-grass, semi-arid region between the 98th meridian and the Rockies, a sixth of the Lower 48.

The most rural parts of the Plains faced long-standing problems—droughts, disappearing topsoil, depopulation, declines in the traditional agricultural and energy economies, and dependence on federal farm subsidies. We argued that in response, the Plains' future would draw on pieces of its past. We called this approach the Buffalo Commons.

To us, restoring a commons for buffalo offered a metaphor for a change to new uses of land that fell between intensive cultivation and pure wilderness, with less emphasis on agriculture and extraction and more on preservation and ecotourism.

For an exercise in social prophecy, the Buffalo Commons has enjoyed surprising impact, accuracy and endurance. Its tenth anniversary provides an occasion to reflect on its reach and ponder its possibilities.

The Buffalo Commons touched off a national debate about the fate of the Plains. Many environmentalists, Native Americans, and preservation- and tourism-based interests were inspired by the Buffalo Commons.

Ranchers, farmers, energy interests and communities often saw it as an assault on their livlihoods, legitimacy and ancestors. Some of our early speeches in the Plains required armed guards, and a 1992 Montana talk got canceled because of death threats.

The Buffalo Commons remains controversial. Reports regularly appear claiming that the Plains economy is booming, but they tend to focus on the Plains state overall or their urban centers, rather than on the most rural parts of the states. For example, in September 1997 the Center for the New West, a Denver group, issued a report celebrating "Economic Resurgence in the Great Plains," and the center's president, Phil Burgess, used the report to write off the Buffalo Commons as "folly." Yet the report included a section titled "Why Growth Slows at the 100th Meridian" and described the area west of that longitude as "still a kind of frontier where development often grinds to a halt and falling birth rates and outmigration take their toll." Exactly!

While conventional development was grinding "to a halt," by the middle 1990s the Buffalo Commons was forming on the ground, particularly in the Northern Plains. Public-land buffalo herds increased markedly.

On private land a noticeable number of Plains ranchers switched from cattle to buffalo and prospered economically and environmentally. Buffalo have nutritional advantages over cattle—less fat and cholesterol, more protein—and They drink less water, trample riverbank areas less and need less ranch work, especially in calving. They survive winter better, as in the 1996-97 Dakota-Montana blizzards and the October 1997 Colorado storm, and they yield higher profits.

The buffalo market, which barely existed 10 years ago, thrives while the cattle (and sheep) markets keep slipping. North Dakota Gov. Edward Schafer sees buffalo production and buffalo tourism as vital to the state's growth. The state's bank lends enthusiastically to buffalo ranchers, and its agricultural extension service offers them technical help. North Dakotans have established a marketing cooperative and a slaughtering-processing facility specially for buffalo and plan another one, which the state is encouraging. In 1996, its agriculture commissioner, Sarah Vogel, told *The New York Times* that North Dakota will someday have more buffalo than cattle.

In 1992, 19 Plains Indian tribes formed the InterTribal Bison Cooperative to reinvigorate buffalo's historically central place in their cultures. The cooperative, based in Rapid City, S.D., trains Indian buffalo producers and tribal land managers; promotes buffalo art and artifacts; operates a joint venture with an Indian-owned farming company; and has plans for a slaughtering-processing facility. The co-op has grown to 42 tribes, including some outside the Plains. The buffalo population on Indian land has tripled since 1992.

Until the 1990s, the Nature Conservancy, the country's leading land-preservation organization, tended to ignore the Plains, but it has recently made many purchases in the region. It often restores buffalo and ecologically associated animals and plants and sets up ecotourism enterprises. The conservancy and other preservation groups plan more such projects.

Federal agencies are also abetting the Buffalo Commons. The Forest Service may allow buffalo to graze on more of its forest and grassland holdings in the Dakotas, Nebraska and Wyoming. And in 1992, the U.S. Interior Department began the Great Plains Partnership, a wildlife protection effort by federal agencies, state governments, and their Canadian and Mexican counterparts. The Clinton administration expanded the program and assigned the Environmental Protection Agency to lead it.

Our research has stimulated other work on the Plains. Anne Matthews' book, *Where the Buffalo Roam* (Grove Weidenfeld, 1992), which focuses on our work (HCN, 12/16/91), was one of four finalists for the 1993 Pulitzer Prize for nonfiction. Since 1993, Lawrence Brown's bimonthly newsletter, *From the Deep Plains,* has presented a South Dakota rancher's friendly attempts to find alternatives to the Buffalo Commons. Ernest Callenbach's B*ring Back the Buffalo! A Sustainable Future for America's Great Plains* (Island Press, 1995) and Daniel Licht's *Ecology and Economics of the Great Plains* (University of Nebraska Press, 1997) support the Buffalo Commons and suggest new ways to achieve it.

Local environmental groups—for instance, South Dakota's Sierra Club chapter and Bring Back the Bison in Evanston, Wyo.—lobby for buffalo.

It has been exciting to watch our metaphor spring to life and acquire the muscle of reality. We see a growing recognition that the idea makes ecological and financial sense—that it offers a plausible option for many places, especially if the other choices are casinos, prisons, hazardous waste, agribusiness or continued slow-leak decline.

The emergence of the Buffalo Commons shows the adaptiveness of the rural West. We confidently await the further return of the buffalo.

Questions for Discussion

For Readings 1 and 2

1. How have farming practices changed in the last 100 years in the Great Plains region?
2. What was the overall purpose of the Soil Conservation Service and how did it accomplish its goals?
3. What are the key elements of the Great Plains Conservation Plan?

For Reading 3

1. Figures 1, 2, and 3 of the Inventory address nationwide trends in soil-related resources. Are conditions improving or getting worse?
2. Use Tables 1, 2 and 3 for the following. Pull out data for one or more of the states in the Great Plains. What land-use changes took place between 1982 and 1992? How do these compare with national trends? (See question 1.)
3. Use Tables 10 and 11 for the following. Using the same states as in question 2, what changes in land erosion occurred between 1982 and 1992?
4. Using the figures and tables in the NRI, determine how soil conditions and land uses are changing in your state or region. If you live in the Great Plains, pick another state or region that you are interested in.

For Readings 4, 5 and 6

1. In your opinion, how practical is the proposal to turn the Great Plains into a buffalo commons?
2. What would be the advantages and disadvantages of such a plan?

Going Beyond the Readings

Visit our Website at <www.harcourtcollege.com/lifesci/envicases2> to investigate soil conservation issues in additional regions of the United States and Canada.

unit 7

Northwest: Wildlife Management

Introduction
Readings:
Reading 1: The Spotted Owl Controversy and the Sustainability of Rural Communities in the Pacific Northwest
Reading 2: Having Owls and Jobs Too
Reading 3: The Birds—The Spotted Owl: An Environmental Parable
Reading 4: Timber Owners Cut a Deal to Preserve Wildlife Habitat
Reading 5: U.S. Considers 80% Increase in Sierra Logging
Questions for Discussion

Managing Timber in the Pacific Northwest

Introduction

In the Pacific Northwest of North America, environmental controversies sometimes rage as monumental and complex as the forests around which they center. Those who cut trees rage at those who would protect them. Those who would protect endangered species rage at those who would sacrifice a few rare plants and animals for reliable, high-paying jobs. Those who favor states rights rage at the federal government. Foresters look at the same data sets, interpret them differently, and rage at each other.

It was not always so. Two hundred years ago, timber in the Pacific Northwest seemed a boundless resource. One of the world's great forests stretched along the coast of North America from Alaska to California and inland to Alberta, Montana, Wyoming, and Colorado. For thousands of years, these forests provided wildlife and indigenous peoples with habitat and sustenance. In the last two centuries, human demands on the forests changed. Fully 25 percent of U.S. forest resources are located in the Pacific Northwest. As needs for lumber and wood products grew, so did the region's timber industry. The harvest of trees became a dominant economic force throughout the region.

Seeds of controversy sprouted around the turn of the twentieth century, when it became obvious that the forests had limits. Whole tracts of trees disappeared as rates of harvest far exceeded the forests' innate ability to replenish. In the United States, the U.S. Forest Service was created to manage federally-owned forests so that trees could be harvested indefinitely. Private forests were left much to their own devices, but, in order to stay profitable, they too had to face the problem of diminishing reserves.

Theoretically, solutions are simple. Forests are renewable. They grow, and if harvest rates match growth rates, sustainable forests are possible. One solution to the region's diminishing forests would be to restrict harvest rates to growth rates. But severely restricted harvest rates eat into industry profits. Again, solutions seem simple. Speed up the forests' growth. Replant harvested trees with seedlings that will in time grow into new timber.

Throughout the region, a technique of timber harvest evolved called "clear cutting." All trees, and indeed nearly all plant life in a particular tract, are cut. All profitable logs are removed, usually by truck on temporary roads that, on the way to sawmills, cross other tracts. Brush and trash trees are carefully burned to minimize risk of accidental, hard to control, expensive fires that could spread to adjoining tracts. The nearby shorn tract is then replanted, largely with trees that grow fast and have high potential market value. What had been a forest then resembles a tree farm-one where sustainable harvests are possible.

Background

The first regional controversies flared like forest fires. Industry and federal foresters could not always agree on what constituted sustainable harvest rates, nor on which species should be replanted. Also controversial were optimal rotation times for particular tracts and the amount of replanting. Difficult and sometimes heavy-handed decisions had to be made, especially on federally-owned forests.

Resentments grew. Lumberjacks in the Pacific Northwest did not like to be regulated.

The issue intensified in the 1970s. Professionally, foresters began to understand that a forest is more than a group of trees. One small example: To be healthy, many plants, including trees in the Pacific Northwest, form intimate associations with fungi growing in soil. The fungi help trees gather nutrients. In return, trees share photosynthetic products. In order to spread from place to place, fungi occasionally grow fruiting bodies (mushrooms) that produce spores. Squirrels find mushrooms tasty and carry them into new locations, thus spreading spores. The fungi are brought to where young trees sprout. This three-way symbiotic (mutually beneficial) relationship, involving trees, fungi, and squirrels, vastly complicates life for foresters. Their chief interest is trees; but to get healthy trees they need to be concerned with fungi, squirrels, and numerous other organisms. Forests are more than trees.

At the same time, the idea spread to the general public. With the environmental movement, people who lived far from forests-some in regions far removed from the Pacific Northwest-suddenly became vitally interested and involved. They found clear-cut areas visually repulsive. They saw that replanted forests did not resemble the magnificent original forests. They worried about how long it would take for new-growth forest to become old-growth forest. What happens to wildlife in the meantime?

Endangered species complicated the picture further. First there were the salmon and trout. When clear-cutting comes too close to streams, soil erosion spoils water quality. Fish die. Into the fray waded fisherman, both sport and commercial. Along with them came Native Americans, who have used these waters for thousands of years. For them, fish in pristine waters are more than food or sport; they are part of their spiritual heritage. These powerful and vociferous groups demanded that their interests-fish and streams-be preserved.

And then there was the northern spotted owl. Often, when confronted with monumental complexity, we focus on one small aspect, and so it was here. In the 1980s, the northern spotted owl became a "poster child" for the entire issue. This small nocturnal bird of prey is found only in old-growth forests. As forests disappeared, owl numbers decreased. In 1986, the owl was listed by the U.S. Fish and Wildlife Service as an officially endangered species. Since then, other endangered species, especially the marbled murrelet, a small sea bird that nests only in old-growth forests of the Pacific Northwest, have complicated things further.

The Problem

The Endangered Species Act prohibits any citizen of the United States from taking action that adversely affects the well-being of any endangered species. In the Pacific Northwest, this requirement had serious implications for both federally- and privately-owned forests. Wherever northern spotted owls lived, their preservation took precedence over timber cutting. The act did not required total cessation of harvesting; it did require that any tree cutting that occurred could not adversely affect the owl.

"These are our trees and our owls," say lumberjacks and timber industry representatives. "We've been cutting trees for generations. It's all we know how to do. It's all we want to do. It sustains our communities. How dare you tell us we can't cut down our own trees."

"They're not your trees and not your owls," respond environmentalists. "As a society, we have a responsibility to preserve our resources. How dare you take our owls and forests for personal greed."

"These forests are located in sovereign states, whose jurisdiction should transcend that of the federal government. How dare you impose outside values on us," cry those who advocate states rights.

"You better not do anything to muddy any more of our fishing streams," demand the fishermen. "Those are our streams and our fish."

"Hey, wait a minute," says the Native American, "these are really ours."

Where is the voice of reason in all this? Environmental issues are difficult enough when they come to us one by one. But in the Pacific Northwest, issues seem to complicate and build on each other. The readings that follow illustrate the magnitude of the problem and point the direction to possible solutions.

reading 1

"The Spotted Owl Controversy and the Sustainability of Rural Communities in the Pacific Northwest"

Daniel Levi and Sara Kocher

Environment and Behavior, Vol. 27, September 1, 1995, p. 631. © 1995 by Sage Publications, Inc. Reprinted by permission of the publisher.

Abstract The viability of rural logging communities in the Northwest is threatened. Even with sustainable forest management, financial and technological changes are decreasing the employment opportunities in rural communities. The contribution of the logging industry to local economics has been decreasing due to lumber competition from other regions, technological change, and the spotted owl controversy. This study reports on the attitudes of people involved in the economic development of the rural Northwest who attended regional economic development conferences held in 1991 and 1992. The participants at each conference rated the economic situation of the region as poor, and primarily blamed federal regulations for their current economic problems. They recognized the need to develop a diversified rural economy and believed the region's greatest economic assets were its environmental quality and quality of life. Attendance at the first regional economic development conference resulted in short-term improvement in attitudes about the future of the rural areas.

The rural communities of the United States are struggling to survive. During the last two decades, many rural communities and counties lost population (Popper & Popper, 1991). As the urban population of this country continues to grow, we are abandoning our rural areas because they no longer seem capable of providing a sustainable economic environment for people.

Sustainability requires that the use of resources in the present not damage their future use. A sustainable rural environment depends on the interrelationship of the physical environment, the economy, and the community (Dixon & Fallon, 1989). Traditionally, rural communities have developed based on the extraction of resources to support the economy. In some instances, this relationship is breaking down because the natural resources have already been used, so there is no longer a resource base for the rural economy (Schallau, 1990). In the last 20 years, the main problem has been changing financial and technological conditions. Farming, logging, and mining operations now require fewer employees. Even when the amount of production remains the same, the ability of natural resources production to support employment and community stability has been reduced dramatically because of technological changes (Gorse & Gorte, 1989; Monteith, 1989).

The problems of rural logging communities are representative of the problems that many rural communities face. The local economy fluctuates wildly depending on the supply and price of wood products (Drielsma, Miller, & Burch, 1990). Competition from other regions, other countries, and alternative products (e.g., plastics) lowers prices and makes constant improvements necessary (Adams & Haynes, 1989). Technological change improves the efficiency of logging and processing wood products, but reduces the number of jobs in the rural communities. All of these changes have occurred while the amount of logging in old growth forests has decreased (Brunelle, 1990).

Policies designed to help alleviate these problems often focus on better management of the natural resources or economic assistance to the natural resources based industry (Sedjo, 1989). Sustainable agriculture and forestry are often viewed by natural resources managers as cures for the problems of rural areas (Schallau, 1990). Although good management of renewable natural resources is an important component of rural sustainability, it may not support as many jobs as did the previous, unsustainable, harvest levels (Daniels, Hyde, & Wear, 1991; Schallau, 1989). Rural sustainability requires the appropriate use of natural resources and a diversified economic system (Schaap, 1989). Alternative types of economic activity are necessary to compensate for reductions in natural resources–based employment due to environmental and technological change.

Rural Communities in the Northwest

The rural Northwest provides a good example of the sustainability problems facing rural communities throughout the United States. The economy of the region depends heavily on logging, which accounts for 44% of the employment in Oregon and 28% of the employment in Washington (U.S. Forest Service, 1988). The logging industry has depended on the vast old-growth forests in the region; however, these old-growth forests are rapidly dwindling. More than 87% of the old-growth forests have already been cut, and another 6% are in parks and wilderness areas (Wright, 1991). Within the next decade, the timber industry will be forced to depend on the region's second-growth forests. This transition to renewable forest practices will allow the timber industry to continue, but on a smaller scale (Heilman, 1990).

The decline of the old-growth forests is just one factor affecting the economy of the rural Northwest, and it may not be the most significant one. The Northwest logging industry has faced increased competition from the Southeast, forcing the industry to modernize to stay competitive (Adams & Haynes, 1989). Although the amounts of timber produced in 1970 and 1990 were similar, the number of people employed by the timber industry dropped by one third during that period because of modernization efforts in the 1980s (Brunelle, 1990). Many older saw mills closed, leaving rural communities with no stable economic base and higher levels of unemployment than in regional urban areas (Weeks, 1990). Unless national economic factors change and technology remains constant, the region will experience at least a 1% to 2% drop in logging employment every year regardless of the resolution of the spotted owl issue (Adams & Haynes, 1989).

The rural communities of the Northwest have not been economically stable during the last decade. The recessions of the early 1980s and 1991 devastated many communities. Technological advances in the milling and wood processing industry changed the size, location, and amount of employmant in the logging industry (Keegan & Polzin, 1987). Japanese markets for raw lumber reduced the need to process timber in the Northwest before a sale. Companies expanded logging operations on private lands to pay off debts from the mergers and acquisitions of the 1980s (Carroll, Bassman, Biatner, & Breuer, 1989; Heilman, 1990). All of these factors placed the stability of Pacific Northwest rural communities in jeopardy.

Amid these severe economic conditions, changing federal regulations due to the spotted owl controversy have added a new dimension of uncertainty and instability. Pushed by the actions of environmental groups, the northern spotted owl has been listed as a threatened species in the Pacific Northwest. The effect of this listing and numerous court challenges have been to remove from timber harvesting much of the remaining old-growth forests on federal lands in the Northwest until a management plan for the owl is approved (Sinclair, 1991). If this removal is permanent, it is expected to lead to a permanent 10% drop in employment in the logging industry (U.S. Forest Service, 1988). Although this is not as severe as the other economic factors related to logging, in the context of an already declining economy it is viewed as a severe blow to the stability of many rural communities.

Response to Sustainability Problems

The economic problems of these rural communities are caused by factors outside of their control. National and international economic trends, a shrinking natural resource base, and technological change all create this challenge to their sustainability (Brunelle, 1990). These rural communities need to try to develop a diversified economic system that will support their communities (Lee, 1989). However, for many people and communities in the region, this has not been their response to their current economic situation. Instead of attempting to develop alternative sources of economic development, they have focused on the spotted owl controversy as the cause of their economic and community problems (Lee, 1990).

There are a number of reasons why residents focus on the spotted owl controversy. Because economic and technological changes are clearly beyond the control of these communities, the controversy is perceived to be caused by federal regulations and actions of environmentalists that can be altered through political action (Lee, 1990). In a manner similar to people's responses to technological disasters, many residents focus on attributing blame for their problems rather than on rebuilding their communities (Cuthbertson & Nigg, 1987; Drabek & Quarantelli, 1967). The spotted owl controversy is a dramatic and highly visible event, which makes it a more cognitively "available" attribution for their difficulties than the other less dramatic changes (for discussion, see Markus & Zajonc, 1985). Finally, there is resistance to occupational change. Many loggers and mill workers do not want to change their occupational identities to work in other industries (Carroll, 1989; Carroll & Lee, 1990).

There is a growing recognition in the region of the need to develop a more diversified rural economic system (Schaap, 1989). Although community leaders do not want to abandon the logging industry, the last two decades have convinced them of the need for alternative sources of employment (Lee, 1989). Tourism, retirement, service industries, and high-technology manufacturing are often suggested as economic alternatives (Bakley, Smith, & Coupal, 1990; Corcoran & Weber, 1990; Monteith, 1989). However, these alternatives can create conflicts with the timber industry because unharvested forests provide the attraction for tourism and some of the lure for attracting new industry (Monteith, 1989). Deciding how to make these trade-off is one of the challenges facing the rural communities of the Northwest.

Study Overview

The rural communities of the Northwest face an economic and a human crisis. The transition to a sustainable economic system will require changes in the beliefs and attitudes of the rural residents. As long as rural residents focus on the spotted owl controversy, they will seek political solutions for their economic problems, even though these political solutions will only delay the need to make a transition to alternative sources of economic activity. The spotted owl controversy has also led to substantial polarization in the rural communities. To foster economic development, the community residents must move beyond this polarization and work together on economic development issues (Johnson, 1993).

Although outside assistance is useful, local community leaders have the greatest impact development (Clark, 1990). Local industrial development people need to work with the existing business community to explore and develop opportunities for expansion and new development (Luke, 1991). Local networks among existing rural businesses need to be formed to provide support. Linkages between the rural communities and state and federal sources of funding and expertise need to be developed. All of these economic development activities are related to the beliefs and actions of community members.

This study examined the beliefs and attitudes of opinion leaders in the region about the sustainability of rural communities in the Northwest. The sample contained people in federal, state, and local government; nonprofit organizations; and private industry who are involved in the economic and community development of the region. The respondents attended a day-long conference in 1991 on the economic development of the rural Northwest. Conference activities included presentations on economic development issues in the region and examples from rural communities of successful local efforts to promote economic development. The participants completed a preconference survey examining the economic conditions and future of the rural Northwest, and an evaluation survey at the conclusion. A second conference was held in 1992 to continue the activities from the first conference. Conference participants completed the same preconference survey at the beginning of the second conference.

The purpose of this study was to examine the beliefs of opinion leaders about the current situation and possibilities of the rural communities in the Northwest, and to study the effectiveness of using regional conferences to change attitudes about rural economic development. The results were expected to show some of the misconceptions and problems that affect these rural communities. Community survival and development require accurate recognition of the existing problems and realistic assessment of alternative futures. This study tries to identify some of the beliefs that support or limit the economic development of the region. In addition, the study shows how effective regional conferences are at changing the attitudes of opinion leaders about economic development issues.

The hypotheses for this study are as follows:
1. The rural communities of the Northwest will be viewed as having substantial, long-term economic problems. The economic problems of these rural communities have been increasing during the last decade. The conference participants are actively involved in rural economic development issues in the region. Their knowledge of the situation should cause them to rate the rural areas as having long term economic problems with no quick solutions.
2. The participants will overestimate the importance of the spotted owl controversy as the cause of their current economic problems, relative to other factors. Although spotted owl protection is a relatively minor economic problem for the region as a whole, it will be rated as a major problem. The controversy is a dramatic and conflict-laden event that has captured people's attention. Even though the conference participants acknowledge that the rural communities are dealing with long-term economic problems, the nature of the spotted owl controversy makes it a convenient cognitive explanation.
3. The participants will recognize that the future of these rural communities depends on a shift away from logging to a more diversified economic base. Resolving the ban of logging due to the spotted owl issue will not rescue these rural communities from their economic problems. Community sustainability will require new sources of economic activity beyond dependence on logging, The conference participants should recognize this and support a variety of types of economic activities for these communities.
4. The participants will recognize that environmental quality and quality of life factors are important economic assets for rural communities. Rural communities cannot directly compete with urban areas on traditional economic factors such as workforce availability and cost of doing business. To attract economic activity, rural communities need to market the assets that are not in urban areas. Environmental quality and quality of life are two assets of rural communities that can be used to attract economic activity.
5. Participation in conferences will improve the participants' attitudes about the economic future of the rural Northwest. One of the goals of the conference was to present examples of successful, small-scale economic development activities in rural communities. These examples should help to improve the conference participants' attitudes about the economic future of the rural communities in the region. However, the effect of these changes in attitudes may not persist and diffuse unless supported by significant changes in the rural communities.

TABLE 1. Global Economic Ratings.

	1991 Preconference	1991 Postconference	1992 Postconference
Current economic situation of rural communities [a]	2.28	2.65	2.35
Assets of the rural communities [b]	2.76	3.45	2.76
Likelihood for significant economic improvement [c]	2.76	3.20	2.92

[a] 1 = very poor to 5 = very good
[b] 1 = very limited to 5 = very abundant
[c] 1 = not at all likely to 5 = very likely

Methods

Subjects

Subjects for the survey were recruited from individuals attending two regional conferences in Vancouver, Washington on September 10, 1991 and October 15, 1992. The conferences focused on issues facing natural resources dependent rural communities in the Pacific Northwest. The advertising and mailing lists for the two conferences were similar. All participants at the first conference were invited to attend the second conference, although attendance at the second conference was substantially less.

At the first conference, 167 individuals completed a preconference survey. Of these individuals, 44% were affiliated with federal or state government, 24% were affiliated with local or regional governments, 22% were employed by private industry, and 10% were members of the media, consultants, or listed multiple affiliations. The majority of the sample (52%) were involved in economic development activities either as business managers or economic planners; but the sample also included individuals active in forest resource management (6%), utility management (11%), education and research (17%), and health and social services (5%). Only 2% of the sample represented environmental interests, and 6% of the sample provided multiple affiliations.

At the second conference, 55 individuals completed a preconference survey. Of these individuals, 42% were affiliated with federal or state government, 54% were affiliated with local or regional governments, and 4% were employed by private industry. The majority of the sample (58%) were involved in economic development, but the sample also included individuals active in forest resource management (14%), education and research (20%), health and social services (2%), and 6% provided multiple affiliations. Most of these participants (69%) did not attend the first conference.

Statistical tests (analysis of variance) were used to examine whether the organizational and professional affiliations of the participants related to their survey responses. None of the survey responses were significantly related (p < .01) to the backgrounds of the respondents.

Surveys

Two surveys were completed by individuals attending the first conference. In the preconference survey, participants provided global ratings of the current economic situation; rated the contribution of several factors to the economic disruption of the rural communities, the importance of various industries to the future of the rural communities, and the importance of various assets to promote economic development; and provided information on their professional affiliations. The lists of economic factors, industries, and assets were developed through reviews of journals and newspaper articles on the economic conditions affecting the rural communities of the Northwest. The preconference survey was administered in the first half hour of the conference. This survey was included in the registration packet, and the conference organizer asked conference participants to complete the survey as part of his opening address.

A postconference survey was handed out at the end of the 1991 conference and, again, the conference organizer asked the participants to complete the survey. This survey asked participants to reanswer three questions related to global ratings of the current economic situation. It was completed by 110 participants.

Participants at the second conference were given the same preconference survey as at the first conference. The method of administration was the same as at the first conference.

Analyses of the data were performed using the Statview Statistical Program. Only results showing p < .01 were considered statistically significant.

Results

Pre- And Postconference Global Economic Ratings

Three global ratings of the economic situation in the rural Northwest were obtained for the two preconference and one postconference surveys. This permitted an analysis of the participants' initial beliefs each year and a measure of the effects of the first conference on those participants' beliefs. The results to these three sets of ratings are contained in Table 1. A t test was used to examine the differences between those individuals who attended both the first and second conferences, and those individuals who attended only the second conference. Among attendees to the second conference, no significant differences were found between those who attended both conferences and those who attended only the second conference, so the subgroups were combined for the following analyses.

The current economic situation in the rural Northwest communities was rated on a 5-point scale, where 1 indicated a very poor situation and 5 indicated a very good situation. The ratings were significantly different across the three measurement periods using an analysis of variance statistical test, $F(2, 329) = 12.77$, $p < .001$. A Scheffe post hoc test showed that the 1991 postconference response was significantly higher than both of the preconference responses.

The assets that the rural Northwest communities have to deal with regarding economic problems were rated on a 5-point scale, where 1 indicated very limited assets and 5 indicated very abundant assets. The ratings were significantly different across the three measurement periods using an analysis of variance statistical test, $F(2, 328) = 21.39$, $p < .001$. Again, a Scheffe post hoc test showed that the postconference rating was significantly higher than both of the preconference ratings.

The likelihood of significant economic improvement in the rural communities in the next 2 to 3 years was rated on a 5-point scale, where 1 indicated that economic improvement was not at all likely and 5 indicated economic improvement was very likely. The ratings were significantly different across the three measurement periods using an analysis of variance statistical test, $F(2, 327) = 7.97$, $p < .001$. A Scheffe post hoc test showed that the postconference 1991 rating was significantly higher than the 1991 preconference rating. The 1992 rating was not significantly different from the other two ratings.

Economic Disruptions, Future Industry, and Assets

The participants provided three sets of ratings evaluating economic conditions in the rural Northwest. These ratings permitted an analysis of their beliefs about the economic disruptions, future industry, and regional assets that will affect the economic and community sustainability of the region. These ratings were only included on the preconference surveys. All of the ratings were made on 5-point scales, 1 indicating not at all important to 5 indicating very important. Because of the smaller sample size, the 1992 results will only be used as a time comparison to the 1991 results.

The participants were asked to indicate the importance of five factors in the economic disruption of the rural Northwest communities. These results are presented in Table 2. Changing environmental regulations received the highest rating followed by technological change in the timber industry. The actions of environmentalists and recession in the economy received similar ratings. Economic competition in the timber industry was rated the least important of the five factors. Some participants listed additional sources of economic disruption, such as poor planning on the part of the government and the timber industry, the lack of funds for development, and the attitudes of local residents toward change.

In the 1991 ratings, there was an overall significant difference across these economic factor ratings using an analysis of variance test, $F(4, 164) = 4.1$, $p < .01$. A Scheffe post hoc test showed that only the difference between environmental regulations and economic competition was statistically significant. There were no significant differences between the 1991 and 1992 importance ratings using t tests.

An analysis of the correlations in the 1991 results showed that those who believed that changing environmental regulations were an important factor in economic disruption also believed that the actions of environmentalists were an important factor ($r = .45$, $p < .01$). Those who believed that technological change was an important factor in economic disruption also tended to believe that economic competition was important ($r = .51$, $p < .01$) and that the effect of the economic recession was important ($r = .26$, $p < .01$).

The participants rated the importance of four types of industries to the future (next 10 years) of the rural Northwest (see Table 3). Participants rated tourism as most important, followed by service industries, manufacturing, and logging. In addition, several participants indicated that agriculture ($n = 14$) and high-technology services and industries ($n = 10$) were likely to be important.

In the 1991 ratings, there was an overall significant difference among the importance ratings of the four industries using an analysis of variance test, $F(3, 162) = 20.3$, $p < .01$. A Scheffe post hoc test showed that logging was rated significantly lower than tourism and service industries, and manufacturing was rated lower than tourism. A comparison of the 1991 and 1992 ratings found one significant difference: The importance of logging was rated significantly lower in the 1992 survey, $t(219) = 2.96$, $p < .01$.

Those who believed that tourism would be an important industry also believed that service industries would be important ($r = .34$, $p < .01$). However, support for tourism and service industries was not negatively rated to logging. Even the strongest supporters of tourism and services industries believed that logging had a role in the future economy of the Northwest.

The participants rated the importance of various assets of the rural Northwest communities that would promote economic development (see Table 4). A high-quality environment and a good quality of life were

TABLE 2. Mean Ratings of Importance of Economic Disruptions

	1991	1992
Changing environmental regulations	3.84	3.94
Technological change in the timber industry	3.58	3.38
Recession in the economy	3.53	3.59
Actions of environmentalists	3.53	3.61
Economic competition in the timber industry	3.38	3.17

Note: Based on "How important are the following factors to the economic disruption of the rural Northwest communities? (from 1 = not at all important to 5 = very important)."

TABLE 3. Mean Ratings of Importance of Future Industry

	1991	1992
Tourism	3.95	3.95
Service industries	3.72	3.78
Manufacturing	3.45	3.62
Logging	3.20	2.75

Note: Based on "What industries will be most important in the future (next 10 years) of the rural Northwest communities? (from 1 = not at all important to 5 = very important)."

TABLE 4. Mean Ratings of Importance of Regional Assets

	1991	1992
High-quality	4.36	4.42
Good quality of life	4.27	4.42
Forest resources	3.51	3.20
Low costs to do business	3.38	3.26
Skilled workforce	3.18	3.09

Note: Based on "What are the most important assets the rural Northwest communities have to offer to promote economic development? (from 1 = not at all important to 5 = very important)."

rated as the most important assets. These were followed (in order) by forest resources, low costs to do business, and a skilled workforce. Availability of funding, community attitudes, and Pacific Rim location were cited as additional assets by some of the participants.

The ratings of the five assets were significantly different in the 1991 survey using an analysis of variance test, $F(4, 160) = 62.4$, $p < .01$. A Scheffe post hoc analysis showed that a high quality environment and good quality of life were significantly different from the other three factors. There were no significant differences between the 1991 and 1992 ratings.

High-quality environment and good quality of life were strongly and positively related in these responses ($r = .52$), $p < .01$. The high-quality environment and forest resources were positively related ($r = .24$, $p < .01$), as were good quality of life and a skilled workforce ($r = .27$, $p < .01$).

There were a few notable intercorrelations among the sets of questions on disruptions, future industries, and assets in the 1991 survey. Those who believed that logging would be an important industry also tended to believe that forest resources are an important asset (r = .41, p < .01) and that the actions of environmental activists are contributing to economic disruption (r = .33, p < .01). Those who believed that tourism would be an important industry also tended to believe that the high-quality environment is an important asset (r = .28, p < .01). Those who believed that manufacturing would be an important industry also tended to believe that skilled workers are an important asset (r = .23, p < .01).

Discussion

The participants in this study, whose jobs are primarily to encourage the economic and community development of the region, recognize the severity of the region's economic problems and are not optimistic about its future. When trying to explain the reasons for the region's economic difficulties, they are more likely to blame federal regulations than the other, long-term, economic problems that have affected employment in the region. When looking into the future, most of the participants speculate that neither logging nor manufacturing will be the most important growth areas for the region. They see a future in which tourism and service industries provide the most potential for growth. This is not a rejection of logging and the use of forest resources but a recognition that these industries will play a decreasing role in the region's economy.

These results provide support for the first four hypotheses in this study. The economic leaders of the rural Northwest recognize the severity of their problems, have realistic views about the future economic direction for the region, and recognize the assets they have to attract new economic development. However, by emphasizing the importance of environmental regulations as a cause of their economic problems, their beliefs are supporting the polarized views that are discouraging the residents of rural communities from working together to solve their economic problems.

The conference these participants attended was designed to help them better understand the economic forces affecting the region, and to provide them with examples of communities that were successful at bringing in new industry. The community examples ranged from small-scale manufacturing to tourism, retirement communities, and other service industries. In addition, the conference encouraged networking among the participants, which included both local economic development people and state and federal representatives.

As was predicted in the fifth hypothesis, the participants left the 1991 conference with a more optimistic opinion of the assets of the region and the ability of the rural communities to improve their economy. Although this showed that the conference was successful in improving people's attitudes toward the future, these more positive attitudes were not found in the next year's conference participants. Given the continuing national recession and stalemate on the logging issue, it is not surprising that attitudes about the economic conditions of the region were similar in 1991 and 1992. The only significant change during the year was an increased recognition that logging was going to be a less important industry in the future. However, it is doubtful that the first conference caused this change in perspective.

The rural communities of the Northwest are dealing with severe economic problems. Even if previous logging levels could be maintained, economic and technological changes are reducing the importance of timber on the local economies. These rural areas cannot compete with the urban areas in the traditional manufacturing sectors. However, there is an alternative future for these rural areas being created by some of the same factors that are now challenging their existence.

Our increasingly urban society places increased importance on rural areas as sources of renewal through tourism and recreation. As our population ages, rural communities are viewed as quality retirement places for urban residents. Information technology has created a variety of new service industries that are the largest growth areas in the economy. The high-technology service industries are less location dependent, and often locate where they will be able to attract highly qualified employees. The high-quality physical and social environments of the rural Northwest may prove to be an important economic asset to attract these economic activities.

Although the value of rural communities will increase as society becomes more urbanized, many rural communities will not be able to wait until they are discovered by urbanites as good places to recreate, work, and retire.

Rural communities in the Northwest are faced with immediate challenges to their survival: They are faced with a trade-off between environmental quality and traditional, extraction-based employment. If these rural communities do not harvest their local timber resources, unemployment will reduce their quality of life and force many residents to leave for urban areas. If they do log their forests, the high environmental quality that makes these communities attractive will be damaged.

This trade-off decision is made more difficult because it is being made within the context of a larger political debate. The spotted owl controversy has diverted people's attention away from their communities' long-term economic problems and focused their attention on an environmental conflict. As shown in this study, many opinion leaders in the rural Northwest blame federal regulations for their economic problems. Like the responses of communities to technological disasters, this has led to the polarization of local opinion and a focus on attributing blame rather than on community rebuilding. This has damaged the ability of the people in these communities to work together to help develop sustainable rural communities.

Solving these economic problems will be a difficult task for the rural communities. The solutions must come from the members of the rural communities, with some assistance from state and federal sources.

Conferences that demonstrate local economic successes and encourage networking are helpful in changing attitudes and providing ideas and connections to resources. However, continued national economic problems and political controversy limit the effectiveness of these community development activities. These rural Northwest community leaders must get beyond the blaming stage to focus their communities' efforts toward developing a sustainable economy.

reading 2

"Having Owls and Jobs Too"
Daniel Glick

National Wildlife, Vol. 33 (8-18-1995), p. 8(6). © 1995 by National Wildlife Federation. Reprinted by permission of National Wildlife Federation.

Sawing down 200-year-old trees was just about the only thing former logger John Dark ever wanted to do. Working in the woods was a dream job for the Oregon native. There were good wages, fresh air and the incredible adrenaline rush of toppling Douglas firs without mangling 4,000 board feet of lumber. Sure, the work was dangerous: Four buddies had been killed over the years in accidents, and the toll on his own knees and back kept rising. But Dark figured the danger was part of the thrill, and he never considered doing anything else.

Then, in 1991, U.S. District Judge William Dwyer declared a moratorium on cutting old-growth-forest habitat that is critical for the northern spotted owl. And Dark lost his job, seemingly a victim of what has been dubbed the "owl wars." But his story does not end there. Like 4,340 of his former colleagues since 1989, Dark opted for formal retraining. He enrolled in business and accounting classes in a mostly federally funded program at a community college near his home in Creswell. The 32-year-old student now plans to build and market ornate wooden dollhouses he has designed. And he says, "Getting laid off was the best thing that ever happened to me."

The optimism of Dark and other former timber-industry workers about finding a place in Oregon's changing economy is shared by a slew of economists, state officials and timber-industry analysts. Since Dwyer's ruling four years ago, the Oregon economy has bedeviled doomsayers who swore the economy would collapse and mill towns would become ghost towns. Like Dark, many former loggers and millworkers, though certainly not all, now credit the bird with rerouting dead-end careers.

Over the past few years—even in Douglas County, the epicenter of the owl conflict—Oregon's unemployment level (4.6 percent as of March 1995) has fallen below the national level (5.5 percent), and property values have risen by nearly 10 percent a year. High-tech industries have poured into the state. In 1994, job growth increased 3.8 percent. Timber companies with second-growth private forest also acknowledge that the last few years have been kind to their bottom lines, since timber prices have risen steadily—in large part as a result of the reduced public-lands supply. "It's been a windfall," says Boise Cascade director of corporate communications Bob Hayes. By way of thanks, argues Andy Kerr, executive director of the Oregon Natural Resources Council, timber companies "should send environmentalists thank-you notes and nice big checks."

Both Hayes and Kerr miss a crucial point. And so do loggers who blame or credit Dwyer's ruling for changing their lives. The dirty little secret of the owl wars is that *Strix occidentalis caurina* is no more responsible for the boom in the state's economy than it is for job losses in the Northwest timber industry. "There's this myth that these changes would not have occurred but for the owl-recovery plan," says University of Oregon economist Ed Whitelaw. "That's simply wrong." Sighs Phil Keisling, Oregon's secretary of state, "People have loaded so much baggage on the shoulders of this 15-inch bird."

Though timber probably will never again reign in the Northwest's economy as it has in the past, it remains the linchpin of many rural communities. The extent of timber's economic importance, however, is a matter of some debate. John Beuter, an industry consultant (and a former Bush administration forest official and forest economist), reported last winter to the Oregon Business Council that the wood-products industry accounts for a third of Oregon's economic base and $1.4 billion in foreign exports. Beuter is convinced that wood products will continue to be a "significant component" of the economy for years to come.

His figures, however, don't seem to hold up against state statistics. By next year, predicts state economist Paul Warner, high-tech jobs will exceed wood-industry jobs for the first time. More important, the economy has diversified so much that the high-tech and wood industries together will yield only 11 percent of the gross state product. The timber industry's importance in the state has declined consistently since the 1950s, when it employed about 14 percent of the labor force; today that percentage has dwindled to 3.8. Top employers in Oregon now include a retail department store, two computer companies and mail-order giant Harry and David's.

At the same time, other forest industries are emerging that can profit without making stumps out of centuries-old trees. Catherine Mater, president of Corvallis-based Mater Engineering, estimates that the sustainable harvest of salal, Oregon grape, sword fern and evergreen huckleberry for the international floral industry could bring in $72 million per year.

The nonprofit Rogue Institute for Ecology and Economy in Ashland has identified a cornucopia of commercial possibilities in the forest: bear grass for basket weaving; Prince's pine, used as a seasoning in cola drinks; huckleberry for funeral wreaths; cedar Christmas boughs; and other plants useful in everything from dyes to medicines. Wild mushroom gathering is already a booming business (so much so that there is concern about keeping the harvest sustainable). Alder, once tossed away as a junk species, is now in demand for furniture.

Former mill towns like Oakridge, which is surrounded by the Willamette National Forest, are promoting "value-added" products that in some way create other jobs, such as making locally produced flower boxes from locally harvested scrap cedar. "We have a lot to learn about how to use the resources we have available to us," Mater says. "The timing is so ripe to look at alternatives."

Some critics scoff at the idea that specialty forest products can ever compete with the timber industry. But others see many small forest-based industries adding up to solid, diversified local economies. Products like mushrooms or ferns "are never going to replace the dollar value that timber has in a one-shot cut," says Bjorn Everson of the Rogue Institute. "But over a 140-year cycle, they make sense."

After 58-year-old Don Walker of Oakridge lost his job as a logger, for example, he first gathered yew bark for experimental cancer drugs. "That was a real good business," he says. Now he grows shiitake mushrooms and tends a Christmas-tree farm. He's still convinced that conservationists overstated the effects of logging on the owl. Still, he says: "Having my own business is a lot more satisfying than working for someone else."

Not all former timber workers are happy with the changes forced on them by the changing economy. After the last mill in Oakridge closed in 1991 and surrounding U.S. Forest Service land was virtually shut down to productive logging, second-generation logger David Sulick "got tired of hunting jobs" and enrolled in courses at Lane Community College in Eugene to become a paramedic.

In his double-wide trailer—festooned with saws, wedges and other tools of his former trade—the 45-year-old ex-Marine describes the frustrations of going back to school after a lifetime in the woods. For three years, he waded through "paperwork 2 inches thick" just to get his applications done and keep up his eligibility for the program that earned him an emergency-medical-technician license last year. Then all he could find for work is a part-time job at $6 per hour driving the local ambulance. Though he would "absolutely" go back to cutting if he could, he's proud of having delivered a baby and helped save several lives, and he begrudgingly appreciates how much he has changed. "Ten years ago, I would've shot an owl in a minute," he says. "Now I guess I'd just take his picture."

Sulick and his peers actually have more choices now than they had after the near-depression in the early 1980s. After that round of layoffs, recalls Ellen Palmer, program director for the Lane County Community College Dislocated Worker Program, "People were saying, 'We are way too dependent on trees here.'" State and local planners began encouraging diversification. In 1978, wood-manufacturing jobs accounted for about 65 percent of all manufacturing jobs in Lane County. Now the figure is closer to 14 percent.

This time, help also comes from Uncle Sam. President Clinton's Forest Plan, announced in April 1993, included $1.2 billion in federal funds over five years to retrain workers and lessen the economic impact on affected communities. At Lane Community College's worker-retraining program, former loggers such as Sulick and log scalers such as 51-year-old Ray Jones have flocked to a cluster of modular buildings on a north Eugene hillside to build new careers.

The bottom line was that protection of spotted owl habitat was far from cataclysmic. "It didn't create the 'Appalachia of the West' that many people predicted," says Clarence Moriwaki of the U.S. Office of Forestry and Economic Development, set up in 1993 to reduce old-growth logging by 85 percent and retrain workers—as promised in President Clinton's Forest Plan.

Despite a loss of 14,500 jobs in the timber industry since 1988 and dire times for some rural mill towns, the Oregon economy more than made up for that sector's job losses with new jobs elsewhere. In recent years, the high-tech sector has blossomed, and the wood-products industry has developed more efficient manufacturing techniques and expanded product lines. The state's timber production still leads the country at around 5 billion board feet a year.

Northern neighbor Washington State, less dependent on timber to begin with, has grown at a slightly less robust pace. To the south, timber country in northern California is economically in far worse shape then Oregon. But again, the reasons—including an industry already in decline—are far more complex than simple owl-habitat protection.

After working for Weyerhaeuser for 22 years and measuring as much as $1.5 million worth of logs a day, Jones is now taking accounting classes. In his gut, he says, he always suspected his line of work would not last until retirement. "My job was a dinosaur," Jones says. "Most of us knew they were just cutting the timber too fast."

Former millworker and now nurse-in-training Darla Hoskins agrees: "I knew that we couldn't continue logging like we had been around here," she says. "The writing was on the wall in the early 1980s." When the retraining program offered a way out, the 44-year-old Hoskins, who began millworking when she was 21, jumped at it. "I figured there wouldn't be any spotted owls getting me out of nursing," she says. "But I still miss the smell of wood."

The original "jobs-vs.-owls" debate focused on whether habitat protection would drive people such as

Jones or Hoskins out of work. Now some economists are arguing that the recent economic boom is partly because of environmental regulations. Economist Whitelaw, also president of the consulting group ECO-Northwest in Eugene, is calculating the dollar value of living in a region replete with giant conifers, steelhead runs and summer blackberries. He argues that such benefits are a powerful draw for skilled workers, despite the fact that the state's wages remain below national averages.

Biologist Rick Brown, a resource specialist with the National Wildlife Federation's Western Natural Resource Center in Portland, agrees with Whitelaw that "we have to recognize the full spectrum of the forest's economic values" but insists the debate about saving ancient forests has to go beyond number crunching. "The forest is home to an unimaginably complex web of living things—what evolutionary biologist E.O. Wilson calls 'the little things that run the world,'" says Brown. "We also have to bear in mind the enduring values of these ancient forests. What about the value of an ancient forest's majesty, or its immense capacity as a source of wonder?"

Such questions are far from academic. The Oregon timber industry holds cultural and political clout far beyond its statewide economic contribution, and efforts to increase cutting on public lands are gaining steam with the new Republican Congress and in Oregon's legislature. The President's Forest Plan, allowing as much as 1.1 billion board feet of harvest annually, went into effect shortly after Judge Dwyer ruled in June 1994 that the plan did not violate the Endangered Species Act. (In comparison, the average annual cut for the owl region was 4.5 billion board feet during the 1980s and 2.4 billion board feet between 1990 and 1992.)

Meanwhile, beyond protected owls and retrained workers, other positive trends have emerged from the owl wars. One example, says Bob Warren of the Oregon governor's forest-policy team, is a recent "quantum leap in forest management."

Clear-cutting has decreased, and no-cut buffer areas around streams are commonplace, though some critics charge that they are still inadequate. The industry has become more efficient and innovative.

Then there are more intangible shifts. "A growing number of people see public lands as something more than just producers of commodities," says Oregon's Secretary of State Keisling. Such sentiments easily apply to other Northwest environmental issues, such as efforts to save several species of endangered salmon. If Oregon's combination of environmental protection and economic success provides a useful model, then more politicians, state planners, economists, foresters and industry representatives might agree with Whitelaw's prediction that "the long-run health of Oregon's overall economy will depend, increasingly, not on sacrificing the environment but on protecting it."

In Lane County, that kind of attitudinal change is coming. The Valley River Inn in Eugene, in the heart of the Willamette Valley, once had its walls decorated with a giant circular mill saw and two 10-foot crosscutting saws. In 1993, during lobby renovations, the hotel took down the two crosscutting saws. And replaced them with a mural depicting an old-growth forest.

"The Birds—The Spotted Owl: An Environmental Parable"
Gregg Easterbrook

The New Republic, Vol. 210(6), March 3, 1994. © 1994 by Gregg Easterbrook. Reprinted by permission of the author.

Recently, I rather casually did something that according to contemporary environmental orthodoxy is inconceivable: I took a hike through the woods and saw lots of spotted owls. Spotted owls are said to be so rare that even an experienced forester spends weeks trying to glimpse one. I saw four in a few hours. The owls were living wild in a habitat where it is presumed impossible for them to exist: a young woodland, not an old-growth forest. And they were living in a place, California, where environmental doctrine holds spotted owls to be rare birds indeed.

In the evolution of political issues there often comes a sequence that runs like this: A new concern arises. For a while the system attempts to deny the claim's validity, but eventually some action is taken. By then advocates have become an interest group, fighting as much for the preservation of their cause as anything else. The fight takes on a life of its own; the specifics of the original issue are discarded.

Today this sequence may be repeating in the matter of the spotted owl. A decade ago researchers warned that the bird was declining toward extinction. Legal gears were set spinning. In 1991 a federal judge suspended most Northwest logging, resulting in the loss of thousands of high-wage jobs. This month the Clinton administration is set to file court documents that make most of those losses permanent. The Clinton owl plan has become a standard Washington lobbying jangle in which business and environmental constituencies drop sixteen-ton weights on each others' heads. The original question of whether the owl is endangered has been discarded. At the White House level, that isn't even discussed anymore. Political and legal maneuvers continue on the assumption that 1980s studies hypothesizing an owl extinction were correct.

They may not be. Research is beginning to suggest that the spotted owl exists in numbers far greater than was assumed when the extinction alarm sounded. Whereas a headline-making 1986 Audubon Society report said that 1,500 spotted owl pairs throughout the United States was the number necessary to prevent extinction, it now seems that as many as 10,000 pairs may exist. "It appears the spotted owl population is not in as bad a shape as imagined ten years ago, or even five years ago," says David Wilcove, a biodiversity expert for the Environmental Defense Fund. Thus Clinton's plan to shut down most Washington and Oregon logging may not only be unnecessary; it may be resting on an illusion.

The illusion of a pending owl extinction is a parable of modern environmentalism, illustrating both its manifest virtues and the internal faults that, uncorrected, could bring it down. The owl fixation has its political virtues: it's an issue easily understood by the public; it graphically illustrates the genuine need for old-growth forest preservation; and it has allowed environmentalism to win numerous battles against the government and the logging industry. Also, as a direct-mail fundraising tool, the little feathered creature cannot be beat; as a symbol of endangered species, the owl can help protect many others that are not so cuddly and not so popular. In all these areas, the movement has its heart in the right place.

Its head is another matter. To serious environmentalists, the owl dispute has become a proxy for the goal of preserving old-growth forests. In private many enviros acknowledge that owl extinction claims have been extensively pressurized with hot air. They justify this on the grounds that a valid goal, old-growth forest preservation, is served. Indeed, old-growth preservation is important for the protection of biodiversity, for conservation of what remains of America's pre-European heritage, and against the prospect that ancient forests may someday be understood to play an irreplaceable ecological role of which men and women are not yet aware. These are ample reasons why logging in the Pacific Northwest should be closely regulated.

Yet an argument based on forest preservation for its own sake would, in the long run, be stronger than specious species arguments. After all, if conservation rules are based on an owl extinction claim that research someday disproves, why shouldn't hell-bent logging resume? Whether spotted owls are really endangered raises the whole question of whether conservation can be placed on a secure, rational foundation that outlives alarmist fads.

"I know they're here," said Lowell Diller as he and I stood in the gathering dusk in a redwood glade outside the forest town of Eureka, California. We had hiked to a spot where Diller had previously marked a nest. For fifteen minutes he hooted to summon the owl pair that lived there. Though we saw no shadows moving in the near-dark, Diller was convinced that the owls were observing the intrusion by genus homo.

Suddenly, no more than fifteen feet away, furry outlines resolved into view. Two spotted owl had conducted their flying approach through a dense forest understory without making any sound audible to us. The owls, who doubtless heard our clumsy footsteps a mile off, regarded us, perhaps wondering, How can these bipeds survive when they make so much noise in the forest?

The owl extinction alarm is predicated on two notions: that spotted owl live only in ancient forests, and that a last, fragile, dwindling population of the northern spotted exists mainly in Oregon and Washington. New research suggests that neither notion is true. California does not end at the Golden Gate; between there and the Oregon border lies a 300-mile corridor of mostly Sierra Nevada forest. This vast woodland, ignored in the owl debate, may contain a profusion of spotted owl.

Diller is a wildlife biologist employed by the firm Simpson Timber. In 1990 he began to survey a northern California tract of medium-age, "managed" timberland owned by that company. Since then, he has found and banded 603 spotted owls. Federal documents assume that only 653 owl pairs exist in the whole of California, and that essentially none lives in private timberland. Diller's 603 owls were found by inspection of a small snippet of the Sierra Nevada. Most California woodlands have never been surveyed for owls.

The primary federal document on which the Northwest logging ban is based assumes that "somewhere between 3,000 and 4,000" pairs of spotted owls exist in the United States. Last year Steven Self and Thomas Nelson, researchers employed by Sierra Pacific, a timber company with a progressive reputation, estimated that California alone is home to between 6,000 and 8,000 spotted owl pairs. Of course, the efforts of Diller, Self and Nelson are backed by industry, which has a financial stake in debunking owl alarms. But then, works of owl pessimism, such as the Audubon report, have been backed by advocacy groups with a financial stake in advancing the same alarms.

Federal spotted owl research has concentrated on Oregon and Washington, the states with the mature, monocultural Douglas fir stands traditionally assumed to be the bird's exclusive habitat. Diller is among the first researchers to look for spotted owl in California. "if research had started in California, the spotted owl would not now be considered endangered," he says.

A significant aspect of Diller's work is that he finds spotted owl reproducing in young woodlands managed by foresters, areas environmental doctrine presumes the bird cannot abide. "The northern spotted owl rarely if ever successfully fledges young from any habitat except old-growth," the Audubon Society declared in 1988. One active nest Diller showed me was not only in a tree glade of medium height and age but was within sight of a logging road where trucks rumble past almost daily. Because Diller has turned up so many owls, his work has backfired on his employer. Simpson Timber has had to file plans that place about 50,000 of its acres into pure preservation status and to restrict company logging in other ways, since tree harvests that might "take" a spotted owl are essentially forbidden even on private land. Meanwhile, Diller's findings have inspired others to begin systematic owl surveys of the vast northern California forest. Agencies such as the California Department of Fish and Game have found spotted owl living and reproducing in several types of non-ancient woodlands, including oak savannas—low-tree habitats unlike any in the Cascade Range of Washington and Oregon.

How can the spotted owl thrive in California forests that are not ancient? One answer may be that young woods are more alive than old ones. Serene Cascade old-growth stands are places of beauty and abiding significance, but they have a little secret, environmentalists would rather not discuss: floors of ancient forests, though rich with diversity, know a relatively low level of life. When tall trees close the "canopy" of an old Douglas fir stand, direct sunlight ceases to reach the understory. Thus plants grow slowly; thus there are fewer seeds to sustain small mammals; thus there are fewer small mammals for predators such as the spotted owl.

Spotteds in Washington and Oregon prey mainly on flying squirrels, whose Cascades population is relatively low. In California, spotted prey mainly on the dusky-footed wood rat. California's managed woodlands have sunlight on the forest floor, because foresters space and trim trees to maximize yield. The warm climate further encourages plants growth. The result is forests with lots of food for small mammals and lots of wood rats.

California woodlands of past centuries may not have had the lush understory now observed. A 1900 Interior Department survey describes natural northern California forests as "rarely if ever dense" and "characteristically open." Though environmental orthodoxy calls timberlands "managed deserts," contemporary timberlands in California have understories so heavily vegetated that they're hard merely to hike through. These dense understories support lots of rodents—and therefore lots of owls. "It's possible there are now more spotted owl in California than before the white man arrived," says Edward Murphy, a forester for Sierra Pacific.

Environmentalists reject such thinking on two grounds. First, they say owls in young California forests are "packed"—driven there by logging of old-growth, sure to die prematurely. This is possible, yet it seems unlikely given that current California owl surveys did not begin until half a century after the timber industry entrenched itself in that state. If owls packed by logging expire, the California spotted should have passed into oblivion by now. Environmentalists also assert that spotteds are an old-growth "obligate," hyperspecialized to the canopy of mature forests. Again, this is possible. Yet it goes against most of what is known about natural selection. A few vertebrates, such as the panda, have become so specialized that they exist solely in a narrow habitat. Most creatures have some genetic ability to adjust to conditions, because environmental conditions have been constantly changing throughout their evolution. The old-growth forests of the Cascade Range did not pop into existence complete with hyperspecialized owls. Those forests evolved from earlier forms. For millennia, some spotted owls must have existed in conditions other than serene old-growth. After all, nature "logs" too, through lightning-induced fires that take down mature trees and replace them with saplings exactly as timber companies do.

Pessimists portray the spotted owl as an old-growth obligate because under this view the bird is a fragile waif, unable to withstand the slightest variation in habitat. But species that cannot withstand change usually perish anyway, even if they are protected by federal law. In the more likely case that the owl has some ability to adapt, the bird lives but the doomsday claims die—a horrible outcome for environmental dogma.

In 1976, shortly after the Endangered Species Act became law, an Oregon State graduate student named Eric Forsman published a master's thesis saying spotted owls of Oregon were "declining as a result of habitat loss." The study caused a sensation among enviros, who were looking for an Endangered Species test case. Also in 1976, in a bonehead decision of epic proportions, clear-cutting was legalized on Forest Service land. Clear-cutting meant rapid reductions of the "unharvested old-growth conifer forests" where Forsman found most spotted owl.

Enviros were right to be incensed. In one of modern government's leading inanities, the Forest Service uses tax funds to subsidize money-losing clear-cutting on public lands that often adjoin private timber tracts harvested at a profit, without subsidies, using selection logging—an ecologically responsible practice that generates more jobs. Though clear-cutting can be defended in a few circumstances, selection logging and the related shelter-cutting, in which trees are removed in clumps so the forest canopy remains, are generally superior both for habitat preservation and sustained timber income.

Forsman's notion of an owl drop in Oregon is unassailable. Although Oregon and Washington today contain about 7 million acres of primal old-growth forest, an area larger than Vermont, ancient forests there were slicked off from roughly the 1940s to the 1980s much faster than they could regrow. Fewer forest acres means fewer forest creatures, including owls. Yet while Forsman's paper is celebrated as a founding text of owl doomsaying, he did not assert that the spotted owl was falling extinct. Indeed, he found some of what Diller has found: birds living successfully in young timberlands, suggesting that "owls could tolerate harvest activity" so long as clear-cutting is not employed. Another reason spotted owls may be plentiful in northern California is that clear-cutting has rarely been practiced there. Most Sierra Nevada logging is by selection or shelter cuts. Can it be coincidence that the owls of the Northwest began the decline detected by Forsman at the same point that the government sanctioned clear-cutting on federal lands there? A fair reading of Forsman's early papers suggests that if clear-cutting were restricted and Northwest logging were reduced, owl populations would rebound naturally as forest acreage did. This reasonable avenue of escape from the owls-vs.-loggers mess is depicted as a sham by green orthodoxy, which holds that the sole hope for the Northwest forest is to have nearly zero commercial activity. Yet forest rebounds in the midst of commercial activity have been the pattern throughout the United States and Western Europe.

Deforestation commenced in the United States roughly two centuries ago in New England, as timber was cut or woods were burned for cropland. About a century ago destructive logging practices began to end in New England, while cropland began to be returned to forest as the stirrings of high-yield agriculture reduced the acres needed for cultivation. New Hampshire was 50 percent forest in about 1850; it is now 86 percent forest, though its population has expanded sixfold. Massachusetts was 35 percent forest in about 1850 and is now 59 percent forest.

Likewise, the Southeast began to be deforested about a century ago and then to reforest about half a century ago. Today the Southeast has far more forested acres than prewar, despite a population boom. Deforestation peaked in Western Europe prewar, then was supplanted by aforestation; today the European Union nations have more forest then fifty years ago, though their human population has nearly doubled. Because the Pacific Northwest was the last place in the United States to which determined logging spread, the deforestation though there was not reached until the 1980s, at the low point for sanctioned clear-cutting. But in that decade Northwest timber firms began planting far more trees than they cut; the reforestation cycle commenced. So long as future logging is held at sustainable levels, the Northwest forest is likely to exhibit the rapid recovery observed everywhere else in the developed world.

Formal warning of spotted owl extinction was not tendered until the 1986 Audubon report said that the spotted owl population was teetering toward the doomsday number of 1,500 pairs. In the report's wake, the northern spotted was "listed" under the Endangered Species Act. Coincident to the listing, a government science panel headed by biologist Jack Ward Thomas concluded that federal policy should ensure that a minimum of 3,000 owl pairs are protected. In 1991 federal Judge William Dwyer banned most logging in Oregon and Washington to assure survival of 3,000 owl pairs. At this point the notion of owl doomsday locked in legally. It has not been questioned since.

There were numerous high-level meetings in the Bush White House on the logging ban, leading to various fumbled initiatives that did nothing to shake Dwyer's order. Shortly after taking office, Bill Clinton staged a theatrical "owl summit" in Portland. He proposed a plan that would allow some logging in Washington and Oregon for a few years (basically through his 1996 re-election bid), then eliminate most Northwest timber commerce in perpetuity. Clinton's final plan is due to be submitted to Judge Dwyer this month.

Forsman's studies, the Audubon warning, the Thomas report, Dwyer's ruling, the Clinton plan—all share something important in common. They all believe Oregon to be the center of the spotted owl universe, with California containing hardly any of the birds.

Thomas, whose owl gloom work won him national standing among environmentalists, was recently named by Clinton as the first biologist to head the Forest Service. Placing a biologist in charge of the agency is an excellent idea, since protection of biodiversity should be a higher government priority than the felling of trees. But will doomsday thinking serve the Forest Service any better than its previous adoration of the chainsaw? "It may well be that there are a significant number of northern spotted owl on private lands in California, but so what?" Thomas told me. "The injunction controls the issue now." Dwyer's injunction discusses the need to preserve old-growth habitat generally, but its legal power derives from the presumption of an owl emergency, a notion neither environmental orthodoxy nor the Clinton administration wishes to disturb with inconveniently positive findings from the field.

Further overlooked is the existence of birds called "California" spotted owls. Since researchers have never surveyed for the California owl in methodical fashion, its population is not well-known. Estimates place its numbers in the low thousands of pairs.

According to environmental doctrine, the fact that the Golden State contains many California spotteds has nothing to do with the "northern" owl extinction alarm because the breeds are disjunct. Yet the birds live in proximity and appear so nearly identical that even ornithologists have difficulty telling them apart. In 1990 ornithologist George Barrowclough, of the American Museum of Natural History in New York, and biologist Ralph Gutierrez, of Humboldt State University in Arcata, California, compared chromosome fragments from the northern and California spotteds. "No genetic difference was found" between the two, their report states. They further noted no significant difference between DNA samples of northern and "Mexican" spotteds, the Mexican spotted being another bird strikingly similar to its Cascade kin. The Mexican spotted roosts in the scrub desert of the Southwest and Mexico, an utterly different place from the moist, old-growth forests doomsayers describe as the sole imaginable habitat for northern spotteds. Barrowclough is now using genome sequencing to determine whether there are subtle DNA distinctions between the northern and California owls that were missed by the first assay. But it's worth noting that Barrowclough calls both bird types "Pacific Coast" spotted owls, reflecting a feeling the two soon may be seen as one and the same.

This seemingly abstract point of ornithology has tremendous bearing on the owl debate. All biologists concur that small, isolated groups are more prone to extinction than large populations spread over a range of habitats. If the owls of Washington and Oregon are a lonely breed, the odds of peril rise. But if those birds belong to a large genetic family existing in a range of habitats spread over the 1,500 miles from Vancouver to Mexico, it becomes unlikely that any local owl population downtrend will lead like a child on a slide to an unstoppable descent. And if the northern and California spotteds are really both "Pacific Coast" birds, there may be 10.000 or more pairs of these raptors in the United States–4,000 to 6,000 northern owls in California, 2,000 to 3,000 northern owls in the Northwest and a few thousand California owls in California.

Does 10,000 pairs of spotted owl sound perilously small? Not only is this far more birds than environmentalist have described as necessary to assure spotted owl survival; it is significantly greater than the nadirs of similar raptors that avoided extinction. The American bald eagle was down to 417 known nesting pairs in 1963 and now has recovered sufficiently that it soon may be "delisted" from endangered status. The peregrine falcon was down to about 1,000 breeding pairs in North America two decades ago and has bounced back to an estimated 5,000; delisting of the subspecies that lives in Alaska is anticipated. Banning of ddt was essential to both birds' comebacks. Recent Forest Service decisions essentially banning most clear-cutting should have a similar positive effect on the owl.

If there are approximately 10,000 pairs of spotted owl of the Pacific Coast type, this number may not differ materially from the level that existed before the white male. Many territorial or top-chain predators, such as owls or grizzly bears, are few in number under natural circumstances because they monopolize a wide prey range. Biologists believe that the ancient forests of the Northwest can support at most one spotted owl pair roughly every 1,000 to 7,000 acres. This suggests that even if no Northwest old-growth tree had ever been felled, the region's spotted owl population might be only a few thousand pairs higher than today, as the natural owls-per-acre factor would impose a low limit on total owl numbers. If a moderate decline in owl numbers has occurred because of logging, it would surely be a cause for concern, but not for sennets of instant doomsday.

Many biologists are uncomfortable with claims of a thriving owl cohort in California. Gutierrez, for one, thinks the reason spotteds are being found in California timberlands is that most "contain a remnant of ancient forest, which sustains the owls. Once those remnants are logged out the owls will die." Gutierrez believes Diller uses techniques that double-count owls. He has also examined the statistical assumptions of the Sierra Pacific study. Applied to a small California old-growth preserve called Willow creek, these assumptions predict 108 spotted owls. Gutierrez's surveys show Willow Creek to be home to seventy-two owls. "This means the optimists are too optimistic by a third," he says.

But suppose Gutierrez is right. Reducing the optimistic estimate by one-third still leaves northern California with 4,000 to 5,400 spotted owl pairs, giving that state more of the birds than Endangered Species Act documents presume exist in the entire country.

In December 1993 a group of owl biologists whose work supports emergency alarms gathered at Colorado State University to debate whether owl estimates are too low. The meeting concluded that for spotted owls, "the population loss rate is accelerating." Environmentalists have since been campaigning to discredit Diller's work on the grounds that it has not yet been peer-reviewed by disinterested scientists (Diller says he will not be ready for peer review until summer) and that it is contradicted by the findings of the better-known researchers who met at Colorado State. But the Colorado State conclusions have not yet been peer-reviewed, either. And the Colorado State report ends by recommending an increase in government funding to researchers whose owl studies are pessimistic—suggesting that they, like industry-backed biologists, are not without a financial stake in their views.

Perhaps anticipating that findings about California owls will be substantiated, environmentalists are shifting their rhetoric. Until recently they spoke mainly of total owl numbers. Now they have begun to assert that owl numbers are less important than the demographic trend; that is, actual birds counted in the laboratory of nature mean less than prospective birds projected by computer model. The Dwyer logging ban accepts this logic, calling computer-projected trends more important than actual owl numbers. This is a quizzical judgment. For more than a decade pessimistic owl studies have been projecting population trends of around –5 percent annually, suggesting that total spotted owl numbers should have fallen drastically by now. Yet actual field surveys continue to find more of the birds than previously counted. Todd True, an attorney for the Sierra Club Legal Defense Fund, says that "numbers of owls in the woods" aren't the issue: "Inevitably as researchers look for owls they will find more than we knew about before anyone was looking. If the demographic trend is negative the species is still in trouble."

True's point has merit. A species might be imperiled even if its population seems profuse, as the human species has uncountable troubles despite robust numbers. But there is a problem for owl doomsday orthodoxy: the northern spotted owl is not a species.

According to the American Ornithologists' Union, which certifies bird types, the "northern" spotted owl is a "subspecies." Only the spotted owl generally is a species, Strix occidentalis to taxonomists. As a species the spotted ranges from northerns that live as far north as British Columbia to Mexicans roosting south of the Rio Grande. Thus even if the spotted ceased to exist entirely in Washington and Oregon, the species line would go forward in many other places.

Yet for the purposes of the Endangered Species Act, the spotted of Washington and Oregon is treated as if it were a species. The Act mandates protection of habitats and "locally distinct" populations, interpreted to mean subspecies living in a manner different from how kin live elsewhere.

Environmentalists hold that even if many spotted owl thrive in California, logging bans must continue in Washington and Oregon because owl populations there are locally distinct; the birds roost in cool climates feeding on arboreal squirrels, a "distinct" Northwest delicacy. This is probably an accurate reading of the Endangered Species Act as written, but it points to a deep logical fault in environmental orthodoxy. If local variations in climate and diet convert creatures to different species, a black man who lives in Seattle, gets rained on and eats salmon would be a different species from a white man who lives in stifling humidity in Louisiana and dines on gumbo. By this theory the human race contains hundreds of species. The sort of people likely to be environmentalists maintain that when it comes to genus homo, all individuals of all origins are exactly the same in genetic heritage. Yet when it comes to animals, the tiniest distinction renders apparently identical living things irrevocably separate species. The typical northern and California spotted owls look and act more alike than the typical black American and white African. But according to politically correct dogma, different people are identical while similar birds are drastically different.

This kind of upside-down logic is important to political environmentalism because Congress will soon debate reauthorization of the Endangered Species Act. If the most famous candidate for extinction is not dying after all, how will the lobbying be kept correctly pessimistic? Two poles of possibility exist for the act. Reactionaries hope to eviscerate it. Progressives would switch the act from its cumbersome creature-by-creature approach to a rationalized system in which blocs of habitat are placed in preservation status but species vacillations within those blocs do not trigger legal panics, since vacillations happen in nature anyway. Environmental orthodoxy wants the act renewed in its effective but panic-oriented form. For this, a continuing owl emergency is a political essential.

Could it be that spotted owls are not endangered but other old-growth species are? Perhaps the replacement of many old-growth forests with young timberlands has caused a wipe-out among creatures harder to count than owls. Current environmental thinking holds that deforestation is the worst form of human activity from the standpoint of biodiversity loss. And there is no doubt that until roughly the past decade, Northwest timber companies slicked off the Cascade Range without regard for conservation. Since the 1991 logging suspension, evironmentalists' legal maneuvers have concentrated on expanding the ban's scope to protect some 1,400 non-owl species presumed to be imperiled old-growth obli-gates. Yet with the exception of Pacific salmon types—whose recent runs have been unequivocal disasters, though perhaps for mainly natural reasons—only a handful of the presumably imperiled obligates has been shown to exhibit worrisome population trends. And thus far there are no known extinctions of animals or vascular plants in the forest regions of Washington, Oregon or California, according to The Nature Conservancy and the Environmental Defense Fund. A half-dozen plants are "missing in action" (not observed recently, though known to prosper elsewhere) and mammals such as the red vole and fisher are believed to be declining in population—but so far, no known postwar extinctions. This in a habitat range that has not only been subjected to extensive logging but also numbers among the most intensely studied in the world and is therefore a place where extinctions are likely to be detected. Combined with the prospect that there are many more spotted owls pairs than previously estimated, this raises the question of whether the owl doomsday, which has cost thousands of honest people their livelihoods and occupied the attention of presidents, is at heart a false alarm.

In contrast, the need to preserve forests is no false alarm. Broad agreement exists among researchers that old-growth woodlands require respite from the indiscriminate logging of the postwar era. Unregu-lated logging left Northwest forests "fragmented"—containing lots of trees, but in blocks chopped into checkerboards. An emerging body of science holds that forest fragmentation imperils biodiversity. For instance, studies suggest that moderate numbers of spotted owl in contiguous "clusters" would be more secure than twice as many owls in fragmented forests. This alone is reason for strict regulation of Northwest logging.

But the clear need for strict regulation of forestry should be argued on its own merits, not by resorting to dubious claims of owl peril. Consider that Jerry Franklin, a University of Washington researcher and the leading proponent of "New Forestry," drove timber companies to distraction in the 1980s by saying that to protect biodiversity and achieve sustainable yield, Northwest logging must decline to one-half the '80s peak rate. Until quite recently, Franklin was the left wing of the debate. Now, responding to the hypothesized owl emergency, Clinton proposes to reduce Northwest logging to 20 percent of the '80s peak.

Consider that as U.S. timber production declines, demand for foreign timber escalates. Many countries will import more wood from nations such as Malaysia and Brazil, where forestry may be summarized by the cry "timmm-burrr!" Today, Japanese firms are slicing off the lush Sarawak rain forest in Malaysia to feed a global wood market energized by U.S. logging bans. In the Sarawak there are no niceties about sustainable yield or protection of species, and there are no well-funded litigators to sue the irresponsible. Moving logging from Oregon to Malaysia may shift the problem out of sight and out of mind. But it is not much of a deal for the environment.

Consider that anti-logging sentiment born of the desire to protect old-growth stands requiring centuries to restore now spills over into activism against logging in young forests easily restored. In 1990 a California ballot initiative nearly banned most logging even on tree plantations. Yet when commercial forestry produces ample tree harvests, the pressure to log out ancient forest declines. "High-yield forestry can work in concert with old-growth preservation," says Michael Oppenheimer, chief scientist for the Environmental Defense Fund.

Consider that from 9,500 (the White House's own number) to 85,000 jobs will be abrogated by the Clinton owl plan. The lost jobs are high-wage employment of the sort that Americans who aren't lawyers or consultants need to send their children to college. Lumber prices have also nearly doubled since the 1991 ban was imposed, adding roughly $5,000 to the price of a new home. This increase in regressive, hitting the working class much harder than the elite environmentalist class.

If it is eventually understood that affluent environmentalists with white-collar sinecure destroyed thousands of desirable skilled-labor jobs in order to satisfy an ideology and boost the returns on fund-raising drives, a long-lasting political backlash against environmentalism will set in. There is still time to avoid this turn of events. Ancient forests can be protected, additional timber jobs restored and the constructive political power of environmentalism sustained. Honesty about owls would be the beginning.

reading 4

"Timber Owners Cut a Deal to Preserve Wildlife Habitat"
Kim Murphy

Los Angeles Times, August 19, 1995. © 1995 by *Los Angeles Times*. Reprinted by permission.

The Murray Pacific Corp. had just closed the book on its portentous encounter with three northern spotted owls, whose wing beats in the night had tied up nearly four square miles of rich timberland—about half the firm's salable trees—in a bid to save the threatened bird.

For two years, the only sound within 1.8 miles of the owls' known nests was the hooting of biologists seeking to lure and count whatever birds might be lurking in the trees.

As lawsuits and agony unfurled throughout the 1980s over the disappearance of the spotted owl from the Northwest's old-growth forests, Murray Pacific spent $650,000 developing a plan to protect its three owls and whatever others might join them.

Strict logging standards were established. Protection buffers around the nests were assured. And the sound of chain saws and logging trucks was about to be heard again throughout Murray Pacific's 53,527-acre tree farm on the Cascade slopes.

That was in October, 1993. A few months later, a biologist documented the sound of whistling wings just before sunrise—the unmistakable calling card of the marbled murrelet, a tiny, diving sea bird that nests in the same old-growth forests.

"We hadn't even broken out the champagne when a marbled murrelet dipped its wing over the west end of the tree farm," company Vice President Toby Murray recalled with a sigh.

Facing what it saw as the prospect of a never-ending parade of endangered species crawling into its forests to make a last stand on life, Murray Pacific got religion. The company and the U.S. Fish and Wildlife Service in June signed an uprecedented all-species protection plan that will guarantee over the next 100 years a measure of safe habitat not only for the owls and the murrelets, but for any red-legged frogs, eagles, goshawks, wolves, grizzly bears, big-eared bats or members of at least 28 other endangered or threatened species that might venture a paw or a claw onto a Murray Pacific forest.

In exchange for a broad array of protection measures that the company estimates will cost about $100 million over the next 50 years, the federal government has pledged to guarantee continued logging operations, even if a new and previously unrecognized endangered species shows up—and even if something like a spotted owl sheds blood under a logger's saw.

Habitat Conservation Plans

These private "habitat conservations plans"—now being prepared for much of the vast Northwest forests and environmentally sensitive landholdings throughout the nation—are the Clinton Administra-tion's attempt to offer some assurance and predictability to landowners who have grown increasingly militant against strict requirements of the Endangered Species Act.

They are part of a growing attempt by Congress to provide financial incentives to private landowners—ranging from estate tax reform, tax credits for endangered species management and voluntary conservation agreements—to turn protection of dwindling species into a cooperative process, not just an exercise in unpopular regulation.

In San Diego County, a habitat conservation plan has allowed a 2,200-unit housing project to proceed in La Costa by setting aside at least 700 acres for preservation of 63 plant and animal species. In Northern California, Simpson Timber Co. has gained clearance to continue logging operations on 380,000 acres of forests that have documented some of the highest densities of spotted owls ever recorded—725 captured and counted to date.

And in Oregon and Washington state, federal officials are negotiating conservation plans for some 5.5 million acres of forest—providing the first, and perhaps the last, opportunity to develop a comprehensive ecosystem plan that will link preserves on federal forests with large tracts of private timberland to guarantee habitats over the next generation, before the old forests fall victim to urban growth.

"Whatever your forest base is going to be for the next generation is [being established] now. With California, Oregon and Washington among the fastest-growing areas in the country, this will be what the forest landscape is pretty much going to look like from 20 years on out," said Curt Smitch, the Fish Wildlife Service's assistant Northwest regional director.

For years, the problem of trying to protect endangered creatures has been that planning is at the mercy of whatever tract of land happens to house a nest or a den at a given time.

"The spotted owl is a classic example. We have just been chasing individuals around the landscape. You can set an area of protection around the owl, but if it moves, the protection goes off that area and goes onto the next place the owl lands. And as soon as the owl goes out of there, somebody goes in there and cuts [the trees]. It's not a long-term strategy," Smitch said.

"The habitat conservation plan allows you to set up more of a landscape approach and...you begin to deal with...the protection of habitat [rather than individual animals]," he said.

The habitat conservation program has become one of the Administration's highest priorities at a time when the Endangered Species Act is under siege in the courts and the Republican-controlled Congress. In Olympia, Wash., Smitch heads a team of administrators and biologists assigned to negotiate conservation plans with landowners ranging from the state of Washington to small tree-farm owners.

Federal officials are hoping that by showing landowners it can make economic sense to strike a deal on behalf of endangered species—winning long-term certainty that they will be able to develop their land—they will avert the threat that an increasingly hostile timber and development industry will convince Congress to simply repeal the Endangered Species Act or hopelessly gut it.

They also hope to offer an alternative to individuals who, fearing their land may soon be designated critical habitat, have in many cases moved to quickly level it and effectively end any discussion.

Economic incentives are the bridge between what we are doing now and what we should be doing for endangered species," said Bob Ferris of Defenders of Wildlife. His group joined in a series of proposals for incentives to private landowners to protect endangered species prepared by the Keystone Center, a nonprofit mediation group, and presented recently to a congressional committee.

The Keystone report suggested measures such as delaying or forgoing estate taxes on large landholdings designated as habitat areas, agreements that encourage landowners to protect species even before they have been officially listed as threatened or endangered, federal tax credits for species management practices, and streamlining the habitat conservation plan process so that it is accessible to small landowners.

Federal officials are in the midst of plan negotiations with many of the region's timber industry giants, including the Weyerhaeuser Co. and the Plum Creek Timber Co., which together hold more than 2.5 million acres in Washington and Oregon.

But they also are seeking to clinch deals with a broad array of small landowners, a prospect seen as less promising because of the enormous complexity and expensive studies required to complete a plan. Murray Pacific spent $1.75 million developing its program, not counting forgone timber revenues.

Proponents are urging the development of a less complicated planning process for small landowners, perhaps a boilerplate conservation plan that could be adjusted for individual parcels. Their participation is key: Small landowners hold a quarter of the forest space in Washington and some 58% of the nation's forests.

Greg Patillo's 700-acre tree farm hugs the coast of southern Washington near Raymond, some of the best hemlock and Douglas fir-growing land in the world. Patillo, and ex-timber company worker, makes his living off the small farm. Three years ago, federal biologists thought that they detected a spotted owl on a neighboring property.

The owl wasn't found again the next year, or the next. But Patillo, who had a patch of 70-year-old timber on his farm, hasn't been able to think of a thing besides the owl since then. Patillo had been harvesting about 10 acres a year, often less. Last year, the owl in mind, he chopped down 70 acres, including most of what might have been owl habitat. To his mind, it's the government's fault.

I'm not looking for a reward. But to use a club and tell me that I may not be able to use my forest simply causes me to panic and causes me to harvest prematurely and harvest more than I would have," he said. "All they're saying to me right now is if I grow that kind of habitat, I may be penalized."

Patillo has no plans to negotiate a habitat conservation plan, even though he has the specter of the marbled murrelet and the coho salmon—whose listing last month as threatened in Oregon and California drove large numbers of landowners to the negotiating table—hanging over his farm.

But Douglas Stinson, who has about 1,000 acres farther inland near Toledo, has already had four negotiating sessions with the Fish and Wildlife Service to prepare a conservation plan for his tree farm.

"We would like to achieve one. I can't say we will or we won't at this point," said Stinson, who worked for years as a forestry employee and bought his small farm up in patches with his spare cash. "About all we can spend is what it takes in our time in writing it up. If it requires a bunch of lawyers and all that, it isn't going to fit for the individual guy."

Stinson, who wants to preserve the farm for his sons against the pressures of urban growth from the booming Interstate 5 corridor a few miles west, already has in place a number of natural conservation measures, including an 80-year harvest cycle. His proposal to the Fish and Wildlife Service is to simply maintain the kinds of habitat he has, rather than tailoring his trees to fit specific species.

"We feel strongly that you can manage a forest and take a product from it and you can still nurture the birds and take care of the habitat," he said. "But if we can't earn a living, there'll be houses here. The highest value for this land today is as a subdivision."

Environmental groups have been supportive of the habitat conservation plan process but fear timber firms may be getting deals that are much too sweet. Interior

Secretary Bruce Babbitt's pledge to guarantee landowners freedom from all future endangered species headaches—the key incentive that brought them to the table—may be overreaching, they say.

"It's as if we know enough today to protect everything that's going to happen over the next 100 years," said Charley Raines of the Sierra Club. While most plans allow extra mitigation measures in the event of unforeseen or extraordinary circumstances, the government has to pay for them.

"In the [habitat conservation plan] process, we're providing the landowner with an officially sanctioned means of continuing to log their land. In return, they protect the fish and habitat to take care of the species that are found on that land. But if there are some changes in the future, we need to be able to deal with that," Raines said.

With the Murray Pacific plan, he said, the company will gain the immediate benefit of being able to log most of its remaining old-growth timber, while the prime habitat benefits will occur much later, as new mature forest land develops in reserve areas.

That is a key feature of most of the habitat plans being negotiated. While significant buffers are set aside along streams and other areas, the plans often rely on the concept of growing habitats rather than preserving them. Thus, habitat may be saved in one area, then logged as new habitat matures on an adjacent tract.

Trees Amid Clear Cuts

Murray Pacific's plan uses modern techniques to accelerate the development of forests into good dispersal areas for young spotted owls over the entire expanse of the tree farm during the next 50 years. In areas being logged, a few mature trees and old dead trees that serve as ideal bird homes are left behind. Whole stands of trees are left amid the clear cuts. Harvests are limited to 1,000 acres a year, and no more than 5,000 acres in any 10-year period.

Murray, the grandson of the founder of the family-owned firm, said company officials told federal authorities that they didn't want to develop a plan if it meant they would become the focus for owl preservation in the region—especially if they didn't get any promises in return.

"We didn't want to be penalized if 50 years from now we've got the owl superhighway of the region and the federal government might come in and say: 'You're now and owl reserve,'" Murray said.

Other timber industry leaders have hung back from signing conservation plans, perhaps hoping Congress will weaken the Endangered Species Act. Some have likewise not been discouraged by the recent Supreme Court decision requiring protection of the spotted owl's habitat, reasoning that the decision will make Congress even more likely to step in.

On the other hand, companies like Murray Pacific see a whole generation of new species on the horizon for endangered status. Recently, the federal government recommended designating 4.4 million acres, including 50,100 acres of private land, as critical habitat for the marbled murrelet in California, Washington and Oregon.

Along with the coho, a large number of other salmon and steelhead stocks are likely to be listed soon, imposing a potential land-use quagmire on areas around rivers and streams throughout the Northwest.

"I don't think the big stuff has even started yet. Wait till they start listing fish," Murray said. In the end, he said, the scale among trees and harvest, birds and fish, revenues and forecasts, appeared to balance.

"We have thought from the beginning that we were willing to make a significant contribution to fish and wildlife," he said, "so long as we could get some kind of understanding that that would be it."

"U.S. Considers 80% Increase in Sierra Logging"
Frank Clifford

Los Angeles Times, August 20, 1996. © 1996 by the *Los Angeles Times.* Reprinted by permission.

The U.S. Forest Service is considering a policy that would increase logging in the Sierra Nevada by as much as 80%, easing restrictions put in place to protect the California spotted owl and the older trees that the species prefers.

Allowing more trees to be cut would provide jobs to hard-pressed Sierra towns and ease the danger of wildfire, without jeopardizing animal and plant species that have been dwindling, Forest Service officials in California have concluded.

But the new policy, which was to be proposed publicly today, is encountering serious opposition from environmental groups. Release of the proposal may be held up so the policy can be further reviewed by Clinton administration officials in Washington, Forest Service sources said.

The proposal would abandon a 1993 strategy to protect the spotted owl and its old-growth forest habitat. Logging has been steadily curtailed—and the biggest, oldest trees placed off limits—since government scientists reported that the loss of Sierra old growth was threatening the California spotted owl's survival.

The new logging policy is being considered dispite recent studies by Forest Service scientists and other government experts that the Sierra's owl population is still declining at a rate of at least 5% a year and that only 15% of the old-growth forests are intact.

Timber industry representatives are hailing the plan to allow more logging in the Sierra—and the taking of some large old-growth trees—as a long overdue return to common-sense forestry.

"This is a significant improvement," said John Hoffman, vice president of public affairs for the California Forestry Assn. "It does not bring back the harvest levels of the 1980s, but it certainly restores a balance between environmental and practical considerations."

The Forest Service's proposal to ease limits on logging almost certainly will reignite the debate in environmental circles about the depth of the Clinton administration's commitment to forest protection.

"Forest protection clearly is Clinton's Achilles' heel, and this will pinch even more," said Carl Pope, the Sierra Club's executive director. The Sierra Club this year called on the federal government to not allow loggers to cut any more trees in national forests.

In May, the administration interceded to prevent the release of a Forest Service plan that would have allowed more logging in the Sierra, but not as much as the latest proposal.

At the time, White House representatives said they wanted any new logging policy to take into account the results of a comprehensive, congressionally mandated study of the Sierra environment. That study, released in June, found that a variety of plants and animals in the Sierra were on the road to extinction and cited heavy logging as a main culprit.

"I can see no scientific justification for cutting more trees, especially large trees," said R.J. Gutierrez, a Humboldt State University wildlife biologist under contract with the Forest Service to study the owls.

The Forest Service proposal would allow sizable logging increases in the Sierra—80% over 1995 levels and about 50% more than what was cut in 1994. More important to the timber industry, it would allow harvesting of old-growth trees now out of bounds.

Insisting that the largest trees in the forest—those more than 40 inches in diameter—would still be protected, Forest Service officials argue that the new formula would spare most of the habitat of owls and other creatures that depend on old growth.

But critics, including scientists who contributed to the recently completed congressional study of the Sierra, say that Sierra old growth consists of complex clusters of mature trees of vary age and size that the Forest Service plan does not protect.

"I have seen no indication that they have a strategy explicitly identifying old-growth reserves and showing how they intend to keep them intact," said Jerry Franklin, a University of Washington professor of forestry who coauthored the section of the Sierra study dealing with old-growth forests.

The forest plan would mean about a 50% increase in jobs and nearly 400% more revenue for Sierra counties, according to a Forest Service analysis. A quarter of the revenue that the Forest Service receives from commercial logging in national forests goes to the counties where the forests are.

Questions for Discussion

For Reading 1

1. According to these authors, what are the economic problems facing the Pacific Northwest? What are the causes of these problems?

2. How are these problems related, directly or indirectly, to endangered species in the region?

3. Given the isolation and small size of many of the communities most affected by timber management problems, what are some reasonable options for developing "sustainable, diversified, local economies"? How might these solutions be implemented?

For Reading 2

1. What are the primary reasons for recent declines in timber industry-related jobs in the Pacific Northwest?

2. What are some of the primary reasons for recent economic improvements in the region?

3. How important are endangered species to the economics of the region?

For Reading 3

1. On whose "side" is this author?

2. A major premise of this author seems to be that change in forestry-management techniques could improve both economic and environmental conditions in the region. What would be the costs and benefits of these changes to the timber industry? To environmentalists? To outdoor recreationists? To endangered species?

For Reading 4

1. To what extent will "habitat conservation plans" solve or exacerbate the region's timber management-related problems?

For Reading 5

1. To what extent will the new policies being considered by the U.S. Forest Service solve or exacerbate the region's timber management-related problems?

Going Beyond the Readings

Visit our Web site at <www.harcourtcollege.com/lifesci/envicases2> to investigate wildlife management issues in additional regions of the United States and Canada.

unit 8

California: Solid Waste Disposal

Introduction
Readings:
Reading 1: The 3 R's of Solid Waste and the Population Factor for a Sustainable Planet
Reading 2: Sonoma County Waste Management Agency 1992-1997 Progress Report
Reading 3: Why Recycle?
Reading 4: Buy Recycled—On Earth Day and Every Day, Shop the Recycled Way
Reading 5: Earth Pledge: Twelve Things You Can Do Today to Fight Global Warming and Environmental Destruction
Reading 6: Elder Earthkeepers Honored With Gifts From the Earth
Reading 7: What Is Bay Area Creative Re-Use?
Reading 8: Environmentalism Begins At Home: Green Building Represents a Vital Earth-Friendly Action
Reading 9: Helping Business Donate/Sell Waste Material
Questions for Discussion

Solid Waste Disposal in Sonoma County

Introduction

On trash day in many U.S. neighborhoods, large green plastic containers on wheels can be found in front of nearly every house. Collectively, an average neighborhood can send 3000 pounds (1.3 metric tons) of refuse to the local landfill. A few days later there is another trash day for the same neighborhood. Millions of neighborhoods throughout North America are all contributing members of the "throwaway society."

Humans have always discarded that which they no longer needed. Discarded arrowheads, skin scrapers, and pieces of pottery tell much of what we know about prehistoric peoples throughout the world. Starting in the 1700s, the Industrial Revolution increased our opportunities to throw away as we accumulated more and more goods and used them less and less wisely. After World War II in North America, the process of mass throwaway accelerated. Producers discovered it was more profitable to produce goods of low quality with projected short lifetimes than to produce durable goods that stayed with consumers for longer periods of time. "Planned obsolescence" became a watchword of modern commerce. Consumers played along. Initial costs for goods were relatively low. When something broke down, it was more convenient to discard and replace than to repair.

Packaging added more waste. As people became more and more health conscious, the popularity of packaged foods increased. Meat wrapped in plastic picked up fewer germs than meat exposed to air. Attractive packaging enticed buyers. Protective packaging was easy to ship and reduced damage to goods.

The Problem

Today the average North American produces around 4 pounds (1.8 kg) of trash per day. In 1960, the average was only 2.7 pounds (1.2 kg). In 1960, North America had fewer people. Then, we produced an estimated 337,500 tons (306,300 metric tons) of trash annually. In 1997, our annual mountain of trash grew to 600,000 tons (544,500 metric tons). What do we do with all that trash?

Most of it goes to landfills. Until recently, we carted our trash off to dumps. Open-air sites at least collected trash all in one place to slowly molder and decay, out of sight and out of mind. In recent years, not only has the sheer volume of trash increased but its nature has changed. Wood, steel, and glass have been replaced with paper, aluminum, and plastics. Added were a whole suite of chemicals—household cleaners, paint solvents, pesticides, automotive products, and the like. As many of these products degrade, unpleasant, flammable, or toxic by-products are produced. The simple open dump evolved into a modern landfill. To keep dangerous substances isolated, the modern landfill is lined with impermeable substances and periodically buried. Today in the United States, three-fourths of our trash ends up in landfills. Problems are created.

Eventually, even the most carefully constructed landfill can leak. Poisonous leachate filters out into underground water, contaminating wells and surfacing in yards, basements, and streams. More than 20 percent of hazardous waste cleanup sites in the United States are municipal landfills. As many organic wastes decay, especially in oxygen-starved environments, methane gas is produced and leaks into the atmosphere. Methane is highly combustible and can produce dangerous fires. It is also a greenhouse gas that adds to global warming. The biggest problem with landfills is simply the land they consume. As municipalities and populations grow, land becomes more precious and limited. No one wants a landfill in their neighborhood.

Incineration, which processes about one-fourth of the trash in the United States, is an alternative to landfills. Typically, truckloads of trash are dumped onto conveyer belts, carried into huge furnaces, and burned. On relatively small tracts of land, huge volumes of trash can be processed.

But incinerators create almost as many problems as they solve. Not all trash is combustible. Incineration reduces trash volume by only 50 to 75 percent. Ash residues are often highly toxic concentrations of heavy metals and must still be disposed of, typically in landfills. Combustible materials are turned into gas and dumped into the atmosphere. Gases include nitrogen and sulfur oxides (both precursors of acid rain), carbon monoxide and carbon dioxide (both greenhouse gases), dioxins, furans, and 28 different heavy-metal-containing gases implicated in cancers, and other public health problems. Filters can reduce but not eliminate undesirable emissions. Water, used to cool hot ash, becomes contaminated with many of the same chemicals and creates its own disposal problems.

Finally, incinerators are extremely expensive to operate. Theoretically, energy produced by incineration could be recovered and put to some useful purpose, but too often it is simply wasted or dumped into the atmosphere. To defray expenses, incinerator operators—both public and privately owned—sell services to neighboring areas. For a fee, trash is trucked in and burned. To the local community, burning extra trash increases air and water pollution and the amount of ash that needs to be landfilled.

No wonder so many communities are opposed to incinerators.

There are other solutions to our mounting trash problem. Many industrialized nations, especially outside of North America, have adopted a waste management hierarchy. First, consumers are asked to reduce volumes of trash produced. The most desirable of these methods is source reduction, followed by direct reuse of products and then recycling. Reducing, reusing, and recycling goods can significantly lower trash volumes available to incinerators and landfills—solutions of last resort.

Unfortunately, most governments continue to focus on managing rather than reducing trash production. Almost invariably they spend money on landfills, incinerators, and recycling programs, in that order. Practically nothing is spent encouraging citizens to reduce consumption or to reuse commodities.

California is an exception. In 1989 its legislature passed AB949, more technically known as the Integrated Waste Management Act. Among other provisions, the act called for an overall, statewide reduction in solid waste production—25% by 1995, 50% by 2000. In the ensuing years, many California localities responded. One of the most exemplary of these is Sonoma County, north of San Francisco. In 1992 the county, along with nine of its principal cities, formed the Sonoma County Waste Management Agency. Under its Joint Powers Agreement, the formal document that established the Agency, it is responsible for reducing, reusing, and recycling as much of Sonoma County's solid waste as possible. The Agency has met with notable success: After only three years of organization, it not only met the mandated 25% overall reduction of wastes, it exceeded it by 12%. In 1996, reduction levels reached 39%. The readings that follow provide additional background related to the seriousness of the problem nationwide, and describe specific projects and other activities of the Agency.

reading 1

"The 3 R's of Solid Waste and the Population Factor for a Sustainable Planet"

Joan Wagner

The American Biology Teacher, February 1995, 57(2). Reprinted by permission of the National Association of Biology Teachers.

Since early times when human beings began aggregating in clusters within protected boundaries, these neo-urban centers became aware of the environmental problems that were to face all overpopulated areas. Every region has a carrying capacity defined by its resources and its ability to sustain those resources. The first solid waste management involved the wastes of humans and domestic animals. In early Rome, with a population of 50,000, ancient laws created public dung heaps. Human feces were used as fertilizer. This "night soil" is still used by Third World countries today. Since there was no common graveyard, families buried their dead in the confines of their homes. Laws forbade the burning of corpses or the disposal of dead bodies in groves of trees or public areas. All major and public holidays contributed to the pollution of Rome. Since dozens of animals were slaughtered at most festivals, the city was filled with street corner sacrificial altars for the use of private citizens. All butcher shops and altars were strategically located near sewer lines in order to facilitate the run of blood and the quick disposal of carcasses. Julius Caesar passed an ordinance that restricted all commercial traffic from sunrise to 4:00 p.m. Besides being noisy and dangerous to pedestrians, it also had to facilitate the clean-up of wastes from the animals used in the traffic. It is also interesting to note that in 1665, when London was stricken by the Black Death, Paris was spared due to the superior hygiene within its city (Crane 1978).

In the United States, until the 19th century, environmental problems were either ignored or accepted as inevitable because much more land and resources existed in proportion to the population (Crane 1978). If people were unhappy about their surroundings, they could move on to a new location. Land and resources seemed too plentiful to foresee any immediate danger to the environment. The effects of a growing population or the onslaught of the "throw-away" generation were not part of the mainstream philosophy of 19th century planners. In 1960, the average American produced 2 1/2 pounds of garbage a day. Since then, we have added more than 70 million people and today we are generating close to four pounds of garbage a day per person (ZPG Fact Sheet 1989). These two trends have created a staggering rise in the amount of resources we lose and the amount of waste we generate. The United States produces more than 160 million tons of waste... more per capita than any other country in the world (ZPG Fact Sheet 1989). Using present day demographics, we will add another 20 million to our population by the close of the century. Those additional people, based on current behavior patterns, will generate 80 million tons of trash by the year 2000. That is enough to fill a convoy of 10-ton garbage trucks wrapped around the world almost three times... about 70,000 miles (ZPG Fact Sheet 1989).

The first Earth Day is responsible for inspiring the formation of a number of environmental government agencies. For example, in New York State, the Department of Environmental Conservation was created. On a national level, the Environmental Protection Agency was formed. In 1970 the newly formed EPA established the main categories of pollution: air, noise, pesticides, radiation, water, and solid waste (Young 1991). Hence, solid waste was now nationally identified as a source of pollution.

Trash generated by the "throw-away-society" was "striking back." Our inability to absorb the steady flow of trash resulted in the construction of more and more landfills and the increased contamination of our ground water (Young 1991). Up until the 1960s, incineration was the choice method for the management of solid waste. However, the constant incineration in these urban areas so added to the air pollution that laws were passed to suspend that management technique and the landfills were further burdened (Young 1991).

When did the "throw-away" society begin? The age of consumerism appears to have commenced after World War II. It was believed then that the total sales of all of the commodities produced by a nation indicated its economic health. In order to encourage large sales, poorly made products that could be sold inexpensively and would need frequent replacement began to dominate the marketplace. The convenience of a product began to eclipse its durability. Often a product was cheaper to replace than to fix because it had to be returned to the manufacturer rather than a local shop for repair. Furthermore, technology caused many products to become outdated. The government helped to perpetuate the low costs of these throw-away products by subsidizing the costs of virgin materials. The consumer was actually paying artificially lower prices because of this government policy. Most manufacturers were never concerned about the disposal of their products because once sold to the consumer they were no longer responsible for the ultimate "resting ground" of the product (Young 1991).

Today, consumers must be better educated about the products they purchase. They should understand how the product affects the environment. For example, if it was manufactured from virgin materials, were these renewable or nonrenewable resources? What effect does this product have on air and water quality? How will the product affect landfills? The education begins in the schools and the topic of consumerism should be approached through interdisciplinary activities. For example, in my school (Burnt Hills–Ballston Lake Middle School), the eighth grade team of teachers has developed an advertising unit in which students design a product that is environmentally sound and develop an advertising campaign to sell it to the community. We must begin to develop a sustainable economy. Products of high quality that last many years need to replace our post war "love affair" with "throwaways." The government needs to provide greater research funds to improve recycling technology. Perhaps there should be tax benefits to encourage the growth of sustainable manufacturing. If the manufacturer is made accountable for the final destiny of its product, the pure economics of this role should dictate the need to manufacture products of improved quality and durability. Add to this the responsibility for the disposal of the packaging of its products, and one should see less stress on our landfills and greater sustainability of our natural resources.

There exists a strategy for the handling of solid wastes. It is called The Waste Management Hierarchy, which is shared by many industrial countries. This approach prioritizes a list of management options. First in the hierarchy is source reduction—it attacks the problem before it becomes one. The manufacturing of products that put stress on landfills must be reduced. Second is the direct reuse of products whenever feasible. Third is recycling. Fourth is incineration (with the recovery of energy), and the last choice would be sanitary landfilling (Young 1991). Finally, since more people translates into more waste generated, all action would be meaningless if the world population is not stabilized.

Though population is growing fastest in Third World countries, the United States has the highest growth rate of all industrialized countries, adding 3 million a year. Although the population density of the Unites States is less than other countries, its population growth degrades the environment more quickly because of the way we use resources and produce waste. If the U.S. growth rate continues, it is predicted that by the year 2050, its population will increase by 133 million people, which is equivalent to adding 38 cities the size of Los Angeles. World population stands presently at 5.6 billion, adding 93 million a year, and has a doubling rate of every 40 years (ZPG Fact Sheet 1993). Thus, individual efforts to recycle and conserve energy will have little impact on the environment unless we factor into the "formula" for sustainable living a policy that will stabilize population growth at both home and abroad.

The United Nations environmental program endorses The Waste Management Hierarchy, as do citizen groups, many industry leaders, and government officials in Europe, North America and Japan. It has been enshrined in the United States law since the passage of the Resource Conservation and Recovery Act in 1976. It is unfortunate that practice has run counter to principle. Most governments faced with waste crisis have tended to focus on waste management rather than waste reduction. As a result, the furthest up the government would get on the Waste Management Hierarchy was the funding of incineration plants. For example, in 1972, New York State's Environmental Quality Bond Act budgeted $215 million for incineration and only $1 million for recycling. Although many state governments are increasingly planning and budgeting for recycling, according to a recent survey 18 in the Northeast and Midwest still expect to spend 8–19 times more on incineration than on recycling over the next five years (Young 1991).

Building new landfills is not the solution. The new environmental regulations make them expensive to construct and many communities are against them because of health concerns. Furthermore, the growing population is putting demands on the land for other uses. The EPA reports that almost 1/3 of the 6500 landfills in our country will be closed within the next few years (Young 1991). It is a particular problem in highly populated areas such as New Jersey and Florida. All of their landfills will be closed by the end of this century—probably sooner for Florida due to its recent disastrous hurricane and cleanup. Presently, the Fresh Kill Landfill on Staten Island, New York, is the highest peak in New York City. The "rolling hills" of parts of Long Island are actually capped landfills. Today, about 75% of the United States' garbage is buried in landfills. The remainder is divided equally between incineration and recycling. The practice of sending garbage to other states is becoming increasingly more difficult. In 1980, the infamous West Islip Barge carrying 3186 tons of waste was rejected by six states before it was finally incinerated (Young 1991).

Presently, about 10% of the waste generated is incinerated in this country. Incineration has certain disadvantages. It does not generate as much energy as would be saved by recycling. For example, recycled paper can save up to five times as much energy as can be recovered by incineration. For high density plastic, recycling can save almost two times as much energy (Young 1991). Incineration is not the final disposal. It is the conversion of primary waste into secondary waste (gases and ashes) which often can be very hazardous.

After incineration, there remains 30–40% of the volume that then must go to a landfill. Incineration requires the use of water to quench hot ash. The contaminated water presents another disposal problem (Young 1991).

Incineration is very expensive. Its high capital costs cancel the lower day-to-day costs of incineration over recycling. It is estimated by the Institute of Self-Reliance in Washington, DC that:

1 ton of garbage costs:
to incinerate $100,000–$150,000
to recycle $10,000–$15,000
to compost $15,000–$20,000
(Young 1991)

Rough calculations reveal that an $8 billion investment in additional incinerators could allow the Unites States to burn 1/4 of its projected solid waste by the year 2000, whereas the same sum spent on recycling and composting facilities could provide enough additional capacity to handle 3/4 of the nation's garbage (Young 1991).

Recycling competes with incineration that must run at near capacity to stay profitable. Effective recycling and waste reduction can cause these facilities to run in the red. For example, waste disposal officials in Warren County, New Jersey, attributed a large part of a local incinerator's weekly loss to the implementation of a state law requiring a 25% recycling rate. The community was forced to reimburse the incinerator builder and operator for its losses (Young 1991).

Using refillables saves energy. Repeated studies have shown that it takes far less energy to wash out an old bottle than to melt it and make a new one, or to make a new one from virgin material. According to a 1982 study, a 12-ounce refillable glass bottle reused 10 times required 24% as much energy per use as a recycled aluminum or glass container and only 9–10% as much as a throw-away made of these materials (Young 1991).

A 1990 study commissioned by a plastics trade group found that a 16-ounce glass refillable bottle used eight times was the lowest energy user of nine containers considered. Only 5% as much energy is needed to recycle aluminum as to manufacture from its ore bauxite. There is a 2/3 energy savings for recycled steel. Newsprint from recycled paper takes 25–60% less energy than from pulp wood. In fact, one ton of recycled paper saves 17 trees, 25 barrels of oil, 7000 gallons of water, and 3 cubic yards of a landfill. Recycling one run of the Sunday *New York Times* would save 75,000 trees (ZPG Fact Sheet 1989). Since paper makes up 50% of the volume and 40% of the weight in a landfill, a large waste reduction could begin with paper. Furthermore, recycled paper reduces air pollution by 74% and water pollution by 35%. Steel produced from scrap reduces air pollution by 85% and water pollution by 76% while eliminating mining wastes (Young 1991). However, there is a caveat. We must restrict ourselves from putting the "cart before the horse" when the topic is recycling. In 1990, New York State passed an important recycling law, but it lacks the foresight that recycled goods need a market. Many recycling companies have reduced their pick up of paper or are charging increased fees for the hauling because they need to cover their costs due to a diminishing market for recycled paper. Mandating recycling without encouraging markets for recycled goods does not address the problem of developing a sustainable economy.

Presently, in the United States, a little more than 10% of municipal waste is recycled—with 25% of the paper being recycled; while Sweden and Switzerland recycle 40% of their waste, and Japan recycles 50%—including 93% of its newspapers (Young 1991). The United States uses more resources than any other country in the world. Though it represents 5% of the world population, it uses 24% of its energy and 20% of its other natural resources such as copper, tin and lead (ZPG Fact Sheet 1993). According to one estimate, the United States consumed more minerals between 1940 and 1976 than did all humanity up to 1940 (Young 1991). Furthermore, it produces more waste per capita than any other country. Having this type of impact on the environment places much of the responsibility of developing a sustainable economy on the United States. We must work with other industrialized countries and the Third World countries to ameliorate the problems precipitated by the "throw-away society" and the population explosion.

The curtain is quickly falling on our "throw-away society." The years of exploiting the land in order to mine our planet's natural resources as if they were infinite must culminate if we are to become a responsible leader in a global, sustainable society.

reading 2

"Sonoma County Waste Management Agency 1992–1997 Progress Report"

Sonoma County Waste Management Agency, August 1998. Reprinted by permission.

Introduction

The Sonoma County Waste Management Agency's (Agency) creation was prompted by California's Integrated Waste Management Act of 1989, commonly known as AB 939. AB 939 sets goals for waste reduction of 25% by 1995 and 50% by 2000. The jurisdiction of Sonoma County worked together early in the planning process and subsequently formed the Agency in 1992 to manage certain programs and waste streams that could most efficiently and cost-effectively be managed on a regional basis.

Joint Powers Agreement

The Agency is a Joint Powers Authority as defined in California's Government Code §6500. Each of the County's ten jurisdictions has one representative, and each representative has one vote. The following list identifies the jurisdictions that constitute the Agency and the 1997 Boardmembers.

City of Cloverdale	Carol Chase
City of Cotati	Marsha Sue Lustig
City of Healdsburg	Barbara Jason-White
City of Petaluma	Gene Beatty
City of Rohnert Park	Angela Fogle
City of Santa Rosa	Marc Richardson, *Agency Vice-Chair*
City of Sebastopol	Paul Berlant, *Agency Chair Pro Tem*
City of Sonoma	Patricia Wagner, *Agency Chair*
City of Windsor	Sam Salmon
County of Sonoma	Ed Walker

Agency Staffing

The Agency has a contract with Sonoma County Department of Transportation and Public Works for staffing and administrative services. The County's Integrated Waste Manager serves as the Agency's Director. The Agency also has partially dedicated legal counsel, program staff, accountants, two fully dedicated part-time clerks, and one part-time administrative aide. Other staffing is available when needed.

Agency Responsibilities

Under the original Joint Powers Agreement, the Agency is responsible for wood waste and yard debris diversion efforts, household hazardous waste management, and countywide waste reduction education. Collection of refuse, wood waste, and yard debris is not within the Agency's scope of work. In 1994 the Agency expanded its scope of responsibilities to include the Recycling Market Development Zone (RMDZ) and solid waste management planning. It further expanded its role in 1996 by becoming a Regional Agency as defined at AB 939, assuming countywide responsibility for achieving the 25% and 50% diversion goals for the local jurisdictions. This reduces the reporting responsibilities of each jurisdiction with the Agency submitting all required reports to the California Integrated Waste Management Board (CIWMB).

Agency Funding

Funding for the Agency's programs comes from four main sources: yard debris and wood waste tipping fee charges; a $2/ton surcharge on each ton of garbage accepted at the Sonoma County Disposal Sites; outside grants; and miscellaneous sources which include profit sharing, interest on reserved funds, and sale of products.

In January 1998, yard debris tipping fees were increased to $27.00/ton from the previous $26.00/ton. Wood waste tipping fees remain at $12.00/ton. These fees cover expenditures for processing, composting, and marketing the various products generated from the accepted materials. The Agency's remaining programs, including household hazardous waste, education, RMDZ, and planning are primarily funded by the $2/ton surcharge. In addition, the Agency receives grant funds from the CIWMB that subsidizes programs focusing on used motor oil and used oil filter recycling.

Demographics

Sonoma County encompasses the incorporated cities of Cloverdale, Cotati, Healdsburg, Petaluma, Rohnert Park, Santa Rosa, Sebastopol, Sonoma, the Town of Windsor, and the unincorporated areas of the county. Sonoma County is the most northerly of the nine counties in the San Francisco Bay Region. The County is bordered by the Pacific Ocean on the west, Mendocino County on the north, Marin County and San Pablo Bay to the south, and Solano, Napa and Lake Counties to the east. U.S. Highway 101 is the major north-south route through the County. Sonoma County consists of a

TABLE 1. Sonoma County Overall Waste Stream: 1991 v. 1995/6

Component	1991 Tons	1995/96 Tons	Difference	1991%	1995/96%	Difference
Paper	141,760	111,652	-30,108	26.2	27.1	0.9
Plastics	37,508	32,815	-5,323	6.9	7.8	0.9
Glass	15,505	14,866	-639	2.9	3.6	0.7
Metals	43,408	30,246	-13,162	8.0	7.4	-0.6
Yard wastes	83,976	29,553	-54,423	15.5	7.1	-8.4
Other organics	174,916	142,567	-32,349	32.3	34.6	2.3
Other wastes	37,858	42,965	5,107	7.0	10.3	3.3
Special wastes	6,576	8,497	1,921	1.2	2.1	0.9
Totals	541,507	412,531	-128,976	100	100	0

mixture of residential, commercial, industrial, and agricultural land use, with an estimated 1997 population of 426,000.

Landfill disposal services are provided at the Central Dispoal Site, located at 500 Mecham Road, Petaluma, California. The site is about 2.8 miles southwest of the City of Cotati and 7.5 miles northwest of the City of Petaluma in an unincorporated area of Sonoma County. The county also maintains transfer stations in Annapolis, Occidental, Sonoma, Healdsburg, and Guerneville.

Diversion

The Agency exceeded the 1995 25% diversion goal by 12% for a total diversion rate of 37%. In 1996 the diversion rate increased to 39%. The Agency has sought to meet the AB 939 goals, and is proceeding with every intent in meeting the 2000 goal of 50%.

Baseline Studies

Waste Characterization Studies

As part of the AB 939 planning process, the Sonoma County jurisdictions conducted Solid Waste Generation Studies in 1991. In 1995 the Sonoma County Department of Transportation and Public Works, Integrated Waste Division (IWD) conducted another study to determine the impacts of diversion programs, which showed a distinct change in the characterization of the waste stream, as summarized in Table 1. The 1995 study will assist the Agency in directing its future efforts toward appropriate waste streams, such as cardboard, wood, and metals. Copies of both studies are available for review or purchase at the Agency's office, 2300 County Center Drive, Suite 216A, Santa Rosa, California.

Public Awareness Study

In early 1996, the Agency conducted a public awareness study to determine the level of awareness in Sonoma County of Agency programs and integrated waste management issues. The overall result was that a minimum of 70% of Sonoma County residents were familiar with recycling, yard debris, and household hazardous waste programs. Copies of the study's summary of findings are available for review or purchase at the Agency's offices, 2300 County Center Drive, Suite 216A, Santa Rosa, California.

Five-Year Budget History

Table 2 details the total Agency budget for five fiscal years, July 1 to June 30, and includes the expenditures and revenues for the Wood Waste, Yard Debris, Household Hazardous Waste, Education, Recycling Market Development Zone, and Planning Cost Centers.

TABLE 2. Total Agency Budget from Fiscal Years 1992-97

	FY 1992-93	FY 1993-94	FY 1994-95	FY 1995-96	FY 1996-97
Summary of Expenditures	265,046	1,289,382	2,278,034	2,479,189	2,475,216
Summary of Revenues	342,362	1,524,736	2,423,539	2,468,495	2,728,809
Net Cost	(77,316)	(235,354)	(145,505)	10,694	(253,593)

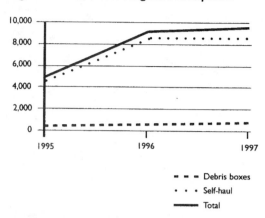

Figure 1. Wood Waste Program Participation

- - - Debris boxes
· · · Self-haul
——— Total

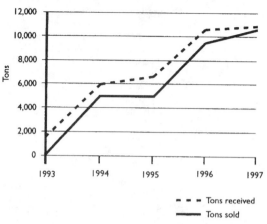

Figure 2. Wood waste delivered and wood chips sold 1993-97 (tons per year)

- - - Tons received
——— Tons sold

Wood Waste

The Integrated Waste Division initiated the wood waste processing program by setting aside an area at the Central Disposal Site. In 1993 the Agency assumed responsibility for the program with a contract to Sonoma Compost Company and Empire Waste Management. Disposal site users are encouraged to separate wood waste with a reduced tipping fee. Wood waste is segregated from the waste stream and chipped for sale as mulch and biofuel.

Reach

Wood waste is collected at the Central Disposal Site and the Sonoma, Guerneville, and Healdsburg transfer stations. Although Figure 1 illustrates participation since 1995, it may not include all participants because users have the option of segregating their loads at the tipping face. When this occurs, the user is recorded as having brought only garbage. Therefore, the number of actual users is greater than demonstrated.

Diversion

Figure 2 illustrates the tons diverted through the program and the total tons of wood chips sold since program implementation.

Five-Year Budget History

Tipping fee charges for wood waste are currently $12/ton. In fiscal year 1997/98, this program accepted an average of 34 tons/day. Table 3 details the total budget for five fiscal years, July 1 to June 30, and includes the expenditures and revenues for the Wood Waste Cost Center.

Looking to the Future

The existing Organics Material Processing, Composting, and Marketing Services agreement between the Agency and Sonoma Compost Company/Empire Waste Management has been extended for two years, until July, 2000. This will enable the IWD to complete the Environmental Impact Report for landfill expansion and make subsequent changes in the operations at the Central Disposal Site. Those changes include plans for relocating the composting operations to a site other than at the landfill. To prepare for this change, a Request for Proposals has been circulated for a program that provides comparable services to the Agency to be provided at a new location.

Yard Debris

In 1993 the Agency initiated the yard debris program with a contract to the joint venture of Sonoma Compost Company/Empire Waste Management. The IWD built the composting facility at the Central Disposal Site, providing it to the Agency for use by its contractor. Most Sonoma County jurisdictions began curbside yard debris collection programs to augment existing curbside recycling programs. The compost processing facility accepts yard debris from self-haul residential, commercial, and these curbside collection

TABLE 3. Wood Waste Diversion Program Budget From Fiscal Years 1992-97

	FY 1992-93	FY 1993-94	FY 1994-95	FY 1995-96	FY 1996-97
Summary of Expenditures	12,909	37,931	48,355	66,975	51,301
Summary of Revenues	2,787	66,609	101,334	97,311	96,223
Net Cost	10,122	(28,678)	(52,979)	(30,336)	(44,922)

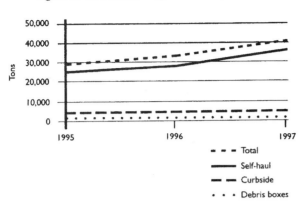

Figure 3. Yard Waste Program Participation

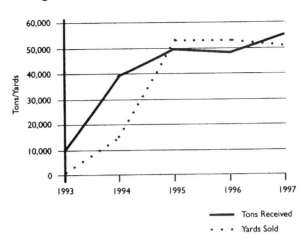

Figure 4. Yard Debris Delivered and Compost Sold

programs. The County's transfer stations also accept yard debris, which is transported to the Central Disposal Site. Yard debris is composted and sold as various products. Disposal site users are encouraged to separate yard waste with a reduced tipping fee.

Reach

Yard debris is collected at the Central Disposal Site and the Sonoma, Guerneville, and Healdsburg transfer stations. Although Figure 3 illustrates participation since 1995, it does not include all participants because users have the option of segregating their loads at the tipping face. When this occurs, the user is recorded as having brought only garbage. Therefore, the number of actual users is greater than demonstrated.

Diversion

Figure 4 illustrates the tons diverted to the program and the total tons per year of compost sold since program implementation.

Five-Year Budget History

Tipping fee charges for yard debris are currently $27/ton. A $1/ton increase was effective January 1, 1998. In fiscal year 1997/98, this program accepted an average of 150 tons/day. Table 4 describes the total budget for five fiscal years, July 1 to June 30, with the expenditures and revenues for the Yard Debris Cost Center.

Looking to the Future

The existing Organics Material Processing, Composting, and Marketing Services Agreement between the Agency and Sonoma Compost Company/Empire Waste Management has been extended for two years until July, 2000. This will enable the IWD to complete the EIR for landfill expansion and make subsequent changes in the operations at the Central Disposal Site. Those changes include plans for relocating the composting operations to a site other than at the landfill. To prepare for this change, a Request For Proposals has been circulated for a program that provides comparable services to the Agency to be provided at a new location.

Hazardous Waste

Household Toxics Roundups (HTR)

Roundups are one day hazardous waste collections for households set up in large parking lots around the county. Prior to the formation of the Agency, many jurisdictions offered annual Household Hazardous Waste (HHW) Collection Events. In 1993 the Agency assumed responsibility for household hazardous waste management and offered 12 Household Toxics

TABLE 4. Yard Debris Program Budget from Fiscal Years 1992-97

	FY 1992-93	FY 1993-94	FY 1994-95	FY 1995-96	FY 1996-97
Summary of Expenditures	55,036	735,922	1,159,726	1,271,131	1,332,517
Summary of Revenues	0	712,048	1,234,866	1,318,254	1,474,171
Net Cost	55,036	23,873	(75,140)	(47,123)	(141,654)

Figure 5. Total participation 1993-97

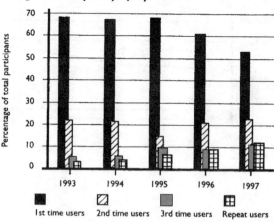

Figure 6. Frequency of repeat users

Roundups (HTR's) a year, with at least one in each jurisdiction (two in Petaluma and Santa Rosa). In 1997, that number was reduced to 11 events a year (one in Petaluma).

Reach

Over the last five years, the Agency's HTR's have served 26,935 participants (Figure 5), an annual average of 3% of Sonoma County households. Participation peaked in 1995, with 7,030 users, and has slowly declined since. Figure 6 demonstrates that half the participants are households that have not previously utilized a Roundup. During the period from 1993 to 1997, approximately 17,650 different households, or 10% of all Sonoma County households, have utilized an HTR.

Diversion

Total pounds managed by the HTR's increased annually until 1996 (Figure 7). While participation decreased in 1996, waste quantities were still increasing. Pounds per participant have increased in 1996 and 1997 (Figure 6). This correlates to a change in the law that allows participants to transport up to 15 gallons of waste, versus the 5 gallons allowed prior to 1996. Some participants obeyed the earlier law by making multiple trips to a single HTR so as not to exceed the transportation maximums. Each trip is recorded as a separate participant.

Five-Year Budget History

Table 5 describes the total budget for five fiscal years, July 1 to June 30, with expenditures and revenues for all of the programs included in the Household Hazardous Waste Cost Center.

TABLE 5. Household Hazardous Waste Program Budget from Fiscal Years 1992-97

	FY 1992-93	FY 1993-94	FY 1994-95	FY 1995-96	FY 1996-97
Summary of Expenditures	168,285	461,616	891,543	835,768	858,420
Summary of Revenues	120,000	505,070	838,117	711,439	780,544
Net Cost	48,285	(43,454)	53,140	124,329	77,876

Program Budget

The HTR's are currently operated by Philips Services Corp. under a two-year agreement that expires January 1,

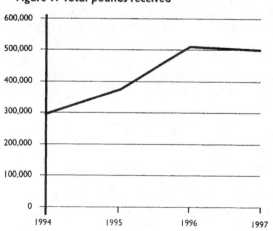

Figure 7. Total pounds received

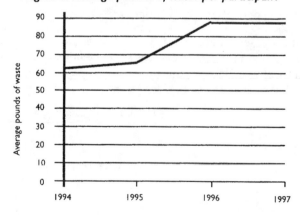

Figure 8. Average pounds of waste per participant

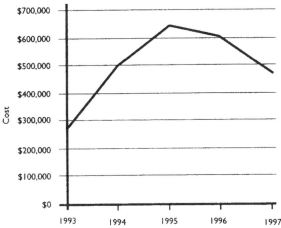

Figure 9. Total expenditures

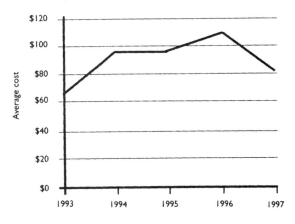

Figure 10. Average cost per participant

1999. The HTR's have two primary cost components: mobilization and disposal. Mobilization is a fixed cost based on the estimated and actual number of participants. Disposal is a variable cost based on the amount and type of waste received. Therefore, to track cost effectiveness, both cost per participant and cost per pound statistics are used. These numbers include both cost factors, but reflect different trends and economies of scale. For instance, for a particular event there may be a high cost per participant with a large waste volume, and with a low cost per pound, or vice versa. In general, the total expenditures over the last five years appear to reflect participation (Figures 8 and 10). However, the cost per pound has constantly decreased (Figure 11). When examining cost per participant (Figure 10), 1996 appears expensive, reflecting old contract pricing and increased pounds per participant. However, the cost per pound for 1996 demonstrates an economy of scale. In 1997, new contract pricing generated a decrease in both cost per pound, $0.93, and cost per participant, $81.

Successes and Future Goals

This program will continue under the current format until it is replaced by the permanent Household Hazardous Waste Facility at the Central Disposal Site, with its mobile collection program and the Toxic Taxi™ service.

Business Hazardous Waste Program (SQG)

In 1993, the Agency started a Small Quantity Generators (SQG) hazardous waste disposal program for businesses that generate less than 27 gallons of hazardous waste per month. These businesses can find hazardous waste disposal prohibitively expensive, since they generate such a small quantity. Business collections occur the day before HTR's at the same location. Businesses must make appointments and are charged disposal fees based on the type and quantity of waste received.

Reach

Between 1993 and 1997, the Agency's Business Hazardous Waste Program has had 666 participants and served 377 different businesses. Since the Agency primarily relies on press releases and environmental inspector to publicize the program, it is difficult to interpret the fluctuations in participation (Figure 12). The program is

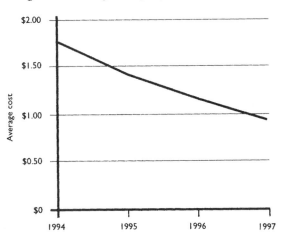

Figure 11. Average cost per pound of waste

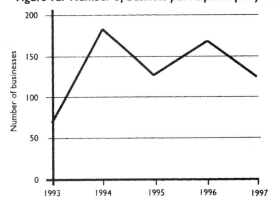

Figure 12. Number of business participants per year

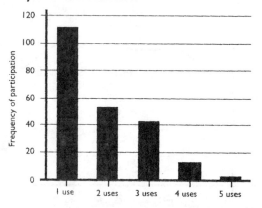

Figure 13. Frequency of participation by individual businesses

apparently popular, with the majority of businesses, 70%, utilizing the program more than once (Figure 13).

Successes and Future Goals

This program will continue as currently operated until the permanent Household Hazardous Waste Facility is built and operational. At that time, the program will be replaced by the HHW Facility, a mobile collection program, and Toxic Taxi™.

Household Hazardous Waste Facility and Mobile Collection Program

The Agency has made progress towards the goal of establishing a permanent HHW collection facility and mobile collection program with completion of facility design, participation in the EIR which is necessary to site, build and operate the facility, and awarding the facility operations contract.

Toxic Taxi™

In 1995 the Agency obtained an HHW Grant from the CIWMB for $30,384 to implement a pilot door-to-door service called Toxic Taxi™. A contract was awarded to Helabilt Environmental, which will provide door-to-door service for a fee when the HHW facility opens. The Toxic Taxi™ will serve residents and businesses and be self-supporting from collection charges.

Oil Recycling Centers

The Agency has received $727,089.39 in Used Oil Block Grants from the CIWMB since 1994. The 1997/97 grant was $146,057. The Agency has utilized some of these funds to expand the used oil collection network in the County. The 1993 Recycling Guide listed only 4 oil recycling centers, while the 1998 Recycling Guide lists 79 centers. Additionally, 29 of those centers accept used oil filters and 19 accept antifreeze.

The County has also expanded the used oil recycling centers at its disposal sites. The Central Disposal Site now collects used oil and used oil filters at the tipping area, as well as at Recycletown and all five transfer stations. Auto batteries are accepted at all disposal sites except Occidental Transfer Station. Additionally, oil grant funds have been utilized in several oil recycling campaigns using every media. Other grant funded programs include:

- Storm drain stenciling
- Maintenance manager training concerning rerefined oil use in fleet vehicles
- Retrofitting of buses and fleet vehicles with oil filter refiners
- GREEN brochure development
- School education programs
- Oil recycling center audits
- Support of city and county oil recycling centers
- Purchase of mobile oil filter crushers
- Purchase of oil tanks, oil filter collection containers, and oil tank sheds
- Support of the Eco-Desk's oil recycling voice mail system (English and Spanish)
- Printing and distribution of "No Toxics" garbage can stickers
- Support of HTR's
- Oil recycling center support (signs and halogen detectors)
- Staff training
- Fair booth displays and give-aways
- Support of an oil recycling intern

Curbside Oil Recycling Pilot

In 1997 the Agency obtained a Used Oil Opportunity Grant from the CIWMB for $103,000 to pilot a new technology in curbside oil recycling. Alternatives To Waste developed a new container, the Safety-Quick Oil Recycling System, that replaces the original drain plug in a vehicle with a new plug that has a valve. The oil container has a hose with the valve connect. Oil is removed by simply plugging in the container. The Agency is testing a prototype of this container in over 300 homes. Should the technology prove successful, the Agency has an agreement with Alternatives To Waste to share in the profits of the venture, allowing Sonoma County jurisdictions to inexpensively implement the program.

Paint Exchange Program

The latex paint exchange program has been expanded: paint is now accepted at the Annapolis, Central, Healdsburg and Sonoma disposal sites. Usable half-full to full latex paint cans are accepted from the public at no charge and redistributed to the public at no charge.

Safe Alternatives Education

The Agency recognizes that the long-term goal of its HHW management programs is to change the public's behavior regarding toxic products by reducing their use and eliminating their improper disposal. To that end, the Agency distributes a variety of brochures on safer alternatives. To strengthen the available information on safer alternatives, the Agency also dedicates staff time and effort to the California Peer Review Project, which accumulates scientific literature to support safer alternative recommendations.

The GREEN Brochure

The Agency worked with 13 other governmental agencies operating within Sonoma County in an informal organization named Governmental Resources Environmental Education Network (GREEN) to produce a poster/brochure that encompasses several environmental messages promoted by the different agencies. The brochure covers used oil recycling, household toxics and pesticides, as well as solid waste, air, and water issues. The brochure was distributed with the 1998 Recycling Guide and by the participating agencies.

Looking to the Future

The infrastructure for recycling of hazardous wastes has been significantly expanded in the last five years. Several new programs have been implemented, and the general awareness of oil recycling and HHW has been greatly increased.

The opening of the HHW Facility at the Central Disposal Site will represent a significant shift in HHW management. The HHW Facility, mobile collection program, and Toxic Taxi™ will represent a collection infrastructure more advanced than any other jurisdiction in California. As the implementation of the collection infrastructure is completed, the focus on education will increase.

Oil recycling funding from the CIWMB appears to be available indefinitely. Funds will continue to support existing collection programs. A new focus will support the development of curbside oil recycling programs. Remaining funds will support programs that transition jurisdictions to more sustainable oil use, such as installation of oil filter refiners and use of rerefined oil in fleet vehicles. Seeking and creating opportunities to work with other agencies to further education of the public about HHW and other environmental issues will continue.

Education

Five-Year Budget History

Table 6 describes the budget for five fiscal years, July 1 to June 30, with the total expenditures and revenues for all of the programs included in the Education Cost Center.

TABLE 6. Education Program Budget from Fiscal Years 1992-97

	FY 1992-93	FY 1993-94	FY 1994-95	FY 1995-96	FY 1996-97
Summary of Expenditures	25,528	53,913	181,356	297,140	264,305
Summary of Revenues	11,884	187,009	241,774	312,097	318,059
Net Cost	13,644	(133,096)	(60,418)	(14,957)	(53,754)

Christmas Tree Recycling

The segregation of Christmas trees for recycling started in 1989 as a volunteer effort coordinated by the Sonoma County Department of Transportation and Public Works and included curbside and drop-off collection. In 1992, the Agency assumed advertising responsibilities. In 1994, curbside collection of yard debris was implemented throughout the county. This collection system was then utilized to collect Christmas trees, rather than the volunteer efforts of previous years. To alleviate concerns that some residents would have difficulty cutting trees into a size that would not interfere with the mechanical, automated yard debris collection system, the Agency approached the Girl Scouts, Boy Scouts, and Middle Way (a local nonprofit supporting the disabled) about creating a fund-raising program to collect whole trees door-to-door. Collection by these nonprofit organizations, for a $5 donation, has been very successful. The Agency provides a recorded phone answering system to take appointments, while the nonprofit organizations manage the publicity.

Target Audience

Curbside yard debris collection is available in every jurisdiction except Cloverdale. Drop-off collection is available in every jurisdiction except Sonoma. A nonprofit collection opportunity is available in every incorporated city. Participation is not tracked due to the difficulty of obtaining accurate data.

Diversion

Christmas trees collected curbside, at drop-offs, or by nonprofits are processed in the Agency's organics processing program. Diversion is not tracked specifically for Christmas Trees due to the difficulty of obtaining the data.

Program Budget

From 1992 to 1996, between $2,000 and $5,000 was spent annually operating the Christmas Tree appointment hotline, printing posters and brochures, and placing ads in the *Press Democrat*. Beginning in 1997, the Agency transferred advertising responsibilities to haulers and nonprofits reducing the estimated annual expenditures for Christmas Tree Recycling to less than $300 (excluding labor).

Looking to the Future

Residents are well informed about Christmas tree recycling options and the program requires little effort or funding to implement. The program will continue without change indefinitely.

Small Change Theatre

Small Change Theatre (SCT) is a professional organization specializing in writing and performing educational programs for children. Actors performing in their programs are experienced in theater, as well as working with children, and there is some student involvement during performances. SCT contracted to provide educational services to elementary schools in Sonoma County for three years, 1994-1996. Their first year contract was funded through a grant from the IWD, with the second two years funded by the Agency.

SCT provided separate scripts, appropriate for kindergarten through third grade and grades 4 to 6, and worked with Agency staff to customize the scripts for Sonoma County. All three scripts provided information on waste reduction, recycling, and reuse and were titled as follows: 1996 "Trashzilla;" 1995 "Showdown at the 3R Ranch;" and, 1994 "Waste in Time." Each performance was approximately 25 minutes in duration. SCT was responsible for scheduling all presentations. Evaluation forms were provided to the schools after the presentations.

Target Audience

Sonoma County elementary schools, kindergarten through sixth grades.

Program Budget

TABLE 7. Three-Year Budget with Small Change Theatre

Funded by	Year	Amount	Number of Schools	Number of Shows	Number of Students
SCWMA	1996	$30,028	69	126	27,204
SCWMA	1995	$32,765	82	139	30,609
COUNTY	1994	$24,905	65	120	25,022

Looking to the Future

This program was very successful in reaching a large number of students and educating them in an entertaining fashion about the importance of waste reduction, recycling, and reuse. At about $1 per student, the program was also cost-effective. However, with comprehensive coverage of nearly all schools, the educational value was reduced since nearly all students in the County had participated. In 1997, the Agency decided to pilot a program emphasizing more diversion from schools, operational changes, and curriculum. However, it is expected that a theatrical-type program will resume in the future.

Public Education Coordinator

The Public Education Coordinator (PEC) position began in 1995 as a one year contract with the Agency combining a variety of tasks including the Recycling Guide, Eco-Desk hotline and resource library, SonoMax (Sonoma County Materials Exchange), and fairs.

Program Budget

TABLE 8. Public Education Coordinator Budget 1995-97

Task	Budget per year		
	1995	1996	1997
Eco-Desk and library maintenance	3,750	2,400	3,000
Eco-Desk telephone services	24,400	7,560	8,190
SonoMax and SonoMax WWW site maintenance[1] (1996 & 1997 only)	16,150	7,200	5,400
Recycling Guide	9,900	8,000	9,300
Fairs	9,300	3,900	6,000
Fairs assistance[2]		1,890	2,225
Meetings	4,700	1,800	1,800
Other	3,500	1,250	2,400
Registration and materials[3]	3,000	2,000	3,685
Total Budget	74,700	36,000	42,000

[1] Printing and mailing of SonoMax quarterly flyer is budgeted seperately (approximate cost for 3,365 = $1,136.00 in 1997).

[2] This cost is based on staff time at $10.00/hour for staffing booths.

[3] Represents fair registration and materials, mileage at $.31/mile, and $20/month for Internet service provider.

Recycling Guide

The Sonoma County Recycling Guide is a comprehensive 24-page hazardous waste, recycling, waste reduction, and reuse directory which has been distributed annually since 1995 to all county residences and businesses by bulk mailing. In 1993 and 1994, the Guide, produced by County staff, was distributed as an insert in the local newspaper. In 1995 the Agency assumed responsibility for the program through the PEC contract. The Guide is printed on newsprint (30% postconsumer recycled) using soy-based inks.

Target Audience

All Sonoma County residents and businesses.

Program Budget

TABLE 9. Printing and Distribution of the Recycling Guide

Year	Number Distributed	Number printed for mailing	Cost for printing/ preparation for mailing
1997	192,255	206,255	13,364.00
1996	192,829	207,829	14,854.59
1995	207,244*	218,000	17,087.00
	Cost for postage	Total cost for printing and distribution	Each
1997	22,044.00	35,408.00	0.17
1996	23,054.22	37,734.52	0.18
1995	25,125.00	42,212.00	0.19

*The number of Guides distributed in 1995 is higher than in other years, with the inclusion of post office boxes in city delivery areas which resulted in duplicate copies at some homes with post office boxes.

Eco-Desk

Background

Using a Microsoft Access database to record the activity, the PEC answers calls from the public from 12:00 p.m. to 3:00 p.m., Monday through Friday, retrieves messages from 15 voice mailboxes, returns calls within one business day, writes monthly and quarterly call tracking reports, and files new resource materials, including educational materials and recycled products information. The Resource Library is open by appointment during normal business hours. The Eco-Desk started in 1992 as a part-time county position. In 1995 the Agency assumed responsibility for the task with the PEC contract. The voice mail system features prerecorded information on a variety of topics including used oil recycling, paint, plastics, HTR's, SonoMax, recycling, and seasonal programs.

Target Audience

All Sonoma County residents and businesses. The following table and figure illustrates commonly asked questions and call summary:

TABLE 10. Commonly Asked Questions at the Eco-Desk 1995-97

Subjects/common questions	% of calls		
	1997	1996	1995
Household hazardous waste	37%	38%	31%
Recycling	30%	31%	23%
Hang-up calls (callers who hang up after listening to pre-recorded information)	12%	13%	N/A
SonoMax	8%	9%	N/A
Hazardous waste collection for businesses	5%	6%	N/A
Other (e.g., illegal dumping, community cleanup, etc.)	4%	4%	9%
Reuse (e.g., furniture donations, etc.)	2%	2%	N/A
Disposal (e.g., directions to the landfill, hours, etc.)	2%	2%	0.6%
Composting	1%	1%	1%

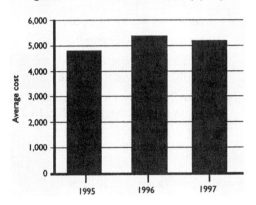

Figure 14. Eco-Desk call summary per year

SONOMAX Sonoma County's Materials Exchange 707.527.3375

The PEC recruits new listings, updates information, designs and produces quarterly materials listing newsletters, maintains the WWW site and coordinates with the statewide program, CALMAX. Patterned after the California Materials Exchange program, SonoMax was started with a County-funded grant to Garbage Reincarnation in 1993. In 1995, the Agency assumed responsibility for the program through the PEC contract. SonoMax is operated through the Eco-Desk Hotline. The goal of the program is to find reuse and recycling opportunities for discards (e.g., plastic bags, buckets, construction debris, etc.). A quarterly newsletter advertises materials *Available* and *Wanted*. Listers are asked to keep track of the number of calls received as a result of the ad and details of successful exchanges. In addition, listings are advertised on the SonoMax WWW site. The PEC updates this WWW site quarterly and when new listings are available.

Fairs

The PEC designs and manages construction of exhibits, recruits staff, and conducts training. The County began exhibiting at fairs and other public events in 1990. Since 1995, fairs have become the responsibility of the Agency within the PEC contract. Fair themes change each year to correspond with current activities in the county and have included 1995 "Be a Smart Shopper;" 1996 "Choose to Reuse;" and, 1997 "Beyond Bottles and Cans."

Fair/events attended in 1997 include:

April	Sonoma Valley Boys and Girls Club Earth Day event
April	Earth Day event at Optical Coating Laboratories, Inc.
July	Sonoma County Fair, July 22-31
October	Harvest Fair, October 3-5
November	America Recycles Day exhibit at WalMart, November 10-16
December	Plan Your Eco-Holiday display at the Santa Rosa Plaza, December 10-22

Target Audience

Sonoma County's residents and businesses.

Looking to the Future

This program has been successful in educating citizens in Sonoma County about the importance of waste reduction, recycling, and reuse and is expected to continue indefinitely.

Bay Area Creative Re-Use (BACR)

Bay Area Creative Re-Use/North Bay (BACR) facilitates the reuse of materials, supplies, equipment and furniture for educational, art, and other purposes by soliciting, collecting and distributing these items, and by developing creative applications for scrap materials. BACR, a nonprofit corporation, was one of four reuse depots in the Bay Area which originated in the late 1970's through a combination of funding from a California Solid Waste Management Board grant and the CETA program. It operated for 5 years in Santa Rosa until the building they were using was sold and the program funding ended.

BACR received a grant from the IWD in FY 1995-95 to reactivate the program. By the end of June 1995, past materials uses and donors had been resurveyed and a temporary (donated) site contained enough materials to initiate limited access in late August 1995. Between July and December 1995, BACR operated on a volunteer basis. The SCWMA awarded BACR a contract for calendar year 1996 to provide a variety of services including access to the warehouse, material pickup, outreach, and administration. The program continues to be funded by the Agency. Volunteers have been integral to this program assisting with the transfer of materials and equipment.

Reach

BACR collects materials from any Sonoma County business whose waste materials are useful to non-profit organizations, schools and city/county departments. The businesses (donors) receive on-site waste assessments, pick-ups on an individual basis, and an annual statement of materials donated with potential for tax benefits.

The recipients are nonprofit organizations, schools, child care programs, camps, convalescent/nursing homes, cultural/arts organizations, city and county departments, and others. Once a month or less, BACR conduct materials exchanges with the reuse organizations in Berkeley, San Francisco, or Sunnyvale.

Compared to 1996, the number of nonprofit organizations/schools/city departments that accessed materials increased 32%, the number of people obtaining materials increased 57%, the number of children and adults using the materials increased 60%, while the amount of materials acquired decreased 22% (Table 11). Compared to 1996, phone requests and correspondence increased 8% and 23% respectively. These increases demonstrate BACR's increased impact in our community.

Diversion

In 1996 a total of 104,868 lbs. of materials/supplies/equipment were collected from 122 businesses,

TABLE 11. Access to Creative Re-Use/North Bay Warehouse

	1996	1997
Access days	52	52
Number of people	838	1,138
Number of organizations	224	296
Number of end users	25,885	41,401

organizations or individuals. During 1997 a total of 82,289 lbs. of materials/supplies/equipment were collected from 130 businesses, organizations or individuals.

Program Budget

TABLE 12. Creative Re-Use/North Bay Budget 1994-97

July 1, 1994–June 30, 1995	25,000
January 1, 1996–December 31, 1996	30,000
January 1, 1996–December 31, 1997	40,000

Looking to the Future

From its inception, BACR has worked with SonoMax, to either refer businesses and individuals to the publication or as a referral when SonoMax did not 'fit' the business' diversion needs or time frame. During 1996, BACR was responsible for diverting a vast amount of fixtures, construction materials, flooring, and other items as a result of the deconstruction of a building in Santa Rosa. The Press Democrat published an article about this project and the publicity boosted BACR's visibility, while raising awareness in the business community about reuse.

In 1997 BACR's presence enabled companies like Lasercraft, Hewlett-Packard, and retail stores such as Home Depot and Riley Street to provide outdated inventory for reuse, rather than have it landfilled. BACR's user groups understand and respect that these items normally need to be destroyed to avoid the chance that they would reappear in the marketplace for sale.

An educational ripple effect is occurring—those obtaining materials are exposed to reuse and resource conservation concepts, and in turn pass their knowledge on to others. BACR's goal is to become fully operational, which includes having adequate space to retrieve more materials and store larger items, purchase a van, increase staff, and expand its programs.

Home Composting Program

The University of California Cooperative Extension is contracted by the Agency for its home composting program, which consists of providing residents with workshops, public education, and informational materials, in addition to maintenance of demonstration gardens. In fiscal year 1996-97, the Agency was approached by two home composting bin vendors to sponsor truck load bin sales, which were held on May 9, 1997 in the cities of Santa Rosa, Petaluma, Rohnert Park, Healdsburg, Sebastopol, and Sonoma.

These events reached two percent of Sonoma County residents, with 3,173 backyard composting bins sold.

During fiscal year 1997-98, the Agency funded a composting pilot program that will research and demonstrate the feasibility of on-site handling and composting of source-separated food wastes at selected commercial and institutional businesses. The CompTainer invessel composter will be moved from site to site to handle the materials generated by the various participants. The first site of this pilot program was the annual 4-H Chicken-Que held at the Sonoma County Fairgrounds. Additional participants include Mistral Restaurant, the Santa Rosa Junior College cafeteria, and the Food For Thought natural food grocery store.

Reach

This program targets residents throughout Sonoma County.

Looking to the Future

The composting pilot program, which has proven successful at all locations, intends to demonstrate that this program can be used by commercial and institutional businesses.

Planning

Planning was added to the Agency's responsibilities in fiscal year 1994/95 to allow the Agency to complete the Non-Disposal Facility Elements for all member jurisdictions. In 1996, the Agency decided to become a Regional Agency as defined as AB 939. As a Regional Agency, all disposal and diversion reporting and annual reports required by AB 939 are prepared and provided by the Agency. All reporting is based on the combined disposal, diversion, and programs of the member jurisdictions, not their individual accomplishments.

Three-Year Budget History

Table 13 includes the total budget for three fiscal years, July 1 to June 31 and includes the expenditures and revenues for the Planning Cost Center.

TABLE 13. Total Budget Planning for Fiscal Years 1994-97

	FY 1994-95	FY 1995-96	FY 1996-97
Summary of Expenditures	1,609	1,179	1,260
Summary of Revenues	5,829	5,258	6,853
Net Cost	(4,220)	(4,079)	(5,593)

Looking to the Future

The planning documents developed between 1993 and 1994 are the guidance documents that the Agency will utilize through 2000 (County Integrated Waste Management Plan (CIWMP), et al.). The Annual Report reviews the progress at implementing programs as outlined in the CIWMP and meeting diversion goals. The Agency will continue to prepare Annual Reports as required by AB 939.

Recycling Market Development Zone (RMDZ)

The primary mission of the RMDZ program is to help local government officials, economic development specialists, and solid waste management professionals develop strategies to attract and establish new businesses and industries that add economic value to secondary materials generated in California. By applying the basic principles of economic development, a RMDZ helps develop industries that will use post-consumer waste materials, thereby helping local jurisdictions meet the Statewide goals for the development of recycling markets.

The Sonoma/Mendocino Counties RMDZ was designed by the CIWMB in 1995. With the Sonoma County Economic Development Board acting as the Zone Administrator, the Agency provided funding to determine successful marketing programs in similar socioeconomic zones; develop and maintain a database of Sonoma County manufacturers who could be potential candidates for RMDZ loans; prepare and circulate press releases to promote workshops aimed at targeted businesses and industries; and work with local lending institutions to inform and create a network regarding the RMDZ.

In 1997, Lake County submitted a redesignation application to the CIWMB to expand the Sonoma/Mendocino Counties RMDZ to include Lake County. Upon approval by the CIWMB, Lake County assumed the Zone Administration responsibilities for the expanded Sonoma/Mendocino/Lake Counties RMDZ. With this change, Agency staff refers requests for information regarding the program to the Zone Administrator and the CIWMB.

Reach

The RMDZ program is designed to reach all businesses and industries located in Sonoma, Mendocino, and Lake Counties that have the potential of using secondary materials generated within the counties for use in the manufacturing of new and existing products.

Diversion

The RMDZ loan program funds up to 50% of the project costs, with a maximum of $1,000,000 for machinery and equipment, working capital, land, and refinancing of current debt. Prior to the addition of Lake County, the Sonoma/Mendocino Counties RMDZ was instrumental in assisting four businesses in acquiring loans for the RMDZ loan program.

Three-Year Budget History

Table 14 includes the total budget for three fiscal years, July 1 to June 31, and includes the expenditures and revenues for the Recycling Market Development Zone Cost Center.

TABLE 14. Total Recycling Market Development Zone (RMDZ) Budget for Fiscal Years 1994-97

	FY 1994-95	FY 1995-96	FY 1996-97
Summary of Expenditures	16,684	6,997	3,677
Summary of Revenues	11,416	24,136	9,646
Net Cost	5,268	(17,139)	(5,969)

Looking to the Future

With the designation of the Sonoma/Mendocino/Lake Counties RMDZ, it is hoped that the program will be able to help more businesses qualify for loans to help them use recycled materials collected from the various programs in their manufacturing processes.

"Why Recycle?"

Sonoma County Waste Management Agency, *Celebrate Earth Day 2000*, an advertising supplement to *The Press Democrat*, April 20, 2000. Reprinted by permission.

In the time it takes you to read this sentence, Californians will buy, consume and throw away 150 tons of raw materials. Before we manufacture these materials into products, we call them trees, rocks, hillsides and canyons. Afterwards, we call them garbage.
—William K. Shireman, President, Global Futures

What single environmental activity has produced all of these benefits and more for California in the last decade?

- Reduced CO_2 emissions by more than 140 million tons—the equivalent of taking 2.8 million cars off the road annually . . .
- Conserved enough energy to supply all of the energy needs for every home in California for 18 months . . .
- Saved more than 600 million trees—equivalent to a forest more than twice the size of Yosemite . . .
- Created 34,000 jobs and contributed over $1.6 billion to California's economy.

Recycling!

In response to concerns over dwindling waste disposal capacity in the late 1980s, California enacted a comprehensive framework of policies aimed at reducing and recycling waste.

The California Integrated Waste Management Act established the goal of reducing waste by 50% by the year 2000. According to the State Waste Board, waste reduction and recycling has already doubled from just 14% in 1989 to 32% in 1997.

California's innovative bottle bill recycling law has utilized market-based incentives to more than double the recycling rate for beverage containers. Since 1991, the Department of Conservation reports, beverage container recycling has exceeded 75%.

Today, the Department of Conservation reports, more than 8 million, or 70% of California households have access to curbside collection of recyclables, compared to less than 200,000, or just 2% a decade ago.

Additionally, California has led the nation in the establishment of minimum recycled content standards for a wide range of products, including newspaper, glass, and even some plastics.

Few environmental policies have produced such immediate results as recycling—Californians are annually disposing of nearly 10 million fewer tons of solid waste today than a decade ago—and fewer still have proven to be as visible and popular with the public. In a statewide opinion survey by J. Moore Methods, 88% of the public said they regularly do some recycling. In another survey, respondents ranked recycling as the most important activity they could undertake to protect the environment.

While California's recycling efforts may have—for the time being—eclipsed any threat of a disposal "crisis," recycling's greatest benefit and potential lies with its ability to conserve resources and reduce pollution and environmental degradation associated with material production.

Today, a participating curbside household will annually recycle 520 lbs. of paper, 180 lbs. of glass, 26 lbs. of plastic, and 2.2 lbs. of aluminum cans. Virtually all of this material is being returned to the economy in the form of new products, thanks to increased consumer demand for Buying Recycled.

More important than the diversion of this material from landfill, this effort will make a substantive contribution to environmental protection: from saving trees to reducing pollution and conserving energy. And recycling may well be one of our most practical and effective tools for combating global warming.

Energy Savings

Recycling saves energy throughout the production cycle because recycled materials require less processing than raw "virgin" materials.

According to a report by the Environmental Defense Fund, recycling a ton of material collected by a typical curbside recycling program saves the energy equivalent of $265 worth of electricity, petroleum, natural gas and coal—that's after deducting the energy used to collect materials.

Conservation of Resources

Every ton of office paper recycled into new paper products saves 17 trees and reduces water usage by 7,000 gallons (Natural Defense Council and others).

According to the Steel Recycling Institute every ton of steel recycled saves 2,500 pounds of iron ore, 1,400 pounds of coal, and 120 pounds of limestone. Americans recycled 1.7 million tons of steel packaging in 1997, enough steel to rebuild the Golden Gate Bridge 20 times.

According to the California Department of Conservation, Californians recycle nearly 400 tons of aluminum cans every day. That's enough aluminum to produce four Boeing 747 jet airplanes.

Reducing Pollution

It's a fact: Recycling reduces pollution. In addition to reducing air and water pollution associated with landfills, using recycled rather than virgin materials reduces the amount of pollutants emitted during resource processing and product manufacturing.

According to a 1992 study by the Tellus Institute for the U.S. EPA, using recycled glass in the manufacturer of glass bottles rather than virgin materials reduces the amount of four criteria air pollutants (CO_2, NO_x, Particulates, and SO_x).

Using recycled aluminum rather than virgin materials decreases environmental management costs attributed to air pollution by 80%.

Pollution Reduction for Recycled vs. Virgin Materials

	Air Pollution	Water Pollution
Aluminum	95%	97%
Paper	74%	35%
Glass	20%	NA

Reducing Greenhouse Gas Emissions

In the past year a global consensus has been reached on reducing greenhouse gas emissions to combat global warming.

According to the U.S. EPA, the largest component of greenhouse gas emissions is carbon dioxide (CO_2) with a third of carbon emissions coming from fossil fuel burning within the industrial sector. Using recycled materials requires substantially less energy in the manufacturing process than virgin materials. As a result, the increased use of recycled materials means far less CO_2 will be released into the atmosphere.

According to the California Integrated Waste Management Board, California is projected to recycle over 25 million tons of material in the year 2000. By recycling this amount, we can reduce CO_2 emissions by 31 million tons, equivalent to the CO_2 emissions produced by 6.2 million cars annually.

According to the U.S. EPA, methane is the second largest source of greenhouse gas in the atmosphere, and landfills account for 37% of methane gas output. By reducing and recycling properly organic materials, including paper, we can divert them from landfill, thereby reducing anaerobic decomposition and the production of methane gas.

Cost Savings

For many businesses, recycling has proven to be cost effective—using less energy to obtain and process materials, and reducing pollution and waste management costs. Here are a few examples.

Every year, the makers of shipping pallets consume half of all hardwood and 10% of all lumber used nationally. At Eastman Kodak, redesigned shipping pallets and altered stacking patterns saved over 7 million pounds of wood and $380,000 in a single year.

Fetzer Vineyards, a worldwide producer and marketer of wines, has reduced its waste by 93% since 1990, saving more than $115,000 in disposal fees.

At Hewlett-Packard's Roseville manufacturing facility, 93% of packaging waste generated is now re-used or recycled, saving the computer giant $1.45 million in disposal costs.

With trash collection and disposal costs exceeding $100/ton in some communities, waste reduction programs, including recycling and composting, can represent a cost effective alternative to traditional waste disposal.

Jobs

According to the California Integrated Waste Management Board, cutting California's waste in half by the end of the decade is projected to create 45,000 new jobs in secondary material collection, processing and manufacturing.

Conclusion

The benefits of recycling go far beyond the materials we keep out of the landfill. According to the Congressional Office of Technology Assessment, the more than 200 million tons of household and commercial garbage generated each year comprises only 1% of the total waste stream in the United States—just the tip of a 7.6 billion ton wasteburg. Nearly 60% of that is industrial solid waste, of which the largest producers are paper and steel manufacturing. Every ton of material recycled and returned to the economy as a raw material may well displace many times its eight in virgin resources, industrial waste and pollution. By recycling and buying recycled products, we cut down on all of the resource, energy and pollution impacts that occur from resource extraction, to material processing and refining to manufacturing.

Whether it's the reductions in pollution and greenhouse gas emissions, the opportunity for cost savings and job creation, or the conservation of energy and material resources, the benefits of waste prevention and recycling are enormous—and go far beyond the simple goal of keeping materials out of the landfills.

Resources:

United States Environmental Protection Agency, Region IX, Office of Solid Waste
Californians Against Waste Foundation
U.S. Environmental Hotline
California Integrated Waste Management Board
California Department of Conservation: Division of Recycling

reading 4

"Buy Recycled—On Earth Day and Every Day, Shop the Recycled Way"

Sonoma County Waste Management Agency, *Celebrate Earth Day 2000,* an advertising supplement to *The Press Democrat*, April 20, 2000. Reprinted by permission.

Thanks to you, recycling is working! The proof is that the paper, plastic, steel, aluminum, and glass that you've been recycling is now made into all sorts of everyday products and packages, which are available on the market right now. There's just one thing left to do. Buy them! That's the "cycle" in recycling. You sort out recyclable materials, your city or town collects them, and manufacturers buy them to make into products again.

You may have already heard the term "Buy Recycled." Why is "Buy Recycled" important? It's simple. In order for the materials that have been collected through community and office recycling programs to have value and get used in the manufacturing of products, there needs to be a demand for those new recycled content products. You create that demand by purchasing products made and/or packaged with recycled paper, steel, aluminum, glass, and plastic.

Look Closely!

Some products made with recycled content are labeled with the above symbol and the words "postconsumer" recycled content. While watching for this symbol helps, not every product made from recycled content is labeled as such. In fact, you may be buying recycled without even knowing it. For example, in California, the average aluminum can is made from up to 55% recycled aluminum.

The following are just a few examples, since hundreds more are available, of recycled products you can buy:

- Building materials—plastic lumber, recycled paint, carpet made with recycled soda bottles
- Garden and yard supplies—hoses, mulch
- Office supplies—copier and printer paper, remanufactured toner cartridges

For more information about how to buy recycled, call the Eco-Desk 565-DESK(3375).

"Earth Pledge: Twelve Things You Can Do Today to Fight Global Warming and Environmental Destruction"

Sonoma County Waste Management Agency, *Celebrate Earth Day 2000*, an advertising supplement to *The Press Democrat*, April 20, 2000. Reprinted by permission.

Every day is Earth Day. By understanding the link between our personal lifestyle and consumption patterns and environmental destruction, we can make better, more sustainable choices. Many of the most environmentally harmful decisions are made in corporate boardrooms, far from our influence—but our individual actions can have tremendous impact when we join together as workers, voters, consumers, and public citizens. Take a few moments to look at how you can make a difference, and then consider making a pledge—to the Earth, to future generations, to the other species, which share our precious ecosystem. Together we have the power to make this the century of ecological balance, restoration, and sustainability.

Transportation

1. Drive Less

The single most important action you can take for the planet is not driving your car. Auto exhaust accounts for about half the smog in urban areas; burning one gallon of gasoline produces 20 pounds of carbon dioxide, a greenhouse gas which causes global warming. Our petroleum dependency affects everything from health insurance costs to U.S. foreign policy. So walk, ride a bike, or take the bus. Riding mass transit uses 25 times less energy than taking the car. Consolidate trips, and carpool or share a ride when you can.

2. If You Must Drive, Drive Efficiently

A tune-up on your car will improve its fuel economy by 6 to 9 percent and save you repair costs in the long run. Slow down—for every mile per hour you drive slower than 65 MPH, you improve your car's fuel efficiency by about 2 percent. Gradual acceleration is far less polluting than rapid, start-and-stop driving, so use a light touch on the gas pedal. And soft tires make the engine work harder, making your car more wasteful, so pump 'em up! Drive a cleaner, more fuel-efficient vehicle: SUV's pollute 2 to 3 times as much as autos, and diesel engines release up to 10 times as much of the most carcinogenic pollutants as gasoline engines.

3. Take the Clean Car Pledge

Clean car technology is here! California's Zero Emissions Vehicle program requires auto makers to begin manufacturing and marketing cleaner, more fuel efficient vehicles. But we still have a long way to go. Nearly half of California's global warming pollution comes from automobile pollution. Car companies are still dragging their feet on production, claiming that consumers don't want to buy clean cars. Prove the car companies wrong by pledging to consider a clean vehicle when you next purchase a car. [See the hybrid electric vehicles on display at the Santa Rosa Earth Day Festival, or check out the Union of Concerned Scientists web site at www.cleanpledge.org to learn more about Zero Emissions Vehicles.]

Energy

4. Conserve Energy at Home

The typical U.S. family spends close to $1,300 per year on natural gas and electricity for a home. Save by turning the thermostat down, and turning off lights and appliances when they're not in use. Switch to compact fluorescent bulbs to use 25% less energy. Weather-strip and insulate your home, wrap your water heater, and buy energy-efficient appliances. If buying a home, consider a smaller, more energy-efficient house. [To find out more about energy conservation and home energy audits, visit the PG&E booth at the Santa Rosa Earth Festival, or check out their web site: www.pge.com.]

5. Switch to a 'Green' Electricity Provider

Every time you switch on a light in your house, you contribute to global warming. Electricity generation is the most polluting industry in the U.S.; about half our electricity is generated from dirty coal plants, and most of the rest is produced by nuclear and natural gas plants. These sources contribute about two-thirds of the emissions associated with global warming, one-third of the pollution that causes acid rain and smog, and over half of our nation's nuclear waste. You can choose to purchase clean, renewable power from an electricity provider that supplies green certified clean power fueled by the sun, wind, geothermal steam, biomass and small-scale hydroelectric generation. Each household that switches to 100 percent green power reduces the carbon dioxide released into the atmosphere every year by three tons! Switching to green power is one of the easiest and most powerful environmental actions you can take for Earth Day 2000. [Visit the 'Green Power Zone' at the Santa Rosa Earth Day Festival for more info, or check out www.cleanpower.org or www.gogreenpower.org to switch "on line."]

6. Ask Governor Gray Davis and Your Local California Legislator to Support Clean Power

Let the Governor hear from you! Two critical programs are at stake this year: continued funding for renewable power generation and the Zero Emissions Vehicle (ZEV) mandate. ZEV requires auto makers to produce and market clean, zero emissions cars. Continued renewable power funding promotes the rapid growth of a clean, renewable power market to help reduce air pollution. Your help is needed to demonstrate strong public support for improved California air quality. Let your voice be heard—write to the Governor today! [Visit the 'Green Power Zone' at the Santa Rosa Earth Day Festival to write the Governor and voice your support for clean cars and renewable power now. You can find your state representative and the Governor's address at the state web site: www.state.ca.us]

Food

7. Eat Less Meat

The way that animals are bred for food pollutes our environment while consuming huge amounts of water, grain, petroleum, pesticides and drugs. The cleanup costs associated with the water and air pollution caused by factory farms are paid for by taxpayers in the form of new water treatment plants and visits to the doctor. 40% of the world's grain is fed to livestock, poultry or fish; decreasing consumption of these products, especially beef, would free up massive quantities of grain and reduce pressure on farmland. Cattle grazing is decimating public lands in the U.S. and rainforests worldwide, while over-fishing is damaging ocean ecosystems. Cutting down on meat is not only good for your health, it's good for the planet. Commit to reducing your meat consumption by at least 10%. [For more info, visit the vegetarian and animal rights booths at the Santa Rosa Earth Day Festival, or check out www.earthsave.org]

8. Buy Organic/Locally Grown Foods

Petrochemiconal and poison-intensive industrial monocrop farming is destroying our topsoil and poisoning our land, water, food, and our bodies. Purchasing organically grown food is voting with your pocketbook for a non-toxic environment. Buying locally grown foods helps support local farmers and keeps your dollars in the local economy. Buying locally and in season also re-connects us to the land we live on, reminding us that all our food started out as water, soil, and sunlight in a particular spot on Earth-and that we must protect that little piece of the planet which sustains us. [For info on local and organic, visit the Community Market booth at the Santa Rosa Earth Day Festival, or check out www.caff.org, and www.ccof.org.]

9. Avoid Genetically Altered Foods

About half of all the soy and one third of the corn acreage planted in the U.S. this year is genetically altered. Much of the canola oil in the U.S. market is from genetically altered plants. These products are in many processed foods, so chances are you're already consuming them without knowing it. In Europe, a coalition of farmers, consumers, and shop owners have successfully fought against these 'franken-foods', and many stores no longer stock them. Last May the British Medical Association called for a moratorium on genetically engineered (GE) foods, warning that the commercialization of untested and unlabeled gene-foods could lead to the development of new allergies and antibiotic resistance in humans. Large corporations like Monsanto have introduced GE foods as part of a larger political and economic campaign to secure patent and intellectual property rights on the genetic blueprints of altered organisms. These firms then 'own' the resulting 'transgene' foods, seeds, or other products, and sell them for profit. Third world farmers are forced to buy new seed each season from these corporations, rather than replanting seeds from their harvest. Many genetic alterations are designed to make the crop more tolerant of pesticides, particularly Roundup, which Monsanto also manufactures, so that greater quantities of pesticides can be u.s.ed. The altered gene is often linked to another, antibiotic-resistant marker gene; researchers warn that these genes may be recombining with disease-causing bacteria in the environment, or in the guts of animals or people who eat GE food, increasing the threat of antibiotic-resistant disease epidemics. Demand that your grocery store label all GE foods, and refuse to buy them. Santa Rosa's Community Market is the first store in the nation to label all foods as GEfree. For more info, visit the Earth First! Booth at the Santa Rosa Earth Day Festival, or check our www.biotech-into.net, www.biointegrity.org, or www.ccof.org/genetic_eng.html.

Waste

10. Reduce, Recycle, Buy Recycled and Compost

Reduce. The best thing you can do for the environment is to reduce consumption. By simply choosing to use less stuff, limited resources aren't consumed and there's less to throw away. Remember to donate items you no longer want to charities as it's always better to reuse something before it is recycled and give others an opportunity to get a benefit from your old things. For a list of local charities and locations to drop off reusable discards, check the Sonoma Country Waste Management Agency's web site, www.recycle.now.org. or call the Eco-Desk at 565-DESK(3375).

Recycle. What single environmental activity saves energy, conserves resources, reduces pollution, reduces greenhou.s.e gas emissions, and creates jobs—recycling! Do you know that waste paper comprises 27% of Sonoma County's overall annual waste stream; that's 111,000 tons of paper, most of it recyclable office paper, junk mail, newspaper, paperboard cartons, magazines and cardboard. Every ton of office paper recycled into new paper products saves 17 trees and reduces water u.s.age by 7,000 gallons. Pledge to recycle all your newspaper, junk mail, cardboard and magazines. For more information call the Sonoma County Eco-Desk 565-DESK(3375) and get a copy of the Recycling Guide 2000.

Buy recycled to close the recycling loop. Collection of recyclables through residential and drop-off recycling programs is just one step in the recycling process. As important is buying the products that are made from the recyclables; a few examples are copier and printer paper, remanufactured toner cartridges, building materials and mulch. For many more examples visit http://www.ciwmb.ca.gov/rcp/.

Compost to save landfill space and combat global warming. Accumulation of greenhou.s.e gases is responsible for global warming. According to the U.S. EPA, methane is the second largest source of greenhou.s.e gas in the atmosphere and landfills account for over a third of methane gas output. By reducing and recycling organic materials, such as kitchen trimmings, leaves and yard debris, we divert material from the landfill thereby reducing anaerobic decomposition and the production of methane gas. For more information on how you can start composting, call the Sonoma County Master Gardener's Information Desk at 565-2608.

11. Keep a Non-Toxic Household & Garden

Reduce the use of pesticides:

- Build healthy soil by adding compost or aged manure and use slow-release organic fertilizer.
- Buy plants that will grow well in this climate and in your yard. Try non-chemical products to control pests such as Teflon tape for root weevils or soap and water solutions for aphids.
- Visit www.recyclenow.org/less-toxic for more information on less toxic ways to manage specific pests.
- Avoid products marked "poison" or "danger."
- Use tools rather than chemicals whenever possible. For example, use a squeegee and vinegar & water.
- Prevent the need for strong chemicals. For example, line your oven with foil to capture drips.
- Purchase latex paint instead of oil-based paint. Latex paint is less toxic and cleans up with water.

12. Dispose of Toxins Properly

Use non-toxic or less-toxic products and tools whenever possible.

- Buy only what you can use.
- Use up what you purchase.
- Give away what you can't use.
- Recycle your used motor oil and filters at one of 74 locations around Sonoma County.
- Bring what is leftover to a Household Toxics Roundup. Call the Eco-Desk, 565-DESK(3375), or visit www.recyclenow.org for Hou.s.ehold Toxic Roundups locations and dates.

reading 6

"Elder Earthkeepers Honored with Gifts From the Earth"

Sonoma County Waste Management Agency, *Celebrate Earth Day 2000*, an advertising supplement to *The Press Democrat*, April 20, 2000. Reprinted by permission.

Respect and sustain the earth so we can long enjoy its gifts. That theme was central to an April 16th Earth Day 2000 celebration honoring 20 Sonoma County Elder Earthkeepers of the Twentieth Century. The event, held at Luther Burbank's Experimental Garden in Sebastopol, was hosted by Earth Elders, a local and global network founded in Sonoma County in 1998 to honor aging, elders and the earth.

Connie Mahoney, Ph.D., founder of Earth Elders explained, "The turn of the century is a perfect time to recognize elder environmentalists as role models who can inspire all of us to continue the important earth-keeping work necessary to sustain life on this planet for future generations."

Honorees were selected from a field of more than eighty nominations submitted by citizens of Sonoma County. A selection committee undertook the difficult task of reviewing the nominations and choosing twenty elders to be honored. Congresswoman Lynn Woolsey presented the awards.

Hand-crafted bowls from fallen native madrone trees (a gift from the earth) were presented to each of the twenty elders honored. Brad Lundborg, a retired physician and member of Earth Elders, designed the bowls exclusively for Earth Elders. "Each bowl is unique and changing. It is a gift from the earth that continues to give. The same can be said for each of the honorees and all of us," Lundborg said.

Lundborg began gathering wood on his Heraldsburg property and hand turning the bowls several years ago because doing so fit in with the relationship he and his family have to the land. Making the bowls and finding "good homes" for them has become more than just a hobby.

In addition to the bowls, live native trees donated by Circuit Rider Productions in Windsor will be planted in the fall at twenty neighborhood schools (a gift to the earth) throughout Sonoma County as a perpetual tribute to the honorees.

"We feel this was an important and appropriate way to thank these elders. Sonoma County is a wonderful place today because of what these people did in the last century," says Mahoney.

Wendy Osmann is general manager of Brighton Gardens by Marriott, a corporate sponsor. She says, "We supported this Earthkeepers recognition because it respects older adults and highlights their positive contributions." The celebration was co-sponsored by Council on Aging, the Interfaith Elder Care Task Force, Retired Senior Volunteer Program (RSVP), New College North Bay and the SSU Gerontology Program.

reading 7

"What is Bay Area Creative Re-Use?"

Sonoma County Waste Management Agency, *Celebrate Earth Day 2000,* an advertising supplement to *The Press Democrat,* April 20, 2000. Reprinted by permission.

Bay Area Creative Re-Use (BACR) is the outgrowth of a cooperative effort of people in the late 1970s from education, industry, government and community organizations throughout the nine San Francisco Bay Area counties. Its mission is to facilitate the reuse of materials, supplies and equipment for educational, art and other purposes by soliciting, collecting, distributing, and developing creative applications for discarded industrial materials. The recipients are nonprofit organizations, parks and recreation departments, schools, day care centers, museums, homeless shelters, senior citizen groups, camps, convalescent/nursing homes and cultural/arts organizations. BACR increases environmental awareness by promoting resource conservation and waste reduction.

Where Do the Materials Come From?

Industries and businesses generate vast amounts of scrap and surplus materials, supplies and equipment which are not currently being used. Manufacturing byproducts, over-runs, out-dated stock and rejects are thrown away when they could be utilized for educational and community service purposes through this system for collection, warehousing, dispersal and education.

What Kinds of Surplus Are Retrieved?

Paper, Formica, corks, thread, wood, yarn, theatre sets, rug samples, leather trimmings, sheet acrylic, tile, Mylar, glass tubes, wire, matte board, zippers, containers, fabric. . . materials as well as finished products, excess inventory, office supplies and equipment. These items currently go to the landfill instead of fueling the community with supplies.

Reuse Benefits Everyone!

Businesses receive on-site waste assessments; pickups on an individual basis; annual statement of materials donated with potential for tax write-off.

Organizations and schools receive much needed materials, supplies and equipment and reduce what has to be purchased.

Industrial discards are used to teach math, science, art, environmental and interdisciplinary subjects.

Resource conservation and waste prevention is promoted through the reuse of materials and equipment.

Businesses reduce disposal costs by diverting from the landfill what was once thought of as "waste."

Cities and counties address AB 939 goals to reduce what goes into the landfill by removing these discards from the solid waste stream.

How to Get Involved

Join the monthly brainstormers group to conceptualize uses for the sometimes wild or not so wacky materials that are retrieved.

Create school curricula incorporating available materials.

Develop and present a class or workshop teaching specific materials' reuse.

Help raise funds or make a personal tax deductible donation.

Call 546-3340 for more information.

reading 8

"Environmentalism Begins at Home: Green Building Represents a Vital Earth-Friendly Action"

Sonoma County Waste Management Agency, *Celebrate Earth Day 2000*, an advertising supplement to *The Press Democrat*, April 20, 2000. Reprinted by permission.

Editor's Note: For more on green building, see the Part V interview with William Browning of the Rocky Mountain Institute in *Environment* 3e by Raven and Berg.

Many Sonoma County residents recycle and compost to help the environment, some carpool or ride bikes, and others purchase paper and other products with recycled content. Yet often it doesn't occur to us to put the object with the greatest impact under the environmental magnifying glass: our homes.

According to a paper by the Worldwatch Institute, buildings account for one-sixth of the world's fresh water withdrawals, one-quarter of its wood harvest, and two-fifths of its material and energy flows. "Green building" gives us an opportunity to construct homes using more environmentally-sound materials, renovate existing buildings with these products, and create structures that require less energy, water and other resources in the future.

A variety of elements can contribute to making a home a "green building." Some examples include designing a house with a shape and orientation that promotes energy efficiency; incorporating lighting and appliances that are energy efficient; building with environmentally-sensitive materials; and including construction and interior finish products that are better for air quality than standard products. A brief overview of these opportunities is provided below.

Housing design—Houses can be constructed in ways that take advantage of their site to save energy. For example, a house can be oriented to get the maximum benefit from sun exposure in terms of heat and natural light. Weatherizing doorways and windows with caulking and adding insulated shutters helps a home retain heat, reducing energy costs. Strategic placement of trees and shrubbery provides a windbreak in the winter and cooling shade in the summer months. Finally, designing a smaller home and yard typically will require fewer resources—construction materials, energy for heating and cooling, irrigation—than a larger home.

Lighting—Installing high-efficiency lighting systems can deliver a fairly short-term return on investment for a homeowner. An analysis by Real Goods shows that the lifetime cost (cost of bulb plus electrical costs) of a compact fluorescent light is often less than half the lifetime cost of a comparable incandescent light—$41.50 for a 25-watt compact fluorescent versus almost $96 for a 100-watt incandescent, for example. Adding strategic solar-powered lights, such as for illuminating driveways or sidewalks, makes home lighting even more environmentally sound-and lowers the homeowner's electric bill.

Sustainable Construction Materials—Homeowners can save both money and resources by choosing reused building products. Salvage businesses often carry home products—cabinetry, sinks, bathtubs, doors and windows—that not only cost less than less than new products, but also can give a home more character. (Imagine a clawfoot tub in your bathroom!) When reused products aren't available, homeowners and builders can choose from a variety of recycled-content products, which help provide markets for recycled materials that are being diverted from California landfills.

Air Quality—Many building materials release toxic gases, such as volatile organic compounds and formaldehyde, which can have a detrimental effect on the health of the building's occupants. Green building materials often are designed to avoid these toxic emissions. Careful selection of construction materials and interior finish products can enhance indoor air quality.

Homeowners who are concerned about environmental issues should encourage their architects, builders and contractors to create a house with these green buildings issues in mind. Homeowners, building professionals and members of the general public are invited to learn more through the Sonoma County Waste Management Agency's Green Building Education Series. Visit the SCWMA Web site—www.recyclenow.org—or contact C2 Alternative Services at c2alts@pacbell.net or by calling 707 568-3783.

"Helping Business Donate/Sell Waste Material: Sonoma County's Materials Exchange, www.recyclenow.org/SonoMax"

Sonoma County Waste Management Agency, *Celebrate Earth Day 2000*, an advertising supplement to *The Press Democrat*, April 20, 2000. Reprinted by permission.

SonoMax is a free service helping local businesses find reuse and recycling opportunities for materials typically discarded, such as empty containers or manufacturing by-products. The goal of the program is to give these too-good-to-toss materials a second chance thereby keeping them out of the landfill.

To facilitate exchanges, a quarterly ad-style newsletter listing local materials AVAILABLE and WANTED is mailed to over 1,000 people throughout the county. It is free to list in SonoMax and many of the items listed are free or low cost. Material categories include construction, containers, durable goods, electronics, glass, metal, organics, paint/wax, paper, plastic, rubber, textiles, wood and miscellaneous. All pricing (if any) and transportation is negotiated between the interested parties.

In addition to the listing flyer, the SonoMax web site, http://www.recyclenow.org/sonomax/ allows visitors to search materials Available and Wanted and to create new ads online. Listings are updated weekly making the web site the best place to view current ads.

All local listings are shared with CALMAX (California's Materials Exchange) and may appear in their statewide catalog and on the CALMAX web site.

Examples of Exchanges:

- **Developer gives away job leftovers and saves on disposal costs.** A local developer listing in the AVAILABLE section of SonoMax donates leftover and scrap materials from new residential construction projects to local non-profits and community groups. Since placing an ad, on-site construction waste has been reduced by 50%, thereby saving money on hauling and disposal fees.
- **Pine crates from LaserCraft are turned into horse troughs.** In addition to listing available horse manure in SonoMax, equestrian Marc Ripens obtains pine crates from LaserCraft which he utilizes on his ranch as horse feeders.
- **Blueprint paper reused for kid's drawings.** The North Coast Builder's Exchange listing in the AVAILABLE section of SonoMax makes used plan sets available to interested teachers and parents.
- **Surplus paint reused by community organizations.** 11,000 gallons of interior and exterior latex paint were recently donated to a Graffiti Fighter's program and to a community church project.
- **Trees from Sonoma County Jail Industries donated all over the county.** Trees grown by adult inmates are donated all over the county to plant on public property. The list of recent donations include: Santa Rosa city streets (7 trees), Bellevue School District (18 trees), Sheppard School (7 trees), Food Link (700 trees).

To place a free ad, to obtain a free flyer, or for more information call the Sonoma County Eco-Desk at 565-DESK(3375) or visit http://www.recyclenow.org/sonomax/.

SonoMax is a business waste diversion program of the Sonoma County Waste Management Agency.

Questions for Discussion

For Reading 1

1. This paper gives additional background information on the magnitude of problems associated with solid waste management throughout North America. Summarize this author's assessment of the seriousness of the problem.

For Reading 2

1. Describe how the Sonoma County Waste Management Agency is organized. How is it funded?

2. How did the solid waste stream change in between 1991 and 1995/6?

3. What happened to these wastes? Describe the various projects and activities sponsored by the Agency. How effective has each activity been?

4. What else could Sonoma County do to further reduce its production of solid wastes?

For Readings 3 Through 7

1. These readings appeared in "Celebrate Earth Day 2000," produced by the Sonoma County Waste Management Agency and distributed to county residents on Earth Day 2000. What was the intent of these articles?

2. Briefly describe the ways, suggested in these articles, that individual citizens can become more involved in recycling.

3. Describe your own recycling efforts. How might you become more involved in recycling?

Going Beyond the Readings

Visit our Web site at www.harcourtcollege.com/lifesci/envicases2 to investigate solid waste disposal issues in additional regions of the United States and Canada.

Northeast: Acid Precipitation

Introduction
Readings:
Reading 1: EPA 1998 Compliance Report: Acid Rain Program
Reading 2: Emissions Trading of Sulfur Dioxide: The U.S. Experience
Reading 3: Acid Rain: A Continuing National Tragedy
Reading 4: A Washington, D.C. Press Conference
Reading 5: Public Service Announcement Campaign A Ringing Success
Questions for Discussion

Minimizing Acid Precipitation in the Adirondacks

Introduction

Appearances are sometimes deceiving. Lakes in the Adirondacks region of New York look much as they always have. They are still crystal blue; still reflective of dawn's first light on a still morning. Jump into one on a hot summer day and its waters are still cool, wet, and bracing. But all is not well with the lakes, ponds, and waterways of the Adirondacks. In many, there are few or no fish. Neighboring trees are dying. The culprit? Acid precipitation.

Far to the west in the Ohio River valley, factories and power plants, largely fueled by coal, have for decades spewed an unpleasant brew of air-polluting smoke. When local citizens and governments complained, these industries built smokestacks hundreds of feet (hundreds of meters) high, to waft smoke high enough that winds carried the problem out of the immediate area. A local problem was solved, but at the expense of those downwind.

Some of the smoke's components, namely sulfur dioxide and oxides of nitrogen, react with water vapor in the atmosphere to form sulfuric and nitric acids. Downwind, over the Adirondacks, these acids precipitate, either dry as tiny crystals or wet dissolved in rain, snow, or sleet. In recent decades, the acidity of rain in the Adirondacks has steadily increased. So have problems.

Background

Acids in rain and melted snow soak into soils and cause several kinds of mischief. Normally, minerals and heavy metals, such as aluminum, mercury, and lead, are held in soils chemically, but acids remove these elements from their binding components.

Free aluminum is taken up by plant roots, especially trees. The aluminum then clogs the trees' vascular systems, diminishing their ability to take up other nutrients and water. Thus weakened, trees lose their ability to tolerate extreme cold, diseases, and insect pests.

Many trees form symbiotic relationships with certain fungi. In exchange for photosynthetic products, especially glucose used for energy, fungi assist trees in gathering nutrients and water. Acidic soil kills symbiotic fungi to the detriment of trees.

Normally, microscopic bacteria, fungi, and other organisms break down leaf litter and recycle nutrients. Acids in soil kill these essential components of the detrital food chain, resulting in nutrient-poor soil. Furthermore, thick layers of leaf litter interfere with the germination of tree seeds and growth of saplings.

Finally, acids in soil free up calcium, phosphorus, and magnesium—key nutrients for trees. Leached from soil by water, they become unavailable to plants.

Acid rain, along with nutrients and metals removed from soil, runs off into waterways. High nutrient levels, especially in ponds and lakes, stimulate growth of algae. Light penetration is diminished in bodies of water, and dissolved oxygen concentration is lowered, especially in deeper waters.

Animals are affected. Sudden snowmelt causes a rush of acid into waterways that causes acid shock to many. High acid concentrations in water may directly affect the young of many animals, including larval stages of insects, tadpole stages of amphibians, and fish fry. Aluminum, leached by acids from soil and dissolved in water, attaches to gills and suffocates aquatic animals, including fish. Mercury bioaccumulates as it is leached from soils and released from bottom muds. First, it gathers in the tissues of aquatic animals. As these are eaten first by small fish, then by larger ones, and in turn by birds and mammals, concentrations increase and may interfere with reproduction.

Humans, too, are susceptible to effects of acid precipitation. Drinking water is affected as high acidity dissolves lead and copper from pipes in houses. Both lead and copper cause health problems, especially in children. Although no effects have been directly linked to Adirondack waters, increasing levels of mercury are a concern to public health officials. Finally, there are economic impacts. Sport fishing is important throughout the region; its loss would be keenly felt.

No wonder there is a public outcry to reduce acid precipitation.

The Problem

Although dying trees are serious concerns in North America, most public attention to date has focused on acid precipitation's effects on waterways. In the United States, the Environmental Protection Agency (EPA) monitors environmental quality and recommends measures and enforces regulations to improve the environment. With respect to acid rain, EPA's overall goal is to reduce emissions of sulfur dioxide (SO_2) and oxides of nitrogen (NO_x) as soon as possible. Regulations focus on public utilities, the biggest emitters. Strict standards on allowable emissions have been set. Noncompliance can result in significant fines and even closure of key facilities.

At first, utilities balked at expenses associated with emission reduction, claiming they had been unjustly singled out. Public utilities in the United States are highly competitive. Consumers demand that reliable electricity be delivered to homes and businesses as cheaply as possible. Every rate hike, no matter how necessary, is resisted vociferously. Utilities claimed that efforts to reduce emissions put them at a competitive disadvantage with other regions where such measures are unnecessary or nonenforced.

Sensitive to these criticisms, the EPA designed an innovative, market-based solution that allows individual utilities to design and implement their own compliance plan. Key to the plan's success in reducing SO_2 emissions is allowance trading. A utility that exceeds its emission goals—that is, emits less SO_2 than is allowed—can sell its excess allowance to other utilities that have not met their goals. This plan creates an economic incentive for compliance. It turns an unpopular expense into an economic opportunity. The plan appeared to be successful. In its first year (1995–1996), total emissions fell significantly below targeted goals.

But not everyone is happy. In the Adirondacks, citizen groups, notably the Adirondacks Council, took exception to the EPA plan. The council was concerned, not with total emission of SO_2 but with the amount of acid precipitation entering local waterways. Aren't the two directly related? Not necessarily, said the council. If utilities in the Midwest buy excess allowances from utilities in other parts of the country, their emissions will remain high and so will the amount of acid precipitation coming into the Adirondacks waterways. They took particular exception to the sale of allowances by utilities in the Adirondacks region.

Has the EPA plan been a success or not? The following readings explain the plan in greater detail, examine pros and cons, and suggest ways in which the plan could be further strengthened.

reading 1

EPA 1998 COMPLIANCE REPORT: Acid Rain Program

U.S. Environmental Protection Agency, July 1999, inside cover and pp. 1–19. Reprinted by permission.

Background

The Acid Rain Program was established under Title IV of the 1990 Clean Air Act Amendments. The program calls for major reductions of sulfur dioxide (SO_2) and nitrogen oxides (NO_x), the pollutants that cause acid rain, while establishing a new approach to environmental protection through the use of market incentives. The program sets a permanent cap on the total amount of SO_2 that may be emitted by electric utilities nationwide at about one half of the amount emitted in 1980, and allows flexibility for individual utility units to select their own methods of compliance. The program also sets NO_x emission limitations (in lb/mmBtu) for electric utilities, representing about a 27 percent reduction from 1990 levels. The Acid Rain Program is being implemented in two phases: Phase I began in 1995 for SO_2 and 1996 for NO_x, and will last until 1999; Phase II for both pollutants begins in 2000 and is expected to involve over 2,000 units. In 1998, there were 408 units affected by the SO_2 provisions of the Acid Rain Program, 235 of which were also affected for NO_x, and an additional 305 utility units affected only by the NO_x provisions.

Acid rain causes acidification of lakes and streams and contributes to the damage of trees at high elevations. In addition, acid rain accelerates the decay of building materials, paints, and cultural artifacts, including irreplaceable buildings, statues, and sculptures. While airborne, SO_2 and NO_x gases and their particulate matter derivatives, sulfates and nitrates, contribute to visibility degradation and impact public health.

The SO_2 component of the Acid Rain Program represents a dramatic departure from traditional command and control regulatory methods that establish source-specific emissions limitations. Instead, the program introduces a trading system for SO_2 that facilitates lowest-cost emissions reductions and an overall emissions cap that ensures the maintenance of the environmental goal. The program features tradable SO_2 emissions allowances, where one allowance is a limited authorization to emit one ton of SO_2. Allowances may be bought, sold, or banked by utilities, brokers, or anyone else interested in holding them. Existing utility units were allocated allowances for each future compliance year and all participants of the program are obliged to surrender to EPA the number of allowances that correspond to their annual emissions starting either in Phase I or Phase II of the program.

The NO_x component of the Acid Rain Program is more traditional, and establishes an emission rate limit for all affected utilities. Flexibility is introduced to this command and control measure, however, through compliance options such as emissions averaging, whereby a utility can meet the standard emission limitations by averaging the emissions rates of two or more boilers. This allows utilities to over-control at units where it is technically easier to control emissions, thereby achieving emissions reductions at a lower cost. Additionally, beginning in 1997, certain Phase II units could elect to become affected for NO_x early. By complying with Phase I limits, these early election units can delay meeting the more stringent Phase II limits until 2008.

At the end of each year, utilities must demonstrate compliance with the provisions of the Acid Rain Program. For the NO_x portion of the program, utilities must achieve an annual emission limitation at or below mandated levels. For SO_2, utilities are granted a 60-day grace period during which additional SO_2 allowances may be purchased, if necessary, to cover each unit's emissions for the year. At the end of the grace period (the Allowance Transfer Deadline), the allowances a unit holds in its Allowance Tracking System (ATS) account must equal or exceed the unit's annual SO_2 emissions. In addition, in 1995–1999 (Phase I of the program), units must have sufficient allowances to cover certain other deductions as well. Any remaining SO_2 allowances may be sold or banked for use in future years.

To the Reader

The Acid Rain Program 1998 Compliance Report summarizes compliance results that, for the fourth consecutive year since the Acid Rain Program began, show 100 percent compliance with both sulfur dioxide (SO_2) and nitrogen oxide (NO_x) requirements. Over the past year there were also a number of significant Program improvements.

First, the allowance transfer deadline, the date by which a unit's allowance account is required to hold enough allowances to account for the previous year's SO_2 emissions, was changed from January 30th to March 1 (Feb. 29 for leap years). This allows affected facilities additional time to determine their previous year's SO_2 emissions and to ensure the availability of sufficient allowances to account for those emissions.

Second, in order to expedite transfers and reduce transaction costs the Acid Rain Program revised its regulations to allow an authorized account representative to specify allowance accounts to which allowances can be transferred without requiring the buyer's signature on each individual allowance transfer form.

Third, to avoid the imposition of extremely large excess emissions penalties for minor, inadvertent accounting errors, the Acid Rain Program now allows for the transfer of unused allowances from unit accounts at the same source to account for the emissions at a unit that lacks sufficient allowances. This leads to a smaller penalty, more in line with the violation, while still ensuring the environmental objective.

Fourth, the monitoring rule was revised to enhance flexibility for industry by reducing monitoring requirements for certain units with low mass emissions, creating new monitoring options for some units, reducing certain quality assurance requirements, and increasing fuel sampling flexibility for certain units. The sum of these changes make the rule more efficient and less burdensome for the regulated community, EPA, and the States.

Finally, the Acid Rain Program permits regulation was revised to make new and retired unit exemptions easier for sources to comply with and simpler for the States to administer. These changes provide States with additional flexibility in meeting public notice requirements in the issuance of Acid Rain permits and allow for "direct/final" issuance of draft and proposed Acid Rain permits. The Program also eased public notice requirements related to the appointment of, and changes to, the designated representative and alternate designated representative.

We will continue to look for ways to improve the Acid Rain Program as we prepare for the year 2000 and the beginning of Phase II, and will work with all interested persons in ensuring that the Acid Rain Program meets its environmental goals with minimum cost and burden for affected sources and States.

Brian J. McLean, Director,
Acid Rain Program

Summary

100 Percent Compliance with Both SO_2 and NO_x Requirements in 1998

All 713 boilers and combustion turbines (referred to as "units") affected by the SO_2 and NO_x regulations of the Acid Rain Program in 1998 successfully met their emissions compliance obligations.

- All 408 units subject to SO_2 requirements in 1998 held sufficient allowances to cover their emissions. Of the 5,300,861 allowances deducted from compliance accounts almost all (5,298,498 or 99.96 percent) were for emissions, but other deductions were also made as required by the Acid Rain Program regulations.
- All 540 units subject to the NO_x requirements in 1998 demonstrated compliance with applicable annual emission limitations. Of those 540 units, 235 were also subject to SO_2 requirements, while 305 units were affected only for NO_x (30 Phase I units and 275 Phase II "early election" units).

1998 SO_2 Emissions of Phase I Units were 24 Percent Below Allowable Level

SO_2 emissions in 1998 were 1.7 million tons (or 24 percent) below the 7 million ton allowable level as determined by 1998 allowance allocations. Since an additional 7.9 million allowances were carried over, or banked, from 1997, the overall number of allowances available in 1998 was 14.9 million, of which affected units consumed only about 35 percent. Actual emissions for the 408 units participating in 1998 were 5.3 million tons, down 180,000 tons from emissions of the 423 units affected in 1997.

1998 Phase I Unit NO_x Emission Rates 41 Percent Below 1990; NO_x Tons 29 Percent Lower Than in 1990

Emission rates for the 265 Phase I utility units dropped by 41 percent below 1990 levels, from an average of 0.70 pounds of NO_x per million Btu of heat input (lb/mmBtu) to an average of 0.41 lbs/mmBtu; this rate is 16 percent below the compliance of 0.49 lbs/mmBtu for these units. NO_x emission levels for these units were 390,254 tons (or 29 percent) below 1990 levels.

1998 NO_x Emission Rates of Early Election Units Even Lower Than Rates for Phase I Units

For the 275 Phase II units which elected to meet Phase I NO_x rates early, emission rates dropped from an average of 0.46 lbs/mmBtu in 1990 to 0.38 lbs/mmBtu in 1998, a 17 percent decrease and 19 percent below the compliance rate of 0.47 lbs/mmBtu for these units. Therefore, while utilization of these units increased by 28 percent between 1990 and 1998, NO_x tons increased by only 8 percent.

Monitoring Performance Excellent Once Again

For the fourth year of the Acid Rain Program, the continuous emission monitors used by participants continue to provide some of the most accurate and complete data ever collected by the EPA. Statistics reflect excellent monitor operation of all units affected by both Phase I and Phase II of the program.

Accuracy: SO_2 monitors achieved a median relative accuracy (i.e., deviation from the

Availability: reference test method) of 3.0 percent; flow monitors, 3.0 percent; and NO_x monitors, 3.1 percent.

Availability: SO_2 and flow monitors achieved a median availability of 99.5 and 99.7 percent, respectively, while NO_x monitors achieved a median reliability of 99.2 percent.

SO_2 Market Active; Volume of Allowances Transferred Between Distinct Entities in 1998 Continues to Increase

Activity in the allowance market continued to increase in 1998. The volume of allowances transferred between unrelated parties in economically significant trades increased from 7.9 million in 1997 to 9.5 million in 1998.

Affected Population in Phase I

Exhibit 1 provides a summary of the affected population of units under the Acid Rain Program from 1995 through 1999. The table illustrates that although the units listed in Table 1 of the CAAA are consistently affected for both SO_2 and NO_x beginning in 1997, the total universe of affected units varies year to year because of the flexibility offered by the program.

SO_2 Program

408 Units Underwent Annual Reconciliation for SO_2 in 1998

There were 398 affected utility units and 10 opt-in units that underwent annual reconciliation in 1998 to determine whether sufficient allowances were held to cover emissions. These 408 units include 263 utility units specifically required to participate during Phase I, 135 utility units not initially required to participate until Phase II, but electing to participate early as part of multi-unit compliance plans[1], and 10 other units that elected to join as part of the Opt-in Program.[2]

1998 SO_2 Emissions Target was 6.97 Million Tons

The number of allowances allocated in a particular year, the amount representing that year's allowable SO_2 emissions level, is the sum of allowance allocations granted to sources under several provisions of the Act. In 1998, the emissions target established by the program for the 408 participating units was 6.97 million tons. However, the total allowable SO_2 emission level in 1998 was actually 14.93 million tons, consisting of the 6.97 million 1998 allowances granted through the program and an additional 7.96 million allowances carried over, or banked, from 1997.

EXHIBIT 1. Affected Units During Phase I of the Acid Rain Program

		1995	1996	1997	1998	1999
SO_2	Table 1	263	263	263	263	263
	Substitution and Compensating	182	161	153	135	Variable
	Opt-in	0	7	7	10	Variable
	Total	**445**	**431**	**423**	**408**	**Variable**
NO_x	Table 1	NA	144	170	171	171
	Substitution	NA	95	95	94	94
	Early-Election	NA	NA	272	275	Variable
	Total	**NA**	**239**	**537**	**540**	**Variable**

[1] During Phase 1 of the Acid Rain Program, a unit not originally affected until Phase II may elect to enter the program early as a substitution unit or a compensating unit to help fulfill the compliance obligations for one of the Table 1 units targeted by Phase 1. A unit brought into Phase I as a substitution unit can assist a Table 1 unit in meeting its emissions reductions obligations. Utilities may make cost-effective emissions reductions at the substitution unit instead of at the Table 1 unit, achieving the same overall emissions reductions that would have occurred without the participation of the substitution unit. A Table 1 unit may designate a Phase II unit as a substitution unit only if both units are under the control of the same owner or operator. Additionally, Table 1 units that reduce their utilization below their baseline may designate a compensating unit to provide compensating generation to account for the reduced utilization of the Table 1 unit. (A unit's baseline is defined as its heat input averaged over the years 1985–1987). A Table 1 unit may designate a Phase II unit as a compensating unit if the Phase II compensating unit is in the Table 1 unit's dispatch system or has a contractual agreement with the Table 1 unit, and the emissions rate of the compensating unit has not declined substantially since 1985.

[2] The Opt-in Program gives sources not required to participate in the Acid Rain Program the opportunity to enter the program on a voluntary basis, install continuous emission monitoring systems (CEMS), reduce their SO_2 emissions, and receive their own allowances.

EXHIBIT 2. Origin of 1996 Allowable Emissions Level

Type of Allowance Allocation	Number of Allowances	Explanation of Allowance Allocation Type
Initial Allocation	5,550,820	Initial Allocation is the number of allowances granted to units based on their historic utilization, emissions rates specified in the Clean Air Act and other provisions of the Act.
Phase I Extension	178,211	Phase I Extension allowances are given to Phase I units that reduce their emissions by 90 percent or reassign their emissions reduction obligations to units that reduce their emissions by 90 percent.
Allowances for Substitution Units	948,708	Allowances for Substitution Units are the initial allocation granted to Phase II units which entered Phase I as substitution units.
Allowance Auctions	150,000	Allowance Auctions provide allowances to the market that were set aside in a Special Allowance Reserve when the initial allowance allocation was made.
Allowances for Compensating Units	15,838	Allowances for Compensating Units are the initial allocation granted to Phase II units which entered Phase I as compensating units.
Opt-in Allowances	97,932	Opt-in Allowances are provided to units entering the program voluntarily.
Small Diesel Allowances	27,656	Small Diesel Allowances are allocated annually to small diesel refineries that produce and desulfurize diesel fuel during the pervious year. These allowances can be earned through 1999.
Total 1998 Allocation	**6,969,165**	
Banked 1997 Allowances	7,959,656	Banked Allowances are those held over from 1995 through 1997 and can be used for compliance in 1998 or any future year.
Total 1998 Allowable	**14,928,841**	

The initial allocation and the allowances for substitution and compensating units represent the basic allowances granted to units that authorize them to emit SO_2 under the Acid Rain Program. Additional allowances for the year 1998 were also made available through the allowance auctions, held annually since 1993. Other allowances issued in 1998 were from special provisions in the Act, which are briefly explained in Exhibit 2 above. In addition, any allowances carried over from previous years (banked allowances) are available for compliance and included in the allowable total.

Beginning in the year 2000 at the onset of Phase II, the volume of allowances allocated annually to the Phase I units will be reduced and the requirement to hold allowances will be extended to smaller, cleaner plants. Nationwide, the cap for all utilities with an output capacity of greater than 25 megawatts will be 9.48 million allowances from 2000–2009. In 2010, the cap will be reduced further to 8.95 million allowances, a level approximating one half of industry-wide emissions in 1980.

SO_2 Compliance Results

Phase I Units Better 1998 SO_S Allowable Emissions Level by 24 Percent

The Phase I units affected in 1998 emitted at a level approximately 24 percent below 1998 allocations, as shown in Exhibit 3. This percentage is about the same as in 1997, with both emissions and allocations registering slight decreases.

Relative to 1997, the 263 Table 1 units decreased their emissions by about 110,000 tons or more than two percent in 1998, while increasing their utilization by just over one half of one percent. The 4.7 million tons emitted by these Table 1 units were still substantially below their 1998 allocation of 5.6 million allowable tons.

Substitution and compensating units in 1998 expended about the same percentage of their annual allocation as in 1997. In 1998, these 135 units were responsible for emitting

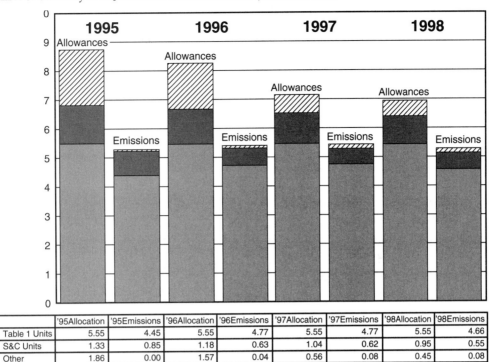

EXHIBIT 3. Summary of SO_2 Emissions versus Allocations (Millions of Tons)

	'95Allocation	'95Emissions	'96Allocation	'96Emissions	'97Allocation	'97Emissions	'98Allocation	'98Emissions
Table 1 Units	5.55	4.45	5.55	4.77	5.55	4.77	5.55	4.66
S&C Units	1.33	0.85	1.18	0.63	1.04	0.62	0.95	0.55
Other	1.86	0.00	1.57	0.04	0.56	0.08	0.45	0.08
TOTALS	8.74	5.30	8.30	5.44	7.15	5.47	6.95	5.29

approximately 550,000 tons of SO_2, about 58 percent of their 950,000 allocation. In 1997, 153 substitution and compensating units emitted approximately 620,000 tons of SO_2, or 60 percent of their 1.04 million allowable level.

Three new opt-in units joined the program in 1998, raising the total allocation to 98,000 allowances and the emissions level to 80,000 tons. The percentage of emissions to allowances allocated to opt-in units in 1998 increased by approximately 1% compared to 1997.

Deducting Allowances for Compliance

The total number of allowances deducted in 1998 was 5,300,861 which represents approximately 76 percent of all 1998 allowances issued. Almost all (99.95 percent) of the deducted allowances were for emissions. Exhibit 4 displays these allowance deductions, as well as the remaining bank of 1995 through 1998 allowances.

At an individual unit, the number of allowances surrendered was equal to the number of tons emitted at the unit, except where the unit shared a common stack with other units. For the purposes of surrendering allowances for emissions at a common stack, the utility was allowed to choose the proportion of allowances deducted from each unit sharing the stack, as long as enough allowances were surrendered to cover the total number of tons emitted. If no apportionment was made, EPA deducted allowances equally among the units sharing the stack to cover total emissions reported by the stack.

Under the Acid Rain Program, certain units applied for and received approval of Phase I Extension plans during the Phase I permitting process. These units fell into two categories: "control units" which were required to cut their emissions by 90 percent using qualifying technology[3] by 1997, and "transfer units" which reassigned their emissions reduction obligations to a control unit. Both kinds of units received extra SO_2 emissions allowances to cover the SO_2 they emitted beyond their basic Phase I allocations during 1995 and 1996. In addition, the control units were given Phase I extension allowances for 1997, 1998, and 1999. A total of 3.5 million allowances was distributed to all Phase I extension control and transfer units.[4]

For 1998, all 19 control units demonstrated meeting the 90 percent reduction requirement and, therefore, did not surrender any 1998 extension allowances. The 1998 tonnage emissions limitation, though, was exceeded by five control units and eleven transfer units and resulted in a surrender of a total of 99,240 vintage 1999 allowances.

SO₂ Allowance Market

The flexibility provided by the Acid Rain Program enabled the 408 units affected in 1998 to pursue a variety

---

[3] Qualifying technology is defined in 40 CFR 72.2

[4] Beginning in 1997, each of the 19 units designated as control units was required to show it had reduced its annual emission by at least 90 percent using qualifying control technology. If a unit could not make this demonstration, all or a portion of the extension allowances it received for the year under the Phase I Extension provisions were required to be surrendered. In addition, also beginning in 1997, each of the same 19 control units and each of the 61 other units designated as transfer units was required to meet a tonnage emission limitation approved in its permit. A unit that exceeded its limitation was required to surrender allowances for the following year.

EXHIBIT 4. SO₂ Allowance Reconciliation Summary

Total Allowances Held in Accounts as of 3/1/99 (1995 trough 1998 Vintage)*	**14,928,841**
Table 1 Unit Accounts	8,585,043
Substitution & Compensating Unit Accounts	1,306,220
Opt-in Accounts	83,962
Other Accounts**	4,953,616
1998 Allowances Deducted for Emissions	**5,298,498**
Table 1 Unit Accounts	4,664,898
Substitution & Compensating Unit Accounts	553,349
Opt-in Unit Accounts	80,251
1998 Allowances Deducted Under Special Phase I Provisions ***	**2,363**
Table 1 Unit Accounts	65
Substitution & Compensating Unit Accounts	1,755
Opt-in Unit Accounts	543
Banked Allowances	**9,627,980**
Table 1 Unit Accounts	3,920,080
Substitution & Compensating Unit Accounts	751,116
Opt-in Unit Accounts	3,168
Other Accounts**	4,953,616

*The number of allowances held in the Allowance Tracking System (ATS) accounts equals the number of 1998 allowances allocated (see Exhibit 2) plus the number of 1997 banked allowances. March 1, 1999 represents the Allowance of Transfer Deadline, the point in time at which the 1998 Phase I affected unit accounts are frozen and after which no transfers of 1995 through 1998 allowances will be recorded. The freeze on these accounts is removed when annual reconciliation is complete.
**Other accounts refers to general accounts within the ATS that can be held by any utility, individual or other organization, and unit accounts for units not affected in Phase I.
***Allowances were deducted for both underutilization and state cap provisions in 1998.

of compliance options to meet their SO₂ reduction obligations, including scrubber installation, fuel switching, energy efficiency, and allowance trading. The presence of the allowance market has given some sources the incentive to overcontrol their SO₂ emissions in order to bank their allowances for use in future years. Other sources have been able to postpone and possibly avoid expenditures for control by acquiring allowances from sources that overcontrolled. The flexibility in compliance options is possible because of the accountability provided through strict monitoring requirements for all affected units that ensure one allowance is equivalent to one ton of SO₂. The program's flexibility enabled all 408 sources to be in compliance in 1998 and significantly reduced the cost of achieving these emissions reductions as compared to the cost of a technological mandate.

The marginal cost of reducing a ton of SO₂ from the utility sector should be reflected in the price of an allowance. The cost of reductions continues to be lower than anticipated when the Clean Air Act Amendments were enacted, and the price of allowances reflects this. The cost of compliance was initially estimated at $400–1000/ton, but was $207/ton at the 1999 allowance auction. Prices have remained in the $205 to $215 range since January of 1999. Some market observers believe lower than expected allowance prices during the first several years of the program were due primarily to lower than expected compliance costs and larger than expected emission reductions, which have increased the supply of allowances and put downward pressure on prices. Exhibit 5 displays the price trend since mid-1994, based on monthly price reports from Cantor Fitzgerald Environmental Brokerage Services, and a market survey conducted by Fieldston Publications.

Activity in the allowance market created under the Acid Rain Program remained strong in 1998, with 1,584 transactions moving about 13.5 million allowances in the Allowance Tracking System (ATS), the accounting system developed to track holdings of allowances. In terms of economically significant transfers, or those between unrelated parties, the volume of allowances transferred rose from 7.9 million in 1997 to 9.5 million in 1998. A record of 70 percent of annual activity consisted of allowances transferred between economically distinct organizations, with more than half representing allowances directly acquired by utilities.

The most active market segment in 1998 in terms of allowance volume was composed of exchanges between brokers/traders and utilities, accounting for 6.3 million allowances. The next most active was the reallocation category, which covered an additional 3.2 million allowances. The category of transfers between unrelated utilities increased to 1.9 million allowances.

EXHIBIT 5. SO_2 Allowance Prices

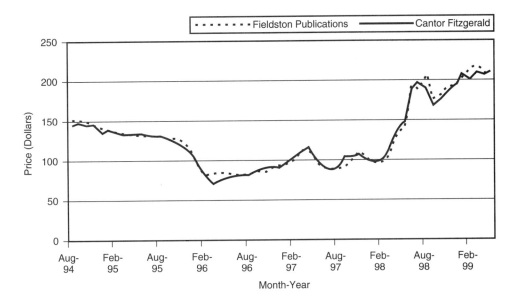

All transactions, along with data on account balances and ownership, are posted on the Acid Rain Division's Internet site (www.epa.gov/acidrain) on a daily basis in order to better inform trading participants. Also available are cumulative market statistics and analysis.

NO_x Program

Instead of using allowance trading to facilitate emissions reductions, the Title IV NO_x program establishes standard emission limitations for affected units. Title IV of the 1990 Clean Air Act Amendments required EPA to establish NO_x annual average emission limits (in pounds of NO_x per million British thermal units of fuel consumed [lb/mmBtu]) for coal-fired electric utility units in two phases.

In April 1995, EPA promulgated 40 CFR Part 76 which established NO_x emission limits beginning on January 1, 1996 for Group 1 boilers that were also part of the Phase I SO_2 program. (Group 1 boilers are dry bottom, wall-fired boilers and tangentially-fired boilers.) Phase I dry bottom wall-fired boilers are subject to a NO_x emission limit of 0.50 lb/mmBtu; Phase I tangentially-fired boilers are subject to a NO_x emission limit of 0.45 lb/mmBtu.

In addition, the April 1995 regulations allowed Phase II Group 1 units to use an "Early Election" Compliance Option. Under this regulatory provision, Group 1, Phase II NO_x affected units can demonstrate compliance with the higher Phase I limits for their boiler type from 1997 through 2007 and not meet the more stringent Phase II limits until 2008. If the utility fails to meet this annual limit for the boiler during any year, the unit is subject to the more stringent Phase II limit for Group 1 boilers beginning in 2000, or the year following the exceedance, whichever is later.

In December 1996, EPA revised the NO_x emission limits for Phase II, Group 1 boilers (0.46 lb/mmBtu for dry bottom wall-fired boilers and 0.40 lb/mmBtu for

EXHIBIT 6. Volume of SO_2 Allowances in Economically Significant Transfers

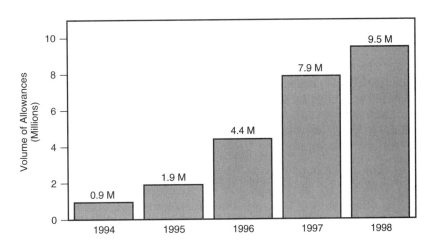

tangentially-fired boilers) and established emission limits for cell burner, cyclones, wet bottom and vertically-fired boilers (referred to as "Group 2 boilers") effective on January 1, 2000. As a result of the April 1995 and December 1996 rulemakings, NO_x reductions were projected to be approximately 400,000 tons per year in 1996 through 1999 (Phase I), and 2,060,000 tons per year in 2000 and subsequent years (Phase II).

Phase I NO_x Units

265 Phase I Units Were Subject to Emissions Limitations in 1998

In 1998, 265 coal-fired utility units were subject to the Title IV Phase I emission limitations for NO_x.[5] The 265 Phase I NO_x affected units include 171 Table 1 units and 94 substitution units whose owners chose to participate in Phase I as part of an SO_2 compliance strategy. This group of units is subject to the Phase I and Phase II. Exhibit 7 shows the number of Phase I NO_x affected units by boiler type.

EXHIBIT 7. Phase I NO_x Units by Boiler Type

Boiler Type	Standard Emission Unit	Table 1 Units	Substitution Units	All Units
Tangentially Fired Boilers	0.45	94	41	135
Dry Bottom Well-fired Boilers	0.50	77	53	130

Phase I NO_x Compliance Options

For each Phase I NO_x affected unit, a utility can comply with the applicable standard emission limitation, or may qualify for one of two additional compliance options which add flexibility to the rate-based compliance requirements:

- Emissions Averaging. A utility can meet the standard emission limitation by averaging the heat-input weighted annual emission rates of two or more units.
- Alternative Emission Limitation (AEL). A utility can petition for a less stringent alternative emission limitation if it uses properly installed and operated low NO_x burner technology (LNBT) designed to meet the standard limit, but is unable to achieve that limit. EPA determines whether an AEL is warranted based on analyses of emissions data and information about the NO_x control equipment.

Exhibit 8 summarizes the compliance options chosen by Phase I affected NO_x units for 1998. As in 1996 and 1997, averaging was the most widely chosen compliance option. For 1998, there were 24 averaging plans involving 204 Phase I NO_x units.

EXHIBIT 8. Summary of Compliance Options Chosen in 1998

Compliance Option	# of Units
Standard Emission Limitation	51
Emissions Averaging	204
Alternative Emission Limitation	10
Total	265

Phase I NO_x Compliance Results

For 1998, EPA has determined that all 265 Phase I NO_x units met the required emission limit through compliance with either the standard emission limitation, emissions averaging, or an alternative emission limitation.

NO_x Emission Rate Reduction

From 1990[6] to 1998, the average NO_x emission rate of the 265 Phase I units declined by 41% (from 0.70 lb/mmBtu to 0.41 lb/mmBtu). As shown in Exhibit 9, on average, both Table 1 and substitution units were below the average Phase I emission limit of 0.49 lb/mmBtu (the heat input weighted average of the applicable limits).

NO_x Mass Emissions Reduction

Exhibit 10 illustrates the change in NO_x mass emissions since 1990 for Table 1 and substitution units. For the 265 units, total NO_x mass emissions in 1998 were 29 percent lower than in 1990, but 3 percent higher than in 1997. While this is the second year total NO_x mass emissions have increased, the ascent can be attributed in part to greater electrical production, as evidenced by an increase in heat input in 1997 and 1998 of 3 percent and 6 percent, respectively, compared to 1996. Without further reductions in emissions rates, NO_x emissions would be expected to rise with increased utilization.

Phase II Early Elections Units

275 Units Were Subject to Early Election Requirements in 1998

Nineteen ninety-eight was the second year in which early election utility units were required to meet the Phase I

---

[5] Compared with 1997, the universe of units remained the same, except that Mt. Storm Unit 2 (WV) was added because its compliance extension expired and Gadsby Unit 3 (UT) was deleted because it was mistakenly identified in previous years as a coal-fired utility unit.
[6] For a more detailed description of the 1990 baseline refer to the Acid Rain Program 1996 Compliance Report.

EXHIBIT 9. Average NO_x Emission Rates for 265 Phase 1 Units

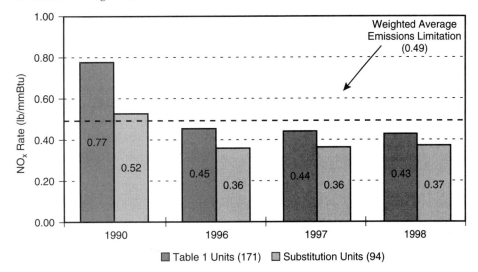

Early Election Compliance Results

For 1998, EPA determined that all 275 units complied with the Phase I, Group 1 emission limitations and have continued eligibility for Early Election in 1999 through 2007.

Average NO_x emission rates for Early Election units have declined by 17% from 0.46 lb/mmBtu in 1990 to 0.38 lb/mmBtu in 1998. This decline is less dramatic than the decline at Phase I NO_x units because 51% of the Early Election units are newer units already subject to the New Source Performance Standards (NSPS) NO_x emission limits. The overall NO_x emission rate for these units is comparable to the average rate of 0.41 lb/mmBtu for all Phase I NO_x units. Exhibit 12 summarizes the NO_x emission rate reductions from 1990 to 1998 by boiler type for the 265 Early Election units that were operating in 1990.

NO_x Mass Emissions Reduction

The total NO_x mass emissions from the operating Early Election units increased by 106,619 tons (or 8 percent) from 1990[8] to 1998, reflecting an increase in utilization (see Exhibit 13). For the 265 Early Election units operating in 1990, heat input increased during the eight year period by approximately 28%.

SO_2 and NO_x Monitoring in 1998

In order to verify the reductions of SO_2 and NO_x emissions mandated under the Clean Air Act and to support the SO_2 allowance trading program, a fundamental objective of the Acid Rain Program is to ensure accurate accounting of pollutant emissions from affected boilers

EXHIBIT 10. NO_x Mass Emissions for 256 Phase I Units

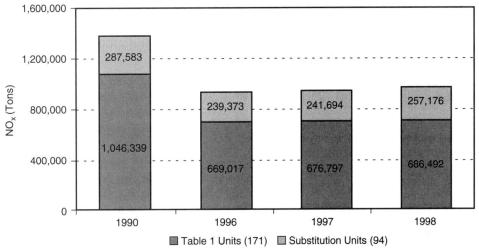

[7] Compared with 1997, the universe of early election units remained the same, except for W C Dale Units 3 and 4 (KY) and H L Spurlock Unit 2 (KY), which were added after being inadvertently omitted in 1997.

EXHIBIT 11. Distribution of 1998 Early Election Units by Boiler Type

Boiler Type	Standard Emission Unit	Operating Group 1, Phase 2 Units	Early Election Units	Percent of Units Electing
Tangentially Fired Boilers	0.45	300	171	57%
Dry Bottom Well-fired Boilers	0.50	314	104	33%
Total		614	275	45%

and turbines. To implement this objective, concentrations of emitted SO_2 and NO_x from each affected until (boiler or turbine) are measured and recorded using Continuous Emissions Monitoring Systems (CEMS)(or an approved alternate measurement method) certified by EPA to meet the high accuracy standards of the Acid Rain Program.

CEMS are used to determine SO_2 mass emissions and NO_x emission rates. SO_2 mass emissions are determined using CEMS to measure SO_2 concentration and stack flow rate. NO_x emission rates, on the other hand, are determined with NO_x and diluent gas (CO_2 or O_2) concentration monitors. These monitors are required to meet strict initial and on-going performance standards to demonstrate the accuracy, precision, and timeliness of their measurement capability.

One measure of the accuracy of a CEMS is the relative accuracy test audit (RATA), which is required for initial certification of a CEMS and for on-going quality assurance. The relative accuracy test audit ensures that the installed monitor measures the "true" value of the pollutant by comparing the monitor to a reference method which simultaneously measures the stack gas pollutant. Thus, the lower the relative accuracy resulting from the test audit, the more accurate the monitor.

All monitoring systems must meet a certain relative accuracy standard in order to be qualified to report emissions to the Acid Rain Program; 10 percent for SO_2 and NO_x and 15 percent for flow (beginning January 1, 2000, the flow standard will also be 10 percent). As a further incentive for high quality maintenance, CEMS that achieve a superior accuracy result, less than or equal to 7.5 percent for SO_2 and NO_x and less than or equal to 10 percent for flow (beginning January 1, 2000, the flow standard for superior accuracy will also be 7.5 percent), are granted a reduced frequency annual RATA requirement in place of the semiannual requirement. Because the RATA determines relative accuracy as an absolute value, it does not detect whether the difference between the reference method values and the readings from the CEMS being tested is due to random error or to systematic bias. Therefore, an additional test is required to ensure that emissions are not underestimated: the bias test. This test determines if the CEMS is systematically biased low compared to the reference method and if so, a bias adjustment factor is calculated and applied to all reported data from that monitoring system to ensure there is no systematic underreporting. Exhibit 14 highlights the relative accuracy results achieved by Acid Rain CEMS in 1998.

Another metric used to determine the effectiveness of a CEMS is the percentage of hours that a monitoring system is operating properly and meeting all performance standards and therefore, able to record and report an emissions value. This metric is defined as the percent monitor availability (PMA). Exhibit 15 shows the monitor availabilities reported in 1998 and indicates that the CEMS used to determine SO_2 mass emissions and NO_x emission rates are well maintained and fulfilling the high performance standards required by the Acid Rain Program.

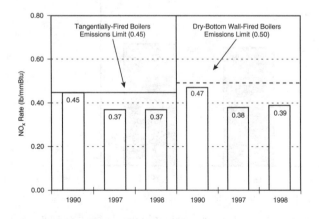

EXHIBIT 12. Average NO_x Emission Rate for 265 Early Election Units (Operating in 1990)

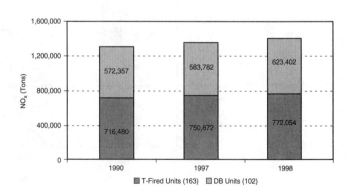

EXHIBIT 13. NO_x Mass Emissions for 265 Early Election Units (Operating in 1990)

EXHIBIT 14. 1998 Relative Accuracy Test Audit (RATA) Results

	SO_2 Concentration	Volumetric Flow Rate	NO_x Rate
Mean Relative Accuracy	4.2%	3.7%	4.1%
Median Relative Accuracy	3.0%	3.0%	3.1%
Percent Meeting Relative Accuracy Standard	95%	97%	91%

EXHIBIT 15. 1998 CEMS Availability

Parameter	Median % Availability at End of 1998	
	Coal-Fired Units	Oil and Gas Units
SO_2	99.5	98.5
Flow	99.7	98.8
NO_x	99.2	98.0

Conclusion

1998 proved to be another successful year for both the Acid Rain Program's rate-based approach to NO_x reduction and cap-and-trade approach to SO_2 reduction. In 1998, all Phase I affected utility units not only met their compliance goals, but exceeded them, achieving an overall reduction of 390,254 tons of NO_x from 1990 levels, and maintaining the extraordinary reductions of more than 5 million tons of SO_2 from 1980 levels, first achieved in 1995. Additionally, the 275 Phase II NO_x early election units had increased emissions of eight percent since 1990, while their utilization increased by 28 percent during the same period.

Exceedance of compliance goals translates into additional environmental and health benefits. For example, the greater and earlier reductions of SO_2 have resulted in a 10–25 percent drop in rainfall acidity in the Northeast in 1995.[9]

One factor mitigating the benefit of the overcompliance in the SO_2 program, of course, is the ability to use banked allowances in the future. The 40 percent of 1995 allowances, 35 percent of 1996 allowances, 23 percent of 1997 allowances, and 24 percent of 1998 allowances that were not retired for compliance purposes can be used to cover emissions in a later year. However, immediate health and environmental benefits are arguably more valuable than a benefit several years from now.

The NO_x program, based on the more traditional rate-based approach, offers less flexibility and displays a lesser degree of overcompliance. It requires each unit to achieve reductions or, at a minimum, for a group of units to achieve an average emission rate equal to or lower than their individual units. This approach does not allow emissions reductions in one year to be used in another year, and as a result, the incentive to overcomply is diminished.

The pattern and certainty of emissions reductions over time will also differ between the two programs. After the year 2000 when both programs are in full implementation, SO_2 emissions are expected to decline steadily to the emissions cap level of 8.95 million tons, whereas NO_x emissions, in the absence of an emissions cap, are expected to rise as existing sources are utilized more and new sources, which are not required to offset their emissions, are built and operated.

Despite these differences, both the SO_2 and NO_x components of the Acid Rain Program are continuing their success in 1998. The significant progress evident at this stage of the program is encouraging. Through the continued efforts of Phase I participants and by additional reductions from Phase II units beginning in 2000, the long term goals of the Acid Rain Program—a 10 million ton reduction of SO_2 emissions and two million ton reduction of NO_x emissions—will be achieved.

---

[9] U.S. Geological Survey, Trends in Precipitation Chemistry in the United States, 1983–94—An Analysis of the Effects in 1995 of Phase I of the CAAA of 1990, Title IV, USGS 96-0346, Washington, DC, June 1996.

reading 2

"Emissions Trading of Sulfur Dioxide: The U.S. Experience"

U.S. Environmental Protection Agency, Acid Rain Program, 1997, pp. 1–5. Reprinted by permission.

Introduction

The overall goal of the United States' Acid Rain Program is to achieve significant environmental and public health benefits through reductions in emissions of sulfur dioxide (SO_2) and nitrogen oxides (NO_x), the primary causes of acid rain. To achieve this goal at the lowest cost to society, the program employs both traditional and innovative, market-based approaches for controlling air pollution.

At the core of the SO_2 program is a cap, or mandatory ceiling, on total SO_2 emissions from utilities. After full implementation of the program, annual emissions will be capped at 8.95 million tons, which represents a 50 percent decrease in SO_2 emissions from electric utilities from 1980 levels. Enforcing this cap is a rigorous monitoring system that allows EPA to ensure accuracy and verifiable emissions reductions.

In a dramatic departure from traditional command and control regulatory structures, this program utilizes a market-based system of tradable emissions allowances. The system allows regulated utilities flexibility in how they attain compliance, so that they can select the most cost effective strategy for their particular facilities.

The program was established in the 1990 amendments to the Clean Air Act, the nation's air pollution control law. As stated in Title IV of the Clean Air Act, the primary goal of the Acid Rain Program is to reduce annual SO_2 emissions by 10 million tons below 1980 levels. To achieve these reductions, the law requires a two-phase tightening of restrictions placed on fossil fuel-fired power plants, which accounted for two thirds of national SO_2 emissions in 1980.

Phase I began in 1995. It affects 263 boilers, or "units," at 110 mostly coal-burning electric utility plants that were identified in the law as the largest SO_2 emitters. An additional 182 units voluntarily joined Phase I of the program early as substitution or compensating units, bringing the total of Phase I affected units to 445.

Phase II, which begins in the year 2000, tightens the annual emissions limits imposed on these large, higher emitting plants. In addition, restrictions come into effect on smaller, cleaner plants fired by coal, oil, and gas. This phase affects existing utility units serving generators with an output capacity of greater than 25 megawatts and all new utility units.

The Clean Air Act also calls for a 2 million ton reduction in NO_x emissions by the year 2000 through a more conventional regulatory program. A significant portion of this reduction will be achieved by coal-fired utility boilers that will be required to install low NO_x burner technologies and to meet new emissions standards.

The Allowance System

Allowance trading is the innovative tool for achieving SO_2 emissions reductions required by the cap. Allowances are the currency with which compliance with the SO_2 emissions requirements is achieved; one allowance authorizes a regulated unit to emit one ton of SO_2 during a specific year or any year thereafter. At the end of each year, the unit must hold an equal or greater number of allowances as the tons of SO_2 emitted that year, i.e., a unit that emits 5,000 tons of SO_2 must hold at least 5,000 allowances that are eligible to be used that year.

The allowance trading system gives utilities flexibility in devising compliance strategies. This flexibility is a result of the fact that a regulated unit need only to hold enough allowances at year end to account for its emissions. Unlike traditional regulatory programs, EPA does not dictate the compliance options. Options might include employing energy conservation measures, increasing reliance on renewable energy, reducing usage, employing pollution control technologies, switching to lower sulfur fuel, or developing other alternate strategies to reduce emissions.

Because allowances have a value and can be purchased and owned, they are sometimes misconstrued as "rights" to a certain amount of pollution. However, the law makes clear that allowances are merely authorizations to emit, since at all times the government retains the

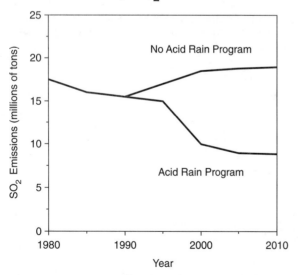

Utility SO_2 Emissions

The Acid Rain Program will result in a 10 million ton reduction in SO_2 emissions from 1980 levels by the year 2010.

authority to require affected sources to comply with the Clean Air Act limits. The concept of authorizing sources to emit is common to all air pollution regulations; no regulations dictate zero air emissions. With more traditional forms of regulations, sources are authorized to emit through emissions limits or through requirements to install specific technologies.

The financial value and enduring validity of excess allowances encourage SO_2 reductions beyond what is required by the present year's allotment. Allowances are fully marketable commodities: units that reduce their emissions below the number of allowances they hold may trade their excess allowances with other units in their system, sell them on the open market to other utilities or through EPA auctions, or bank them to cover emissions in future years.

Banking

Banking encourages early emissions reductions by allowing participants to store their excess allowances and by assuring that those allowances will retain their compliance value in future years. Utilities may use banked allowances to help comply in later years, or they may sell them. In the first year of the Program's implementation, Phase I affected units collectively achieved 40 percent greater emissions reductions than was required by law. This dramatic result demonstrates that banking leads to SO_2 reductions earlier than if banking were not permitted. These extra reductions result in environmental and health benefits that begin sooner.

Banking also reduces compliance costs by allowing utilities more flexibility in the timing of their pollution control investments.

Phase I affected sources are likely to be banking allowances in anticipation of the more stringent emission limits and higher costs in Phase II. It is expected that rather than a sharp drop in emissions at the start of Phase II, reductions will move toward the ultimate cap more gradually as banked allowances from previous years are used for compliance. Although the reduction rate may slow down after Phase II, the environmental and health benefits sparked by the initial overcompliance will have been accruing for several years.

Regardless of how many allowances a unit holds, it is never entitled to exceed the source-specific limits set under Title I of the Clean Air Act to protect public health.

Allocations

EPA allots allowances to each affected unit for each year, beginning with Phase I units in 1995. These allocations were determined prior to the start of the trading program, based on selected emission rates and each unit's representative fuel utilization level. In Phase I, the specified emission rate was 2.5 pounds of SO_2/mmBtu (million British thermal units). This rate was applied to units with existing rates greater than 2.5 pounds/mmBtu and was multiplied by the unit's average fuel use from 1985 through 1987. These allocations are listed in the Clean Air Act and codified in the Allowance System Regulations.

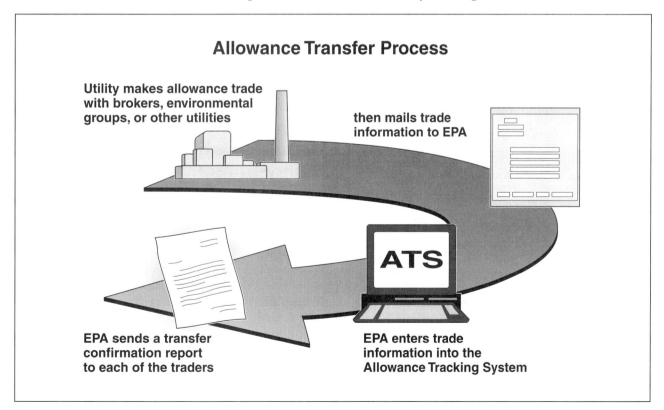

In Phase II, the limits imposed on Phase I plants will be tightened, and emissions limits will be imposed on smaller, cleaner units that were not included in Phase I. EPA allocated allowances to each unit at the lesser of its existing emission rate or 1.2 pounds of SO_2/mmBtu, multiplied by the unit's baseline fuel consumption.

Equity Issues in Allowance Allocations

In addition to the standard formulas for allocating emission allowances, the Clean Air Act Amendments set forth special provisions to address equity concerns raised by some states. For example, in some cases, states that had already reduced the emissions of their electric utilities to levels well below the national average were given extra allowances. Similarly, a state with high population growth in the 1980s was given bonus allowances for its electric utilities to compensate for this growth. In all cases, these redistributions of allowances were done without increasing the overall level of national emissions. In other words, increases in allowance allocations in certain states or at certain electric utilities were offset by decreases in the allowance allocations to other states or emissions sources.

Permits

Permitting under the Acid Rain Program is simple and flexible. For example, the Phase II permitting form consists primarily of a statement of standard legal requirements in the Clean Air Act that companies certify they will meet. Utilities may select from several compliance options, devise the most cost-effective compliance plan, and revise plans easily without needing government review or approval. The standard permit forms ensure consistency on a nationwide basis.

Allowance Transfers

EPA is responsible for recording the transfer of allowances that are used for compliance and for confirming that utilities hold at least as many allowances as tons of SO_2 emitted by the end of the year.

To fulfill this role, EPA designed and maintains the Allowance Tracking System (ATS), a computer program that is the official record of allowance holdings and transfers. Every utility unit, corporation, group, or individual holding allowances has an account with ATS, held in the name of a designated individual. EPA established accounts for utility units affected by both Phase I and Phase II. In addition, any person or group wishing to purchase allowances may open an ATS account.

The structure of the Acid Rain Program simplifies the transfer of allowances, thereby lowering the cost of transactions. The transfer process is comprised of four steps for utilities. A utility makes an allowance trade with another utility, a broker, or environmental group. They enter information about the trade on a one-page form and mail it to EPA. Next, EPA enters this information into ATS and then sends a transfer confirmation report to each of the traders. Because there is no need for case-by-case approval of trades, allowance transfers are straightforward and fast. Only transfers of allowances to be used for compliance require EPA notification; notification of other transfers is voluntary.

ATS also makes the movement of allowances easy to track. ATS records the issuance of all allowances, the holdings of allowances in accounts, the deduction of allowances for compliance purposes, the transfer of allowances between accounts, and the number of allowances held in EPA reserves. Information on the ATS accounts is available to the public via the Internet.

Determining Compliance

At the end of the year, units must hold in that year's compliance subaccount an amount of allowances equal to or greater than the amount of SO_2 emitted during that year. By January 30 following the compliance year, units must finalize and report to EPA allowance transactions that were used in attaining quantities of allowances necessary for compliance. The amount of emissions is contained in the Emission Tracking System (ETS), which is operated by the Acid Rain Program and records each hour of each unit's hourly emissions throughout the year.

After the January 30 deadline, EPA deducts allowances from each unit's compliance subaccount in an amount equal to its SO_2 emissions for that year. If the unit's emissions do not exceed its allowances, the remaining allowances are carried forward, or banked, into the next year's subaccount.

Ensuring Environmental and Health Benefits through Stringent Monitoring

Tradable allowances are intended to achieve environmental goals at the lowest cost to society. However, even if no trades were to occur, these environmental goals would still be met by the Acid Rain Program. The Program uses several mechanisms to ensure that its environmental and health goals are met. During Phase II, the Clean Air Act places a cap of 8.95 million on the total number of allowances issued to units each year. This effectively caps emissions of 8.95 million tons annually and ensures that the mandated emissions reductions are maintained over time. This reduction represents a 50 percent decrease in SO_2 emissions from 1980 levels.

To enforce the integrity of the emissions limit, EPA tracks on a continuous basis each unit's emissions of SO_2, NO_x and CO_2, as well as volumetric flow and opacity. In most cases, a continuous emission monitoring (CEM) system must be installed. Through the CEM system, units report their hourly emissions data to EPA on a quarterly basis. The reporting process is evolving from the submittal of diskettes to the use of modems, which has greatly increased the efficiency of reporting. The data is then recorded in the Emissions Tracking System, which serves as a repository of emissions data for the utility industry.

Strong enforcement measures deter noncompliance. If annual emissions exceed the number of allowances held, the owners or operators of delinquent units must pay a penalty of $2,000 (adjusted for inflation) per excess ton of SO_2 or NO_x emissions. this fee is substantially higher than the cost of compliance, e.g., an allowance. In addition, the number of exceeded allowances is deducted from the violating utility's account for the next year to fully offset the environmental impact.

The emissions monitoring and reporting systems are critical to the program. Monitoring ensures, through stringent accounting, that the SO_2 and NO_x emissions reduction goals are met. They also instill confidence in allowance transactions by certifying the existence and quantity of the commodity being traded.

Emissions data is available to the public in several formats, including over the Internet. Such public access provides the program transparency that assures integrity and public trust in the system.

Results

The Acid Rain Program's environmental goals are already starting to be met. After the first year of implementation, the Acid Rain Program has witnessed large reductions in SO_2 emissions. In 1995, all 445 Phase I affected units were in compliance, and utilities reduced emissions 40 percent more than required by the cap. Annual emissions from these Phase I units dropped by more than half between 1980 and 1995, from 10.9 million tons to 5.3 million tons.

Environmental effects corresponding to these early reductions are starting to emerge. A United States Geological Survey study determined that 1995 wet sulfate deposition declined in the eastern United States by 10 to 25 percent. Other anticipated benefits include avoided health costs of $12 to $40 billion per year by 2010; improvements invisibility amounting to $3.5 billion by 2010; fewer acidic lakes and streams; and reduced damage to buildings and outdoor cultural artifacts.

At the same time, the Acid Rain program has cost significantly less than the benefits, and estimates for the program's long term costs continue to drop. As of 1994, the estimated cost of the program was $2 to $2.5 billion per year by 2010. The cost is half of what a command and control regulatory program would cost.

The cost of reducing a ton of SO_2 from the utility sector has been much lower than expected; scrubber costs have dropped, removal efficiencies have improved, and expected increases in costs associated with the increased use of low sulfur coal have not materialized.

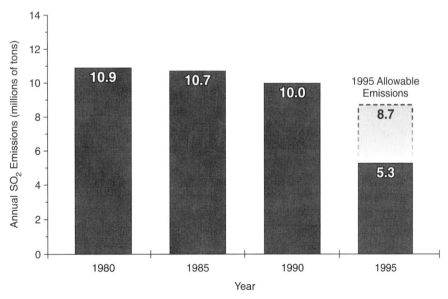

SO_2 Emissions
445 Phase I Affected Utility Units

Emissions at Phase I affected utility units were 3.4 million tons below their required level in 1995.

These reductions in cost are reflected in allowance prices. In just two years, allowance prices have dropped from $150/ton to less than $100/ton.

Important Roles Played by Participants

Successful implementation of the Acid Rain Program hinges on straightforward and well-defined roles for each participant. EPA monitors utilities' emissions to ensure compliance, and measures the environmental effects of the program. Industry's role is to reduce emissions in the most cost effective manner. Brokers provide information on the market and facilitate allowance transfers for utilities and other parties, thus reducing transaction costs. Finally, environmental organizations monitor the overall performance through access to emissions and allowance data. In addition, these groups may purchase and retire allowances, a small but direct action individuals can take to reduce pollution.

Just as each participant has a role in ensuring the integrity of the Program, each element of the Program works in concert to bring about efficient emissions reductions. The allowance trading system capitalizes on the power of the marketplace to reduce SO_2 emissions in the most cost effective manner possible. The permitting program allows sources the flexibility to tailor and update their compliance strategy based on their individual circumstances. The continuous emissions monitoring and reporting systems provide the accurate and standardized accounting of emissions necessary to make the program work, and the excess emissions penalties provide strong incentives for compliance. Each of these separate components contributes to the effective working of an integrated program that harnesses market incentives for the benefit of the environment.

"Acid Rain: A Continuing National Tragedy"

© 1998 by The Adirondack Council. Reprinted by permission.

"We still have a very major problem with acid rain. That is scientific fact. In that regard, the 1990 Clean Air Act Amendments have not worked very well."
Acid rain scientist Dr. Gene Likens, *Boston Globe*, February 8, 1998

Despite the Clean Air Act Amendments of 1990, acid rain continues to degrade ecosystems in high-elevation forests and waters from New York's Adirondack Park, Taconic Ridge, Catskill Park and Hudson Highlands; through New England's Green and White Mountains and the mid-Appalachian range in the Carolinas and Tennessee. Associated nitrogen-based air pollution is ruining aquatic habitat in Long Island Sound and the Chesapeake Bay and even Colorado's Rocky Mountains are now showing air-pollution-related damage.

The air pollution that causes acid rain has been falling on some areas of the United States for nearly a century. But the damage acid rain causes can take a long time to develop. In many of the most heavily damaged regions—such as the forests of New York and New England—scientists have been documenting ecological damage since the 1970s. Now, other regions are discovering that their health and environment are suffering too.

In 1990, Congress amended the Clean Air Act and instructed the U.S. Environmental Protection Agency to create the nation's first acid rain control program. In 1992, the Bush Administration boasted that the new program would "end acidity in Adirondack lakes and streams." But many recognized right away that the program would be inadequate to stop the destruction in the Adirondack Park and the nation's other most sensitive ecosystems.

In 1993, the NYS Dept. of Environmental Conservation, the Natural Resources Defense Council and the Adirondack Council sued the EPA over the new program. In a partial settlement of the suit, the EPA agreed to complete a 1996 report to Congress on whether the new program would have the desired effect. The report confirmed our fears.

The EPA noted that the current federal acid rain program could only slow the rate of damage done to the Adirondack Park. More lakes would die. Meanwhile, acid rain and the air pollution that causes it are damaging other areas of the nation at an alarming rate.

Acid rain has been harming the ecosystems of the Catskill Mountains for as long as the Adirondacks. While the Catskill Park is one-tenth the size of the Adirondack Park and has far fewer lakes, its legendary rivers and trout streams have lost much of their vitality.

Farther south, the mid-Appalachian Mountains are being devastated. Spruce forests are dying, streams are losing their fish. Insect infestations in forests threaten to wipe out entire species.

In New England, studies by acid rain research scientist Dr. Gene Likens showed that the hardwood forest of New Hampshire's Hubbard Brook area has stopped growing. Sugar maples are at a particular risk—bad news indeed for furniture and syrup makers.

Up north, the Canadian government estimates that by 2010, even with full implementation of the Canadian and American acid rain programs, an area the size of France and Britain in southern Canada will continue to receive harmful levels of acid rain. As many as 95,000 lakes will remain damaged, they stated in 1997.

Out west, scientists in the Rocky Mountains are finding that power plant emissions are saturating high-elevation watersheds in Colorado with acid-causing nitrogen. Evergreen forests are losing their needles and tree health is declining throughout the Front Range.

Acid rain damage is not limited to forests and aquatic ecosystems. In Pennsylvania, the monuments at the Civil War battlefield in Gettysburg are deteriorating far more quickly than similar structures in places not affected by acid rain. Throughout the Northeast, stone, brick and block buildings, as well as automobile finishes, show signs of more extensive and rapid weathering than counterparts in other regions of the country.

In the Chesapeake Bay and Long Island Sound, nitrogen-based pollution is overloading the water with nutrients. This contributes to an overabundance of algae, which when they die and decay, deplete the water of precious oxygen needed by all aquatic animals. The condition is known as hypoxia.

Closer to our homes, acidity in water supplies is leaching poisonous metals such as lead into the drinking water. Copper is killing the beneficial bacteria that make septic systems function. Airborne particles of sulfur—the chief component of acid rain—also cause and worsen lung diseases.

The few fish species that can survive in acidic waters are accumulating mercury in their body tissue. Now, mammals and birds that live on those fish are showing signs of mercury contamination. More than 500 lakes and ponds (out of 2,800) in the Adirondack Park are already too acidic to support the plants and aquatic wildlife that once existed in them. Each spring, an entire winter's acidic snowpack melts into the Park's waters, jolting them with a huge jump in acidity known as "acid shock." It could not happen at a worse time. Many of the Park's plants, animals and insects are at their most vulnerable at the beginning of the growing season.

Red spruce forests on the western-facing slopes of the Park's High Peaks region are stunted and dying at a rapid pace. Those forests receive extremely high levels of polluted precipitation that blows in from the coal-fired smokestacks of the Ohio Valley and beyond. Day after day, even when it doesn't rain or snow, the pollution hangs in acid clouds that shroud the mountains in a caustic fog.

Adding insult to a long list of injuries, Canadian studies show that the larvae of black flies—the bane of spring outdoor activities in the Northeast and southern Canada—seem to thrive in acidic waters. Consequently, their populations are exploding as pollution changes the chemistry of the waters from which they hatch.

The Adirondack Park is suffering the worst damage in the nation from acid rain. And because nearly all of the utility plant pollution that causes acid rain in the Adirondacks comes from outside the state, New Yorkers alone can do little to prevent the onslaught.

The good news is that the current acid rain program is costing utility companies far less than they predicted when Congress was contemplating the Clean Air Act Amendments of 1990. As a result, the total costs of finishing the job Congress intended to do in 1990 would still be less than original estimates.

Appalachian Damage

Often, a tree weakened over time by acid rain will die from what—on the surface—appears to be an unrelated disease or other stress factor. Species such as spruce and fir, for example, can succumb to severe cold, drought, insect infestations or diseases that may have caused little damage to a tree located where acid rain is less severe.

Reports by the U.S. Forest Service indicate that death rates for many tree species have doubled or tripled in parts of the greater Appalachian range since the 1960s and 1970s. The greatest proportion of dead trees are located in areas that receive the highest doses of sulfur- and nitrogen-based air pollution.

The Mechanics of Acid Rain

Acid-rain-causing pollution is carried on prevailing winds and can drift for hundreds of miles before it is deposited by precipitation. Adirondack, Catskill and Appalachian mountain regions are the hardest hit because prevailing winds carry the pollution from several other states onto those mountain ranges. As the winds rise over the mountains, the moisture they contain cools and condenses into clouds, which reach the point of saturation. The resulting rain, snow, sleet and/or fog has high concentrations of sulfur and nitrogen pollution. The sulfur-dioxide becomes sulfuric acid.

The nitrogen becomes nitric acid. Alkaline minerals such as calcium and magnesium in soil and the air (a.k.a. base cations) help buffer acidity. But acid rain washes those minerals out of the soil faster than weathering can replace them by breaking down rock.

[Editor's Note: See Figure 20-14 in Raven/Berg *Envrionment* 3e for an illustration of the mechanics of acid deposition.]

Mercury Kills

Loons require pristine shorelines and seclusion to successfully nest and breed. They find vast areas of suitable habitat within the Adirondack Park. But their habitat is shrinking in most other areas of the Northeast due to human encroachments. As a result, the loon has become a symbol of the health and solitude of the Adirondack wilderness.

Unfortunately, the fish that loons eat are becoming increasingly contaminated with mercury—one of the deadliest toxic metals associated with acid rain.

Cranberry Lake and Stillwater Reservoir are two of the largest and most popular water bodies in the western Adirondack Park. They lie at the northern and southern boundaries of the Five Ponds Wilderness Area, which contains the largest contiguous virgin forest remaining in the Northeastern United States. In the mid-1990s, both lakes were found to have substantial mercury contamination. Like aluminum, mercury can be leached out of soil by acidity, but it is also found in the same smokestack pollution that causes acid rain.

In 1996, something new and alarming happened. New York State officials advised women of child-bearing age, the elderly, and young children to avoid eating yellow perch and smallmouth bass from either lake. Both fish are less susceptible to aluminum-related gill damage than most species of trout and salmon. But that resistance allows them to live longer in acidic waters, where they slowly accumulate mercury in their fatty tissues.

The NYS Health Department has now issued similar mercury warnings for more than a dozen Adirondack lakes. Every single Northeastern state now has mercury consumption warnings for fish taken from its waters.

The poisoning of these valuable game fish is a tragic blow to the Park's tourism industry. And the contamination is growing deadly for bird and mammal species that rely on fish for food, since the accumulated mercury in fish is transferred to whomever, or whatever, eats the fish.

An Adirondack Case Study

Fifty years ago, Big Moose Lake was teeming with life. For a half-century, tourists flocked to the lake from near and far to escape city life and relax on the shore of a pristine lake, fifty miles from the bustle and pollution of the nearest urban center.

Trophy-sized brook trout, white fish, landlocked salmon and lake trout abounded, beckoning anglers from throughout the world to ply its remote, chilly waters. Acid rain has exterminated those fish species. Former natives such as crayfish, freshwater shrimp, frogs, hooded mergansers and otters are rarely seen anymore.

By 1980, the tourist hotel operators had given up on Big Moose Lake's fishing as a means of attracting tourists. They watched helplessly as millions of dollars in potential revenues slipped from their collective fingers. Worse yet, one lakeside business was about to discover that acid rain can make people sick, too.

Covewood Lodge owners Diane and C.V. "Major" Bowes were dumbfounded when their children began complaining about the taste of their drinking water, which was drawn from a well next to the lake. When one of their young daughters developed stomach cramps and diarrhea, the Boweses had their water tested.

Test results showed the water contained five times as much lead as is deemed safe for human consumption. The water contained copper as well. Both metals were being leached out of the inside of their pipes and plumbing fixtures due to the corrosive water. Lead is highly toxic to humans. But copper also kills the beneficial bacteria that allow septic systems to break down wastes and purify wastewater. The Boweses now treat their water to make it safe for drinking.

In 1996, the NYS Dept. of Health began conducting tests of water supplies throughout the state. Health officials reported that nearly all of the lakes, ponds and reservoirs they tested for toxins in the Adirondacks and Catskills were at least slightly acidic. Often, the tests showed no chemical contamination in the source of the water; but high levels of lead and copper coming out of the taps in people's homes.

How Do We Solve This Problem?

Given all of the damage done to our health and environment by acid rain and the toxic chemicals associated with it, it is unconscionable that this destruction is allowed to continue. But utility companies that burn coal and produce the most acid rain also produce the cheapest electricity in the country. That is a strong incentive to keep doing what they are doing. In 1992, the U.S. Environmental Protection Agency created an acid rain control program to reduce sulfur-dioxide emissions by 50 percent nationwide. We now know that this goal is too low. In addition, the federal government has done little to control nitrogen-oxide emissions from electric plants. Research has demonstrated that nitrogen-based air pollution is a main culprit in the destruction of many life forms in lakes, rivers and coastal estuaries. The good news is that there is a solution to the acid rain problem. Scientists in Northeastern states and southern Canada agree that sensitive areas of the country would have a fighting chance at recovering from decades of pollution if utility companies reduce their sulfur-dioxide and nitrogen-oxide emissions by 70 to 75 percent below 1990 levels. Better yet, the solution is simpler and less expensive than most people think.

1. **Make the existing program work.** Private industry has already grown accustomed to the current sulfur-dioxide control program. Making the program work better would be more practical than trying to tear it down and start over. The program sets a cap on the total amount of pollution allowed nationwide and allows individual companies to buy and sell the rights to that pollution. This allows those who can make the largest and least expensive cuts to do so right away. They are then free to sell the rights to the pollution they don't emit. Over time, the total number of pollution rights (a.k.a. allowances) issued by EPA will drop by 50 percent below 1990's pollution levels. **Action: Reduce the cap another 50 percent. This would bring the total reduction in pollution to 75 percent below 1990 levels.**

2. **Create a new pollution-trading program for nitrogen-oxides.** Creating a federal allowance-trading program similar to the sulfur-dioxide program would give utility companies a financial incentive to make deep cuts in emissions. **Action: Create a nitrogen-oxide pollution trading program that reduces nitrogen-based air pollution by 70 percent below 1990 levels.**

3. **Keep monitoring the results.** Cuts in smokestack emissions have caused a drop in the amount of acid rain chemicals hitting the ground, but some watersheds have been so hard-hit for so long, deeper cuts will be needed—nationwide or in a targeted geographic region—to ensure that they recover. In addition, any new cuts in sulfur, nitrogen and mercury should be coupled with monitoring of the effects of the cuts on the ground. **Action: Biological surveys and chemical tests should be performed on a regular basis at least through the year 2020 to ensure that the pollution cuts made in the meantime have the anticipated effect.**

4. **Give EPA the authority to keep making cuts.** Congress expected EPA to fix the nation's acid rain problems when it approved the Clean Air Act Amendments of 1990. That did not happen and EPA says it has no authority to order deeper cuts on its own. **Action: EPA's Administrator should have explicit authority to order new cuts to protect human health and sensitive ecosystems without further Congressional action.**

What Is All Of This Going To Cost?

In 1992, utility companies said it would cost them a total of $6 billion a year to comply with the federal acid rain program. A recent study by the Massachusetts

Institute of Technology showed that the actual cost of compliance is less than $800 million annually. That is less than one-seventh of the projected price. EPA predicted in 1990 that allowances would trade for $1,200 to $1,500 per ton by now. The average cost in 1998 was less than $120.

Creating an allowance trading program to control both sulfur-dioxide and nitrogen-oxide emissions would be very attractive to utility companies and securities brokers, who can help us convince Congress to act right away—before more lakes and forests die.

For the same amount that we expected to spend as a nation just on sulfur-dioxide, we can reduce both sulfur-dioxide and nitrogen-oxides by 70 to 75 percent below 1990 levels and continue monitoring the results.

reading 4

"A Washington, D.C. Press Conference"

The Adirondack Council Newsletter, August 1999, p. 4. Reprinted by permission.

On June 16, the Adirondack Council hosted a press conference on acid rain at the U.S. Capitol, joining forces with a U.S. Senator and two Congressmen, as well as historic preservation organizations and groups striving to protect outdoor artworks. The message to the rest of Congress was simple: Pass the *Acid Deposition and Ozone Control Act.*

Joining the Council at the press conference were U.S. Sen. Charles Schumer and U.S. Reps. Sherwood Boehlert, R-Utica, and John Sweeney, R-Saratoga. Also at the press conference were representatives from the National Trust for Historic Preservation, the Save Outdoor Sculptures Project (Heritage Preservation), D.C. Preservation League, Potomac Heritage Partnership, Historical Society of Washington D.C., National Audubon Society, Trout Unlimited and Citizens Campaign for the Environment.

The press conference focused on the damage acid rain has done to historic buildings and monuments in the nation's capital. The Adirondack Council is attempting to broaden the coalition of organizations and politicians who are interested in stopping acid rain by showing people from various parts of the nation and various interest groups what they have at stake.

Susan Nichols of Save Outdoor Sculptures captured the attention of the media with her discussion of the damage done each year to priceless works of art, including the Statue of Liberty and any outdoor sculpture made of limestone, marble, copper or bronze.

"Acid rain is the leading cause of damage to outdoor statuary today," Ms. Nichols told reporters from the Associated Press, CNN, C-Span and NBC, who had gathered on the lawn overlooking the U.S. Senate at the Capital.

From the Capitol building itself to the Lincoln and Washington memorials, acid rain destroys the surface of carved limestone and marble by washing away the calcite. Pits develop on the smooth surfaces at the base of the columns. Anyone who brushes across the surface comes away covered with a chalky, white powder.

On copper and bronze objects, sulfur dioxide pollution combines with the copper to form copper-sulfate, which runs down the surface as a liquid. Holes and streaks develop in the metal, while smooth surfaces turn rough.

Whether the damage is done to stone or metal, the loss of fine details and intricate carvings happens much more quickly in acidic precipitation than in untainted rain. The damage is irreversible. Faces become smooth and amorphous. Names, dates and other inscriptions disappear.

The *Acid Deposition and Ozone Control Act* (S.172/H.R.25) is sponsored by Senators Daniel Patrick Moynihan, D-NY, Charles Schumer, D-NY, and James Jeffords, R-Vt., Jack Reed, D-R.I., Joseph Lieberman, D-Ct., John Kerry, D-Mass., Dianne Feinstein, D-Ca., and Barbara Boxer, D-Ca. House sponsors include Sherwood Boehlert, R-NY and John Sweeney, R-NY, and more than 40 of their colleagues.

The bill calls for an additional 50 percent cut in sulfur dioxide pollution—on top of the 50 percent cut ordered by Congress in 1990, for a total cut of 75 percent below 1990 levels. It also would require a 70 percent reduction in nitrogen oxide pollution (which causes both acid rain and smog) from electric power plants.

reading 5

"Public Service Announcement Campaign a Ringing Success"

The Adirondack Council Newsletter. August 1999, p. 5. Reprinted by permission.

In July, the Adirondack Council released nearly 1,000 copies of radio and television public service announcements on the need to stop acid rain. Calls have been streaming into the Council's Acid Rain Hotline from Florida, California, West Virginia, Rhode Island and the Carolinas, where stations appear to be giving the ads excellent air time.

The voices of musicians Bonnie Raitt and Natalie Merchant are featured in the national campaign.

"New York's Adirondack Park is the one place in the nation hardest-hit by acid rain, but we are by no means alone in suffering extensive damage. The Adirondack Council has been fighting acid rain in the Adirondacks for two decades. Now we are organizing and funding the national campaign to stop acid rain across America," said Adirondack Council Executive Director Timothy J. Burke. "We are pleased to have the assistance of the nation's most respected environmental and historic preservation organizations, who joined us in a *New York Times* ad calling on Congress for deeper pollution cuts. And we are grateful that Bonnie Raitt and Natalie Merchant are lending their well-known voices to this effort."

Questions for Discussion

For Readings 1 and 2

1. Basically, how does EPA's plan work to reduce SO_2 and NO_x? How are these emissions different?

2. How does the EPA measure success?

3. According to the EPA, how successful has the plan been in reducing SO_2 emissions?

4. According to the EPA how successful has the plan been in reducing NO_x emissions?

For Reading 3

1. According to this reading, how serious is the problem of acid precipitation in the Adirondacks and in other regions. What are the sources of acid precipitation?

2. In addition to reducing emissions of SO_2 and NO_x, what are some other possible solutions to the problem? Would they be short-term or long-term solutions?

For Readings 4 and 5

1. Describe how acid precipitation can adversely affect urban environments.

2. The more people are concerned with acid precipitation, the more likely it is that Congress will enact further legislation. What other citizen groups might join the Adirondack Council in voicing their concerns? Describe how the Adirondack Council is soliciting their cooperation.

3. In your opinion, has the plan been successful? How might it be improved?

Going Beyond the Readings

Visit our Web site at <www.harcourtcollege.com/lifesci/envicases2> to investigate acid precipitation issues in additional regions of the United States and Canada.

unit 10

Southwest: Air Quality

Introduction
Readings:
Reading 1: Smog City Case Studies
Reading 2: Cleaner Fuels and Cleaner Cars
Reading 3: The False Promise of Electric Cars
Reading 4: New Vehicles Now Less Efficient Than Junked Ones
Questions for Discussion

Improving Air Quality in Los Angeles

Introduction

If there is one thing we would like to take for granted, it's the air we breathe. In a typical day, the average adult takes in more than 4000 gallons (15,000 liters) of air. Oxygen is absolutely essential to all of us and there is no substitute. We expect the air we breathe to be clean and free of impurities.

For a growing number of us, such is not so. Worldwide, more than half the human population living in urban areas breathes unclean air at least some of the time. Transportation and industrial processes as well as meteorological conditions conspire to pollute the air of urban areas with unpleasant, sometimes dangerous substances. The list of pollutants is long and troubling. Major pollutants include:

- Sulfur dioxide   from gas- or oil-fueled factories and power stations; heating plants; waste incinerators; diesel engines
- Nitrogen dioxide   from coal-, oil-, or gas-fueled furnaces; waste incinerators; motor vehicles
- Carbon monoxide   from motor vehicles; combustion of gas, oil, or coal
- Volatile organic compounds (VOCs)   from gas-, diesel-, or oil-fueled engines; leakage at gas stations; paints
- Ozone   from sunlight triggering chemical reactions between VOCs and nitrogen dioxide

- Suspended particulates from coal- and oil-fueled furnaces; waste incinerators; gas- and diesel-fueled vehicles
- Greenhouse gases (carbon dioxide and methane) — carbon dioxide (CO_2) from fuel combustion; methane (CH_2) from leakage at gas stations; coal mining; landfills

Under most weather conditions, especially where human numbers are low, these pollutants simply diffuse into the air causing little harm. Not in cities. High-pressure weather systems may set up inversions, in which relatively warm air masses are trapped beneath blankets of cooler air. When winds are low and conditions are stable, pollutants are trapped and concentrations build up to harmful levels.

Mountains, such as those around Mexico City or Los Angeles, worsen air-pollution problems. In the worst cases, these conditions persist for months and human health is affected. Young children are particularly susceptible. Infants' lungs are not fully developed; their airway passages are relatively narrow. The rate at which they take in air is, pound for pound, nearly double that of adults. As infants grow into children, they spend time playing outdoors, vigorously gulping in air. Breathing through open mouths avoids the filtering action of nasal passages. Polluted air damages young lungs and air passages and sets the stage for lifelong, chronic respiratory problems. In the United Kingdom, one child in seven now experiences chronic wheezing.

To people suffering chronic respiratory problems, air pollution intensifies and worsens their condition. Pregnant women, because of increased oxygen needs during pregnancy, are also at risk, as are their unborn children. Older people are even more susceptible. Decreasing lung capacity is a normal part of aging. Pollutants in air hasten the process, often shortening life.

Together, these groups may make up 20 percent of an urban population. They are not the only ones affected by air pollution. Serious smog episodes cause burning eyes, running noses, throat irritation, headaches, tiredness, and nausea, even to healthy people. Those who work or exercise outdoors strenuously, such as laborers and joggers, increase exposure and experience increased rates of respiratory illnesses. Air pollution may be causing a worldwide increase in asthma and allergies. Indeed, air pollution is a problem that affects all those living in or near cities.

Point sources—sites such as factories or power plants that produce significant amounts of air pollution—are relatively easy to identify and remedy, albeit not always effectively. Tall stacks on such facilities place emissions relatively high in the air, where winds carry them away; however, sulfur dioxide and oxides of nitrogen react with water vapor to produce sulfuric and nitric acids and return to earth as acid rain. Solving a local problem creates a regional one. Scrubbers placed on smoke stacks trap and reduce but do not eliminate emissions. Conversion to alternate fuels—usually oil, natural gas, or coals containing low levels of sulfur—reduces emissions in some cases. Major obstacles to pollution reduction measures at point sources are primarily economic and political.

Nonpoint sources of air pollution, mainly motor vehicles, are more difficult to address. This is the single most important source of air pollution worldwide. An estimated 675 million vehicles produce copious quantities of sulfur dioxide, nitrogen dioxide, carbon monoxide, carbon dioxide, VOCs, ozone, particulates, and toxic metals, particularly lead and cadmium. The problem is only going to get worse as demand for vehicles increases, particularly in developing nations. Currently, the worldwide auto industry is producing vehicles at the rate of more than one per second, or 34 million a year—a rate of increase significantly greater than the growth in overall population. At that rate, the world's already crowded roadways will be carrying more than a billion cars by the year 2030. As poor people become affluent, their first cars are usually older, cheaper models, which are often the worst polluters. A worst-case offender can produce more pollutants than 50 normal vehicles.

Solutions are few. Attempts to get people to use alternate, less-polluting means of travel have met with only limited success. Catalytic converters—devices placed on vehicle exhaust systems to trap or chemically change emissions—reduce but do not eliminate emissions. Those typically in use today take several minutes to warm up, during which time emissions are uncontrolled. Cars equipped with converters use relatively richer fuels, whose pollutant levels are potentially greater than low-grade fuels. Better converters that require less warm-up time are being designed. Recent attention to reduce vehicle emissions is focusing on more efficient cars—low-emission vehicles that use less-polluting fuels such as alcohol-enriched gasoline or natural gas, and zero-emission vehicles, such as electric cars.

Few cities have worse air pollution problems than Los Angeles. The southern California climate has long periods of warm, stable, still conditions, especially in summer. Temperature inversions are common. With mountains to the north, south, and east, and weather systems approaching from the west, air is trapped in a bowl above the city. There, 14 million inhabitants operate 9 million cars, and 40,000 sources of industrial air pollution. Los Angeles is more spread out than most cities, which increases driving distances. In a typical Los Angeles year, one day in three has unhealthy levels of air pollution, and in 1988 there were 266 such days.

The city has launched a vigorous campaign to lessen air pollution. One of its more controversial measures is to require use of zero-emission cars, even though the technology to produce them does not exist. The reasoning behind such a plan is that creating a huge market for electric cars will stimulate car manufacturers. The following readings are about Los Angeles's air pollution reduction plan and its controversial electric car requirement.

reading 1

"Smog City Case Studies"
D. Elsom

Smog Alert, published by Earthscan, London, 1996. Reprinted by permission on behalf of Kogan Page Limited/Earthscan Publications, Limited.

Illustrating the Seriousness of the Urban Air Pollution Problem

Most of the world's cities are suffering some serious air pollution problems which are being tackled with varying degrees of success. Four case studies are examined here: Los Angeles, London, Mexico City and Athens. Los Angeles was chosen because it is the city which first experienced serious photochemical smogs and where the authorities have applied some of the most radical and stringent pollution control policies in the world in an attempt to rid the city of air pollutants. London is where the term smog was first coined, being used to describe the polluted smoky fogs it had been experiencing for centuries. London highlights the fact that the successful elimination of one major air pollution problem does not necessarily mean that others will not emerge requiring even more stringent pollution control policies and measures. Unlike the other case studies, Athens is not a megacity (commonly defined as urban agglomerations with a population exceeding 10 million), but it highlights how even a medium-sized city generating sizeable pollutant emissions in an unfavourable topographical setting and climate can suffer pollution problems as severe as any megacity. Mexico City is the final case study explored. It offers a dire warning of the unhealthy air quality that many other cities in rapidly developing countries may have to cope with in the near future—if they are not already having to do so—unless economic, social and planning policies can be followed which check the phenomenal surge in pollutant emissions that has characterised most urban development in the world until now.

Los Angeles: Car City, Smog City

Los Angeles—car city, smog city. To many people, linking these two descriptions of Los Angeles explains why the region experiences such poor air quality. Indeed, emissions from 9 million motor vehicles in the Los Angeles basin are responsible for many of its air pollution problems but industry, businesses, urban planners and the climate have all also played a part in creating the serious air quality problem facing Los Angeles today.

Los Angeles first experienced the yellow pall of photochemical smogs in the 1940s. Ever since, these smogs have been stinging eyes, burning throats and lungs, and causing tightness in the chests of residents and visitors to the city. Moreover, the effects can be long term as children growing up in the city have been shown to have suffered a 10–15 per cent reduction in lung function. Autopsies of more than 100 Los Angeles County teenagers and young men who died suddenly between 1987 and 1989 showed that three out of four had some lung damage. The autopsies did not identify causes, but air pollution was believed to be a factor. The human health costs of the basin's air pollution have been estimated at around $10 billion. Even if concentrations of only one pollutant, PM_{10}, could be reduced to meet federal standards, this would prevent 1600 premature deaths annually among those who suffer from chronic respiratory disease.[1]

The Los Angeles basin or South Coast Air Basin (comprising four counties) contains nearly 14 million people, 9 million vehicles and 40,000 industrial sources of pollution. Road transport accounts for 44 per cent of reactive organic gas emissions (VOCs), 55 per cent of nitrogen oxides and 87 per cent of carbon monoxide emissions. Around 70 per cent of the surface of Los Angeles is devoted to the car in the form of roads, driveways, parking and petrol stations. This includes the Santa Monica Freeway (I-10) from Los Angeles to Culver City which is the busiest highway in the US with 288,000 vehicles a day. Pollution emissions become trapped within the basin, which is bounded by mountain ranges on three sides and with the Pacific Ocean forming the fourth side. Air pollution problems are made worse in the basin by the presence of frequent temperature inversions, whereby cool air from the Pacific Ocean underlies warm air aloft. This inversion layer traps pollutants and limits the amount of clean air into which pollutants can dilute while, at the same time, the mountains prevent the pollutants from escaping the basin. Long periods of intense sunshine and high temperatures favour photochemical reactions which produce ozone. Land and sea breezes cause the pollutants trapped in the basin to be recirculated around the basin. The effect of the sea breeze circulation is evident in the spatial distribution of the number of days when ozone exceeds federal standards in that the highest exceedances occur well inland away from downtown Los Angeles, which is the major source of

---

[1] Hall, J V, Winer, A M, Kleinman, M T, Lurman, F W, Brajer, V M and Colome, S D (1992) 'Valuing the health benefits of clear air' *Science*, vol 2655, pp 812–816; Krupnick, A J and Portney P R (1991) 'Controlling urban air pollution: a benefit-cost assessment', *Science*, vol 252, pp 522–528; Lents, J M and Kelly, W J (1993) 'Clearing the air in Los Angeles', *Scientific American*, vol 269 (4), pp 18–25.

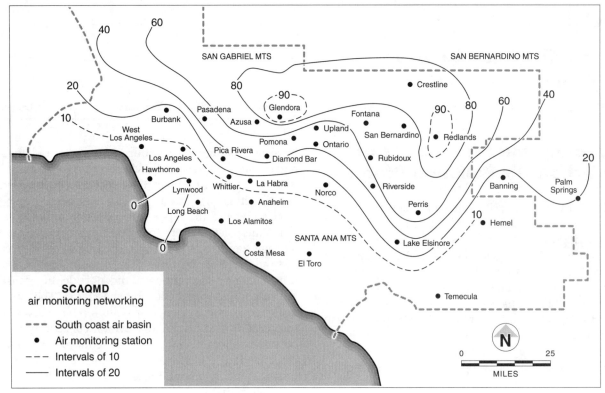

Figure 9.1 Number of days on which the federal one-hour ozone standard of 120 ppb (240µg/m³) was exceeded in the South Coast Air Basin in 1993.
Source: South Coast Air Quality Management District (1994) Final 1994 Air Quality Management Plan: Meeting the Clean Air Challenge, SCAQMD, Diamond Bar, ch 2, p 10.

the precursor emissions (Figure 9.1). A similar spatial pattern occurs for peak hourly ozone concentrations.[2] Some pollution does escape from the basin. For example, of the 90 days in San Diego in 1993 when state standards were exceeded, 51 days (57 per cent) were caused by the transport of pollution from Los Angeles, about 190 km (120 miles) to the north.

The seriousness of California's air quality problem, not just Los Angeles but Sacramento, San Diego and San Francisco, has long allowed it to adopt stricter standards and control policies than those applied nationwide. California claims to have the strictest air quality standards and control programme in the world. A range of standards and alert levels has been adopted for health protection purposes. The state standard for average hourly ozone concentrations is 90 ppb (75 on the Pollutant Standards Index; PSI) whereas the federal standard is 120 ppb (100 PSI). California established a health advisory reporting level of 150 ppb (138 PSI) in September 1991 after medical research showed that ozone posed a health threat at this lower concentration (Figure 9.2). Sensitive individuals are advised to avoid all outdoor activity and athletes are advised to avoid strenuous outdoor activities at this level. A stage 1 episode (commonly called a smog alert) is issued at 200 ppb (200 PSI) and sensitive people are advised to go indoors, and the general public are told to avoid vigorous outdoor activities. A stage 2 episode, when all physical activity should stop, is issued at 350 ppb (275 PSI). Once stage 2 alerts were commonplace (e.g., 15 in 1980), but none has occurred since 1988. A stage 3 episode, last called in 1974, when everyone should stay indoors, is set at 500 ppb (400 PSI).

Despite several decades of pioneering and increasingly stringent control legislation, Los Angeles experienced the worst air quality in the US in 1993 with respect to carbon monoxide, nitrogen dioxide and ozone, together with very high levels of fine particulates. Table 9.1 highlights that Los Angeles–Long Beach and Riverside–San Bernardino are very much worse than other cities in the US with regard to the number of days when ozone exceeded the federal standard from 1984–1993. Interpreting long-term trends from this table is difficult due to the confounding factors of meteorology and emission changes. For example, just as the worsening in 1988 can be attributed in part to meteorological conditions being more conducive to ozone formation than previous years, the 1992 decrease was due,

---

[2] Blumenthal, D L, White, W H and Smith, T B (1978) 'Anatomy of a Los Angeles smog episode: pollutant transport in the daytime sea breeze regime', *Atmospheric Environment*, vol 12, pp 893–907; Elsom D M (1992) *Atmospheric Pollution: a Global Problem*, second edition, Blackwell, Oxford, pp 41–42 and 226–231; GEMS (1992) *Urban Air Pollution in Megacities of the World*, Blackwell for WHO/UNEP, New York, pp 25, 135–146.

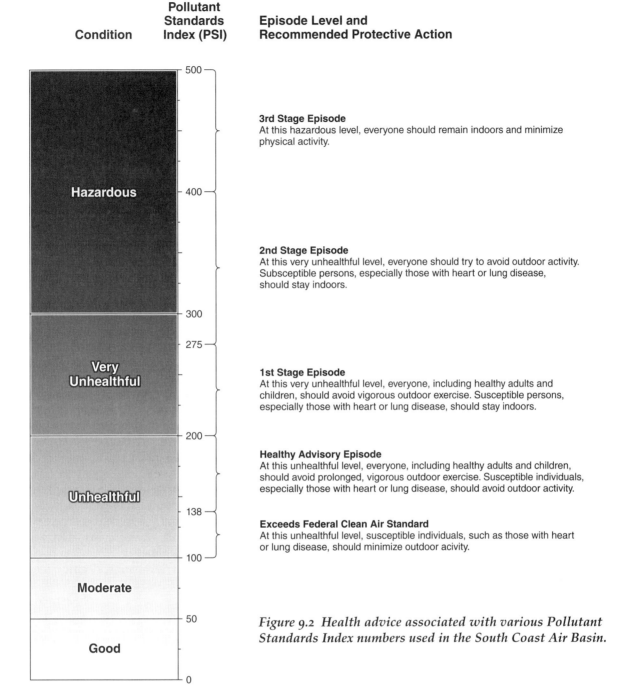

Figure 9.2 *Health advice associated with various Pollutant Standards Index numbers used in the South Coast Air Basin.*

in part, to meteorological conditions being less favourable for ozone formation than in other recent years. Also, since the worst year of 1988, the volatility of petrol has been lowered by federal and state regulations, so leading to a reduction in emissions of ozone precursors. Trends which attempt to adjust for meteorology do show a downward trend during the past decade. For example, the number of days in the Los Angeles basin exceeding the federal standard showed an overall decrease of 22 per cent over the period 1976 to 1991 (from 155 to 121 days). Similarly, the number of basin days with stage 1 episodes showed a 64 per cent decrease over the same period (from 111 to 40 days).[3]

Almost like a boxer throwing one last heavy punch aimed at dealing with its pollution problems once and for all, the South Coast Air Basin adopted the radical

---

[3] Davidson, A (1993) 'Update on ozone trends in California's South Coast Air Basin', *Air and Waste,* vol 43, pp 226–227; California Air Resources Board (1992) *Ozone Air Quality Trends in California 1981–1990,* California Air Resources Board, Sacramento; South Coast Air Quality Management District (1994) *Air Quality Trends, 1976–1993,* South Coast Air Quality Management District, Diamond Bar.

TABLE 9.1. Number of days when hourly ozone levels exceeded the federal standard of 120 ppb (240 mg/m^3) for the period 1984–93 for selected locations in the US

Area	1984	1985	1986	1987	1988	1989	1990	1991	1992	1993
Houston	49	48	43	52	48	34	48	40	30	26
Los Angeles	152	154	159	145	165	138	118	111	130	102
New York	17	19	7	15	32	10	13	19	3	6
Philadelphia	22	25	19	32	34	17	11	24	3	20
Riverside	164	142	151	149	163	146	121	119	126	112
San Diego	51	50	42	40	45	54	39	25	19	14
Washington	12	12	10	20	35	4	5	16	2	12

Source: Extracted from US Environmental Protection Agency (1994) *National Air Quality and Emissions Trends Report, 1993*, US EPA Report 454/R-94-026, Research Triangle Park, NC, pp. 124–125.

Air Quality Management Plan initially in 1989, with the final version being approved in 1994. This remarkable plan has three tiers or stages based on the anticipated availability of levels of technology. The first tier relies on existing technology, the second on breakthrough technologies and the third tier relies on technology that has not yet been fully developed, but is considered attainable. Through this plan the South Coast Air Basin is intended to attain federal standards for nitrogen dioxide by the end of 31 December 1995, carbon monoxide by 31 December 2000, fine particulates (PM_{10}) by 31 December 2001 with provisions for a five year extension (i.e., a target of 2006 being most likely) and ozone by 15 November 2010. The target date for achieving state standards for nitrogen dioxide is the end of 1997, carbon monoxide the end of 2000, PM_{10} post-2010 and ozone post-2010.[4]

One of the key measures in the plan is to require that 40 per cent of cars, 70 per cent of freight vehicles and all diesel buses convert to cleaner fuels such as methanol by 1998. However, the most radical requirement is that 2 per cent of cars sold in the state in 1998 and 10 per cent of 2003 must be zero emission vehicles (ZEVs). With subsequent re-examination this requirement may be extended to 20 per cent by 2005, 35 per cent by 2007 and 50 per cent by 2009. By 2003, not only will 10 per cent of all new vehicle sales have to be ZEVs, but 15 per cent will have to be ultra-low emission vehicles and 75 per cent low-emission vehicles. Emission limits are given in Table 9.2. By 2007 all new cars must use clean-burning fuels such as methanol or electric power. Major motor and oil manufacturers have expressed doubts that technology can meet these demands, but after several years of trying to get the pollution control agencies to weaken and delay their proposals they appear to have accepted the need to try to meet the stringent fuel and vehicle requirements. It may have helped some manufacturers to have accepted the decision, albeit reluctantly, because California is the largest car market in the world and failure to comply would mean large fines or being barred from selling in California. Chrysler has opted for advanced lead-acid batteries for its ZEVs, although it considers that nickel-metal hydride batteries have shown much promise when production costs can be reduced. Using 27 connected batteries, Chrysler's minivan can travel 96–112 km (60–70 miles) before recharging is needed (eight hours for full recharge using conventional household current). Each battery pack will cost between $4500 and $6000 and last about three years. Chrysler must produce about 1500 electric vehicles in California in 1998, rising to 7500 by 2003.[5] General Motors launched its $35,000 two-seater EV-1 in 1996. It uses lead-acid batteries (nearly 40 per cent of the car's weight) and has to be recharged every 140 km (88 miles). California's decision to insist on sales of a small but growing number of ZEVs has boosted interest and investment in developing electric cars, not only in other US states such as New York and Massachusetts, but throughout the world. Many other cities in the US and elsewhere in the world may follow the technology forcing policy example set by California. There were 12,600 clean-fuelled vehicles in California in 1994. The LA Metropolitan Transit Authority operates 330 methanol-powered buses and has ordered 196 compressed natural gas buses. The Orange County Transit Authority plans to purchase 70 propane buses. Smaller electric shuttle buses are growing in numbers throughout the state.

Tighter federal and state emission standards for vehicles are only fully effective in tackling pollution problems if vehicle owners ensure that their vehicles meet these standards. In 1984, vehicle inspection tests

---

[4] Lloyd, A C, Lents, J M, Green C and Nemeth, P (1989) 'Air quality management in Los Angeles: perspectives on past and future emission control strategies', *Journal of Air and Waste Management Association*, vol 39, pp 696–703; South Coast Air Quality Management District (1989) *Air Quality Management Plan*, South Coast Air Quality Management District, El Monte; South Coast Air Quality Management District (1994) *Final Report Air Quality Management Plan: Meeting the Clean Air Challenge*, South Coast Air Quality Management District, Diamond Bar.
[5] Nauss, D W (1995) 'Chrysler to sell electric van battery choice', *Los Angeles Times*, 6 April; Reeves, P (1994) 'LA recharges its batteries', *The Independent*, 17 May.

TABLE 9.2. Californian emission standards for motor vehicles (g/mile, US test cycle)

Type of vehicle	Carbon monoxide	Nitrogen oxides	Reactive Organic gases*
Conventional	3.4	0.4	0.250
Transitional low emission	3.4	0.4	0.125
Low emission	3.4	0.2	0.075
Ultra-low emission	1.7	0.2	0.040
Zero emission	0	0	0

*Non-methane hydrocarbons (VOCs)

were introduced to ensure they did. These smog checks are required for motor vehicles once every two years or after a change of ownership. However, an undercover investigation found that only one in four of all vehicles was receiving a reasonably thorough inspection. This has prompted California to introduce centralised inspection stations (test only) as well as the current decentralised system of garages (test and repair) licensed to undertake smog checks. Older vehicles or vehicle makes with a higher than average frequence of smog check failures may be assigned to centralised stations to ensure the highest standard of test is applied. As found in many other cities around the world, it is the 10 per cent high-emitting vehicles that contribute more than half the total vehicle emissions. To help identify gross polluters there is the 1-800-CUT-SMOG line to report polluting vehicles so as to encourage these vehicle owners to clean them up. Vehicle scrapping programmes have been explored as a means of dealing with some of the older vehicles which have emissions. Payments are made to scrap these cars or loans are offered for replacement (5000 were scrapped in 1993 and 1994).

The pollution potential of petrol has ben examined critically in recent years and reformulated petrol has been introduced. Reformulated petrol is petrol whose composition has been changed to reduce emissions of VOCs and benzene and it contains an oxygenate such as ethanol or MTBE. Whereas oxygenated fuels are intended to reduce carbon monoxide levels (and to a lesser extent hydrocarbons) during cold days, reformulated fuels are aimed at reducing smog-forming emissions in warm weather. Phase 1 reformulated petrol became available in 1992 and is intended to reduce reactive organic gases by 7 per cent. Further reductions are expected with the improved phase 2 reformulated petrol from 1996. Reformulated petrol is compatible with modern car engines and fuel systems, although in poorly maintained cars it may lead to a slight reduction in distance travelled and increased hesitations after start-up. The federal Clean Air Act 1990 mandated the use of reformulated petrol in all areas that failed to meet air quality standards for ozone. Conventional petrol may not be sold in those areas. The federal Environmental Protection Agency classified non-attainment areas for ozone into five classes according to the extent of their problem. These five classes ranged from marginal (relatively easy to clean up) to extreme (which will take a lot of work and long time to clean up). Los Angeles is the only area to be classified as extreme. The 1990 act uses this new classification to tailor clean-up requirements to the severity of the pollution and set deadlines for reaching clean-up goals.[6] California adopted its own Clean Air Act two years earlier (in 1988). It has often been the case that federal pollution control legislation follows California's lead. This Act sets guidelines for the strategies and specific measures to tackle its extreme air pollution problems and aims to reduce emissions of carbon monoxide, nitrogen oxides and reactive organic gases (VOCs) by 5 per cent every year.

Control of emissions from industry and businesses is through the use of operating permits. to provide some flexibility and market incentives an emissions trading scheme was set up in the Los Angeles area in 1994 for 1000 of the biggest polluters. It is called RECLAIM (Regional Clean Air Incentives Market). Permits are issued for a specified value in emissions of hydrocarbons and nitrogen oxides and, more recently, for VOCs. Companies that reduce emissions below the permitted level can sell the credits so gained to others for whom similar action would be uneconomic. Each year the emission value of the permit for each company is reduced by 5–8 per cent, forcing companies to reduce their emissions and so leading to a steady improvement in air quality from year to year. By 1997, if the scheme continues to expand, it may have saved companies about $670 million.[7]

Vehicle emissions are being cut down by reducing the number of solo commuters and increasing the number of workers who take transit, car pool, van pool,

---

[6] Griffin, R D (1994) *Principles of Air Quality Management*, CRC Press, Boca Raton, pp 283–297; Klausmeier, R and Kishan, S (1995) 'Worldwide developments in motor vehicle inspection/maintenance (I/M) programs', *Proceedings of the 10th World Clean Air Congress, Espoo, Finland*, vol 3, Finnish Air Pollution Protection Society, Helsinki, paper 546.

[7] South Coast Air Quality Management District (1995) *Annual Report 1994: Clean Air is Everybody's Business,* South Coast Air Quality Management District, Diamond Bar, 16 pp; 'Smog exchanges', *Acid News,* No 2, May 1992, p 14; 'South Coast Air Quality Management District', *Journal of Air and Waste Management Association*, vol 42, 1992, pp 704–705; 'Smog trading gets green light', vol 42, 1992, p 402.

walk, cycle or telecommute (eliminating 3 million work-trips by 2010 is the telecommuting target). Employers with 100 or more employees at a work site are required to take part in the trip reduction programme. An average vehicle ridership (AVR) is calculated by dividing the number of employees reporting for work in working hours by the number of vehicles driven by the employees over the working week. If there are 300 employees reporting for work each day making 1327 journeys by vehicle during the week, the AVR is 1500 divided by 1327, which equals 1.13. In 1988, the regional AVR was about 1.15. AVR targets are set which are 1.75 in the downtown area, 1.5 in intermediate areas and 1.3 in outer areas. It is likely that these targets will be raised in the coming years. Employers are required to have a plan approved which outlines how the AVR target will be achieved and maintained. Trip reduction plans are expected to include a list of incentives such as preferential parking for ride sharers, subsidies for car pools or use of public transport and facilities to encourage the use of bicycles. Adoption of non-standard work schedules (e.g., 40 hours over four days or 80 hours over nine days) may also help reduce the number of trips. Employers failing to submit plans have been fined such that total fines amounted to $2 million in 1992. By 1994, around 6000 companies employing nearly 2 million commuters had set trip reduction plans. Although this tackles commuters employed by large companies, it does not tackle other work-related journeys, shoppers and many other journeys.[8]

Four out of five employees throughout the US enjoy free parking either through parking spaces made available to them or through public parking costs being paid by employers. Los Angeles intends to limit free parking and require employers to offer employees cash equivalent to the value of their parking perks so that more employees may choose to share transport, walk or cycle, or use public transport. Los Angeles once had an extensive tram system (using the Red-cars), but the motor manufacturers bought it and closed it down. In 1990 a new light railway network began with the Blue line running between downtown Los Angeles and Long Beach for 32 km (20 miles) and in 1993 the 27 km (17 miles) Red Line was added, running north from downtown to Hollywood. Expansion of the network will add the Busway and the Green Line such that in 30 years time there may be 590 km (370 miles) of light rail network. The 150-mile metro network being developed at a cost of $5 billion may encounter problems in persuading some people to travel underground in an earthquake-prone area. Nevertheless, public transport is at last receiving the attention it has needed for several decades. Plans for a fully integrated public transport system for southern California are estimated to cost up to $100 billion. Some of the potential benefits of a public transport system will not be realised because the Los Angeles Basin is so vast, about 95 km (60 miles) wide. This means that mass transit systems can serve only a limited number of commuters, especially given that many commuting journeys in the basin are not to or from downtown Los Angeles.

Los Angeles is attempting to improve air quality using a wide range of measures. Although technofixes involving the introduction of less polluting fuels and vehicles are an essential part of the long-term air quality control strategy for the basin, increasing attention is being given to try to modify the lifestyle of residents, such as through ride sharing and telecommuting programmes, staggering working hours, encouraging greater participation in park and ride commuting, and even moving businesses closer to residential areas. Air quality is better than it was in the 1950s and 1960s, but it is still poor even after implementing the most stringent air quality standards and pollution control policies in the US Rapid population growth, the expanding size of the urban area and the distance travelled by vehicles continue to partially offset progress towards clean air. Consequently, doubt remains as to whether the Los Angeles basin will attain federal and state air quality standards within the next 10 years as the Air Quality Management Plan intends. The decades that have passed during which Los Angeles residents have had to suffer poor air quality with only limited improvements have been achieved during that time offer a warning, or perhaps a lesson, to other megacities both in the Northern and the Southern world.

---

[8] Bae, C-H C (1993) 'Air quality and travel behaviour; untying the knot', *Journal of American Planning Association*, vol 59, pp 65–74; Grant, W (1994) 'Transport and air pollution in California', *Environmental Health and Management*, vol 5, pp 31–34.

"Cleaner Fuels and Cleaner Cars"

D. Elsom

Smog Alert, published by Earthscan, London, 1996. Reprinted by permission on behalf of Kogan Page Limited/Earthscan Publications Limited.

Introduction

Smog alert systems can alleviate the worst excesses of the pollution problem but cannot prevent smogs from re-occurring during stagnant weather conditions. The solution to the problem of urban populations experiencing poor air quality is a long-term commitment to reduce pollutant emissions to achieve healthy air quality. Reducing the emissions per motor vehicle can be achieved through the introduction of more fuel-efficient engines, the use of cleaner fuels and improved technology. Where private cars remain the principal means of travel within an urban area, a technological approach to pollution control would be to replace existing diesel- and petrol-engined cars with zero-emission vehicles. The speed with which cleaner fuels and cleaner vehicles are introduced will be influenced greatly by the commitment of city and national governments to improving air quality as they can set stringent technology-forcing emission and fuel efficiency standards as well as using economic instruments to influence adoption (e.g., tax incentives to encourage vehicle owners to use less polluting fuels and buy zero-emission vehicles).

Exhaust Emission Standards

Emission standards for vehicles can be tightened as improved engine technologies, pollution control equipment and fuels become available. Alternatively, emission standards may be set to force the automobile and oil industries to develop improvements at a faster pace than they would have otherwise. The US and Japan have been leaders in successive tightening of emission standards but the European Union is beginning to catch up. Vehicle emission standards in the European Union were tightened significantly with new cars from 1993 (stage 1), requiring the fitting of catalytic converters (Table 6.1). From 1996 a further reduction (stage 2) takes place in petrol-engined vehicle emission limits of 30 per cent for carbon monoxide and 56 per cent for hydrocarbons plus nitrogen oxides over the 1993 levels. This European Union directive applies to new vehicle types from 1996 and to all new vehicles from 1997 and brings the European Union broadly in line with 1994 US federal standards. The European Union plans to promote further reductions in emissions by the year 2000 (stage 3) with a joint auto-oil research programme examining the potential improvements of cleaner engine technologies and fuels to produce ultra-low emitting vehicles. Stage 3 emission standards will equate approximately with California's ultra-low emission vehicle standards.[1]

TABLE 6.1. European Union (EU) emission standards for cars

Type	CO	HC + NO$_x$	Particulates
EU stage 1 (1993)			
Petrol and diesel	3.16	1.13	0.18
EU stage 2 (1997)			
Petrol	2.20	0.50	—
Diesel, indirect injection	1.00	0.70	0.08
Diesel, direct injection	1.00	0.90	0.10
EU stage 3 (2000) as proposed by the European Parliament			
Petrol	1.00	0.10 (HC) 0.10 (NO$_x$)	
Diesel	0.50	0.10 (HC) 0.10 (NO$_x$)	0.03

Catalytic Converters

To meet new new emission standards for cars, catalytic converters have been required to be fitted in cars in the US and Japan since the 1970s, in the European Union since 1993 and in four southeast Asian countries (Malaysia, Singapore, Taiwan and Thailand) also since 1993. Catalytic converters are of two basic types. The simplest is an oxidation catalyst which consists of a canister fitted with a porous ceramic element, coated with a thin layer of platinum or other noble metal of the platinum group (the catalyst), in which air and the exhaust gases are mixed to convert carbon monoxide and hydrocarbons to carbon dioxide and water vapour. Its disadvantage is that it does not reduce nitrogen oxides. Oxidation catalysts are appropriate for diesel engines but function better if the fuel contains as little sulphur as possible.

The three-way catalyst is the type usually fitted to petrol-engined vehicles. It is a similar canister but with a different mixture of platinum metals: platinum catalyses reactions to remove hydrocarbons and carbon monoxide whereas rhodium removes nitrogen oxides. It

---

[1] Royal Commission on Environmental Pollution (1994) *Eighteenth Report: Transport and the Environment*, HMSO, London, pp 120–124. Comparison between, say, European Union and US emission standards is difficult because each use different test cycles. As well as setting standards for hydrocarbons (HC) + NO$_x$ the US sets standards for NO$_x$, total HC and non-methane HC.

needs careful control of the fuel/air ratio. Too little air (oxygen) and the engine produces more carbon monoxide and hydrocarbons. Too much air and the engine emits more nitrogen oxides. For optimum efficiency it must operate around an air to fuel ratio of 14:7:1, known at the stoichiometric ratio. To regulate the ratio, a device called a Lambda sensor fitted to the exhaust system. It measures the oxygen content of the exhaust gases and feeds information electronically to the fuel injection control or carburettor to keep the air/fuel mixture at the optimum level. the three-way catalyst converts carbon monoxide, hydrocarbons and nitrogen oxides to carbon dioxide, water vapour and nitrogen (Figure 6.1).

The use of catalytic converters requires unleaded petrol as lead poisons the catalyst, preventing it from functioning correctly. Regular vehicle inspections are needed to ensure the catalyst is working correctly as it can be tampered with, removed, poisoned by lead or damaged in some way. The catalyst system plus Lambda sensor is referred as a regulated, controlled or closed-loop system. Whereas the controlled three-way catalyst can achieve emission reductions of 80–95 per cent in test conditions or 75–80 per cent in all driving conditions (i.e., taking into account its initial poor performance from cold starts), the uncontrolled three-way catalyst reduces emissions at best by 50–70 per cent. One of the disadvantages of three-way catalysts is that as the engines are constrained to operate very near the stoichiometric ratio, the additional fuel savings offered by lean-burn engines cannot be used. According to UK research a car fitted with a catalytic converter has a fuel consumption 3–9 per cent higher than one without, and carbon dioxide emissions can be 9–23 per cent greater.[2]

One problem with catalytic converters is that they take several minutes to warm up to their operational 'light-off' temperature of around 300–400°C during which emissions go unchecked. The length of many urban journeys may be only a few kilometres (miles) so this can be a serious limitation of such pollution control equipment. Additionally, a catalyst can 'go out' when the engine is idling in congested traffic as it falls below the temperature at which it operates effectively. Conversely, at high driving speeds, if temperatures exceed 800°C, a catalyst can fail. Engine misfiring or 'bump starting' can also cause failures as can malfunction of the Lambda sensor used to determine the air to fuel ratio. Perhaps 5 per cent of all catalytic converters may fail and have to be replaced in the first three years. In Sweden, which has had catalytic converters fitted in cars since 1989, the failure rate after five years was found to be up to 15 per cent in some makes of cars. Consequently, especially on cold days of calm and stable weather conditions, carbon monoxide will remain a significant problem, especially along busy streets, at road junctions, in tunnels and at traffic lights until the motor industry tackles the problems of 'cold start' emissions by speeding up the time taken for the catalyst to begin operating.[3]

Technology improvements should soon lead to pre-warmed or quick 'light up' catalysts becoming standard (with a dashboard light to indicate when it is functioning correctly). By 1995 there was a growing number of car manufacturers offering some method of quickly warming the catalysts to their operating temperature. However, existing cars will still give rise to excessive cold start emissions (Figure 6.2). Emission performance deterioration of cars fitted with catalytic converters is another issue that manufacturers need to address. European research suggests that on average cars fitted with catalysts deteriorate such that carbon monoxide emissions increase from 1 g/km at new to 3 g/km at 80,000 km (50,000 miles) and hydrocarbons plus nitrogen oxides from around 0.4 g/km at new to 1.0 g/km after 80,000 km. California has required improved warranties on catalytic converters from 1990 cars onwards, extending the seven year and 110,000 km (70,000 miles) warranty to 10 years and 160,000 km (100,000 miles). This will improve the reliance of the systems and ensure that repairs are undertaken instead of being waived as too costly as has been permitted under the 'smog check' (emissions testing) programme. Since 1994 California has required new cars to be fitted with computerized systems with dashboard warning lights to indicate any malfunction in the pollution control equipment which needs repair. It is possible that the technology may eventually become available to measure the total amount of pollutant emissions a vehicle produces in a year, which could then be the basis for applying the polluter pays principle, taxing vehicle owners according to their pollution impact. It would certainly encourage car owners to keep their vehicles better maintained.

Vehicle Emissions Testing and the Identification of Gross Polluters

Periodic vehicle exhaust emissions testing is intended to identify vehicles that fail to meet mandatory emission standards and so require repair (or scrapping if repair is not possible). For example, one badly maintained car can produce as much pollution as up to 40 cleaner vehicles. Emissions tests are usually part of a larger vehicle inspection/maintenance (I/M) programme which examines whether a vehicle is safe and roadworthy. Checks can be made to assess whether a vehicle's pollution control equipment is functioning

[2] 'Catalytic converters study published', *The Environmental Digest*, 56, Feb 1992, pp 13–14.
[3] Holman, C, Wade, J and Fergusson, M (1993) Future Emissions from Cars 1990 to 2025: the Importance of the Cold Start Emissions Penalty, World Wide Fund for Nature UK, Godalming.

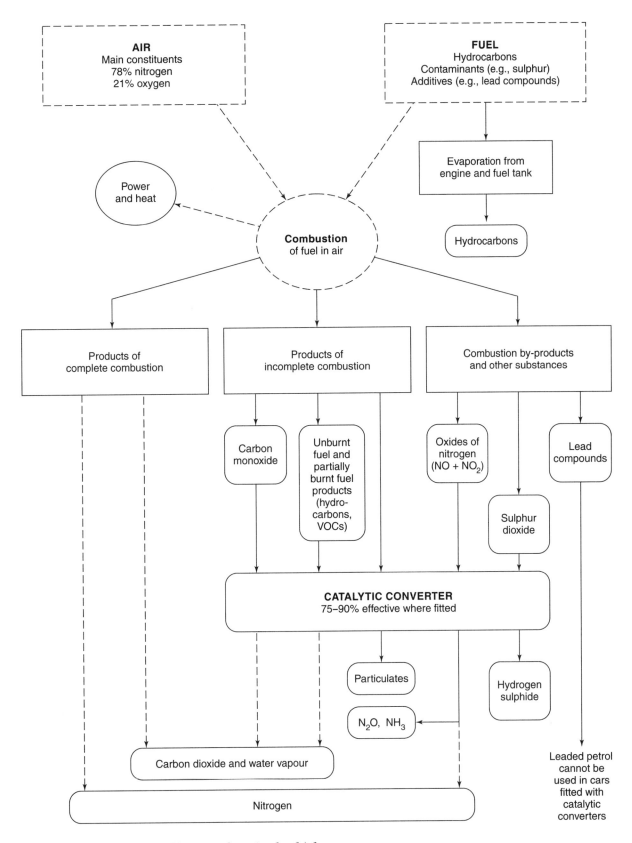

Figure 6.1 Pollutants emitted by petrol-engined vehicles
Source: Royal Commission on Environmental Pollution (1994) Eighteenth Report, Transport and the Environment, HMSO, London, p 22.

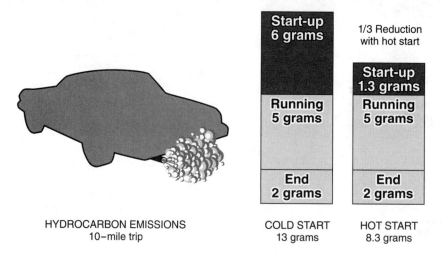

Figure 6.2 Comparison of the amount of hydrocarbons emitted from a petrol-engined car fitted with a catalytic converter during a cold start and a hot start. Around half of the hydrocarbons emitted during a short trip happen during the cold start.
Source: Data based on an assessment by the Californian Air Resource Board.

correctly and has not been tampered with or, in the case of older vehicles, whether engine maintenance has been adequate to meet the less stringent emission standards applied to older vehicles. The 'smog check' has been implemented in the US for many years and has been introduced more recently in Europe. From September 1995, the amount of carbon monoxide emitted by petrol-engined cars in the UK must not exceed 4.5 per cent for 1973–1986 vehicles and 3.5 per cent for post-1986 vehicles. For diesel-engined vehicles the level of smoke opacity must not exceed 2.5/m for non-turbocharged engines and 3.0/m for turbo-charged engines. However, these emission standards need extending to include other exhaust pollutants. The European Union is proposing to standardise vehicle inspection tests, starting when a vehicle is four years old and then continuing at two year intervals. This compares with Germany's annual test beginning when a car is one year old, France's two year interval test not beginning until the car's seventh year, and the UK's test which is compulsory every year from the car's third birthday.[4]

Inspection and maintenance programmes usually measure exhaust emissions while the vehicle is stationary, with its engine at idle for several minutes. The US Clean Air Amendments of 1990 require states with serious or worse ozone or carbon monoxide non-attainment areas to implement an enhanced I/M programme. This requires vehicles to be tested under load conditions over a simulated urban driving cycle using a treadmill device termed a dynamometer. It also requires inspection of the small carbon canisters fitted to vehicles, intended to control evaporative emissions of hydrocarbons, to ensure that these have not been disconnected or are not malfunctioning. Inspection station networks can be either centralised (high volume test facilities with no repair facilities) or decentralised (licensed vehicle repair facilities which can perform repairs as well as inspect the vehicle). California's enhanced I/M inspection programme intends to assign vehicles to test-only centralised stations where there is a likelihood that they are a high polluter. The I/M programmes have been claimed to be very successful in reducing vehicle emissions. For example, the I/M programme for cars together with fleet renewal in New York reduced emissions of nitrogen oxides by 38 per cent and carbon monoxide by 34 per cent between 1980 and 1987. California's smog check introduced in 1984 was aimed at reducing vehicle emissions by 25 per cent by 1994.[5]

Given that newer vehicles tend to be much less polluting than older vehicles, the average age of the vehicle fleet and the rate at which it is renewed can contribute significantly in lowering total vehicle emissions in an urban area. The average age of vehicles in many countries may be around ten years. In the US the average age of cars on the road is eight years but 25 years ago it was only five years. This trend is thought to be due to the relatively higher prices of new vehicles and increased durability of older cars. Governments can speed up fleet renewal by introducing vehicle scrapping schemes whereby older vehicles are scrapped in exchange for a government grant towards the cost of purchasing a new (and cleaner) car. In 1990, California offered $700 for pre-1971 models and eventually retired over 8000 vehicles. The Spanish government's offer of $750 to owners who scrapped their old cars produced a 19 per cent rise in sales of new vehicles in 1994. In 1993, Budapest city council offered the 120,000 owners of grossly polluting two-stroke vehicles grants and loans for four-stroke vehicles or public transport passes. If

[4] Royal Commission on Environmental Pollution (1994) op cit, p 141.
[5] Klausmeier, R and Kishan, S (1995) "World-wide developments in motor vehicle inspection/maintenance (I/M) programs', *Proceedings of the 10th World Clean Air Congress, Espoo, Finland*, vol 3, Finnish Air Pollution Prevention Society, Helsinki, pager 546; Lawson, D R (1993) 'Passing the test—human behaviour and California's smog check programme', *Air and Waste*, vol 43, pp 1567–1575.

owners chose the free public transport passes they received a two-year pass for giving up a Trabant and a three-year pass for a Wartburg.[6]

Emissions testing using remote sensing of vehicles as they are being driven along roads (rather than using laboratory analysers in pre-arranged static tests) is gaining increased attention from governments. This is because it is cheaper and quicker, being able to test thousands of vehicles in a relatively short time. One of the remote sensing devices most widely used is the system called the Fuel Efficiency Automobile Test (FEAT) developed by Donald Stedman at the University of Denver in 1987. It uses infra-red spectrometry to provide an analysis of carbon monoxide and hydrocarbon emissions from a vehicle as it passes through an infra-red beam directed across a road. An ultraviolet-based system has now been added to measure nitrogen oxides and opacity (smoke).[7]

If remote sensing replaced periodic static testing of all vehicles it would have to be located in several places across an urban area to ensure that few vehicles escaped detection. However, the accuracy of remote sensing is not as good as the static tests, especially for hydrocarbons and nitrogen oxides. This has prompted many governments to use remote sensing as an excellent measure to support rather than replace existing periodic I/M inspections as it can identify gross polluters between the periodic tests. This will enable the enforcement agencies to discover vehicles whose owners have tampered with pollution control equipment (i.e., reset controls) after having passed an emissions test, those who possess fraudulent emission test certificates, and those vehicles whose poor maintenance has allowed the efficiency of the vehicle engine to deteriorate since the test. whether remote sensing leads to on the spot fines for polluting vehicles or simply a requirement that the owner takes the vehicle for a full inspection test within a set time period may be determined by how accurate remote sensing devices are considered to be compared with testing at I/M stations.

Remote sensing has been particularly useful in recent years in highlighting the fact that some vehicles are responsible for excessive emissions such that one 'gross polluter' can produce as much as 40–50 times the exhaust emissions of a 'clean' car. Tests in four cities in the UK highlight this situation (Table 6.2). The highest emitting 10 per cent of the vehicles contributed 40–59 per cent of total vehicle emissions for carbon monoxide and 52–67 per cent of total vehicle emissions for hydrocarbons. In contrast, the cleanest 70 per cent of the vehicles contributed only 8–23 per cent of total carbon monoxide emissions and 7–21 per cent of total hydrocarbon emissions. The four cities produced a range of values because the vehicles sampled were characterised by differing proportions of cold-starting vehicles (with their higher emission levels), differing speeds and driving conditions (low or high speed, cruising or decelerating), and differing vehicle age distributions and composi-

TABLE 6.2. Contribution of exhaust emissions from the dirtiest 10 per cent and cleanest 70 per cent of vehicles to total emissions of carbon monoxide and hydrocarbons in four UK cities measured using remote sensing techniques

City	Percentage of total CO emissions contributed by dirtiest 10 per cent of vehicles	Percentage of total HC emissions contributed by dirtiest 10 per cent of vehicles	Percentage of total CO emissions contributed by cleanest 70 per cent of vehicles	Percentage of total HC emissions contributed by cleanest 70 per cent of vehicles
London	57	66	11	10
Edinburgh	41	52	23	21
Leicester	40	53	22	20
Middlesborough	59	67	8	7

Source: Muncaster (1994)[8].

[6] Centre for Exploitation of Science and Technology (1993) *The UK Environmental Foresight Project, vol 2, Road Transport and the Environment*, HMSO, London, p 97; 'American cars older than ever', *The Environmental Digest*, 86, Aug 1994, p 15; 'DDR cars; getting them off the streets', *Acid News*, 5 Dec 1993, p 5.
[7] When a vehicle breaks the FEAT infra-red beam a measurement is taken of the absorption in front of the vehicle and as the vehicle exits the beam. The detector converts the incident infra-red radiation to a voltage signal, thus allowing the carbon monoxide to carbon dioxide and hydrocarbon to carbon dioxide ratios to be calculated from the measured voltage changes both in front and behind the vehicle. The carbon monoxide/carbon dioxide and hydrocarbon/carbon dioxide ratios are the only valid measurements that can be made because the instrument cannot distinguish the intensity or position of the exhaust plume. However, using appropriate known relationships, values can be derived for the effective air/fuel ratio, the emissions of carbon monoxide or hydrocarbons in grams per gallon (or litre) of fuel, the emissions of carbon monoxide or hydrocarbons per mile (km) and the percentage carbon monoxide or hydrocarbons. The system is calibrated daily with a certified gas mixture. A video camera can be focused on the rear of the vehicles as they pass so that their license plates can be recorded.

tions (petrol and diesel, catalysts and non-catalysts).[8] These remote sensing surveys clearly indicate the enormous air quality benefits that could be gained by using legislation and enforcement measures to clean up the gross polluters. targeting gross polluters offers a simple and cost-effective measure. In general, legislative and economic measures which encourage proper maintenance of vehicles should improve air quality significantly.

Some diesel-powered vehicles produce excessive emissions of particulates, especially fine particulates (PM_{10}), which can be a serious threat to health. The worst polluters are the poorly maintained vehicles, those using poor quality fuels and those whose emission controls have been tampered with. Periodic vehicle testing programmes may not always be effective in identifying these gross polluters. Many of the mechanical problems that cause excessive smoke emissions are very expensive to repair. this leads to vehicle owners taking temporary measures to disguise high smoke emissions when the vehicle is tested. Actions include using higher grade diesel fuel during the test or adding a smoke-suppressant additive to the fuel. In Chile, it is reported that owners have reduced smoke opacity by adding a handful of gravel or some water to the vertical exhaust pipe to suppress the smoke temporarily. Such problems point to the need for random roadside inspections to catch the gross polluters in addition to periodic opacity testing. In some countries, the public are encouraged to play a part in identifying 'dirty diesels' by passing on details of smoky lorries, buses and even taxis they have spotted, via a telephone hotline or pre-paid reply cards, to the government vehicle inspection departments for possible follow-up inspections.[9]

Increasing Fuel Efficiency in Vehicles: Lean-burn Engines

The air to fuel ratio has a marked effect on the emission of pollutants. If the mixture is rich, that is too little air, combustion is incomplete, giving rise to high levels of carbon monoxide and hydrocarbons, but as combustion becomes more efficient the combustion temperature increases, so generating more nitrogen oxides. The stoichiometric ratio of 14:7:1 is the value at which conventional engines run most efficiently. However, if engines employ a higher ratio, the additional air reduces the cylinder temperature and decreases nitrogen oxide emissions. At very high ratios the benefits of lower emissions of nitrogen oxides are offset by increases in hydrocarbons because of the more difficult conditions for combustion. Lean-burn engines are designed to run at an air to fuel ratio of around 18:1 (even 20:1 in theory), a ratio which would cause conventional engines to stall. Lean-burn engines offer potential fuel savings of 5–10 per cent. Emissions from lean-burn engines are improved over a standard car, but are less effective than a car fitted with a catalytic converter, especially when operating at high speed or under strain (e.g., during acceleration, climbing steep hills, carrying heavy loads) when they revert to a richer mixture and then emit similarly high levels of nitrogen oxides to a standard car without a catalyst. Lean-burn engines can be fitted with an oxidation catalyst (which does not need to operate at the stoichiometric ratio), but they cannot be fitted with a three-way catalytic converter for which the air to fuel ratio of 14.7 is critical.[10]

Fuel consumption can be reduced by various improvements in engine and vehicle design. Governments can encourage such technological improvements by setting increasingly stringent average vehicle fuel consumption targets at, say, 5–10 year intervals. Penalties would then be imposed on manufacturers failing to achieve the target, whereas tax benefits could be offered to those who better the targets. Sales of less fuel-efficient vehicles could be discouraged by raising taxes on them or simply increasing fuel prices generally (which are low in the US) and undertaking a commitment to increase them each year (as has happened in the UK).

Cars sold in the US on average (sales weighted and applied to each car manufacturer) in 1993 consumed around 28 miles per gallon (mpg), nearly twice as efficient as cars sold in the mid-1970s, but with little improvement in recent years. President Clinton originally pledged during his campaign to increase the fuel efficiency required by law (the corporate average fuel economy or CAFE standard) from the 1990 level of 27.5 to 40 mpg by 2000 and 45 mpg by 2015. Major US car manufacturers (Chrysler, Ford, General Motors) are currently co-operating with the government on a programme using advanced technology (e.g., lightweight materials, improved petrol engines, fuel cell engines) aimed at achieving a fuel consumption of 80 mpg.[11]

Reducing the weight of the vehicle body (e.g., using aluminum, structural plastic) can save fuel by reducing the power required and making it economical to use a smaller engines. Shedding 10 per cent of a car's weight may reduce fuel consumption by about 7 per cent. Fuel efficiency can be increased by using better tyres. For example, in 1992 Michelin developed a tyre using new polymers which is claimed to reduce surface resistance by 35–50 per cent, so cutting up to 10 per cent in fuel consumption. Fuel efficien-

---

[8] Muncaster, G (1994) 'Experience with remote sensing', paper presented at the *National Society for Clean Air and Environmental Protection (NSCA) seminar of Targeting Traffic Pollution—Options for Local Air Quality Management, 8 December 1994*, NSCA, Brighton.
[9] Klausmeier and Kishan (1995) op cit.
[10] Mitchell, C G B and Hickman, A J (1990) 'Air pollution and noise from road vehicles', in European Conference of Ministers of Transport, *Transport Policy and the Environment*, OECD, Paris, pp 46–75.
[11] 'US plans 80 mpg car' and 'US car technology initiative', *The Environmental Digest*, 75, Sep 1993, p 12.

cy can also be improved by better engine lubrication such as the 'zero viscosity' oils currently being developed which may increase fuel efficiency by 4–6 per cent. Finally, it should be realized that many of the technological improvements are most effective when a vehicle is new and that deterioration of the vehicle with age leads to reduced fuel efficiency. Moreover, poor maintenance, bad driving habits and congestion all also worsen fuel efficiency.[12]

Vapour Recovery: Large Canisters Versus Petrol Pump Nozzle Recovery

Evaporative emissions of VOCs can arise when tankers deliver petrol to filling stations, when cars are being filled from petrol pumps, and from vehicle fuel tanks and fuel systems when running. In addition, evaporative emissions can continue when a warmed-up engine has stopped and from the effects of high air temperatures on the fuel tank. In 1992, evaporative emissions of VOCs from the fuel tank and fuel system of vehicles accounted for about 10 per cent of total emissions of VOCs in the European Union. to control these emissions, vapour collection systems can be fitted either to petrol pumps or to the vehicles (carbon canisters).

The greatest public exposure to VOCs such as the carcinogen benzene tends to occur at petrol stations, but usually only for short periods. Attempts to reduce exposure at petrol stations include so-called stage I controls which are aimed at preventing vapour escaping during tanker deliveries to underground storage tanks. Such controls have been in force in the US since the 1970s and are currently being phased in throughout Europe, Austria, Sweden and parts of Germany (e.g., Munich and Berlin in 1987) have long required their installation and they achieve about 80 per cent recovery of toxic vapours. Installation of petrol vapour recovery devices in service stations in Los Angeles reduced emissions equivalent to 19 per cent of total hydrocarbons in California in 1980.

Stage II controls require petrol pumps to be fitted with vapour recovery equipment (vapour recovery sleeves on the nozzles of the pump hoses) as is the case in California. Currently, the European Union is debating whether to implement stage II controls or to require each new vehicle to be fitted with its own vapour recovery system, namely a large carbon canister. Petrol pumps could be fitted with recovery equipment relatively quickly, but at a cost to industry of $2 billion. In contrast, it would be less expensive to require all new vehicles to be fitted with large carbon canisters, but this would take several years before the majority of vehicles on the road have such equipment to curb VOC emissions. In 1993 Germany passed a law requiring all new petrol stations to be fitted with vapour recovery equipment. Existing stations have three to five years to comply. By 1995, Denmark, Luxembourg and Sweden had introduced similar national requirements.[13]

Carbon canisters fitted to a car allow the fuel system to breathe but absorb VOCs. A sealed cap is used on the fuel tank and a line runs from the top of the tank to a charcoal filter which absorbs and stores fuel vapour. When the engine is running, fresh air is drawn into the canister and mixed with the petrol vapours which have been absorbed onto the charcoal filter and the vapour/air mixture flows to the engine. Benzene is largely removed with a carbon canister. small carbon canisters, standard on US cars and recently introduced in the European Union, are intended to absorb vapour from the fuel tank 'breathing', running losses when driving and from hot engines after use, whereas large carbon canisters (1–5 litres capacity) also recover vapours when refuelling.

Diesel Particulate Traps (Filters)

Particulates from diesel engines have long been of concern for their potential impact on health. Trapping exhaust particulates using a particulate trap or filter may be attempted, but if the filter cannot be regenerated (i.e. if the soot is not burned to form carbon dioxide) the particulates build up, causing back-pressure problems in the engine leading to possible 'blow off' of the particulates. To burn off the particulates a temperature exceeding 550°C is usually required, but the normal road temperature of diesel exhausts is only around 150°C.

Various regeneration methods have been attempted. External energy sources such as fuel or electric burners are used to burn off the particulates. Catalytic trap oxidisers employ a catalyst to regenerate the filter by lowering the temperature required to burn off the particulates (oxidise the soot). One problem with using the catalyst is that this can lead to sulphate formation from fuel containing sulphur. This highlights the fact that plans to introduce diesel catalyst technology need to be anticipated by ensuring that low sulphur fuels are available, or that all diesel fuels have a lower content in general. Catalytic fuel additives are another approach. Tests on 110 buses in Athens claimed up to a 90 per cent reduction in particulates using a fuel additive which lowers the particle combustion threshold from around 550 to 480°C. This demonstration project proved so successful that the Greek Government has mandated the system for all buses in Athens. Generally, particulate traps can reduce particulates by more than 50 per cent, but the high temperatures used in some filter regeneration processes increase the formation of nitro-

---

[12] Royal Commission on Environmental Pollution (1994) op cit, p 131; 'Mitchelin's green tyre to cut fuel consumption', *The Environmental Digest*, 58, Apr 1992, p 13.
[13] 'Vapour recovery' *Acid News*, 2, Apr 1993, p 10; 'Stopping Petrol Fumes', *Acid News*, 4, Oct 1995, p 9.

gen oxides. Currently, particulate traps are relatively expensive and may cost up to 25 per cent of the vehicle engine costs (e.g., traps typically cost $4500 in 1992). Given this high cost there is clearly a need for governments to offer financial incentives for their installation on buses and heavy goods vehicles.[14]

Petrol-Engined Versus Diesel-Engined Vehicles: Which Pollutes Less?

Diesel-engined cars and vans offer better fuel economy and efficiency. In a petrol engine a mixture of air and fuel is drawn into the cylinder, compressed and ignited by a spark. In a diesel engine the fuel is injected into air which is already very hot as a result of compression and thus no spark is needed. As a diesel engine works with a surplus of air, the combustion process is more efficient. This results in a reduction in the amount of fuel consumed by at least 25 per cent by volume and 15 per cent by mass compared with a petrol engine of comparable power. Petrol engines tend to be better than diesel engines in terms of performance (e.g., acceleration).[15]

Assessing which type of engine is less polluting is difficult. Not only does it depend on the type of pollutant being considered, but also on whether pollution control devices (e.g., catalytic converters, carbon canisters) are fitted (Table 6.3). The level of engine maintenance also plays a key part. Not surprisingly, in terms of carbon monoxide, hydrocarbons, nitrogen oxides and particulates, the petrol car fitted with a three-way catalytic converter (when functioning correctly) produces lower emissions than one without a catalytic converter. Diesel emits less carbon monoxide than a car fitted with a catalytic converter, a similar amount of hydrocarbons, more nitrogen oxides and considerably more particulates. If a diesel car is fitted with an oxidation catalyst, then emissions of carbon monoxide and hydrocarbons are even better. The advantages of petrol cars fitted with a three-way catalytic converter may disappear with the degrading of catalysts with age after about 60,000 km (37,500 miles). When comparing the emissions of particulates from cars using leaded petrol with diesel cars we should recognise that they are chemically different such that the health risks associated with each group of particulates may differ. diesel particulates also have a soiling factor of around seven times that of petrol particulates. Emissions of PAHs such as the carcinogen benzo(*a*)pyrene is about 50 per cent greater from diesel than petrol cars without catalysts although this reduces when oxidation catalysts are

TABLE 6.3. Comparison of emissions from petrol and diesel cars

Pollutant	Petrol without three-way catalyst	Petrol with three-way catalyst	Diesel without oxidation catalyst	Diesel with oxidation catalyst
Regulated pollutants				
Nitrogen oxides	****	*	**	**
Hydrocarbons	****	**	***	*
Carbon monoxide	****	***	**	*
Particulates	**	*	****	***
Unregulated pollutants				
Aldehydes	****	**	***	*
Benzene	****	***	**	*
1,3-Butadiene	****	**	***	*
Carbon dioxide	***	****	*	**
PAHs	***	*	****	**
Sulphur dioxide	*	*	****	****

Asterisks indicate which type of car typically has the highest emissions: *, lowest emissions; **/***, intermediate; and ****, highest emissions.
The table indicates only the relative importance of emissions between the types of vehicles. The difference in emissions between, say, *** and **, may be an order of magnitude, or much smaller.

Source: Quality of Urban Air Review Group (1993) *Diesel Vehicle Emissions and Urban Air Quality*, Second Report, Department of the Environment, London, p 15.

---

[14] Holman, C (1992) *Cleaner Buses*, Friends of the Earth, London, p 7; Quality of Urban Air Review Group (1993) *Diesel Vehicle Emissions and Urban Air Quality*, Second Report, HMSO, London: 'Cleaner diesel buses claimed', *The Environmental Digest*, 61, Jul 1992, p 14.
[15] Royal Commission on Environmental Pollution (1994) op cit, p 24; Quality of Urban Air Review Group (1993) op cit.

fitted to diesel engines (e.g., from 1996 in the European Union).[16]

Fuel Improvements

Pollutant emissions can be reduced by changing the composition of the fuel. The lead, sulphur, aromatic hydrocarbons and benzene content of fuels can be reduced or eliminated. The volatility of the fuel can be lowered to reduce evaporation of VOCs. Oxygenated compounds can be added to improve combustion.

Unleaded Petrol

Traditionally, lead compounds have been added to petrol to prevent knocking as this is a cheaper alternative to refining the petrol to a higher octane number. Knocking is the distinctive 'pinking' noise which occurs when there is spontaneous ignition of the petrol and air mixture in the corner of the cylinder furthest away from the spark plug. The lead allows engines to operate at higher compression ratios and under greater loads before knocking occurs. Lead also acts as an engine lubricant, reducing the wear in some parts such as valve seats. Concern over the potential health effects of airborne lead together with the recognition that catalytic converters, fitted to cars to reduce exhaust emissions of carbon monoxide, hydrocarbons and nitrogen oxides, do not function properly when lead compounds are present (the catalyst is 'poisoned' by lead) resulted in the US, Canada, Japan, then Europe and more recently many countries around the world reducing the lead content of petrol as well as making unleaded petrol available.

Health concerns focused on young children who lived near busy roads and experienced high levels of lead in their blood which may have affected their behaviour (e.g., attentional disorders, learning disabilities, emotional disturbances), caused dysfunction of the central nervous system (irritability, clumsiness) and affected educational characteristics (lower average IQ). Around 1970 the lead content of most European petrol was 0.84 g/l but by the late 1980s the lead content had been reduced in many countries to 0.15 g/l, the lowest level usable in petrol engines existing at the time without special adaptations. At the same time the availability of unleaded petrol increased in preparation for all new cars from 1993 which were required to be fitted with catalytic converters. These cars were designed with hardened valve seats and operated without knocking using unleaded petrol which was of a lower octane number (95) than most leaded petrol (97–98). Airborne levels of lead fell dramatically as a result of the reduced lead content of leaded petrol and with the rise in sales of unleaded petrol.

The success of government intervention in encouraging the greater use of unleaded petrol was demonstrated in the UK. In 1986, when unleaded petrol was first introduced, public demand was small even among owners whose cars could run on unleaded petrol because it cost more than leaded petrol. It was not until the government introduced a tax differential in favour of unleaded petrol, and increased this differential each year, did sales escalate. By the start of 1993, when new cars had to be fitted with catalytic converters and therefore had to run on unleaded petrol, so ensuring sales would continue to increase, unleaded petrol sales accounted for half of all petrol sales. By 1994, unleaded petrol deliveries reached 57.6 per cent in the UK (Table 6.4). throughout the European Union, unleaded fuel accounted for 62.4 per cent of total petrol deliveries. Unleaded petrol accounted for more than 99 per cent in Austria, Finland and Sweden (all of whom joined the European Union in 1994 and so increased the percentage use of unleaded petrol). Motorists in the European Union's northern Member States tend to use higher percentages of unleaded petrol than southern motorists.[17]

TABLE 6.4. Unleaded petrol as a percentage of total petrol deliveries in the European Union (EU)

Member State	1993	1994
Austria	98.1	99.8
Belgium	57.4	64.8
Denmark	75.6	98.1
Finland	87.0	99.9
France	40.8	50.0
Germany	88.7	92.3
Greece	22.9	27.6
Ireland	38.5	48.6
Italy	23.8	32.6
Luxembourg	69.0	75.6
Netherlands	75.0	80.1
Portugal	20.9	30.0
Spain	14.8	22.2
Sweden	79.5	99.4
UK	52.0	57.6
Total EU	56.1	62.4

Source: European Press Office, Luxembourg.

---

[16] In the past decade, sales of diesel-engined vehicles in Europe have increased markedly and they now account for 20–25 per cent of vehicle sales. In the late 1980s and early 1990s, sales were encouraged by some car manufacturers advertising diesel-engined vehicles as 'green' or even 'pollution-free' by stressing the advantage of being more fuel-efficient and producing lower emissions of some pollutants, especially compared with petrol vehicles not fitted with catalytic converters. However, by 1994 the concern for fine particulates (and carcinogenic compounds) as well as nitrogen oxides (linked by the public with rising asthma incidence) from diesel vehicles compared with petrol engines with catalytic converters (required on all new cars since 1993) changed the assessment.

[17] Elsom, D M (1982) 'The phasing in of unleaded petrol in the United Kingdom', Clean Air, vol 22, pp 226–232.

Oxygenated Fuels

Oxygenated fuels (oxyfuels) used to lower winter carbon monoxide emissions were pioneered in the state of Colorado and led to the US Clean Air Act 1990 requiring the use of oxygenated fuels in all areas which have not attained the federal air quality standard for carbon monoxide. Oxygenated fuels with an oxygen content not less than 2.7 per cent must be used during those times of the year that are considered 'high carbon monoxide conditions'. In the US, Albuquerque, Denver, Las Vegas, Phoenix, Reno and Tucson are among those cities in which oxygenated fuels have been introduced during the winter months in an attempt to improve combustion efficiency and to reduce carbon monoxide emissions by as much as 20 per cent. In 1995 oxygenated fuels cost about 10–15 cents per gallon more than conventional fuel.

Oxygenated fuels contain small amounts of ethanol or methanol derivatives. These include ethyl-t-butyl ether (ETBE) derived from ether (and produced from corn) and methyl-t-butyl ether (MTBE) derived from methanol. The oxygen-rich additives help the fuel's hydrocarbons to burn more efficiently at low temperatures, thereby converting more of the carbon monoxide to carbon dioxide and also reducing hydrocarbon emissions. Nitrogen oxides are unaffected. The minimum 2.7 per cent oxygen by weight in US oxyfuels is achieved by adding 15 per cent by volume of MTBE or 7.7 per cent by volume of ethanol.

In those US cities where oxygenated fuels were introduced, carbon monoxide levels have fallen by 10–15 per cent (e.g., Denver claims a 12 per cent reduction). They benefit cars with catalytic converters (as well as older cars without) because the catalysts take several minutes to reach their operating efficiency, during which carbon monoxide and hydrocarbon emissions are uncontrolled. If the additives burn incompletely, they can form aldehydes. MTBE and methanol produce very small amounts of formaldehyde, whereas ETBE and ethanol produce acetaldehyde, both of which may cause cancer. A causal link between some people who suffer headaches, insomnia, nausea, dizziness, skin problems and coughs when driving cars using oxygenated fuels is claimed, but not yet confirmed or rejected.

In Finland, oxygenated fuels were introduced in 1991 and now account for 95 per cent of all fuel sold in the country. To achieve an oxygen content of 2 per cent, Finland uses 11 per cent of MTBE or t-amyl methyl ether, the latter being favoured. Finland is the only European country to manufacture oxyfuels, which are exported to Sweden for use in their cold winters. Emissions of carbon monoxide in Finland have fallen by 10–20 per cent and hydrocarbons by 5–10 per cent as a result of introducing oxygenated fuels. The European Union may consider the wider adoption oxygenated fuels as one way to reduce the benzene content of fuel as oxyfuels, by helping hydrocarbons to burn more efficiently, offer an alternative means of maintaining a high octane content.[18]

Other Reformulated (Cleaner-Burning) Fuels

Reformulating the composition of petrol and diesel that can be used in all vehicles can reduce vehicle emissions and improve air quality. Lowering the volatility of fuel so it does not evaporate as readily can reduce emissions of VOCs, which react with nitrogen oxides in sunlight to produce ozone. Following the worst period of ozone pollution in the US in 1988, national regulations lowered the national average summer Reid vapour pressure of petrol, which is a measure of its volatility, by 11 per cent. A further reduction of 3 per cent took place between 1989 and 1990. A modelling analysis of New York City conditions estimated that the impact of the Reid vapour pressure reductions was a 25 per cent reduction in VOC emissions. The UK Government had intended to reduce the volatility of petrol sold during the summer months by 1993, but was delayed until 1995 due to resistance from the oil industry.[19]

However, lowering the volatility of petrol through a reduction in its butane content may give rise to other pollution problems if the octane level lost through this process is regained by increasing the aromatic content (i.e., giving rise to more benzene) or by the use of oxygenates (i.e., perhaps increasing formaldehyde depending on the additive used).[20]

The US is increasingly making the use of reformulated petrol in nonattainment air quality areas compulsory. Californian drivers began using less polluting (and slightly more expensive) phase 2 reformulated petrol all year round from March 1996 (conventional petrol no longer permitted to be sold in the area). This fuel has a minimum of 2 per cent oxygen, a maximum of 1 per cent benzene, reduced aromatic hydrocarbons (25 per cent less than before), negligible sulphur dioxide, no heavy metals and it is not allowed to result in an increase in nitrogen oxides. In addition, the emissions

---

[18] Burke, M (1995) 'Are oxyfuels good for us?', *New Scientist*, 15 Jul, pp 24–27; Kierman, V (1994) 'US says drivers must use more alcohol', *New Scientist*, 9 Jul, p 7; Miller, S S (1992) 'Carbon monoxide and oxygenated fuels in US cities', *Environmental Science and Technology*, vol 26, p 45; in 1995, 100 petrol stations in Vienna, Austria were selling petrol containing TBE, *The Environmental Digest*, 1995/8, Aug, p 12.

[19] US Environmental Protection Agency (1994) *National Air Quality and Emissions Trends Report, 1993*, Report EPA 454/R-94-026, US EPA, Research Triangle Park, NC, pp 44 and 47; Her Majesty's Government (1992) *This Common Inheritance, Second Year Progress Report*, HMSO, London; 'Oil industry delays controls on petrol volatility', *ENDS Report*, 234, Jul 1994, p 33.

[20] House of Commons Transport Committee (1994) *Transport-related Air Pollution in London*, Sixth Report, vol 1, HMSO London, p 26.

of ozone-forming VOCs and toxic pollutants must be reduced by 25 and 20 per cent, respectively. It is estimated that a reduction in the carcinogenic compounds will result in a fall in potential cancer cases of 35 per year from 1996 to 2010.[21] Because of its carcinogen effects, benzene has recently received much attention in Europe. Currently, the European Union has a limit of 5 per cent by volume, but there is growing pressure to lower the limit to 1 per cent. Diesel has also received attention from the European Union with the sulphur content of diesel being reduced from 0.3 per cent by weight in 1989 and 0.2 per cent in 1994 to 0.05 per cent in October 1996.[22]

Reformulated fuels can lead to small but significant improvements in air quality in an urban area. One of its attractions for air quality managers is that it can reduce emissions from all vehicles, irrespective of age. It can be used all year round, tackling summer air quality problems (e.g., reducing emissions of ozone-forming VOCs) and winter problems (e.g., reducing carbon monoxide by its increased oxygen content). There is no need to modify existing vehicles to be able to use reformulated petrol or diesel. It should not affect a vehicle's performance unless the vehicle is in a poor mechanical condition, when drivers may notice a slight increase in fuel consumption and hesitations after start-up. Unlike alternative fuels (e.g., methanol, electric battery), air quality improvements do not have to wait for several years until the sales of new vehicles designed specifically to use alternative fuels have begun to replace existing vehicles in significant numbers. There is no need to convert storage tanks, service station pumps or workshop equipment as when using alternative fuels. Given their higher costs of manufacturing and retail price, reformulated fuels can be used selectively if necessary—that is, in a city with poor air quality rather than throughout the country.[23]

Alternative Fuels: Biofuels

Biofuels have received increasing attention in recent years as a possible replacement for petrol or diesel or to be mixed with conventional fuels. Bio-fuels may be alcohols such as ethanol and methanol to replace petrol or esters such as rapeseed methyl ester to replace diesel There are about 50–60 million cars—about 10 per cent of the global total—powered by some form of alcohol fuel such as pure ethanol or methanol or using a mixture of up to 20 per cent with petrol. As ethanol and methanol are liquids they offer a potentially convenient replacement for petrol. Alcohols emit pollutants, albeit less than petrol, and so are considered as transitional fuels in, say, California's long-term commitment to introducing cleaner vehicles.

Ethanol

About 4 million cars in Brazil are powered by ethanol produced from sugar cane (hence the sickly smell of burnt sugar sometimes associated with car exhausts) using fermentation followed by distillation to recover the ethanol. Typically, as a fuel in its pure form, ethanol produces 20–30 per cent less carbon monoxide, about 15 per cent less nitrogen oxides and insignificant amounts of sulphur dioxide (petrol contains more than three times as much sulphur, although the amounts are very small anyway). Carbon dioxide release by ethanol-powered vehicles is balanced by its absorption in new sugar cane. Vehicles need alteration to run on ethanol. To prevent corrosion the inside of the fuel tank is coated with tin, the fuel lines with copper and nickel, and the carburettor with zinc. The piston needs strengthening because ethanol has a higher detonation temperature, which means the fuel-air mixture in the combustion chamber needs higher compression. The energy content is only two-thirds that petrol so the fuel tank needs to be larger to provide the same range for a vehicle. Of significance for cold climate countries is the fact that ethanol cars are poor starters at low temperatures.

The Brazilian experiment in using ethanol (pro-alcohol) began in 1975 as a response to the high oil prices in the 1970s. Initially, ethanol was used as an additive to petrol, but all-ethanol was introduced in 1979. However, although 80–90 per cent of all new cars sold in Brazil between 1985 and 1988 were equipped to use ethanol, there has been a dramatic decline since then with only 20 per cent of all new cars being ethanol users in 1991. The reasons for this change were a combination of nationwide shortages of ethanol in 1989–1990 due to sugar cane farmers being dissatisfied with the government payments to them, and the fall in world oil prices which made ethanol too expensive without substanial government subsidies. The cost of ethanol production is up to twice that of petrol. The national trend towards using gasohol (78 per cent petrol, 22 per cent ethanol) rather than ethanol was accelerated after 1992 when the government stopped subsidising the pump price of ethanol. Brazilian car manufacturers have also put pressure on the government because

---

[21] Griffin, R D (1994) *Principles of Air Quality Management*, Lewis, Boca Raton, pp 300–301; 'California to have cleaner petrol', *Acid News*, 2, May 1992, p 6.
[22] Royal Commission on Environmental Pollution (1994) op cit, p 124; some fuels contain detergents to stop the build up of sooty particles in the engine.
[23] Holman, C (1994) *The Effects of Petrol Quality on Emissions for Passenger Cars*, World Wide Fund for Nature, Godalming; Holman, C (1995) 'Reformulated fuels—a quick fix solution', paper presented at the *National Society for Clean Air and Environmental Protection (NSCA) Seminar on Greener Fuels for Cleaner Air?*, Birminghan, Feb 1995, NSCA, Brighton.

they would prefer to produce only one type of car engine to run on gasohol.[24]

São Paulo illustrates some of the successes of using ethanol either in its pure form or as an additive to produce gasohol. The city has 1.1 million ethanol-powered and 1.2 million gasohol-powered vehicles. All petrol cars were phased out in the mid-1980s during a complete ban on car imports (this ended in 1990), after which high import taxes have kept imported car numbers low, although these can be converted to run on gasohol. Table 6.5 presents the air quality changes that are predicted if all the city's 2.3 million vehicles were powered by petrol, gasohol or ethanol. Significant air quality improvements are predicted if all the vehicles use ethanol, whereas the replacement of the current ethanol/gasohol vehicles of São Paulo with petrol-powered vehicles would result in a major deterioration in air quality. One additional advantage of ethanol is that it does not contribute to airborne lead. Airborne lead levels in Brazilian urban areas fell from 1.6 mg/m^3 in 1978 to 0.3 mg/m^3 in 1983 following widespread ethanol adoption.

TABLE 6.5. Percentage air quality changes compared with the present air quality if all vehicles in São Paulo were powered by one type of fuel

Fuel	Carbon monoxide	Hydrocarbons	Nitrogen oxides
Petrol	+120	+100	+10
Gasohol	+40	+35	No change
Ethanol	−20	−20	−10

Source: Compiled from information given by Homewood (1993)[24].

In the US, 95 per cent of ethanol is produced from corn (maize) and about 5 per cent from sugar cane or other biomass or organic matter such as wheat, potatoes and sugar beet. It is used as an additive to increase the octane rating of petrol and to reduce carbon monoxide emissions (refer to the earlier section discussing oxygenated fuels). This oxygenated fuel or gasohol, compulsory in winter months in many cities with carbon monoxide air quality problems, is subsidized by the federal government and some states exempt gasohol from their fuel taxes. In Europe, the European Union allows 5 per cent ethanol derived from cereals (wheat, maize), potatoes or sugar beet to be added to fuels (petrol and diesel). Even ethanol derived from surplus wine was permitted in 1991. Similarly, when Stockholm faced a shortage of ethanol in 1995, the city authorities were given permission by the European Commission to import 5000 tonnes of surplus red wine from Spain, from which to produce ethanol.

Methanol

Methanol, also known as wood alcohol, can be produced from wood, coal or natural gas. Its high octane rating has resulted in it being used in racing cars such as the Indy 500 race cars since 1965. Unlike petrol-powered vehicles, methanol-powered vehicles emit only a few compounds, primarily unburnt methanol and formaldehyde. Unburnt methanol is much less photochemically reactive than the organic compounds emitted by petrol-engined vehicles, so ozone formation ought to be reduced when using methanol. However, the overall ozone reductions depend on how much formaldehyde is produced as this is very reactive and has a high ozone-forming potential. Formaldehyde is also a suspected carcinogen. Recognizing this concern, in 1989 California set standards of 15 mg/mile of formaldehyde as the ultimate standard needed to be achieved by vehicles using methanol. Providing formaldehyde emissions can be kept low, methanol is considered to offer useful air quality benefits compared with petrol, but only as a transitional step towards the long-term aim of the widespread use of zero-emission vehicles.

There are many practical considerations in methanol replacing petrol. Methanol produces only half the energy as the same volume of petrol, so the fuel tank would have to twice as large to enable the same range to be driven. Methanol is highly corrosive, so petrol stations would need new storage tanks and cars would need stainless-steel fuel tanks and corrosion-resistant fuel lines and carburettors. As it is more toxic than petrol, self-service methanol pumps would pose potential problems. Pure methanol-powered cars are difficult to start in cold weather, even at 10°C. To overcome this problem M85 mixtures are used, that is 85 per cent methanol and 15 per cent petrol. Mehanol's overall environmental impact would depend in part on whether it is produced from natural gas or coal. Production from natural gas would produce similar carbon dioxide emissions as petrol, but production from coal would increase emissions of this greenhouse gas by 50 per cent. The cost of production of methanol is similar to that of petrol, but the cost of converting filling stations and cars points to the need for financial support from the government if methanol is to displace petrol use significantly.[25]

---

[24] Boels, L B M M (1995) *The Potential of Substitute Fuels for Reducing Emissions in the Transport Sector*, European Federation for Transport and Environment, Brussels, pp 22–23; Homewood, B (1993) 'Will Brazil's cars go on the wagon?', *New Scientist*, 9 Jan, pp 22–23; 'Brazil may abandon alcohol', *The Environmental Digest*, 70, Apr 1993, pp 14–15.

[25] Boels, L B M M (1995) op cit, pp 20–22; Nadis, S and MacKenzie, J J (1993) *Car Trouble*, World Resources Institute, Beacon Press, Boston, pp 63–66.

Biodiesel: Rapeseed Methyl Ester

The growing surplus of farmland needed to produce food crops in the European Union has increased farmers' interests in growing crops to produce fuels, using the 15 per cent of land ('set aside') required to be taken out of food production by large farms since 1992. The main focus of attention has been on producing biodiesel from oilseed rape (called canola in Canada and the US). Oil is extracted from rape simply by crushing, with 3 t of rape yielding one tonne of oil. Most diesel engines can run on unblended rape oil without modification, but they become clogged after several days. To prevent this, glycerine must be removed by separation from the oil. Each tonne of oil is mixed with 110 kg methanol in the presence of a nitrogen hydroxide catalyst and heated to 40–50°C. The glycerine settles out, leaving a clear thin liquid, rapeseed methyl ester.

Many European countries have begun building rapeseed methyl ester production plants. Austria has more than 100 petrol stations selling biodiesel. Several trials using bus fleets have been attempted. In the UK, Reading conducted trials in 1993 using buses fuelled by rapeseed methyl ester. No engine modification was necessary and emissions of sulphur dioxide, particulates, carbon monoxide, nitrogen oxides and carbon dioxide were reduced for a power loss of 2–5 per cent compared with a conventional diesel engine. The trials suggested there may be problems with lubricants and the rapeseed methyl ester damages rubber pumps and hoses more than diesel. Despite the overall success of this pilot experiment, wider adoption of rapeseed methyl ester by European bus and lorry fleets is hindered by the higher cost compared with conventional but more polluting diesel fuels (e.g., twice as costly as diesel in the UK). This points to the need for tax concessions (e.g., cutting duty on biofuels to one-tenth that of diesel) and government subsidies if biodiesel is to become competitive with fossil fuels. However, producers in Austria claim that they can produce rapeseed methyl ester for the same price as diesel, so the penetration of this biofuel into the market may increase in the coming years.[26]

Compressed Natural Gas (Methane) and Liquid Petroleum Gas

Gas can be used as a fuel for vehicles in the form of compressed natural gas or liquid petroleum gas. Whereas compressed natural gas is methane gas, liquid petroleum gas consists mainly of propane and butane produced as by-products from oil refineries. Liquid petroleum gas is used extensively in the Netherlands where it accounts for 15 per cent of vehicle fuels. Compressed natural gas is the cleanest fossil fuel and currently provides power for 700,000 vehicles worldwide. There are 300,000 such vehicles in Italy and 230 filling stations (the Italians have been using natural gas for 40 years), 200,000 vehicles in Russia, 100,000 in New Zealand, 100,000 in British Columbia, Canada, and 100,000 in Argentina. Currently, there are 100,000 natural gas vehicles in the US, but with the stringent emission standard targets applied in California being increasingly adopted by other states it is possible that the number of natural gas vehicles could reach 4 million by 2010. All new buses in Buenos Aires began using compressed natural gas in 1990, whereas Sydney has 250 natural gas buses operating. Budapest in Hungary intends to replace its 400 diesel buses with compressed natural gas vehicles at a cost of $6 million, but this is expected to pay for itself in four years due to the difference in petrol and compressed natural gas prices. The US Parcel Service plans to convert 2700 delivery trucks to compressed natural gas in Los Angeles and eventually all its 50,000 vehicles. Brussels introduced a fleet of 20 buses in 1994 powered by natural gas. In 1995 in the UK, there are over 300 vehicles running on natural gas and eight fast-filling stations. Mexico City has plans to convert all public transportation vehicles to run on natural gas. Almost all such vehicles have modified petrol or diesel engines that run on natural gas stored in high pressure tanks located in the boot of a car or on the chassis or roof of a van, bus or lorry.[27]

The advantages of compressed natural gas are that it is often cheaper than petrol (e.g., half the price of petrol in the US due to no federal tax being applied) and it produces lower pollutant emissions. Carbon monoxide is reduced by 90 per cent, reactive hydrocarbons by 50 per cent and there are virtually no particulates emitted. Unlike petrol, no benzene is emitted. However, on the negative side emissions of nitrogen oxides may be higher. It emits less carbon dioxide, but methane is a greenhouse gas so this offsets this advantage with respect to the problem of global warming. It is considered a safe fuel with no fires or explosions arising from using compressed natural gas in vehicles. It is better suited to buses and trucks than cars because it offers only 25 per

---

[26] Devitt, M, Drysdale, D W, MacGillray, I, Norris, A J, Thompson, R and Twidell, J W (1993) 'Biofuel for transport: an investigation into the viability of rape methyl ester (RME) as an alternative to diesel fuel', *International Journal of Ambient Energy*, vol 14, pp 195–218; Elvington, P (1993) 'Bio-diesel fuel: merits questioned', *Acid News*, 3, Jun, p 15; McDiarmid, N (1992) 'British uses to run on flower power', *New Scientist*, 3 Oct, p 18; Meyer, C (1993) 'Rough road ahead for biodiesel fuel', *New Scientist*, 6 Feb; Patel, T (1993) 'France placates farmers with plant fuel plan', *New Scientist*, 27 Feb.

[27] Compressed natural gas (85 per cent) is also mixed with hydrogen (15 per cent) to form hythane (i.e., 'Hy' from hydrogen, 'thane' from methane.) It emits less hydrocarbons and nitrogen oxides compared with compressed natural gas. Biogas, like natural gas, consists of methane, but whereas natural gas is a fossil fuel, biogas comes from renewable sources. It is produced from the decomposition of organic matter, such as sewage sludge, but it can also be salvaged from waste dumps. Linkopping in Sweden, Colorado Springs in the US and Tours in France all have buses running on biogas. Elvingson, P (1994) 'Transportation fuels: exploring the alternatives', *Acid News*, 5, Dec 1994, pp 6–7.

cent of the energy provided by a similar amount of petrol, so needs much larger storage tanks, causing the vehicles to be heavier and/or some loss of space.

Zero-Emission Vehicles

Electric (Battery) Vehicles

Electric vehicles using batteries have existed for many decades but they were unattractive to most potential users because the early electric motors were inefficient, the electronic devices to control the motors were unreliable and the batteries could not store enough energy to propel the vehicle very far, not helped because the batteries made the car very heavy. Today the situation has improved considerably. Current electric cars offer the greatest potential for use in urban areas where frequent stops and starts associated with commuting and deliveries (e.g., post and parcel deliveries) lead to a high consumption of conventional fuels.

Batteries cannot store nearly as much energy, or generate as much power, as internal combustion engines. The two key parameters which need to be considered are power density and energy density. Power density is the amount of power per kilogram of battery weight that can be extracted from a battery. Low power density translates into poorer vehicle acceleration, a reduced ability to climb hills and a slower recharge time. Energy density measures the amount of electrical energy that can be stored in each kilogram of battery. A low energy density results in a reduced range between recharges. An analysis of power density and energy density reveals the advantage of internal combustion engines over batteries. Modern engines typically have a power density exceeding 400 W/kg and an energy density of more than 200 W-h/kg. In comparison, conventional lead-acid batteries offer a power density of less than 100 W/kg and an energy density less than 40 W-h/kg. The challenge is to develop a battery that approaches the qualities of the combustion engine.[28]

The conventional lead-acid battery uses alternate lead and lead oxide plates suspended in dilute sulphuric acid. Advanced lead-acid batteries (power density of 200 W/kg) are used in General Motors' successful Impact. This sporty lightweight (plastic-bodied) car offers 0 to 95 kph (60 mph) acceleration in just eight seconds—better than many sports cars—but it needs a two to eight hour recharge. In 1992, General Motors, Ford and Chrysler signed an agreement to co-operate in the development, design, testing and possible manufacture of electric vehicle components that could be used in each of the companies' electric vehicles. Other companies are testing alternative batteries. BMW uses sodium-sulphur batteries (the batteries contain liquid sodium and sulphur at a temperature of around 300°C) propelling its EI prototype introduced in 1991. It requires 12 hours for a full recharge from domestic electricity supplies and offers a range of 240 km (150 miles) and a top speed of 120 kph (75 mph). Peugeot is developing the nickel-cadmium battery and Chrysler the nickel-iron hydride battery. The nickel-cadmium battery in Nissan's Future Electric Vehicle introduced in 1991 offers a 15 minute fast recharging time, a range of 250 km (150 miles) and a top speed of 130 kph (70 mph). Zinc-air dry cells were powering 40 postal vans in Germany in 1995, using a three minute recharging time by changing electrodes for a range of 300 km; the results of this pilot study have convinced the German Post Office to use electric vehicles for 80 per cent of its fleet. Other batteries being considered by car manufacturers include zinc-bromine and lithium batteries. generally, alternatives to the conventional lead-acid battery are more expensive, less reliable, harder to maintain and, in some cases, potentially dangerous (toxic contents).[29]

PSA (Peugeot-Citroen) and Renault are running large-scale trials in cooperation with Electricite de France which generates much of its electricity using nuclear power stations. In 22 towns, including Aignon, Bordeaux, Douai, La Rochelle and Strasbourg, battery recharging stations are being set up at the roadside and in public or private car parks. This will provide electric cars (e.g., models based on the Peugeot 106, Citroen AX and Renault Clio) with convenient range-extending opportunities during the day. To maintain the momentum towards electric vehicles, the French Government announced in 1995 that it would grant a subsidy of about $1000 to anyone buying an electric car, whereas the state-owned electricity company is expected to offer a $2000 subsidy per car to manufacturers or importers of electric vehicles. these measures are aimed at securing a target of 100,000 electric cars in France by 2000.[30]

There are still many practical difficulties before electric battery vehicles gain strong favour among the wider public. The batteries have a limited lifetime and need to be replaced every two or three years. They take a long time to recharge compared with the simplicity of filling a vehicle with petrol or diesel. Typically, the rate of recharging is slow, say six to eight hours (i.e., overnight at home or during the day when parked at work) for a fully depleted lead-acid car battery. Energy density remains a problem, with a typical maximum range of 190 km (120 miles) before the batteries are fully depleted. The batteries, operating in series, take up significant space within a vehicle and they may represent one-third

---

[28] Howard, G (1992) 'Flat out for the car of the future', *New Scientist*, 7 Nov, pp 21–22; Pratt, G A (1992) 'EVs: on the road again', Technology Review, Aug/Sep, pp 50–59; Sperling, D (1994) *Future Drive: Electric Vehicles and Sustainable Transportation*, Island Press, Washington.

[29] Lossau, N (1994) 'German's post vans go electric', *New Scientist*, 22 Jan, p 20; 'BMW and Nissan launch electric cars', *The Environmental Digest*, 49/50, Jul/Aug 1991, p 17; 'Rapid recharge for batteries in Nissan's electric car', *New Scientist*, 7 Sep 1991.

[30] 'French boost for electric cars', *The Environmental Digest*, 1995/4, p 15; 'Electric vehicle test', *Acid News*, 3, Jun 1993, p 14.

of the total weight of the vehicle and a similar proportion of its total cost. Given the need to replace batteries every few years, renting or leasing rather than buying may make more sense. Electric vehicles cost more than conventional vehicles, but this should reduce greatly as the scale of production increases. Generally, the challenge for battery manufacturers is to produce small lightweight batteries which can store large amounts of energy and can be recharged in a few minutes.

Although the benefits of electric vehicles for urban air quality are immense, it is important to recognise the huge increase in the number of batteries that would need to be manufactured for use in electric vehicles. Clearly a commitment to the recycling of these batteries, some containing toxic substances such as cadmium, needs to be adopted at the outset so that new environmental problems are not created. Electric vehicles will still cause some air pollution problems. They produce emissions of particulates from tyre wear, brake pads and clutches. A battery is not a power source: it merely stores electricity generated elsewhere. Electric vehicles do not emit pollutants directly, but the electricity to recharge the batteries does. This electricity is generated elsewhere, sometimes in power stations well outside urban areas, but pollutants are still generated and it represents a trade in emissions from one place to another, as happened with the shift from domestic coal burning to electric and gas heating. the widespread use of electric vehicles may require increased electricity generating capacity. If these power stations are coal- or oil-fired, significant emissions of nitrogen oxides, sulphur dioxide, carbon dioxide and particulates may be produced. as a sustainable approach to tackling the air quality problem in urban areas, using electric vehicles should not be at the expense of increasing carbon dioxide emissions and so adding to global warming. In that context, an electric vehicle ought to be recharged with electricity generated by solar, water, wind or geothermal energy sources.

Hybrid vehicles are being developed as a means of reducing emissions in urban areas while offering motorists the performance advantages of existing cars. They enable less polluting electric power to be used within urban areas, but a small petrol or diesel engine offers the advantages of a conventional car for long distance travel. The Volvo ECC (environmental concept car) introduced in 1993 has a tiny gas turbine, running on diesel and driving a lightweight generator at very high speed. Even though one-fifth of the car's weight is batteries, it is not particularly heavy, has a top speed of 175 kph (108 mph) and would go 670 km (416 miles) on a tankful if you did not exceed 90 kph (56 mph). Battery power alone would take it only a fraction of that distance at a lower speed. The Canadians have developed a petrol-electric hybrid vehicle using electric motors in each wheel of the car, with the batteries being topped up by a small petrol engine which runs for a maximum of 30 minutes in every hour of driving.[31]

Hydrogen Fuel Cells

Hydrogen is almost the optimum fuel in that when burned it emits only water vapour and, depending on the process, minute quantities of hydrocarbons (from the crankcase) and nitrogen oxides. Hydrogen can be used in a liquid form (requiring storage at –250∞C), as a pressurised gas (but the storage tanks are heavy and bulky) or as metal hydrides. The latter consist of powdered metal mixtures including iron, manganese, nickel, titanium and vanadium. Hydrogen gas is pumped into the tank containing the metal catalysts and, cooled by water pumped in alongside the hydrogen, the metals absorb the hydrogen to form a hydride. When the hydride tanks are heated, hydrogen is released to the engine. The system is safe and cannot explode or cause explosive fires. Unfortunately, metal hydride systems are heavy, which so far limits the range of cars using this system to around 160 km (100 miles). Refuelling takes the form of feeding hydrogen gas into the storage tank through a hose. This can take 10–15 minutes, which is slow compared with filling up with petrol. A water hose is needed to remove the heat generated when hydrogen is absorbed by the powdered metals. Mercedes-Benz have 20 prototype cars and vans limited to a range of about 200 km (125 miles) between refuelling, but plan to have hydrogen-powered Mercedes buses operating in Hamburg in 1997.

Hydrogen fuel cells offer possibilities in a decade or two, especially if legislation promoting ultra-low and zero emission vehicles becomes more widely adopted in Europe, North America and Japan. A fuel cell converts chemical fuel into electricity at up to 50 per cent efficiency with no moving parts. The fuel cell was developed in 1839 by Sir William Grove in London and until the 1990s its main application for transportation has been to provide a source of electrical power and drinking water for the Apollo moon missions in the 1960s and the Space Shuttle. The principle of the fuel cell is the reverse of the electrolysis of water in which electricity is passed through a water-based electrolyte and produces hydrogen and oxygen at the electrodes. In the fuel cell hydrogen is fed to a catalysed anode and oxygen (air) to a catalysed cathode. The electrolyte is typically a membrane capable of conducting ions. The products of this process are electricity, water and a little heat. Unlike batteries, fuel cells are not dependent on recharging, but simply use hydrogen as the fuel. Fuel cells can be powered by hydrogen which is stored or produced on-board by steam reforming from hydrocarbons or alcohol fuels (e.g., methanol) or ultimately directly by alcohols or hydrocarbons. The fuel cell used in

---
[31] 'Canadians develop hybrid car', *The Environmental Digest*, 89/90, Nov/Dec 1994, p 15.

a commuting vehicle could be little more than the size of two conventional batteries, but space is needed to store the fuel on-board. Prototype buses using only a solid polymer fuel cell powered by gaseous hydrogen stored on-board were introduced in the US in 1993 (e.g., the Los Angeles airport shuttle bus). Increasing numbers of buses and vans powered solely by fuel cells, offering double the efficiency of the internal combustion engine, are likely to be produced by the end of the century and cars next century.[32]

Solar Cars

The ultimate electric-powered vehicle would use solar energy. Currently, solar panels on a roof do not provide enough power for all a vehicle's daily travel needs, although it could be used to extend an electric battery's life between, say, nightly recharging at home. In 1991, 10,000 solar panels were added to the roofs of conventional petrol-engined cars in Europe, mainly to power electronic accessories such as the ventilation of car interiors on hot days and charging batteries when sitting idle. In 1992 a demonstration project in California involved solar carports (parking spaces) which were equipped with plugs to charge electric car batteries from an array of solar cells. In 1995 the two-seater solar–electric battery hybrid car called the Hawaiian Sunray became available on the market. This comes with solar panels for use either at home or for bolting on-board. This egg-shaped mini-car is fitted with ten lead-acid batteries and has a range of up to 160 km (100 miles) between charges, achieved by either using household electricity outlets or from the solar panels. Its top speed is 110 kph (70 mph).[33]

The potential for solar-powered cars is beginning to be demonstrated. For example, the three-wheeled solar-powered car which won the annual 3000 km (1875 miles) race across Australia in 1993 managed to achieve a solar energy efficiency (the proportion of the sun's energy converted into motive power via photovoltaic cells) of 20 per cent. This was an improvement of 5 per cent over the winning car in 1987. Its average speed for the race was nearly 85 kph (53 mph). Unfortunately, the solar cars competing in the Darwin-Adelaide race were cramped, at times unstable and the drivers often had to suffer temperatures of 50°C.[34]

Conclusions: Promoting Cleaner Fuels and Cleaner Cars

The initial costs of developing and using reformulated and alternative fuels and cleaner vehicles can be high.

This points to the need for government subsidies and tax incentives to encourage the adoption of these cleaner technologies. Governments can influence the take up of less polluting fuels by introducing fuel duties/tax according to their environmental impact. They can offer grants and subsidies to the car industry to develop and manufacture zero-emission vehicles and for the public to buy them. For example, during the mid-1980s Germany raised the annual duty for cars meeting their existing emission standards, but lowered it for cars which met the new stringent European Union standards which would come into effect in 1993 and which would require catalytic converters to be fitted. By 1988, more than 75 per cent of new cars sold in Germany already met the 1993 emission standards and by 1990 this had exceeded 90 per cent. Consequently, Germany was five years ahead of, say, the UK, and its citizens also gained five years of air quality benefits.[35]

California offers the greatest stimulus for introducing cleaner cars as it required at least 2 per cent of sales from motor manufacturers selling more than 30,000 vehicles a year to be zero-emission vehicles—assumed to be electric—in 1998 if they wished to continue selling in the state. This amounts to about 35,000 vehicles a year on Californian roads. This requirement will increase to 5 per cent by 2003 and 10 per cent of sales in 2005 (Table 6.6). This legislation is providing a model for other states to follow. California's commitment is not only fopr zero-emission vehicles but also for a phased strengthening of emission targets for all new vehicles. This progresses from transitional low emitting vehicle targets through low emission vehicle targets and ultra-low emission vehicle targets to zero-emission vehicles. Encouragement and requirements to introduce cleaner fuels and cleaner cars can take many forms. In 1990, Denver stipulated that fleets of 30 or more vehicles in the city must convert at least 10 per cent of their petrol-engined vehicles to alternative fuels by the end of 1992. In 1994, the UK Government made a commitment to increase road fuel duty by at least 5 per cent each year, so providing an incentive to economize on fuel and to choose more fuel-efficient vehicles. In January 1993, the European Parliament urged governments to use subsidies and tax incentives to stimulate the sales of electric cars. In this way, it suggested electric cars could account for 7 per cent by volume of total urban vehicle traffic in the European Union by the year 2002. However, Renault considers that it will not be until 2015 that electric cars will represent 10 per cent of

---

[32] Fisher, S (1994) 'Milestone on the road', *Financial Times*, 15 Apr, p 16; Lloyd, C (1994) 'Electric van gets clean away from batteries', *The Sunday Times*, 24 Apr, section 3, p 12.
[33] Essoyan, S (1995) 'Something new under the Hawaiian sun: a solar car', *Los Angeles Times*, 5 Apr.
[34] Anderson, I (1993) 'Solar dream car comes through', *New Scientist*, 20 Nov; 'Solar car race successes', *The Environmental Digest*, 77/78, Nov/Dec, p 19.
[35] House of Commons Transport Committee (1994) op cit, vol 2, p 55. A similar situation applied in the Netherlands where incentives were also offered.

TABLE 6.6 Sales requirements for zero-emission vehicles as passenger cars and light-duty vehicles in California. Targets listed after 2010 have yet to be finalized in legislation. In 1996 the Californian Air Resources Board suggested it might relax its 1998 target but would keep rigidly to its goal of 10 per cent by 2003.

Year	Percentage of sales
1998	2
1999	2
2000	2
2001	5
2002	5
2003	10
2204	10
2005	20
2006	20
2007	35
2008	35
2009	50
2010	50

Source: South Coast Air Quality Management District (1994) Final 1994 Air Quality Management Plan: Meeting the Clean Air Challenge, SCAQMD, diamond Bar, chapter 4, p 17

the total European Union market and 40 per cent of cars used in towns. Using a variety of incentives, the Japanese government aims to have 200,000 electric vehicles in operation by 2000.[36]

Traffic management in congested, polluted cities can encourage greater adoption of cleaner cars such as by banning diesel- and petrol-engined cars from city centres and requiring these to be parked at park and ride sites on the outskirts of the city. From such sites buses, trams or trains could be used to reach the city, or perhaps electric cars could be rented to travel into the city centre.[37] Free parking and the permitted use of high-occupancy vehicle lanes even when there is only one person in the car are other incentives that can be offered to drivers of electric vehicles. The widespread use of less polluting alternative fuels and zero-emission vehicles is likely to remain a long-term objective, but reducing emissions by improved engine technologies, fitting pollution control equipment and changing fuel composition can produce a small but significant improvement in urban air quality.

Many people would like technology to solve their pollution problems and not have to change their lifestyles, habits or behaviour. They would like to continue using their car, preferably as the sole occupant, and not have any restrictions imposed on its use. Unfortunately, technology can only help the problem, not solve it. Technological solutions to air quality problems can even worsen the situation by causing other environmental pollution problems and resource depletion. Fitting cars with catalytic converters is a major step adopted by many countries, but the rising numbers of cars and the increasing distances they are forecast to be driven each year will offset some of the benefits of the expected pollution reduction. This suggests traffic-generated pollution will remain a problem despite current and planned stricter exhaust emission limits for new petrol- and diesel-engined vehicles. The limitations of technological solutions to air quality problems facing urban areas places greater responsibilities on the authorities to tackle the problem of vehicle-generated pollution in other non-technological ways. Some of these, traffic management and control policies and measures, are explored in the next chapter.

---

[36] 'Dutch support for electric cars' and 'Toyota and Nissan co-operate', *The Environmental Digest*, 67, Jan 1993, p 16.
[37] For motorists willing to park on the outskirts of Coventry, UK, there is a proposal to make electric hire cares available for driving into the city centre or for other uses within the city. It will use a 'dial a car' system involving mini-electric Peugeot cars.

reading 3

"The False Promise of Electric Cars"
Eric Peters

In late 1989, California's air quality bureaucracy passed rules effectively requiring the sale of electric cars beginning in 1998. (Though the state does not specifically mention them, electric cars are the only vehicles that currently meet its ultra-strict "zero-emission vehicle" standard.) Eleven Northeast and mid-Atlantic states are considering or have already approved electric vehicle mandates patterned after the California edict.

Some 20% to 33% of the entire U.S. new car market (depending on whether all the interested states pass the necessary legislation) would be covered by these mandates. All told, if the mandates come through, automakers would be required to supply some 70,000 electric vehicles within three years and nearly one million through the year 2003, as requirements call for more electrics.

Consequently, automobile manufacturers have been faced with the daunting prospect of having to design and build a practical, marketable—and safe—electric vehicle (EV) for retail sale by California's 1998 deadline.

Why is the EV mandate "daunting"? Mainly because electric vehicle development is still in its infancy—despite a lot of rosy talk by electric vehicle advocates. Major technical and engineering hurdles remain to be overcome before EVs are expected to be a reasonable alternative to internal combustion automobiles.

Essentially, the available evidence reveals that electric vehicles are being forced into the marketplace before they're anywhere near ready—and that consumers will reject them overwhelmingly. Problems with EVs include limitations of current battery technology and vehicle performance, uncertainty about vehicle safety, and prices significantly higher than comparable conventional cars—all issues that may be disquieting to potential EV owners. In addition, EVs may also represent a new threat to the environment that could turn out worse than anything one could fairly attribute to gasoline-burning automobiles.

Meanwhile, many alternatives to electric cars—natural gas, propane, even super-clean conventional vehicles—might prove more palatable to consumers and less threatening either to the environment or public safety. But they have been shunted to the sidelines while electric-vehicle mandates occupy center stage.

Obviously, the effort to build a viable electric car must be successful—or it will be a disaster both for the industry and consumers. Millions of dollars have already been spent in research and development—to say nothing of eventual tooling and other related production costs—by each of the automobile manufacturers. When all is said and done, the total industry commitment to electric vehicles is likely to exceed several billion dollars. To put this in perspective, $1 billion is about what Chrysler spent on bringing the Neon, the company's hugely successful new economy car, from the designer's sketchpad to dealer showrooms. In other words, for the money being spent on EVs, the manufacturers could have designed, built, and sold several entirely new conventional car models.

If electric cars don't sell—or more importantly, *if they can't be sold at prices which reflect their true cost to manufacture*—the automobile industry will be faced with two equally unpalatable choices: The car companies will have to subsidize the "sale" of electric vehicles—selling the cars below cost and raising prices of conventional cars—to make them more attractive to consumers. Or manufacturers may simply unload fleets of the unsaleable electrics to commercial users (mainly utilities) at tremendous discounts. Either way, the money lost will have to be made up from purchasers of conventional, gasoline-burning cars and trucks.

In other words, you and me.

According to informal conversations with industry representatives of all the major domestic automobile manufacturers (none want to speak out publicly against electric cars at this juncture, for obvious reasons), buyers of gasoline-powered cars can expect to see sticker prices of conventional vehicles escalate anywhere from $500 to $1,000 or more should the EV mandate stay in place.

Call it an "electric vehicle surcharge," if you like. Such a significant and dramatic price increase could easily push many buyers out of the market. Ironically, air quality could actually worsen as people cling to their older—and higher polluting—cars and trucks in lieu of buying the newer models, which they can no longer afford.

All of the electric vehicles being evaluated for eventual production—the GM Impact, Ford Eco-star/Ranger pick-up truck, Chrysler TEV electric minivan—are expected to have retail prices in excess of $20,000 when they hit the market. (That figure represents an estimate of what a mass-produced EV is likely to cost—and includes cost savings attributable to economies of scale, etc.)

However, according to a recent report by the U.S. General Accounting Office (GAO), "The initial purchase price of vehicles that meet the reasonable demands of consumers will most likely remain at least two and three times higher than comparable internal combustion engine vehicle prices in the near term."

If these figures can't be brought down considerably, it's doubtful any but a few zealots will find electric cars attractive given their limitations—and consumers will have to subsidize their sale.

Advanced Technology?

So what's wrong with electric cars? Aren't they the promising new technology of tomorrow that will liberate us from dirty and unhealthful internal combustion? Sadly, they are not.

Electric cars have a number of serious liabilities that haven't yet been overcome, despite nearly a century of trying. Chief among these are poor and unreliable performance.

To understand the engineering problems with electric cars, you have to get at their basic flaw—the power source. After an extensive review of the topic, the GAO report notes that "limitations in the range, power, recharging capabilities, and life of batteries remain the largest technical obstacles for the commercialization of EVs."

All electric vehicles currently being readied for production rely on primitive (and gigantic) lead-acid battery packs for power. These cumbersome units, weighing 800 pounds and up, store only enough power for a very limited range as compared to conventional vehicles. To travel 100 miles, GM's Impact would require 5.67 liters of gasoline weighing 10 pounds under conventional power or 880 pounds of a lead-acid battery, the GAO notes.

Technologically speaking, these batteries are essentially the same as the battery packs that powered early electric vehicles—like the 1908 Baker Electric—90 years ago. As a result, the over-the-road performance of today's "high tech" electric cars is only slightly better than the electric cars back then.

The Baker, for example, ran about 30 miles before it needed to be recharged; the typical "model" electric car nowadays has a real-world range of 70 to 100 miles—and, like the Baker, needs eight or more hours to juice back up. Even in 1908, when internal combustion automobiles were in their infancy, this limited range was unacceptable and cars like the Baker soon went out of production.

On today's electric vehicles, use of even basic accessories such as the air conditioner, headlights, windshield wipers, etc., draws power from the battery pack and can decrease the usable range by 25% to 50% or more depending on conditions. City driving also shortens the driving range (30 to 50 miles in current vehicles).

Cold weather hurts battery performance even more. When the outside temperature falls below freezing, usable range plummets to 20 miles or less—making EVs practically unusable in colder regions during winter. (While frigid weather isn't generally a problem for Californians, it is a fact of life in the Northeast, where many EV mandates are being pondered.)

Electric-vehicle advocates dismiss concerns about lead-acid batteries, stating they are only an "interim" technology. Indeed, news reports frequently highlight one or another advanced battery design that promises to overcome range and performance problems. But these more advanced batteries being tested (e.g., sodium sulfur, lithium, nickel-cadmium) are extremely expensive (Ford's Ecostar prototype uses a sodium sulfur battery, which cost tens of thousands of dollars all by itself), are often hazardous, and are not likely to be ready for commercial use for many years. According to the GAO, "each battery type has its individual positive and negative attributes." For example, lead acid, with its performance limitations, is plentiful and relatively cheap; nickel cadmium has high power, which is good for acceleration, but both nickel and cadmium are expensive. Cadmium, moreover is quite toxic. Lithium is considered among the most promising designs for performance and cost, but there are safety and recharging questions and its development is not expected before 2010.

The limitations imposed by battery technology are probably sufficient to sink the electric car as a practical alternative form of transportation. The GAO notes, for example, that even as future research improves battery design, vehicles powered by batteries "will most likely always have shorter ranges and longer refueling times than comparable" gasoline-powered vehicles. But, poor battery performance is just one of the electric car's many weaknesses.

Take maintenance costs. Though electric-vehicle advocates like to point out that EVs require no periodic oil changes or tuneups, they frequently neglect to mention the issue of battery pack replacement. Like the small 12-volt battery which starts your conventional car in the morning, an electric vehicle's battery pack must be replaced every three years or so because it is no longer able to hold a charge. The cost of replacing the

Real World Comparison

To put the electric car in better perspective, consider the real-world performance of the Ford Ecostar—which I had opportunity to test drive for a week recently.

The Ecostar is among the most technologically advanced and best-performing prototype electric vehicles out there. It features a state-of-the-art battery pack, direct-drive motor, and the latest electronics to back it all up.

The Ecostar is the working prototype upon which Ford Motor Company's eventual production model electric vehicle—very likely a converted Ranger pick-up truck—will be based. When that vehicle reaches showrooms, it is expected to have a price in excess of $20,000.

I put the Ecostar—which is based on Ford's European Escort—up against my second car, a "parts chaser" 1974 VW Beetle—which I purchased for $900.

What did I find? A twenty-year-old, $900 used VW is superior in every category of meaningful performance save braking distance to the "advanced" Ecostar. (My primitive four-wheel drum brakes are no match for the Ford's modern disks.)

- Range. The Bug can go about 200 miles on a full tank and takes maybe three minutes to refuel at any gas station. The Ecostar, with its 75 hp AC motor, has a "safe" range of about 60 miles (you don't want to push it), and takes eight hours or more to recharge on 220 volt current. On household 110 volt current, recharge time doubles or triples.
- Acceleration and Top Speed. Shifting hard, the Bug can make it to 60 in about 10 seconds; top speed is around 95 mph at redline with the wind at your back. Passing tractor trailers isn't easy, but it's possible. With a fresh charge, the Ecostar does 0–50 mph in 12 seconds; top speed is about 70 mph. As the battery loses its charge, however, the Ecostar slows down appreciably. Slight inclines become challenging; passing anything is out of the question.
- Passenger and Luggage Capacity. The Bug seats four tightly, with a little room left for stuff in the trunk; Ecostar is a two-seater, but has a rated payload capacity of 880 to 1,021 lbs. The Ecostar does have air conditioning (but that draws power away from the battery and eats into the range), a nicer ride, and even handles better. But the point here is a decades-old used car that was considered "bottom of the barrel" even when new can do most of the important things better—and for a lot less money—than an allegedly "state of the art" electric vehicle. –E.P.

battery pack is expected to be in the several thousand dollar range for the cheap lead-acid units; much more for the "advanced" battery pack designs. For electric car owners, this is the equivalent in expense and hassle of having to replace the engine and transmission in a conventional vehicle after three years of service.

EVs are also poor performers. Compared to even the slowest and least powerful current-year economy cars, electric cars are so feeble they border on being dangerous. A vehicle with a 0–60 time of more than 15 seconds, for example, will have trouble merging safely with faster traffic—or pulling out from an intersection. Most modern cars—even the real dogs—can accelerate from a stop to 60 mph in under 12 seconds; the best electric cars can take as long as 30 seconds. Ford's Ecostar, for instance, does 0–50 in 12 seconds. The GM Impact is slightly faster—but not by much. And the Chrysler TEV van needs almost 30 seconds to hit 60 mph.

On the highway, most electric cars have top speeds (with a full charge) of 70 mph or less. While it's true that is more than the national speed limit, it leaves little in terms of reserve power for passing or maintaining speed on hilly terrain. And when the battery begins to lose its charge, this already marginal performance often diminishes to the point of being genuinely hazardous.

Only by drastically cutting weight, size, and carrying capacity have engineers managed to get decent performance out of an electric vehicle. That's why most of them are so small. They usually seat only two and have no extra room for luggage or cargo.

When you compare the performance of EVs costing $20,000 plus to conventional new cars in the $10,000 to $15,000 range, the magnitude of the looming electric car fiasco becomes apparent. For $15,000, one can buy a nicely loaded Ford Taurus, Chevy Lumina, or dozens of similar models that don't need half a day to recharge, can carry a family and its stuff, have decent performance, and go hundreds of miles before refueling. Even $10,000 will get you a Geo Metro, Hyundai Accent, Ford Aspire, or several other models that offer performance superior to any electric car likely to be on the market by 1998.

There are problems with traveling far from home. The major impediment to electrics after batteries, according to the GAO review, is infrastructure support. That is, how do you charge your battery when traveling? A 1993 survey found that 76% of respondents wouldn't buy an EV until quick-recharging stations became widely and publicly available. Little has been done in terms of establishing roadside chargers, setting standards, addressing safety concerns (i.e., which standards for plugs and which system is safest to charge in the rain?), and the like.

Given all these limitations and uncertainties and barring a major technological breakthrough, consumers are not apt to respond favorably to electric vehicles. As John B. Heywood, of Massachusetts Institute of Technology's Sloan Automotive Laboratory comments: "Most auto buyers do not care what propulsion system is under their hood so long as it is the cheapest available and requires minimum compromises in vehicle comfort, performance, and aesthetics. It is the cost, weight, and bulk size of the total propulsion system that are the critical issues. The gasoline-fueled engine now dominates auto use because it is much better than its competitors in these areas."

Safety Questions

Safety of electric vehicles is a matter of critical importance for consumers. Though EVs are fundamentally different in their basic design from conventional vehicles, with large quantities of caustic materials on board (e.g., sulfuric acid), as yet no special safety or crashworthiness standards have been issued by the National Highway Traffic Safety Administration (NHTSA).

EVs merely have to meet the standards which apply to conventional vehicles that are in place at the time of their manufacture—and waivers may be granted by the government under a loophole in the existing law. "In all likelihood," the GAO notes, "new or revised regulations will be required to ensure EV crashworthiness." For example, Evs made of more light-weight materials may be "less able to absorb and direct the energy of a collision," which could result in less protection for occupants. Battery fasteners and enclosures "are likely to require special attention to minimize hazards associated with high voltage and reactive chemicals."

Leaks in the battery pack—and the possibility for the release of explosive hydrogen gas (a by-product of all lead-acid batteries)—are a particular concern. All EV batteries present some safety hazards. The advanced sodium sulfur batteries appear to present the "most serious hazard," the GAO reports, threatening high voltage electric shock, fire and toxic gases. Ford parked its entire sodium sulfur fleet last summer to investigate two fires in the battery packs.

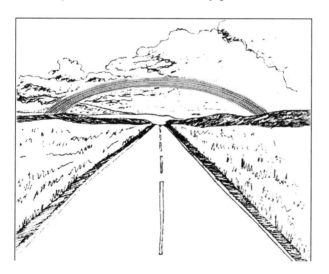

Interestingly, consumer advocacy groups haven't been vocal on the issue. Clarence Ditlow of the Center for Auto Safety observes that some of the smaller, independent electric car companies (those not affiliated with the Big Three) have put their prototypes through crashworthiness testing. But consumers should be concerned about the electric vehicles to be produced by the major manufacturers—the vehicles they will see in showrooms in a few years.

Ford and GM have put their respective electric vehicle prototypes through informal crash testing and claim their EVs will meet 1998 NHTSA standards for conventional passenger cars. However, the fact remains that no specific standards which address the unique safety issues associated with electric cars have yet been promulgated by the federal government—and consequently we have no objective way of determining how safe electric vehicles really are. The real point here is whether safety standards written for conventional cars have any applicability to electric vehicles.

Bruce Zemke, a staff development engineer with GM, says that discussions are under way between NHTSA, the Society of Automotive Engineers (SAE), and car industry representatives to ascertain whether special crashworthiness/safety standards should be issued for electric cars.

"We need to engineer the product so that safety is not an issue," he explains. "We have to think about how electric vehicles should be evaluated—and whether they should be judged on an equivalent basis with internal combustion cars."

The manufacturers are doing what they can, but they don't have very much time to work with and are under the gun to have electric cars ready for sale by 1998.

Pollution Problems

Of course, the issue behind all the mandates is whether EVs do, in fact, represent the "environmentally sound" technology advocates have claimed. If they do represent a way to clean up the air, perhaps these burdens are worth shouldering. But, if not, then why should Americans have to pay for them?

Those unfamiliar with the subject tend to view the electric car issue as a contest between "zero emissions" (i.e., environmentally responsible energy) and irresponsible, outdated, and "dirty" internal combustion engine technology. But the fact is that electric cars produce pollution, too—it's just in a different form and sometimes originates in a different place.

It turns out that more and more studies are coming to the conclusion that electric cars may not be all that "earth friendly" after all. The most recent of these, published in the journal *Science*, concludes that lead-based electric cars actually would increase the threat to public health and the environment. The analysis by researchers at Carnegie-Mellon (with funding from the National Science Foundation) finds that the mass production of electric cars using lead-acid battery packs would exponentially increase the public's exposure to lead pollution. According to the study, electric cars would create more than 60 times the amount of lead pollution as comparable vehicles burning leaded gasoline.

"These lead discharges would damage ecology as well as human health," the researchers write. "Even with incremental improvements in lead-acid battery technology and tighter controls on lead reprocessors, producing and recycling these batteries would discharge large quantities of lead into the environment."

Lead is a more serious environmental threat than urban smog, which is why it was gradually removed as an octane-enhancing agent from motor vehicle fuels beginning in 1975.

"Electric vehicles will not be in the public interest until they pose no greater threat to public health and the environment than do alternative technologies, such as vehicles using low-emissions gasoline," the researchers conclude, adding that nickel-cadmium and nickel metal hydride batteries also "do not appear to offer environmental advantages."

John Undeland of the American Automobile Association (AAA) expresses concern that if the Carnegie study is correct in its conclusions, it could mean the introduction of electric cars might negate all the air quality gains which have been made through the elimination of leaded gasoline.

"This would put us back at square one. We've done a lot of work to remove lead from the environment, and now we're faced with something that might undo all our efforts. This is just another indication that we haven't got to the point where electric cars are viable," he says. "Electric cars are not 'zero emissions' cars—they're 'elsewhere emissions' cars."

In point of fact, studies suggest that the added demand for electricity from coal-fired utility plants will result in a manifold increase in so-called "stationary source" pollution. This is especially true in the Northeast, where the coal burned is of the high-sulfur type that is by nature the most polluting.

Estimates of the increase in sulfur dioxide emissions—which cause acid rain—vary from 17% to 2,100%, according to figures compiled by the National Conference of State Legislatures.

Yet big electric utilities—which hope to find a captive audience for their excess generating capacity—continue to advocate EVs and EV mandates. And for the most part, "environmentalist" groups have been strangely silent on the question of electric vehicles' potential for contributing to worsening air quality.

The clincher is that electric cars probably aren't necessary to improve the quality of the air we breathe anyway. Advances in emissions control technology for gasoline-fueled cars have come within a hair of making them free of harmful pollutants. "Zero emissions" electrics, in this respect, make only small improvements over current "low emission" vehicles. And while electrics totally eliminate certain smog-causing emissions from tailpipes, most of these emissions will continue to come from the stationary sources—factories and such.

"The proportional contribution of stationary source emissions has been getting larger and larger over the past three decades while the contribution of mobile source [cars and light trucks] has been getting smaller and smaller, despite a substantial increase in the total number of miles driven each year," says AAA's William Berman.

Overall, the car industry has managed to eliminate 90% to 98% of the harmful emissions coming from the tailpipes of their new cars. Catalytic converters and computer-controlled engine management systems that precisely regulate what's happening under the hood are to be credited for these improvements.

AAA's 1994 study, "Clearing the Air," for instance, states that conventional cars and light trucks contribute less than one third of the contaminants that combine to form urban smog—and that passenger cars and light trucks are no longer the single biggest contributor to regional air quality problems. By 1996, according to the study, just 24% of the total output of Volatile Organic Compounds (VOCs)—precursors to urban smog—will come from passenger vehicles, down from 71% on 1970. A similar and equally dramatic decline in nitrogen oxide emissions is expected to occur as well, with 1996 vehicles estimated to produce just 20% of this pollutant.

The study concludes: "While ground-level ozone continues to be a pervasive problem in many U.S. cities, automobiles and light trucks are no longer the primary or even secondary cause of summertime ozone 'smog' in the 10 cities studied."

There may be a glimmer of hope for car buyers, however. According to Reuters, "Officials from the California Air Resources Board (CARB) and Gov. Pete Wilson's office say they are still not convinced that a commercially viable electric vehicle will be available by the 1998 deadline set by the state." Similar concerns have been raised recently by officials in Massachusetts and New York, raising the possibility that the mandates may at least be delayed.

But given the inertia behind the EV mandates—and the commitments that have been made by politicians and bureaucrats—it's doubtful the "zero-emissions vehicle" mandates will be modified. Consumers, meantime, may have to get used to continued false promises.

"New Vehicles Now Less Efficient Than Junked Ones"

Sierra Club Action Newsletter, Vol. II, #147, August 14, 1997. Reprinted by permission.

Sometime irony can be delicious. Here's an example, however, that will make you cringe. Years ago, opponents of more restrictions on auto emissions argued that fuel efficiency and air quality would continue to improve without new regulations because every year old polluting vehicles would be replaced with new, more efficient, cleaner-burning ones.

Not any more.

According to the Environmental Protection Agency, for the first time on record, old cars and trucks now being retired are more efficient, on average, than new vehicles being sold. (*New York Times*, Aug. 11, 1997)

This is because gas-guzzling minivans, light trucks and sport utility vehicles have doubled their market share over the past two decades. They now make up nearly 40 percent of the new vehicles sold.

The corporate average fuel economy (CAFE) standards for new cars is 27.5 miles per gallon, but light trucks are only required to average 20.7 mpg.

For years, the introduction of new vehicles did improve overall fuel economy, albeit too slowly for Sierra Club tastes, because the cars and trucks being replaced were gas guzzlers from the early 1970s.

But those inefficient vehicles have long since been taken out of service. The cars going to the junkyard today tend to be from the mid-1980s, when efficiency was at its peak. For the last three years, new vehicles have averaged 24.6 mpg, compared to a high of 25.9 in 1987 and 1988.

This decrease in overall efficiency, combined with more driving and lower gas prices, has resulted in record consumption of gasoline, half of which comes from imported oil.

The Sierra Club supports raising CAFE standards to 45 mpg for cars and 34 mpg for light trucks. It's the biggest single step we can take to curb global warming, cut our dependence on fossil fuels, and protect sensitive areas like the Arctic National Wildlife Refuge and the new Utah national monument from oil drilling.

Questions for Discussion

For Reading 1

1. According to this author, how serious is the air pollution problem in Los Angeles?

2. Briefly describe how Los Angeles intends to reduce vehicle emissions. How effective do you think each of these measures will be? Specifically, for each measure, what are its strong points and what are its weaknesses?

For Reading 2

1. What are the prospects for developing cleaner fuel for vehicles? For each method, what are its strengths and what are its weaknesses?

2. What are the prospects for developing cleaner vehicles in the near future? What are the relative strengths and weaknesses in low-emission vehicles and zero-emission vehicles?

3. What could be done to improve the marketability of zero-emission vehicles?

For Reading 3

1. According to this article, what are the major drawbacks to electric vehicles?

2. One of the major problems with electric cars today is their cost. Who should bear these extra costs? Do you think drivers in general would be willing to pay more for such cars? Assuming that they were safe, efficient, and environmentally friendly, how much more would you be willing to pay for an electric car?

3. If electric cars are not the answer to improving urban air quality, what are the alternatives?

For Reading 4

1. According to this article, why aren't we making steady progress in reducing vehicle emissions? How might we get "back on track"?

2. What are the implications of this article for Los Angeles?

Going Beyond the Readings

Visit our Web site at <www.harcourtcollege.com/lifesci/envicases2> to investigate air quality issues in additional regions of the United States and Canada.

unit 11

Southeast: Wildlife Management

Introduction

Readings:

Reading 1: Bluefin Tuna in the West Atlantic: Negligent Management and the Making of an Endangered Species

Reading 2: Historic Rationale, Effectiveness, and Biological Efficiency of Existing Regulations for the U.S. Atlantic Bluefin Tuna Fisheries: A Report to the United States Congress

Questions for Discussion

Managing Bluefin Tuna in the Western Atlantic

Introduction

As little as 20 years ago, futurists predicted that marine fisheries would provide salvation for a starving world. Human populations would continue to grow, so the reasoning went, and arable land would become exhausted or transformed into cities. We would become more and more dependent on the vast, unlimited resources of the sea. "Let them eat fish" seemed like a palatable solution to food shortages everywhere.

Unfortunately, in today's world, ocean resources are neither vast nor unlimited. Fishery resources are beginning to show signs of overuse. Fishers worldwide expend ever-increasing efforts with ever-decreasing results. In 1976, an estimated 64 million tons (58 million metric tons) of fish were harvested worldwide. Yields increased dramatically and reached a peak in 1989 when 97 million tons (88 million metric tons) were harvested. The peak was followed by a rapid decline, which leveled off at around 88 million tons (80 million metric tons) harvested annually.

Modern fishers are no longer restricted to their nation's offshore waters. For example, fishing fleets are sufficiently large and seaworthy that Japanese fishers can and regularly do ply potentially productive waters anywhere in the Pacific, Indian, and even Atlantic Oceans. Japanese buyers can and regularly do pay top price for fish anywhere in the world, air freight them to Japanese markets, and sell them still fresh from the sea. It is often more likely that an Atlantic salmon caught in the North Sea will feed a Japanese family halfway around the world than a Scandinavian family living near the catch site.

In recent years, steady fish harvests have been sustained through rapid expansion of fishery efforts in the Indian Ocean. In other oceans, notably in the Atlantic and Pacific, commercial harvests have declined steadily.

Everywhere the pattern is the same. A new fishery resource is discovered, exploited, and overexploited in rapid succession. As fish populations dwindle, prices increase, encouraging fishers to continue exploitation even as harvests decrease. Finally, the market crashes. World attention focuses on some other newly discovered resource, leaving locals to deal with often severe economic losses.

Why must it be this way? Why can't fishery resources be managed? Why can't harvests be restricted at the first signs of population decrease? Why can't fishery resources be harvested in a sustainable manner? They probably can be, but there are problems. The fishing industry is more interested in making profits today than in long-term fishery management. Populations of fish are extremely difficult to measure and monitor. Fishers and regulators regularly disagree on numbers of fish until it becomes clear that few remain. Painfully little is known about fish numbers or what controls reproduction. Regulations are difficult to come by and even more difficult to enforce, especially offshore and when dealing with fishers who are citizens of other countries. Many commercially important fish are wide ranging, present in one country's waters for only part of the year. Successful management requires international cooperation.

A case in point that directly involves fishers in North America is the bluefin tuna in the Western Atlantic. Here is a truly magnificent fish. Adults weigh up to 1500 pounds (680 kg). Ecologically they are a top carnivore found in discrete populations in many of the world's oceans. Bluefin tuna are among the fastest of fish, capable of sustained speeds of 30 mph (48 km/h) with bursts exceeding 50 mph (80 km/h). Western Atlantic Ocean bluefin tuna migrate annually from tropical waters in winter to near Arctic waters in summer. Sport fishers crave their speed, beauty, and fighting ability. Commercial fishers crave their profitability, for they are also delicious, and Japanese diners pay up to $350 per pound ($770 per kg) for bluefin tuna at a sushi bar. Commercial fishers can expect to sell a large adult for more than $30,000. No wonder their numbers have decreased more than 90 percent from an estimated 250,000 to 22,000 in less than 25 years.

The plight of the bluefin tuna has drawn international attention, concern, and cooperation. In 1969, the International Commission for the Conservation of Atlantic Tuna (ICCAT) was established by treaty and granted overall scientific and management responsibility for "tunas and tuna-like species," which include marlins and swordfish. Its stated goal is to manage these fish on a sustained harvest basis. Harvest quotas and other regulations will be implemented, if necessary, to impede further declines in population, promote population recovery, and allow for continued and, if possible, expanded harvest rates indefinitely into the future. Today, the commission consists of 22 nations, mainly countries that rim the Atlantic Ocean plus Japan, a major harvester and consumer of bluefin tuna.

ICCAT consists of two bodies. A governing body is comprised of delegate commissioners named by each of the member nations. They meet periodically to review catch statistics and other data, establish harvest quotas, and monitor compliance and progress. Their responsibilities and actions require a delicate balance between economic needs of the fishing industry and recovery needs of the various fish populations. The industry argues for the highest possible harvest quotas to sustain profitability. Often the welfare of the fish is taken up by scientists and conservationist organizations who often argue for harvests to stop.

A second ICCAT body is a scientific advisory committee established and largely funded by the commission. Its task is to assess population size and dynamics as precisely as possible and recommend courses of action that would facilitate recovery and sustainable harvests. When their work started, little was known about the population dynamics and basic biology of these species. Numerous field studies were designed and conducted to answer basic questions.

In 1981, harvest quotas for the bluefin tuna were set by ICCAT and divided among various participating nations who were expected to monitor and enforce restrictions among their citizens. Ostensibly, harvests were allowed only for purposes of scientific monitoring—that is, to supply scientists with data. For example, extensive tag and release studies were initiated to provide data on seasonal movements. In 1981, fishers were allowed to harvest 1280 tons (1160 metric tons) of bluefin tuna to facilitate tag recovery and similar studies. Fishers were expected to supply commission scientists with specific data about any fish caught. In subsequent years, additional quotas were allowed.

In the United States, management of marine fisheries falls under the responsibility of the National Marine Fisheries Service (NMFS), a branch of National Atmospheric and Oceanic Administration. Its job is to manage, through regulation, inspection, and enforcement, the bluefin tuna harvest allotted to U.S. fishers by ICCAT. They too must carefully balance harvest allocations among a bewildering group of potential fishers. For example, without careful regulation, the entire U.S. quota could be harvested early in the year in the Gulf of Mexico and off the coasts of Florida and the Carolinas, precluding any harvesting by mid-Atlantic and New England fishers. In all regions, sport fishers vie with professional fishers such as purse seiners, who vie with long-liners and so on.

How well ICCAT and NMFS manage bluefin tuna is not an easy question to answer. At stake is the future of one of the ocean's most magnificent creatures. At stake too is a cherished way of life shared by all those who take and consume this resource. Continuation of both requires wise and careful international management. The readings that follow point to answers and solutions.

reading 1

"Bluefin Tuna in the West Atlantic: Negligent Management and the Making of an Endangered Species"
Carl Safina

Conservation Biology, Vol. 7, No. 2, June 1993. Pp. 229–233. Reprinted by permission of Blackwell Science, Inc.

The bluefin tuna (*Thunnus thynnus*) is a creature of superlatives. Growing to 1500 pounds (700 kilos), traveling on transoceanic migrations, and reputedly capable of swimming 50 miles (90 km) per hour, it is one of the largest, most wide-ranging, and fastest of animals. To anyone who has seen this saber-finned giant explode through the surface of the sea, it is among the most magnificent. The bluefin is also one of the most valuable and most over-exploited of creatures. Its west Atlantic breeding population has plummeted 90% since 1975, from an estimated quarter million to 22,000 animals (ICCAT 1991). The east Atlantic adult population is currently estimated to be half what it was in 1970 (ICCAT 1992). A highly prized delicacy in Japan's most exclusive sushi restaurants, a single bluefin can bring fishermen up to U.S. $30,000, sell at auction in Tokyo for more than U.S. $60,000, then cost diners U.S. $350 per pound. Most fishing for bluefin tuna is driven by the Japanese market and this international trade.

Built, like most tunas are, for speed and endurance ("tuna" is from Greek meaning "to rush"), its fins retract into slots during high-speed acceleration. The bluefin tuna is "one of the most highly developed of the tuna species" (Cort & Liorzou 1991). Through a highly developed circulatory thermal exchange system known as the rete mirable or "miraculous network," bluefins maintain body temperature of 24° to 35°C, though they inhabit waters ranging as low as 6°C (Cort & Liorzou 1991).

The bluefin tuna inhabits both sides of the Atlantic and the Pacific. Highly migratory, western Atlantic bluefin range from waters off Labrador south at least to Brazil and breed almost exclusively in the Gulf of Mexico (Mather 1974; Clay 1991). There is approximately a 3% west-to-east exchange of adult bluefin across the Atlantic Ocean (Suzuki 1991). The eastern and western populations are considered distinct for fishery management purposes (ICCAT 1989). Southern bluefin tuna (*Thunnus maccoyii*), whose adult population has also declined continually and is now "remarkably low" (ICCAT 1992), migrate circumpolarly in the southern hemisphere, apparently breeding only off Java and northwest Australia but ranging into the south Atlantic.

The body responsible for stewardship of Atlantic tunas is the International Commission for the Conservation of Atlantic Tunas (ICCAT, pronounced "eye-cat," hereafter referred to as the Commission). The Commission comprises roughly 20 member Atlantic-rim countries, plus Japan, a major fisher, importer, and consumer of Atlantic tunas. Founded in the late 1960s, the Commission assumed scientific and management authority for "tunas and tuna-like species." In practice this has included both tunas and such taxonomically distant species as marlins (*Istiophoridae*) and swordfish (*Xiphias gladius*).

The Commission's scientific committee, composed of scientists from several countries, complies catch statistics and models population trends. The Commission's managers are responsible for setting fishing management policy. The managers often have strong industry ties. For example, at this writing one of the three U.S. commissioners works full time as executive vice president for a national seafood industry advocacy and lobbying firm. Another works for the U.S. Department of Commerce.

The Commission's charter mandates that it manage for maximum sustainable yield. But though several species are declining under the Commission's purview, the west Atlantic bluefin tuna is the only one for which the Commission has ever recommended catch quotas. These quotas have long exceeded maximum sustainable yield. An independent panel convened in 1991 by the U.S. National Marine Fisheries Service (NMFS) found that the west Atlantic bluefin population "could provide substantially greater yield and spawning potential if fishing mortality was reduced." The panel also noted that the bluefin population modeling that NMFS contributes to the Commission "used state-of-the-art methodology and is generally of high quality" (NMFS 1991).

But the Commission's managers have repeatedly ignored their own scientists' advice. In 1981 the Commission's Standing Committee on Research and Statistics concluded that the western Atlantic's bluefin tuna population was depleted and that catches "should be reduced to as near zero as feasible." The Commission's managers set a 1160 metric ton annual quota, ostensibly for "scientific monitoring." (The Commission's U.S. advisory committee chairman remarked publicly 10 years later, "It isn't actually a scientific quota and we never really believed it was.") In 1983, the "scientific" quota was raised to 2660 metric tons. Throughout the 1980s this annual "scientific" quota remained unchanged, though the Commission's own data showed the breeding population declining each year, as fishing mortality increased several-fold (Fig. 1). In 1990, the Commission's Standing Committee on Research and Statistics noted that existing catch quotas "will cause the decline of the age 8+ group [breeders] to continue" and "is expected to result in an increase in the estimated fishing mortality rate and a corresponding decline in the estimated stock size" (ICCAT 1990).

In 1991, the Commission's scientific committee reported that the west Atlantic breeding population had declined 24% in the preceding 12 months. The scientific committee also projected an 8% decline between 1992 and 1995 under current fishing pressure, but projected a 47% increase between 1992 and 1995 if the quota was cut by 50%. By that time, Sweden had announced that it would seek to list the western Atlantic bluefin on CITES Appendix 1, and Swedish observers were present at the 1991 meeting of the Commission (CITES, the Convention on International Trade in Endangered Species of Wild Flora and Fauna, is a treaty with approximately 120 party countries at present. Appendix 1 includes species threatened with extinction that are affected by international trade). Apparently feeling some pressure, the Commission's managers cut quotas 10% that year. No recovery target or recovery schedule for bluefin tuna, or any other species, has ever been adopted by the Commission.

The Commission is not in a strong position to regulate or manage much of the international bluefin tuna fishing. Countries that are not Commission members make signficant catches. Japan Fisheries Agency officials have said that catches by non-ICCAT countries may exceed 80% of the Commission's members' catch in the Atlantic, and "an increasing number of boats have been reported flying flags of convenience of ICCAT non-member countries so as to target bluefin tuna in the Gulf of Mexico and the Mediterranean Sea [the only known spawning areas], and thereby fish without any restrictions" (Miyake 1992).

These problems have drawn the attention of conservation organizations. Based on the Commission's management record and on its 1990 statistical report, and in an attempt to make the Commission accountable for its actions, the National Audubon Society in the spring of 1991 proposed to the U.S. Fish and Wildlife Service that the western Atlantic population be included in CITES Appendix 1. Such a listing would have suspended export to Japan of bluefin tuna from the listed population. The U.S, under heavy pressure from tuna exporters, decided not to seek listing. Consequently, Audubon sought in the World Wildlife Fund's assistance in furthering discussions Audubon had initiated with Sweden, whose once-productive bluefin fishery had vanished in the last two decades. In October 1991, Sweden announced its proposal that the western Atlantic population be listed in CITES Appendix 1 and the eastern Atlantic population in Appendix 2 (the latter listing would mandate monitoring of international trade).

A month after Sweden's announcement, the Commission held its annual meeting in Madrid. The

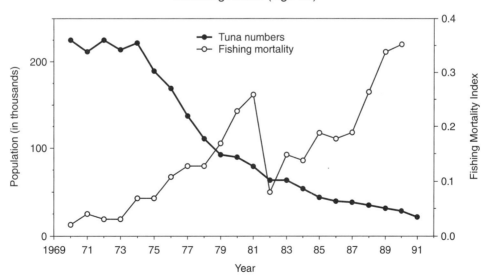

Figure 1: *While managers have ignored their own Commission's science, west Atlantic bluefin tuna have been in prolonged decline, and pressure on the animals has increased significantly. Data from ICCAT 1991.*

U.S. position at that meeting was to pursue a 50% reduction in catch quota for the western Atlantic population. This position, the first catch reduction for bluefin tuna proposed by a Commissoin country in nearly a decade, had been hard-won by conservation advocates working against bitter industry lobbying. This lobbying had escalated to the White House when tuna exporters hired a lobbying firm partnered by a former White House political director and by the man who would become chairman of the Republican National Committee during the Presidential campaign. Conservationists countered with their own White House contacts and prevailed, and the policy to seek a 50% catch quota reduction, based on the U.S. National Marine Fishery Service's scientists' population projections, emerged.

At the 1991 Commission meeting, the three member countries fishing for bluefin in the western Atlantic were encouraged to decide fishing and quota policies among themselves. However, two of the three U.S. Commissioners told members of the U.S. delegation that they did not believe such a catch reduction was necessary or politically realistic (the third commissioner later reported that he had been excluded by the chief U.S. commissioner from attending a closed session with Canadian and Japanese delegates). Initial Japanese opposition to the U.S. position for halving allocations shifted to support of a 50% phased reduction over four years. Japan's desire to avoid a CITES listing and Canada's strong resistance to reduced quotas resulted in continued negotiations throughout the meeting.

By the Commission's adjournment, a conditional four-year phased 25% reduction was agreed to. Over the four-year period this phased scheme would reduce the allowed catch by 17.5% compared with the current allocation scheme. By year four, the quota would be reduced by 25%. However, it was agreed that this phase-in will be automatically abandoned after the 1993 or 1995 scientific reports of the Commission, unless these reports "indicate otherwise." No quantitative targets for population recovery were set. Japan had also offered a trade resolution which would have banned trade in bluefin tuna from non-ICCAT countries, but this resolution was resisted by the European Community because it would have affected some of their non-ICCAT members.

CITES' 100-plus member nations and a host of non-governmental organizations and observers convened in March 1992 in Kyoto. Japanese seafood representatives picketed the convention. Listing west Atlantic bluefin as proposed would have suspended only 1% of Japan's total tuna species imports, and roughly 15–20% of its bluefin tuna imports (up to 85% of Japan's bluefin imprts come from the east Atlantic, Pacific, and Indian Oceans—including southern bluefin—and this international trade would not have been interrupted by the proposed listings). But Japanese delegates said repeatedly to members of the conservation community, "This is the same sort of thing you did to us on whales, sea turtles, and driftnets, and we will not let this happen with bluefin tuna. If you succeed here, you will go on to the next species and the next until you destroy our food culture."

Japan, with Canadian and U.S. assistance, worked feverishly to force Sweden to withdraw the proposal prior to a vote. One Swedish delegate said, "The Japanese and Canadians are applying the worst kind of pressure. We will have to do as instructed." Fierce conflict over the bluefin proposal was evident in the constant flow of position papers and rebuttals distributed throughout the CITES meeting by the Japan Fisheries Association, the Japan Tuna Federation and their American consultants, the Japanese government's Fisheries Agency, the World Conservation Union (IUCN), and a wide variety of international conservation groups. The paper debate raging in the conference lobby was reflected in the attentions of reporters from around the world; only the fight over elephant ivory received more attention.

A peculiar bluefin tuna "briefing book" appeared, containing a six-page statement by "The Federation of Japan Tuna Fisheries Cooperatives, in agreement with the International Commission for the Conservation of Atlantic Tunas, the Inter-American Tropical Tuna Commission, the National Fisheries Institute of the United States [the seafood industry lobbying firm employing one of the U.S. tuna Commissioners], the U.S. East Coast Tuna Association [bluefin exporters], the International Coalition of Fisheries Associations . . . " This did little to dispel impressions that the Commission for the Conservation of Atlantic Tunas has conflicts of interest.

When the bluefin proposal was formally taken up on the floor of the CITES conference, Sweden advocated the need for a 50% quota reduction, adding that they would withdraw the proposal if the Commission's countries agreed to pursue quota reductions. The U.S., Canada, Morocco, and Japan eagerly agreed, and Sweden withdrew their proposal.

Although floor statements were permitted on all other proposals that were withdrawn by their proponents during the two-week conference, an attempt at floor debate by the World Conservation Union (IUCN) was not permitted on the bluefin issue. Eyewitnesses to closed meetings that had occurred over the bluefin proposal reported that, apparently to squelch debate, Canada and Japan had insisted that no floor statements be permitted from anyone besides the fishing countires, and Sweden, under severe duress, acquiesced. Sweden's delegation head said at a press conference after formally withdrawing its proposal, "We had come expecting to debate the biological merits of our tuna proposal, but it soon became evident that other concerns would decide the issue." TRAFFIC USA (Trade

Record Analysis of Flora and Fauna in Commerce; a subsidiary of the World Wildlife Fund) later reported "official discussion over the merits [of Sweden's proposal] never even took place, due to backroom politicking . . . In a well-orchestrated presentation designed to foreclose any open debate . . . Sweden, under extreme pressure . . . withdrew its proposal before any views could be heard" (Hemley 1992). An African delegate later wrote "I was sickened by the manipulations on the Bluefin Tuna."

A special Commission meeting was convened in May 1992 in Tokyo in accordance with the promises made by the fishing countries to the CITES assembly prior to Sweden's withdrawal of its proposal. Although Japan seemed ready to ban imports of tuna from non-ICCAT countries pending consensus by Commission members that it do so, France and the European Community (which includes several non-ICCAT countries) again blocked any attempted trade controls. Although Japan supported in principle the U.S. proposal for halving catch quotas, Japan's delegates emphasized that non-ICCAT countries must be dealt with first. By adjournment, the Commission appeared more incapable than ever of addressing conservation needs, even those it had identified itself.

The Commission's mismanagement is perhaps best highlighted by the U.S. National Marine Fisheries Service's recent analysis showing that if the Commission had simply not raised the catch quota from 1160 to 2660 metric tons in 1983, the adult population would by now have been approximately 3.4 times what it is, and would have been steadily increasing, rather than declining (Powers 1992). When the catch level of 2660 metric tons for Atlantic bluefin tuna was put in place in 1983, 2660 metric tons was roughly 15% of the breeding adult biomass. By 1991, 2660 was approximately 50% of the breeding biomass. Under the system whereby a fixed tonnage (rather than a percentage) quota is taken from an ever-diminishing adult population, pressure on the remaining fish increases even if the number of animals taken remains constant. The National Marine Fisheries Service states that "the objective of stemming the decline in adult population size will not be achieved under the management program now in effect. In order to stem further decline in the adult spawning stock . . . it is necessary to reduce the allowable take 50% or more" (NMFS 1992). (The U.S. connot unilaterally reduce its catch quota because, through a fishing-industry–sponsored 1990 amendment to the United States' Atlantic Tunas Convention Act, the U.S. has legally prohibited itself from setting a domestic catch quota that is less than the quota provided to the U.S. by the Commission; this appears to be the only law that suspends U.S. discretion over management of its own natural resources.) The American Fisheries Society's 1991 resolution on western Atlantic bluefin tuna concluded that "all directed harvests should be immediately prohibited . . . ." In 1992 the Society stated that "The present management regime (A) will not allow the stock to recover, (B) poses an unacceptable risk of there not being enough adult fish to spawn new generations of tuna, and (C) is counter to the long-term interest of both fishery producers and consumers . . . Although the threat to the biological integrity of the stock is of most concern, the economic losses incurred to date are staggering and cannot be recovered" (AFS 1992).

Despite this increasing international scrutiny from conservation groups and fishery scientists, the Commission has attempted to remain insulated. At its 1992 meeting in Madrid, the Commission refused observer status to the World Wildlife Fund. The Commission's scretariat also refused to distribute to its delegates a modestly worded joint statement by World Wildlife Fund, Audubon, and the Center for Marine Conservation, which requested rebuilding targets and timetable, catch reductions that reflect the Commission's scientists' population projections, and a certificate-of-origin system for bluefin (and similar measures for swordfish, whose population trends have also been a source of serious concern).

Despite the Commission's attempts to remain behind closed doors, scrutiny of the Commission by conservation organizations and fishery scientists seems likely to increase. Audubon, World Wildlife Fund, and the Center for Marine Conservation have formed "ICCAT Watch" to track and publicize the Commission's activities. And just prior to the Commission's 1992 meeting, the American Fisheries Society's president warned the U.S. delegation head that "Failure to implement strong conservation measures would be a serious mistake which could cause concerned countries to find solutions through . . . CITES and the Endangered Species Act" (Fetterolf 1992).

The Commission may be changing its approach. At its November 1992 meeting the Commission adopted some modest but potentially constructive measures. It recommended that beginning 1 September 1993, all frozen bluefin imported into any of the Commission's member countries be accompanied by an "ICCAT statistical document." The original proposal by the U.S., Canada, and Japan was for a "certificate of origin," but after three days of intense negotiating with the European Community, France, Portugal, and Spain, the countries agreed to replacing the term "certificate of origin" with "bluefin statistical document," allowing use of a Commission-accepted document or its equivalent, instead of a certificate that would have required validation by a government official. The countries also agreed to delete any language about whether bluefin unaccompanied by appropriate documents would be denied entry into any member country of the commission. Each country may determine the disposition of undocumented bluefin depending on its interpretation of obligations under the General Agreement on Tariffs and Trade. Prior to implementation of this program for imports of fresh bluefins (the high-value animals prized for sashimi), several practical problems regarding handling by customs officials must be resolved. Despite its dilutions and potential implementation problems, this arrangement could help distinguish non-Commission international trade from that of member nations. This could give Japan, by far the major importer, the discretion to prohibit imports of bluefine tuna from non-Commission countries, depending on its interpretation of trade obligations. The Commission also directed its scientific committee to provide and evaluate "various management measures that could be implemented on the east and west bluefin tuna stocks and provide scientifically based target options for the rebuilding of the stocks in a reasonable period of time." The scientific committee will conduct a new assessment of the west Atlantic bluefin population in 1993.

These movements give rise to the hope that the Commission's long quiescence may be changing. Real progress will not have been made, however, until Japan prohibits non-Commission imports, catch quotas are halved, and recovery plans are implemented to rebuild the breeding population to its early-1970s level within a decade (a recovery rate equivalent to the depletion rate that the Commisson allowed). If the Commission's conservation and rebuilding measures remain inadequate after its 1993 meeting, a CITES Appendix 1 listing for west Atlantic bluefins may be necessary. It also seems that a CITES Appendix 2 listing, which would require uniform monitoring of all international bluefin trade, might offer the most efficient and cost-effective way of accomplishing the monitoring of non-Commission countries' catches, something the Commission itself has identified as necessary.

Catches of other fishes under the Commission's purview have been in excess of maximum sustainable levels, including blue marlin (*Makaira nigricans*), white marlin (*Tetrapturus albidus*), swordfish, east Atlantic bluefin tuna, east Atlantic yellowfin tuna (*Thunnus albacares*), albacore tuna (*Thunnus alalunga*), and bigeye tuna (*Thunnus obesus*) (ICCAT 1992). Currently, nothing prevents systematic repetition of the west Atlantic bluefin scenario for these species, several of which have already declined significantly. Until the Commission becomes serious about complying with its charter mandate to manage for sustainable yield, the acronym ICCAT will appear to represent International Commission to Catch All the Tuna. If suspension of international trade in west Atlantic bluefin tuna eventually occurs, whether through CITES, another legal mechanism, or through commercial extinction, it might seem the result of the Commission's long history of failing to heed its charter and its scientists.

reading 2

"Historic Rationale, Effectiveness, and Biological Efficiency of Existing Regulations for the U.S. Atlantic Bluefin Tuna Fisheries— A Report to the U.S. Congress"

National Marine Fisheries Service, 1996, pp. 9–52. Reprinted by permission of the National Marine Fisheries Service.

Historical Review of the Fishery

History of the Fishery

In the eastern Atlantic Ocean and Mediterranean Sea, bluefin tuna have been exploited for thousands of years. Before the 20th century, there was little directed bluefin tuna fishery existing in the United States. There was no market for bluefin, and giant fish (≥310 pounds) were regarded as a nuisance because of the damage they caused to fishing gear. In the early 1900's, a sport fishery developed for small (up to 135 pounds) and medium (135–310 pounds) bluefin off New York and New Jersey, and for giants off Prince Edward Island (Canada) and in the Gulf of Maine. This rod and reel fishery expanded rapidly following World War II, and continues today from Cape Hatteras to the Canadian border. In addition, it is locally important in the Straits of Florida. Occasional sport catches are also made in the Gulf of Mexico.

Until 1958, the U.S. commercial fishery employed mostly harpoons, handlines, and traps. Much of the catch occurred when bluefin tuna were encountered during operations targeting other species. Commercial purse seining began with a single vessel in Cape Cod Bay in 1958, and expanded rapidly into the region between Cape Hatteras and Cape Cod in the early 1960's. The purse seine fishery between Cape Hatteras and Cape Cod was directed mainly at small medium bluefin, and at skipjack tuna, all for the canning industry. North of Cape Cod, purse seining was directed at giant bluefin.

High catches of juvenile bluefin were sustained throughout the 1960's and into the early 1970's and, along with the intense longline fishery pursued by Japanese vessels in the 1970's (in the late 1970's, approximately 10,000 giant bluefin tuna were taken in one year alone out of the Gulf of Mexico), are believed to be responsible for the decline in abundance of larger bluefin in subsequent years. During the 1970's, a new market was developed for giant bluefin tuna (310 pounds or greater), and fresh bluefin is now flown directly to Japan for processing into sushi or sashimi. This resulted in a sharp increase in ex-vessel prices from $0.20 per pound to recent averages of appoximately $10 to $16 per pound, providing a virtually irresistible incentive for additional fishing effort.

The peak yields of bluefin from the western Atlantic (about 8,000–19,000 mt) occurred in 1963–1966 when much of the catch was taken by longlines off Brazil; since then catch rates off Brazil have been very low. During the late 1960's and 1970's yields averaged about 5,000 mt. In 1982 a catch restriction of 1,160 mt was imposed by ICCAT (based on the SCRS recommendation at the November 1981 meeting to hold catches as near zero as possible); the catch limit was increased to 2,660 mt for 1983 (because of uncertainty about the proper scientific quota monitoring level) and was held at that level through 1991. Yields generally have been within 15 percent of the target catch levels since 1982. The U.S. generally caught 40–60 percent of the total yield during 1960–1975, about 30 percent during 1976–1981 and has taken about 60 percent of the yield since 1982. During the 1960's and 1970's a North American purse seine fishery for juveniles and the longline fishery usually took 70–80 percent of the yield and recreational fisheries usually took 10 percent. The value of medium- and large-sized bluefin tuna increased substantially during the 1980's with the increased importance of the Japanese market. As a result, most of the large-medium and giant-sized fish now caught by rod and reel are sold.

By 1973, the United States and other nations concern at ICCAT about the decrease in the abundance of bluefin tuna in the North Atlantic. In response to this concern, ICCAT recommended, in 1974, a minimum size limit. With the passage of the Atlantic Tunas Convention Act (ATCA) in 1975, the United States took action to comply with the ICCAT recommendations and limited the United States harvest by imposing quotas and size limits. In spite of the ICCAT recommendations and United States compliance with these recommendations, west Atlantic bluefin stock abundance continued to decline. After conducting a series of stock assessments, ICCAT's SCRS recommended in 1981 that catches from the west Atlantic bluefin stock be reduced to as near zero as possible to stem the decline of the stock. Based on this recommendation, allowable landings of western Atlantic bluefin have been restricted since 1982.

The development of the Japanese market for giant bluefin blurred the distinction between sport and commercial fishing. The traditionally recreational catch for medium and giant bluefin was being sold for shipment to Japan. However, the 1992–93 rule banned the sale of small medium as well as school bluefin. Recreational anglers often purchase a General category permit so they can sell any bluefin larger than 178 cm (70") straight fork length which they might catch. This has resulted in a considerable number of General category permits being issued although only a small fraction of the permitted vessels actually catch and sell fish in that category. Given that there also has been a "trophy fish" Rod and Reel Incidental catch permit which allows anglers to retain one giant fish per season, the preference for the General category permit clearly indicates an economic interest in commercial-sized fish in addition to a "recreational" interest.

Management Regime

Background

The International Convention for the Conservation of Atlantic Tunas (Convention) was signed in 1966. The Convention was ratified by the United States in 1967 and entered into force on March 23, 1969. Twenty-two countries are currently parties to the Convention.

The objective of the Convention is to maintain "the populations of these fishes at levels which will permit the maximum sustainable catch for food and other purposes." The Convention established ICCAT, with two primary responsiblities: scientific assessment of Alantic tuna and tuna-like fishes; and recommendations, applicable to tuna and tuna-like fishes that may be taken in the area subject to the Convention at levels which will permit the maximum sustainable catch." The Convention provides that every conservation recommendation shall become effective for each member country unless a member country files a formal objection with the Commission and thus is exempt.

Congress enacted ATCA, 16 U.S.C. 971 et seq., in 1975, to provide the framework for the United States' participation in ICCAT and, because the Convention is not self-executing, to provide ICCAT with suggested recommendations. Under a 1990 amendment to the Magnuson Fishery conservation and Management Act (MFCMA), the Secretary of Commerce is delegated the authority to adopt regulations necessary "to carry out the purposes and objectives of the Convention" and the ATCA, and to promulgate regulations "as may be necessary and appropriate to carry out" the recommendations of ICCAT. In November 1990, Congress amended the ATCA to require, inter alia, that "no regulation promulgated under this section may have the effect of increasing or decreasing any allocation or quota of fish to the United States agreed to pusuant to a recommendation of the Commission."[1]

Regulations governing the conduct of the U.S. Atlantic tuna fisheries are currently under the authority of ATCA. Implementing regulations are found at 50 CFR part 285. The Fishery Conservation Amendments of 1990 (FCA), Pub. L. 101-627, also authorize management of Atlantic tuna under the MFCMA. The Secretary will continue to issue regulations governing the tuna fisheries under the authority of the ATCA until a Fishery Management Plan (FMP) is developed and regulations are issued under the Magnuson Act.

At the 1981 ICCAT meeting, the Commission considered recent stock assessments that showed a continued decline of bluefin in the western Atlantic ocean. The SCRS recommended that harvest levels of bluefin be as near zero as feasible for two years, with the small amount of catch for scientific monitoring purposes only, and ICCAT adopted management measures for the western Atlantic which significantly limited the United States and the total western Atlantic harvest. See 47 FR 17086 (April 21, 1982). Officials of the United States, Japan, and Canada, the three ICCAT member nations most actively fishing for bluefin in the western Atlantic, consulted and recommended to their governments measures to implement the ICCAT recommendations.[2]

The purpose of the 1982 quota was to provide SCRS scientists with catch statistics to allow continued monitoring of the western Atlantic bluefin stock. Some ICCAT members felt that the quota was set too low to provide adequate fishery-dependent data. During the 1982 annual meeting, citing a need for improved data from the fishery and uncertainty in stock assessment results and appropriate monitoring levels, ICCAT increased the allowable annual harvest from the western Atlantic to 2,660 mt where it remained for the fishing years 1983 through 1991. See 47 FR 25350 (June 11, 1982). The recommendation from the 1982 ICCAT meeting for the level of the scientific monitoring quota increased the United States quota to 1,387 mt, to provide more complete and accurate biological information for stock monitoring purposes. This quota remained in effect through 1991. See 48 FR 27745 (June 17, 1983). Although there have been modifications to the regulations, the present gear allocation scheme, with slight modifications, has been in place since 1983, with the exception of the purse seine allocation, which has been reduced by 51 mt.

Domestic regulations to carry out the ICCAT recommendations were implemented in 1982 and 1983. At that time, there were several user groups participating in the

---

[1] The Secretary of Commerce has delegated responsibilities under the ATCA to the Administrator of the National Oceanic and Atmospheric Administration (NOAA). See Department Organization Order 10–15, 3.01(aa). This authority has been further delegated to the Assistant Administrator for Fisheries (Director of NMFS), which is the division within NOAA that is responsible for management of the nation's fisheries. See NOAA Directives Manual 05–60, 2.

[2] These measures included a scientific monitoring quota of 1,160 mt for the entire Western Atlantic and an allocation of the quota between these three countries (U.S.–605 mt; Japan–305 mt; Canada–250 mt).

fishery, which were described in the 1982 Environmental Impact Statement. That document discussed the expected impacts of the ICCAT-imposed quota reductions on the environment with particular focus on the participants. The National Marine Fisheries Service (NMFS) used several guiding criteria, or objectives, to ensure consistency with the ATCA and other applicable U.S. law.

Recent Regulations

At the November 1991 meeting of ICCAT, the Commission recommended additional measures intended to enhance recovery of the stock. As a member of ICCAT, the United States is obligated to adopt domestic regulations to comply with the ICCAT recommendations. The measures taken included: (1) that the Contracting Parties whose nationals have been actively fishing for bluefin in the western Atlantic (Japan, Canada, and the United States) institute effective measures to limit the quota for the biennial period 1992–1993 to 4788 mt, but not to exceed 2660 mt in the first year; (2) that the biennial quota be taken by the Contracting Parties in the same proportions as previously agreed to for 1990; (3) beginning with the 1992 catch, if a Contracting Party exceeds its annual or biennial quota, then in the biennial period or the year following reporting of that catch to ICCAT, the Contracting Party will compensate in total by reducing the quota of the domestic catch category responsible for the overharvest; (4) that the three Contracting Parties will prohibit the taking and landing of bluefin weighing less than 30 kg (66 pounds) or having a straight fork length less than 115 cm (45"), to no more than 8 percent by weight of the total bluefin catch on a national basis; and (5) that the Contracting Parties institute measures to preclude economic gain to fishers from landing bluefin less than 115 cm or 30 kg.

In 1992, NMFS promulgated regulations to require annual renewals of required permits, to impose additional restrictions on the Incidental category, and to reduce the boat and bag limits in the Angling category.[3] Final regulations for the 1992–1993 season were published (See 57 FR 39205, July 24, 1992). The primary elements of the 1992–93 rule included:

1. determination of new quotas for all user categories, set at 90 percent of historical catch levels over the period of 1983–91 in response to the ICCAT recommendation of a 10 percent cut in fishing quota;
2. prohibition of the sale of school, large school and small medium-sized tunas (<178 cm (70") straight fork length);
3. more restrictive Angling category bag limits; and,
4. prohibition on the retention of fish <66 cm (26") straight fork length (young schools).

At the November 1992 meeting of ICCAT, member nations recommended a statistical documentation program to enhance the scientific monitoring of the Atlantic bluefin tuna catch. Additionally, the SCRS reported on the status of Atlantic yellowfin and bigeye stocks indicating that they are currently fully exploited or possibly overfished.

In 1993, NMFS again promulgated revisions to the regulations governing the Atlantic tuna fisheries (See 58 FR 45286, August 27, 1992). These actions were limited in scope and were intended to help meet existing ICCAT obligations and to improve the efficiency of the domestic fishery management program. The specific regulatory changes were to:

1. require Atlantic bluefin tuna dealers to submit daily reports via fax and a bi-weekly report instead of the present weekly report;
2. require permits for vessels fishing in the Angling category;
3. require at-sea observer coverage on vessels taking Atlantic tunas, if so requested by NMFS;
4. prohibit the use of non-authorized gear in the Atlantic tuna fisheries except pursuant to an experimental fishing exemption;
5. allow the Assistant Administrator for Fisheries, NOAA, to make inseason transfers of potentially underharvested quota between fishing categories;
6. raise the amount of General category set-aside for the late season fishery from 40 mt to 65 mt;
7. allow for inseason adjustments to the Angling category bag and boat limits for private, party and charter boats; and,
8. make technical changes to enhance administration, management and enforcement.

The amendments were needed to improve efficiency in collecting data necessary for monitoring the stock, to further the goal of minimizing economic displacement while preserving traditional fisheries, and to make maximum use of the available resource. Additionally, a number of technical corrections were needed to clarify the language of existing regulations. Finally, pursuant to the 1992 ICCAT SCRS reports concerning status of other tuna stocks and the need for data upon which to assess the potential impacts of new fishing gear and methods, NMFS took action to prohibit unauthorized gear from the Atlantic tuna fisheries and to implement experimental fishing provisions for new gear.

In addition to the regulatory changes, NMFS also took action in 1993 to adjust quotas as required under the regulations at 50 CFR 285.22(h). This action, under

---

[3] Formal definitions of the five fishing categories for Atlantic bluefin tuna are provided in the regulations (50 CFR 285).

existing authority established for the Assistant Administrator, was needed to adjust the 1993 catch quotas for underharvest/overharvest in certain fishing categories. Also, pursuant to the implementing regulations at 50 CFR 285.22(f), the Assistant Administrator took action to make inseason transfers from the reserve to make up for overharvest in 1992 and to improve scientific monitoring of the stock in 1993.

In an effort to address increased participation in the fishery, NMFS published a control date for Atlantic tuna fisheries on September 1, 1994. The purpose of this control date is to advise current and future participants that access to the fishery may be limited at some point in the future, and that access after the control date is not assured. In fact, access for vessels already in the fishery prior to the control data could also be based on past participation and/or catch history. Nevertheless, a total of 2,261 new bluefin tuna permits have been issued since the control date; some of these may represent multiple permit holders (e.g. General and Angling), however it is estimated that at least 1,500 permits are new entrants. Thus, the control date does not appear to be discouraging entry into the fishery.

On April 14, 1994, NMFS published an interim final rule for the 1994 fishery. The interim final rule: (1) establishes fishing category quota allocations for the 1994 bluefin fishing season; (2) amends the specified amount of other species to be landed as a condition for landing an incidental bycatch of bluefin in the southern longline fishery (1,500 pounds during the months of January through April, and 3,500 pounds for the months of May through December; (3) adjusts the line that separates the northern and southern regulatory areas for vessels using longline gear and possessing an Incidental Catch permit for bluefin; (4) requests comments on the future use of curved length measurements for bluefin and alternate means to provide closure notices; and (5) makes technical corrections to the regulatory text.

In a separate action, NMFS proposed to amend the regulations at 50 CFR 285 to implement a statistical documentation program for international trade in Atlantic bluefin tuna. The documentation program will allow the ICCAT Secretariat to monitor exports of non-member countries and improve estimates of total catch.

In July of 1995, NMFS issued a rule to revise the regulations governing the Atlantic tuna fisheries to: set Atlantic bluefin tuna (ABT) fishing category quotas for the 1995 fishing year; control fishing effort in the ABT General category; extend vessel and dealer permitting and reporting requirements to additional Atlantic tunas fisheries; adjust angler bag limits; and make amendments to clarify the regulations, facilitate enforcement and improve management efficiency. These regulatory amendments address scientific monitoring and allocation issues in the ABT fisheries and simplify rules applicable to recreational fishing for tunas. The permitting and reporting provisions enhance data collection and enforcement of catch restrictions in the Atlantic tuna fisheries and enable the United States to collect fishery information needed by ICCAT to produce stock assessments. These actions are necessary to begin implementation of the 1993 recommendation of ICCAT regarding fishing effort on yellowfin tuna, and to implement the 1994 recommendation of ICCAT regarding fishing quotas for ABT, as required by ATCA.

By final rule published June 14, 1996, NMFS amended those regulations governing the Atlantic tuna fisheries to: set Atlantic bluefin tuna (ABT) fishing category quotas for the 1996 fishing year (Table 5; Appendix I), revise allocations to monthly quota periods and establish the effort control schedule in the ABT General category, allow the partial transfer of quotas among Purse Seine category permit holders and amend landing requirements, and increase minimum sizes for Atlantic yellowfin and bigeye tunas.

Description of the Fishery

Background

United States landings of Atlantic bluefin tuna for the 1983–1995 period are provided in Table 6. The historical level of landings has generally been determined by the quotas since 1982, although the Angling category, which was managed with bag limits rather than quota monitoring until 1992, was a notable exception prior to implementation of survey methods to monitor catches more closely.

Fishing Grounds, Seasons, Gear, and Regulations

Table 7 summarizes the traditional gear, area, size of fish, and seasonality of the U.S. bluefin tuna fishery. Handgear includes rod and reel, harpoon, kegline, and handline. The handgear fishery for small and medium tuna (almost exclusively rod and reel) is distinct enough from the handgear fishery for giant tuna (all types of handgear) that separate statistics are reported for these fisheries. Since the smaller fish are no longer sold, dealer landing reports do not record landings of Angling category fish. Angling catches are monitored through the NMFS Large Pelagic Recreational Fishery Survey. Bluefin tuna are still occasionally captured by traps and gillnets, subject to strict bycatch regulations, but the contributions of these types of gear to the total U.S. bluefin catch have become very small.

Under current regulations, any boat or vessel wishing to fish for bluefin must obtain a permit from

NMFS. Depending on the vessel, gear type and size of bluefin targeted, the permit may be Angling, General, Harpoon, or Incidental (limited entry is in effect for the Purse Seine category). Any giants and large mediums which are sold must be immediately reported by permitted dealers and marked with identifying tags, so catches of these fish are monitored quite closely.[4]

U.S. Purse Seine Fishery

U.S. vessels fishing for Atlantic bluefin tuna with purse seine gear originally operated from several ports in the northeastern United States, California, and Puerto Rico. The fishery traditionally targeted small and medium tuna in nearshore waters (rarely outside 200 km) between Cape Hatteras and Cape Cod in the summer, and giant tuna in the Gulf of Maine in late summer and early fall for the cannery industry. A combination of quota regulations and over-investment in fishing capacity has severely contracted the duration of the fishing seasons.

In 1982, a limited entry system with non-transferable individual vessel quotas (IVQ's) for purse seining was established, effectively excluding any new entrants to this category. Equal quotas are assigned to individual vessels by regulation; the IVQ system is possible in this category given the small pool of ownership in this sector of the fishery. Currently, only five boats comprise the Atlantic bluefin tuna purse seine fleet. Larger, distant water seiners from the U.S. Pacific coast in the past sometimes diverted operations from the yellowfin and skipjack fisheries to fish for bluefin, but this practice is no longer permitted. These larger vessels were actually less efficient in the shallow shelf waters of the northwest Atlantic than are the smaller vessels presently involved in the fishery.

Bluefin tuna purse seining operations use typical purse seining procedures. Spotter aircraft are used to locate fish schools. The vessels themselves may not even leave the docks until suitable concentrations of fish are located. Although an official starting date of August 15 is set by the regulations, the start of the purse seining fishing season coincides with availability of fish in schools large and dense enough to warrant attention. Once sufficient densities do appear, catch rates are so high that the annual quotas are usually met within weeks for the large mediums and giants. Current regulations for the bluefin tuna purse seine fishery allocate a total catch quota of 250 metric tons (divided evenly into five IVQ's of 50 mt), of which a minimum of 90 percent are giants (>206 cm or 81" curved fork length) and 10 percent may be large mediums (185–206 cm or 73–81" curved fork length). In addition, purse seiners are limited to a 1 percent bycatch limit on undersized bluefin (<185 cm or 73" curved fork length) which cannot be sold. Any bycatch of undersized bluefin tuna by these same vessels when targeting yellowfin or skipjack is included in this 1 percent limit.

U.S. Handgear Fishery

The U.S. handgear fishery is a summer and fall fishery. Fishing for small bluefin tuna with rod and reel generally begins in early summer off Virginia, and the center of activity moves northward into the New York Bight as the season progresses. Giant bluefin are caught in Cape Cod Bay, the Gulf of Maine, and other New England waters during summer and early fall with all types of handgear. In February of 1995 and again in 1996, large concentrations of tuna (migrating south) appeared off North Carolina, generally holding on sunken ships and similar structure. Sporadic rod and reel catches of giants have been reported in late spring from the Gulf of Mexico. Fishing usually takes place between 8 and 200 km from shore. Beyond these general patterns, the availability of fish at a specific location and time can be quite unpredictable.

The handgear fishery for bluefin is composed of a diverse collection of boats and fishers. Most of the boats are greater than 7 m in length and are privately owned by individual fishers. Charter and party boats also are involved in the fishery. Small bluefin are most typically caught by trolling with artificial lures, although "chunking" (a form of chumming with large pieces of fish rather than using a slurry of ground fish) has become popular in some areas, using rod and reel. Giants are harpooned, or are caught by trolling, or by chumming and drifting with several types of hook and line gear. Mackerel, whiting, mullet, ballyhoo, and squid are the usual choices for bait.

Bluefin tuna is the intended target on many fishing trips. Other tunas and large pelagics are taken on trips directed at bluefin. Bluefin catches are also reported from trips directed primarily at sharks.

Details of operations, frequency and duration of trips, and distance ventured offshore probably vary as widely as the equipment, skill and enthusiasm of the individual fishers vary. Effort increases sharply with reports of fishing success. Intense efforts with rod and reel may also be mounted during tuna "tournaments."

Since 1975, catch and effort data for the U.S. rod and reel fishery have been collected through the Large Pelagic Survey (LPS), as indicated above. This program concentrates on the rod and reel fishery for offshore pelagic fish between Cape Hatteras and Cape Cod, and in recent years, through Maine. Estimates are made of the geographical breakdown of catch by size, number of vessels, hours fished, and number of lines. This pro-

---

[4] All large medium and giant bluefin which are caught but not sold (e.g. with a Rod and Reel Incidental category permit) must also be tagged and reported (50 CFR 285.30 (c)(2)).

gram collects information about fishing for both small and large bluefin, but because of its geographical scope, coverage has been more complete for the small fish rod and reel fishery. This survey is used to estimate the Angling category catch for quota monitoring purposes, as sales data do not exist for these fish. For the past two years, a supplementary survey has been used to collect basic socio-economic data on the large pelagic recreational fishery (see Appendix IV).

Incidental Longline Fishery

The longline incidental fishery consists of about 600 permitted vessels. The fact that bluefin represents an incidental catch for these vessels suggests that while elimination of the incidental quotas (without major season-area closures) may reduce total revenues to the longline fleets, it should not substantially reduce total employment. Again, longline vessels may increase fishing effort on other species to make up for lost revenue.

The full impact of closing the fishery would be difficult to gauge. Even with zero quota, it is unlikely that it would be possible to reduce fishing mortality in the longline fisheries to zero. The problem appears to be more severe outside the Gulf of Mexico, where logbook data indicate that numbers of bluefin discarded may have exceeded the numbers landed by more than 200 percent throughout the period 1989–1992. Data from the same source indicated that discards within the Gulf of Mexico have been highly variable, ranging from slightly more than 200 percent of landings in 1989 to about 20 percent of landings in 1992. The extent to which longliners can or do target bluefin tuna in each area is unknown.

The only way to ensure zero incidental mortality is to completely close down all longline fisheries; however, this is unlikely to be practical either within or outside the Gulf of Mexico due to the magnitude of the season-area closures that would be required, and the relative importance of associated fisheries (see FEIS, para 4.3.2 (C) for a more complete analysis of the relative importance of associated fisheries in the Gulf of Mexico). In terms of volume of longline landings, bluefin tuna is surpassed by yellowfin tuna, swordfish, bigeye tuna, albacore and sharks in the Northeast, and yellowfin tuna, swordfish and dolphin fish in the Gulf of Mexico. In terms of value of landings, bluefin tuna may form a significant portion of total revenues locally in both regions due to the high prices received.

Socio-Economic Aspects of the Fishery

Prices and Markets

The ex-vessel price of Atlantic bluefin tuna in the United States has increased substantially from roughly $0.20 per pound during the early 1970's to nearly $16.00 per pound dressed weight in 1993. This increase is largely attributed to increased demand for fresh bluefin tuna from Japan, the principal consumer of U.S. Atlantic bluefin tuna. The sensitivity of U.S. ex-vessel bluefin tuna prices to Japanese wholesale market prices is demonstrated in Figure 3. The close correlation between average Japanese wholesale market monthly prices and average U.S. ex-vessel monthly prices clearly demonstrates that the U.S. dealers and fishers are "price takers" on the fresh bluefin tuna export market. Since the U.S. share of total bluefin tuna supplies to the Japanese market is relatively limited, it is unlikely that reductions in U.S. landings and exports will have a significant impact on Japanese prices. However, prices could vary considerably in certain months in which the United States is a major supplier of fresh bluefin and if quota changes affect supply from other countries as well (Canada, Japan's own landings).

Prices paid to U.S. fishers and exporters thus depend on the purchasing power and preferences of Japanese consumers, supplies of competitive product in Japan, packing and transportation costs, the United States–Japanese exchange rate, and other factors. It should be noted that the U.S. dollar depreciated considerably against the Japanese yen in the 1970's and 80's, further stimulating the expansion in U.S. bluefin tuna exports to Japan. Yearly average exchange rates show that the U.S. dollar depreciated relative to the Yen over 14 percent from 1992 to 1993. This indicates that even if Japanese market prices remained steady, prices paid to U.S. processors and fishers would increase. Prices paid to U.S. fishers at the ex-vessel level also depend upon the primary market structure, as discussed below.

In addition to market factors, characteristics of the individual bluefin tuna itself can significantly affect the price such that each fish is evaluated and priced. Size can be an important determinant of price per pound. According to industry sources, bluefin tuna prices offered for individual fish peak in the range of 500 to 700 pounds (round weight) due to the costs (including risk) of investing upwards of $10,000 or more for individual fish. As the wholesale price of bluefin tuna rises, the dealer incurs interest costs if the fisherman is paid prior to the sale of the bluefin on the Japanese market (unless on consignment).

Fish quality and condition are also important determinants of price. Industry sources indicate that bluefin tuna are evaluated by expert graders on the basis of four criteria: fat, freshness, color and shape of the fish. Graders accord individual bluefin a letter grade (A, B, or C) for each criteria. In addition, stomach lining, body temperature, and tears and bends in the flesh are factored into the final price paid to fishers. Detailed data on how quality and condition affect prices are known only by dealers for their own transactions. However, because Atlantic bluefin tuna gain weight and fat content while resident in the northeast during the late summer months, time period of catches should roughly reflect fat content and thus can serve as a proxy for quality in predicting prices.

Although it is generally acknowledged that a higher fat content yields a better price for bluefin tuna, other factors, such as world bluefin tuna market conditions, also influence price. Variablity in supply and prices demonstrate the volatility of the Japanese market and the difficulty in predicting even general price trends on a monthly basis. Although a fatter fish may fetch a higher price whatever the market conditions, prices for lower quality bluefin in another month may be higher if total supplies to the Japanese market are relatively low.

U.S. fishers can sell their catch dockside or hire an agent who sells the fish on consignment. There is anecdotal evidence that consignment prices paid to U.S. fishers can be greater than dockside prices. Again, the new dealer reporting form implemented in 1994 includes the recording of whether or not the fish was sold on consignment or dockside.

Preliminary analyses clearly show that consignment prices are significantly higher than dockside, particularly for Harpoon category fish. On a daily basis, the level of risk faced by the consignment seller depends greatly on fluctuations in supplies to Japan and exchange rates.

Prices paid to U.S. fishers are also a function of the structure of the ex-vessel market. It is interesting to note that the number of bluefin tuna dealer permits issued has increased from 56 to 474 permits from 1990 to 1994. The rapid jump in the number of dealers may be attributable to increasing restrictions in other New England/mid-Atlantic fisheries. It would be expected that a growing number of bluefin tuna dealers would increase the competition for bluefin tuna at the ex-vessel level and yield the best possible price for the bluefin tuna fishers. In 1994, there were 1,177 vessels selling bluefin tuna to some 92 dealers who actually bought bluefin. Only a limited portion of licensed dealers actually purchased bluefin tuna (less than 20 percent of dealers in 1994), and over a fourth of these dealers only purchased one fish.

Expect for some direct sales to retailers, U.S. exports of Atlantic bluefin tuna to Japan are auctioned at several wholesale markets, particularly the large Tsukiji Central Wholesale Market in Tokyo (Weber 1990). The average nominal wholesale price of bluefin tuna on the six major markets was around $14 per pound in 1991 and 1992, and jumped to over $16 per pound in 1993 (although a large share of this increase can be attributed to the change in the exchange rate).

Ex-Vessel Gross Revenues

Ex-vessel revenues from recorded sales of bluefin tuna in the commercial fishery, by category, for 1992–1994 are presented in Table 8. These figures are presented in constant (1994) dollars, so that cross-year comparisons are more meaningful. All vessel categories exhibited an increase in real gross revenues from 1992 to 1994. Overall real gross revenues increased nearly 10 percent from 1993 to 1994, although total volume of landings increased only 5 percent. Average prices were relatively strong in 1994, particularly in the Harpoon category.

Before examining trends in gross revenues by category, it should be emphasized that this discussion focuses on gross revenues only, and not net revenues. Individual vessels may have experienced a decline in net revenue even with higher gross revenues reported for their fishing category. Thus, trends in gross revenues can only indicate the average trends in gross income and the effect on fishers' net revenues if their costs remained relatively steady over the period examined.

Net revenues can only be calculated if adequate cost information is collected, through a logbook or survey research project. In any case, to the extent that fishing costs are reflected in the producer price index, costs may not have increased substantially from 1992 to 1993, and from 1993 to 1994, with respective (annual) changes in the index of 1.5 percent and 1.3 percent.

Angling and Charter Boat Revenues

In most fisheries in the United States a clear distinction is possible between "commercial" and "recreational" fishers. This distinction is not always obvious in the bluefin tuna fishery. Even after the ban on the sale of fish under 178 cm (70") straight fork length that went into effect with the July 1992 rule, anglers who may have otherwise been considered recreational fishers could land a fish 178 cm (70") straight fork length or above and elect to sell the fish. In fact, the large number of permit holders in the General category may be explained by the purchase of permits by recreational anglers "in case" they land a commercially-sized fish. Recreational anglers who never sell bluefin tuna can obtain the "Incidental Rod and Reel" permit, which allows them to harvest one bluefin tuna of commercial size, per year, for trophy purposes only.

New regulations implemented in 1994 required recreational fishers (anglers) to obtain a permit for Atlantic bluefin tuna. Although it is still possible to hold both the Angling and General category permits, a number of recreational fishers opted for the "Incidental Rod and Reel" permit. In fact, the clear distinction now possible between recreational and commercial permit holders implied that all of the General category permittees would have to pass Coast Guard fishing vessel commercial safety requirements. An estimated 1,600 recreational anglers opted for the "Incidental Rod and Reel" permit rather than make investments to meet the safety requirements invoked by having a General category permit.

Given the ban on the sale of fish under 178 cm (70") straight fork length the direct income associated with the Angling category is limited to charter/party boat operations. As with the commercial fishing categories, the ideal analysis would include calculation of costs and revenues to charter vessels such that producer surplus could be estimated. The economic importance of the recreational fishery is not limited to charter vessel producer surplus, however, nor does it necessarily depend upon the value of the landings which are sold, but rather the participants' willingness to pay for recreational fishing. These non-market values are difficult to estimate, and involve either direct questioning (contingent valuation) or indirect survey techniques such as the travel cost method, as a basis for estimating demand (and thus consumer surplus) for recreational fishing. The economic importance of the recreational bluefin tuna fishery, including non-market benefits, should thus be kept in mind when examining the gross revenue figures from other categories, despite the difficulty in attaching a dollar value to the recreational fishery.

Direct charter/party boat income may be estimated using results from a previous study. Based on sampling from 1988–1990, an average of 2,500 charter boat trips per year targeted on bluefin tuna. This estimate, however, is tenuous. It was derived from data from New Jersey and Virginia and expanded coastwide by NMFS estimates. Assuming that charter boats charge about $800 per day, the gross revenues from bluefin fishing would be about $2 million. These direct revenues represent nearly 10 percent of the total gross revenues to the other bluefin tuna categories, and is an underestimate of revenues accruing to charter boats because some of the bluefin landed are probably caught by the captain or mate and sold (only large mediums and giants after the 1992 rule). Additionally, tips which are typically given to the mate (about $100 per trip) are not included. As an alternative estimate, consider the 1994 LPS estimate of 10,119 charter boat trips taken in the entire recreational large pelagic fishery (Table 11). Assuming that approximately 35 percent of these trips are targeted specifically at bluefin tuna,[5] this figure provides a revised estimate of $3.19 million for 1994 gross revenues. The producer surplus component of the value of the recreational fishery would thus be these gross revenues minus costs incurred in providing the charter boat services.

Once again, it should be emphasized that these net revenues would be only a part of the value of the recreational fishery, since angler consumer surplus (ACS) is another important component as well. Angler consumer surplus is generated from charter and party boat services as well as from private vessel participation in the recreational fishery. Preliminary estimates of ACS in the private bluefin tuna fishery are $1,132 per fishing trip. Given that total catch of school and small medium bluefin tunas in 1994 was half that of 1993, it is likely that ACS fell as well.

Participation

Participation in the bluefin tuna fishery is examined to provide information on the relative activity in the fishery by sector (Table 9). While there are a large number of fishing vessels holding bluefin tuna permits, only a fraction of these actually land and sell bluefin tuna. As discussed in Section 3.4.2(2), a control date for the Atlantic tuna fisheries was published on September 1, 1994; however, new permit issues continue to rise. In 1994, only 7 percent of those vessels with a bluefin tuna permit actually sold a bluefin. Since those fishers not selling a tuna may in any case consider themselves participants in the fishery, the percentage of vessels landing fish is referred to as the "success rate." This should not, however, be interpreted as synonymous with economic success or performance, as costs are not being compared with returns.

General Category

The General category is characterized by a large number of fishers attempting to harvest a relatively small number of fish. In 1994, only 8 percent of General category permit holders landed any fish, and over 50 percent of those who did land a bluefin in the General category landed only one or two fish. Nevertheless, the total number of vessels landing bluefin increased from 1993 to 1994. Often a General category permit is purchased as an "insurance policy" in the unlikely event that a commercial-sized bluefin is landed (e.g., by "anglers"). The permit fee (actually zero in 1994) pales in comparison with the potential payoff for a high quality large medium or giant bluefin. For those vessels that landed a bluefin tuna in 1994, the average number of fish per vessel was 4.1 fish for the season (for about $20,000 per year), slightly lower than the 4.4 fish landed in 1993. The average weight per fish declined from 1993 to 1994 from 396 pounds to 367 pounds, respectively (fish worth about $5,000 each).

Harpoon Category

With only 37 vessels landing fish in 1994, the Harpoon category is the smallest of the directed fishery categories in value and volume of landings. The average number of bluefin sold per Harpoon category vessel landing bluefin is nearly 10 fish per year (valued at about $50,000 per year), and over half of the fleet landed more than five fish in 1994. This is in contrast with the General category, which averages just over four fish per "successful" vessel and in which only 25 percent

---

[5] This estimate is based on a study by Figley (1992) of the New Jersey large pelagic recreational fishery. This same proportion (35%) is used in estimates of expenditures, costs, and employment in the recreational fishery throughout this document.

landed over five fish. Thus, "successful" vessels in the Harpoon category were "more successful" on average than General category vessels. The General category figures are "distorted" somewhat, however, by the presence of recreational anglers holding General category permits. The average weight of each Harpoon category fish declined slightly from 1993 to 1994 from 405 pounds to 362 pounds, respectively.

Purse Seine Category

The purse seine fleet, as indicated above, consists of five vessels, each of which holds an equal amount of bluefin tuna quota (60.2 mt each in 1994). The number of fish, as well as gross revenues, would be similar for each vessel. The average number of bluefin harvested by each vessel in this category in 1994 was 341 fish worth about $1,750,000. Each fish weighed an average of 389 pounds in 1993, up slightly from the average of 380 pounds in 1993.

Incidental Catch Category

In 1993, 435 bluefin were caught incidentally to other fishing operations, primarily in the longline yellowfin and swordfish fisheries. These fish averaged 493 pounds (down from 516 in 1993). Bluefin tuna were landed by 184 Incidental category permit holders in 1994. One noticeable change over these three years is the relative decrease in the number of vessels landing higher numbers of fish. In 1994, only 3 percent of those vessels landing under the Incidental category landed more than five fish, in contrast with 21 percent in 1991. As discussed previously, additional restrictions on the incidental catch of bluefin tuna have made it more difficult for those participants to capture large numbers of fish. These regulations require a minimum targeted catch volume in order to land one bluefin tuna.

Angling Category

Estimates of angler participation are reflected in the fleet size estimates reported in Table 10. Nearly 8,000 vessels fished for bluefin tuna (along with other species) in 1993 in the nine states included in the Large Pelagic Recreational Fishery Survey. Participation is also reflected in the total number of trips, by vessel type and by state, as shown in Table 11.

Employment

Table 12 presents estimates of direct and indirect full time equivalent (FTE) employment in certain portions of the Atlantic bluefin tuna fishery. An estimated total of over 1,200 direct and indirect full time equivalent jobs are attributed to the Atlantic bluefin tuna fishery (NMFS, 1995).

Costs

Cost information is difficult to obtain in most fisheries. For the most part, there are no cost data collection systems. The collection of cost data in the bluefin tuna fishery is made more difficult because of the seasonal nature of the fishery and the varying motivations of the participants (profit or fun or a combination of both). Each category is discussed separately. For each category, the variable costs of fishing for bluefin tuna are estimated. Fixed costs are not included in these calculations, for various reasons. Firstly, it is assumed that commercial vessels will continue to fish as long as variable costs are covered, at least in the short run, since fixed costs are incurred whether or not the vessel engages in fishing. Secondly, for both commercial and recreational vessels, it is assumed that a number of species may be targeted, and the relevant "decision" is which species the vessel operator chooses to target. The level of capital investment in vessels, gear, and other equipment is considerable in both the recreational and commercial fisheries for Atlantic bluefin tuna; however, no estimates of these values are currently available.

General Category

Until the July 1992 ban on the sale of bluefin tuna under 178 cm (70") straight fork length, there was considerable overlap between the General and Angling categories. The distinction is somewhat less blurred, however, since the ban on the sale of school and small medium fish has been implemented. However, there still is some overlap, such as anglers and charterboat operators who purchase a General category permit in the event they hook a large medium or a giant.

Some idea of expenses associated with trips made by General category fishers who are very dependent on the fishery (incomes over $50,000) was obtained through conversations with fishers. The variable cost per day of fishing was estimated at $466 (1992 dollars). This crude estimate is roughly comparable to average costs found in previous studies of the Virginia and New Jersey recreational fisheries (approximately $375 per day). The professional commercial fisherman has the extra expense of a paid crew member, but less expense for food (only two people on the vessel) and for items like lodging and entertainment. For the skilled fisherman who might land one fish every three days, variable costs per fish could be approximated at $1,400.

The survey of a sample of members of the General Category Tuna Association indicated that average variable costs per fishing trip were approximately $391. When combined with the number of trips necessary to land a fish (for those able to land a fish) the average variable costs per fish are approximately $2,623 (the value of the average fish being about $5,000).

Angling Category

Studies in Virginia (Lucy et al. 1990) and New Jersey (Ofiara and Brown 1987) reported costs associated with bluefin tuna fishing. In both studies, data were collected from surveys of anglers (including charter boats) fishing for "big game" fish. Average expenses, in 1992 dollars, are $375 per trip as reported by the vessel owner. Undoubtedly, trip expenses were shared in some part by the other passengers on board. In the New Jersey study, there were an average of 4.7 people on board; in the Virginia case, there were 4.1 anglers per trip.

That these studies combined data for private and charter boats clouds the financial picture. Charter boats are, in general, larger and more expensive to operate than private boats. At a minimum, charter boat variable costs per trip will include private vessel costs, plus wages for the mate of about $80 per trip—a total of $392 (Virginia and New Jersey averaged). Expenses for anglers on a charter boat (assuming the charter boat fee of $800 per trip is split between 6 anglers) would be the charter fee, meals and lodging expenses (estimated at about $60 per person), plus tips to the mate of $100 per boat trip—about $210 per person.

The supplemental socio-economic survey of the LPS in 1993 indicated that average variable costs for a private vessel targeting Atlantic bluefin tuna were $315 per trip. Travel costs were estimated based on mileage between the home and the point where the vessel is moored, and averaged $27 per angler. The number of trips averaged 10 per angler and a mean of 8.63 fish were landed per trip. Of the 29,883 private vessel trips targeting large pelagics, as estimated in the 1994 LPS, approximately 35 percent may be assumed to be targeting bluefin tuna. Based on this estimate, total expenditures are estimated at $3.3 million.

Harpoon Category

Currently no cost information is available on this category. Costs might be similar to those of the General category vessels except for the occasional use of spotter planes and the share for a third crewman. Several of the vessels included in the General Category Tuna Association survey indicated that they use harpoons.

Purse Seine Category

Through the cooperation of several vessel owners, data were obtained for 1990 seasonal fishing costs for the Purse Seine category. Variable costs, including crew wages and payroll taxes, fish spotting services, fuel, supplies, food, travel, lodging, and unloading, were estimated to be slightly over $1 million per vessel. Fixed costs of insurance, professional fees, and office fees averaged a little over $100 thousand per vessel. Purse seine owners contacted in early 1993 felt that this estimate of their operating costs ($1,100,000) was still reasonably accurate. This assessment is corroborated by the very slight increase in the producer price index from 91–94 (3 percent). These costs do not give a complete picture of the vessel's operations, however. Depreciation, opportunity costs of capital, drydocking, and activities in other fisheries are unknown. There is anecdotal evidence that the bluefin tuna seining fleet was severely affected by lower catches of other tunas in the 1992 season.

When contacted in 1994, purse seiners indicated that their variable fishing expenses when targeting Atlantic bluefin tuna average some $1,750 per day, plus crew share costs. Given an averge of 30–40 days for each vessel to fill its quota, and a share of 55–60 percent to the crew members, an estimate of $10,581 of variable harvesting costs per metric ton was calculated for bluefin tuna landed in the Purse Seine category. Again, this estimate excludes fixed costs, which can be very high in the purse seine fishery.

In sum, purse seiners' fixed and variable costs are a minimum of $1,100,000 per vessel per year. In 1994, their landings were worth about $1,750,000 per vessel.

Incidental Category

If the "incidental" catch of a bluefin tuna is truly incidental (that is, if fishers would have made the same trip and fished in the same manner) then the cost of catching a bluefin incidentally is essentially zero. Only handling costs can be directly attributed to the catch of bluefin tuna, and are assumed to be minimal.

Beginning in 1992, new rules requiring minimum landings of a targeted species for every bluefin and a limit of one incidental bluefin per trip should have reduced the motivation to catch bluefin. As discussed above, there is evidence that the incidental fishery is becoming somewhat more "incidental." Although the possibility of catching a valuable bluefin will still be a factor, there are insufficient data to predict fishing practices in this category. Therefore, in this analysis the basic assumption is that bluefin catches are truly incidental and that the associated costs of incidental catch are zero. This assumption simplifies the calculation of producer surplus for the Incidental category, since it is precisely the gross revenues from the sales of bluefin tuna.

Processing and Export

To maximize fish quality, much of the processing of export-quality Atlantic bluefin tuna in the General, Harpoon, and Incidental categories takes place on board. Fishers maintain freshness by gutting and bleeding the fish and protecting it from heat and sunlight, preferably by immersing it in ice or an ice brine (Kojima and Gaw 1990). Following these procedures in their entirety can be more difficult for purse seiners due to their large harvests in one trip. In fact, industry sources indicate that most Atlantic bluefin tuna were landed whole in the purse seine fishery in 1993.

Once landed, most Atlantic bluefin tuna are immediately graded and prepared for export to Japan's fresh fish market. Export-quality fish are either refrigerated or placed into an ice water bath until ready for export. Fish are then placed individually in foam-lined wooden crates filled with ice for transport to an airport and flight to Japan.

U.S. fishers may either sell their landings to federally-licensed dealers or hire a dealer to sell the tuna in Japan on consignment. There is evidence from dealer reports that roughly half of U.S. landings are sold on consignment for fishers. Agents earn a commission ranging from 4 percent to 9 percent, and fishers also pay expenses for shipping, handling, tariffs, and customs (Weber 1990). Weber reported that in 1986, costs for shipping and handling a 400 pound giant tuna amounted to $1.81 per pound, including $0.28 per pound for labor and handling. Because Japan and the United States are signatories of the General Agreement on Tariffs and Trade, U.S. exports of fresh or frozen tuna to Japan qualify for only a 5 percent tariff. In addition, there is a $50–85 customs clearance charge per shipment, depending on weight, and a $40 tax on each tuna sold on consignment (Bellows, personal communication).

Although the United States is the world leader in imports of frozen and prepared tuna, it exports virtually all landings of large Atlantic bluefin tuna to Japan, which is itself the world leader in imports of high-valued tuna species (Sonu 1991). Industry reports that roughly 90 percent of all giant and medium bluefin are exported fresh to Japan for auction in a wholesale market, usually the large Tsukiji Central Wholesale Market in Tokyo. The percentage of landings which are exported is lowest at the start of the season when fat content is low, increasing to nearly 100 percent in late summer and early fall.

According to the East Coast Tuna Association, virtually all United States exports are conducted by U.S. companies or agents; that is, Japanese ownership of U.S.-origin tuna begins after first sale in Japan.

Since 1986, U.S. exports of bluefin tuna (including small amounts of Pacific bluefin tuna) fluctuated between 800–1,000 metric tons. The vast majority is exported in fresh form, which commands a higher price, for the raw tuna market (sashimi and sushi). The United States is the leading exporter of fresh bluefin tuna to Japan, supplying around 30 percent of the Japanese market since 1987. Canada, Spain, Tunisia, and Australia also export appreciable and roughly equal amounts of fresh bluefin tuna to Japan (southern bluefin in the case of Australia; Sonu 1991).

The Japan Marine Products Importer's Association reported imports of about 954 metric tons (2.1 million pounds) of Atlantic bluefin tuna from the United States in 1992 (all fresh) valued at approximately $22 million (Jan–Nov). These figures are roughly consistent with U.S. landings of giants and mediums after adjusting for dressed weight (factor of 0.8) and the percentage of bluefin which are exported throughout the season (90 percent). Preliminary U.S. data for 1993 indicate that U.S. exports of fresh bluefin tuna to Japan jumped by 30 percent in quantity but only 12 percent in value in 1993: some 1,603 mt of fresh bluefin tuna were exported at a value of $15.2 million.

Australia is by far the leading exporter of frozen bluefin tuna to Japan, providing 60 percent of total imports since 1987. Since 1987, U.S. exports of frozen Atlantic bluefin tuna to Japan have been negligible.

Summary

The SCRS estimates that current biomass levels of western Atlantic bluefin tuna are less than 20 percent of 1975 levels. The commercial and recreational fisheries for Atlantic bluefin tuna in the United States are of considerable economic importance, particularly on a regional level, providing over 1,200 full-time equivalent positions in direct and indirect employment. Total commercial landings in 1994 are estimated at an ex-vessel value of nearly $21 million. Recreational expenditures are estimated at $3.19 million in the charterboat industry and over $3.3 million in the private vessel recreational fishery. The Atlantic bluefin tuna fishery is characterized by a wide diversity of user groups, each with varying motivations, gear types, seasons and areas fished.

Principal Bluefin Tuna Management Measures

The United States, as a member of ICCAT is obligated to implement its recommendations. It does so by following management measures stipulated by ICCAT and by implementing domestic fishery management measures that are consistent with the objectives of ICCAT. These measures are briefly described below.

Scientific Monitoring Quota

U.S. Allocation from ICCAT

As noted above, the ABT allocation for the western Atlantic bluefin tuna fishery is defined by ICCAT as a scientific monitoring quota. ICCAT recommended in 1982 that quotas should be established for the primary purpose of providing the most complete and accurate biological information for stock monitoring purposes. This was the basis for the increase in quota from 1981 to 1982, and the quota has varied little since that time.

Scientific Information from the Fishery

Scientific information collected from the ABT fishery includes biological data, migration patterns, and CPUE indices. All fisheries provide biological information through tissue samples, otoliths, blood samples, and other tests, which contribute to studies on age and growth, reproduction, physiology, and other critical research. Migration patterns are studied through tagging of bluefin tuna, in combination with biological sampling.

The CPUE indices, two of which are provided through the Large Pelagic Survey (LPS) of vessels in the Angling and General categories, are critical elements in the stock assessment of ABT. The LPS is an annual survey of the large and small fish rod and reel/handline fishery from North Carolina through Maine during May–October. The goals of this survey include catch monitoring, biological sampling, and most importantly, the construction of annual indices of relative abundance for Atlantic bluefin tuna ABT. The LPS, which includes a dockside and telephone survey, underwent extensive review in 1995 in the light of a number of problems. Some of these have been addressed in the 1996 LPS, while others will require longer term solutions. Because the CPUE index is generated from data collected through a survey procedure, the quality of the estimates provided by the survey is directly related to sample size.

Research and Monitoring

A comprehensive report on research and monitoring activities for HMS was completed earlier this year for Congress. In addition, a report on a research and monitoring program for highly migratory species (HMS) has been drafted by NMFS. This report, prepared to conform with Section 302(b) of the Fisheries Act of 1995, outlines a research and monitoring program to support the conservation and management of ABT and other HMS. A summary of the report will be published in the Federal Register, and copies of the report will be made available to the public.

The research and monitoring report addresses the work conducted on identification of the range of ABT and other HMS stocks, tagging studies, genetic studies, aerial surveys of ABT, observer coverage and port sampling of commercial and recreational fisheries, improving permitting and reporting programs, computer modeling, and other research programs underway. NMFS has included in this report the most recent figures on the allocation of budget and personnel to HMS research.

National Research Council Review and Recommendations

An independent external peer review of the bluefin tuna stock assessments was commissioned by NMFS in 1994. This review was conducted by the National Academy of Sciences' National Research Council (NRC), which challenged several assumptions in the SCRS assessments and found an error in the 1992 U.S. rod and reel CPUE indices. As a result, the SCRS convened its bluefin tuna working group to reassess the western component of the bluefin tuna stock ahead of the previously-agreed schedule. The SCRS adopted some, but not all of the NRC recommendations. The new SCRS assessment resulted in a more optimistic view of the status of the western Atlantic bluefin tuna stock for several reasons, including that an extra year of data provided evidence that the stock may be beginning to rebuild.

The NRC report includes six major research recommendations, which are listed below. Actions taken to implement these recommendations are listed under each item:

1. Conduct more rigorous tests of the one-stock hypothesis using the most appropriate technologies capable of detecting contemporary population genetic structure.

In addition to in-house research, NMFS has funded a two-year university research project on genetic studies of stock structure.

2. Microconstituent analysis and archival tags be used to provide information on spawning fidelity.

NMFS is conducting background studies on the utility of microconstituent analysis prior to implementing a two-year research program. NMFS will host a workshop, planned for fall of 1996, to lay out a research plan on microconstituent analysis, including priorities for research and appropriate tools for determining stock structure, migration, and spawning fidelity. In August 1995, NMFS held a workshop on tagging methods and produced a report detailing a tagging research program and priorities for action. NMFS also continues to conduct extensive in-house tagging research and to fund outside efforts.

3. Spawning biomass, sex ratio, age at maturity, and fecundity in the spawning ground be estimated and that larval performance, as affected by environmental conditions, be studied.

Research on life history characteristics is ongoing by NMFS, focusing on age and growth, as well as spawning dynamics.

4. Tagging programs be undertaken with appropriate combination of conventional, acoustic, and archival tags.

NMFS has expanded its conventional tagging programs, particularly in the winter North Carolina bluefin tuna fishery, in which over 2,700 ABT were tagged over the past three years. To date, recoveries in each year have ranged from 2 to 6 percent. NMFS funded research on sonic tags in North Carolina and New England. NMFS is collaborating with the private sector in supporting research using new archival tags. NMFS will also fund research on improving the design and behavioral characteristics of archival tags.

5. Synthesize analysis of existing data on distributions of bluefin tuna in relation to spatial and temporal dynamics of major oceanographic features.

NMFS has funded aerial survey research for three years, which includes analysis of oceanographic features such as water temperature and productivity relative to schooling behavior and distribution of ABT.

6. Review of all research proposals and resulting manuscripts, including external peer review.

NMFS is designing a process for consistent peer review of research proposals for bluefin tuna in order to ensure that relatively scarce research funds be allocated to the most capable researchers conducting research in priority areas. NMFS scientists routinely submit their research results to peer review.

ICCAT Measures

Quotas

A quota is the maximum number or weight of fish that can be legally landed in a time period. It can apply to the total stock, a nation's fishery on a portion of such a stock, or to its subsets, which might be composed of allocations of different sized individuals (e.g., small, medium and large) to different gear types. The purpose of the quota is to allow some level of fishing while protecting a stock's ability to reproduce sufficiently to maintain its population—optimally at the level of MSY.

ICCAT determines the appropriate total biannual quota for a particular stock taking into consideration the advice of its scientific advisory panels (SCRS) and its most recent biannual stock assessments and population projections. Such decisions require the agreement of all participating states, and if one disagrees, the measure is rejected. However, if all ICCAT states can reach a consensus, it recommends biannual quotas for each participating member state based on the agreements made by their representatives and generally following each state's historical participation in the fishery. For example, if a member state historically had 40 percent of the landings of a particular stock, it could expect to be allocated a like percentage of the stock's next biannual quota. However, this same percentage share might be substantially less in total weight if the stock had declined in abundance compared to prior years. From ICCAT's perspective, the sole purpose of any quota for landing Atlantic bluefin is to provide information to monitor the stock and its continuing decline or recovery.

Each member state may harvest no more than its allocated quota, and overages and underages are considered and adjusted by ICCAT in arriving at future biannual quota allocations for each member state. ATCA requires that ICCAT's recommendations, including biannual quotas (as well as minimum size limits and other such management measures), be followed by U.S. regulatory actions issued as final rules or regulations as published in the Federal Register. Once a quota is reached (whether for the fishery or a subset), the fishery or subset is closed generally by Federal regulation, until a new quota allocation is made and implemented as above.

Minimum Size

Minimum size limits are recommended by ICCAT to provide a disincentive against harvesting the very youngest fish in order to give them some level of protection and to at least delay harvest until they had grown larger. However, ICCAT'S minimum size limit recommendations are not directly related to the age of maturity. Doing so would protect juveniles until they could spawn at least once—the most desirable situation to maintain a population's health. For example, ICCAT's minimum size limit for Atlantic swordfish is 41 lb, a juvenile in its second year and about 3 years prior to its first spawn at a weight of about 150 lb for females. Female bluefin tuna, however, do not mature until they weigh about 300 lb. ICCAT's minimum size limit is 14 lbs for bluefin tuna. The United States implements ICCAT's size limit recommendations by regulation the same as it does for quotas.

8 Percent Limit on School Fish

In 1992, NMFS implemented a recommendation by ICCAT to prohibit the taking and landing of bluefin weighing less than 30 kg (41 lb) or 115 cm (45 in) fork length with discretion to grant tolerances of no more than 8 percent by weight of the total bluefin catch on a national basis. (This limit is in addition to ICCAT's minimum size limit mentioned above.) This recommendation was adopted by ICCAT in order to enhance recovery of the bluefin stock and stem its continuing decline. The Angling category is the only domestic category that traditionally harvests and retains fish that small. The Angling category was also affected by the measure adopted at the same time prohibiting the sale of fish less than 178 cm (70 in), the only fish that category may catch and retain.

No Directed Fishing on the Spawning Stock

ICCAT prohibits directed fishing on the spawning stock of Atlantic bluefin tuna. Until 1992, the lack of a requirement that bluefin be landed in conjunction with other species and the short distance from port to the fishing grounds in the Gulf of Mexico made it possible for vessels to direct their fishing on bluefin despite a limit of two fish per trip. In 1992, NMFS published regulations to prevent directed fishing for Atlantic bluefin tuna in the Gulf of Mexico and to enhance enforcement of the current regulations. These regulations: (1) prohibited retention of Atlantic bluefin tuna harvested from the Gulf of Mexico, except for vessels permitted in the Incidental Catch category; (2) reduced the incidental catch limit for Atlantic bluefin tuna from two to one fish per vessel per trip in the longline fishery operating south of 36°N. latitude (southern area); (3) conditioned the incidental catch limit of one Atlantic bluefin tuna on the landing of at least 2,500 pounds of other species in the southern area; and (3) made other technical revisions to the regulations.

Domestic Measures

Permit Requirements

Permits or certifications are required of all vessels fishing in the Atlantic bluefin tuna fishery, including the Angling category (for recreational fishers). To monitor the catch in the Angling category with its hundreds of boats, it is essential to have an accurate count of the universe of angling vessels. Catch per unit of effort estimates are extrapolated by NMFS to estimate the total recreational catch. This procedure is much more timely and accurate if the true number of recreational vessels involved is already known. Permits were issued first to party and charter vessels in 1993 and to the owners of private recreational fishing vessels in 1994. Angling category permits distinguish recreational from commercial fishers. NMFS considers that General category permit holders are commercial fishers who must therefore abide by Coast Guard regulations pertaining to safety equipment requirements. Those anglers who do not want to sell their catch, but who want to retain a trophy fish (one large medium or one giant bluefin annually) taken incidental to fishing under the regulations applicable to the Angling category, may apply for an Incidental Catch (Rod and Reel) category permit. Permits are also required of Harpoon, Purse Seine and Incidental category commercial vessels (see Appendix 1).

Reporting Requirements

Reporting requirements are necessary to ensure that the ICCAT-recommended biannual quotas for the United States are not exceeded. Thus it is essential to have a system in place to provide data to monitor the stock and its harvest by all domestic user groups. In 1993, NMFS required Atlantic bluefin tuna dealers to submit daily reports by FAX, as well as by mail. This was done, at the suggestion of fishers and dealers, to give NMFS better information for quota monitoring and to allow NMFS to make decisions on seasonal closures or adjustments on a real-time basis. Weekly dealer reports were replaced in 1993 with biweekly reports to enhance the usefulness of the information collected.

Quota by User Group

After it receives its annual quota allocations from ICCAT, the United States allocates these annual quotas among the user categories authorized for this fishery. These include the General Category (rod and reel, handline, harpoon or bandit gear), Angling Category (rod and reel or handline), Charter/Headboat Category (rod and reel or handline), Harpoon Category, Incidental Category (longline, purse seine, fixed gear traps) and the Purse Seine Category. See Appendix I, Atlantic Bluefin Tuna Brochure for details by category. Except for large medium or giant bluefin caught by the recreational fishers permitted in the Angler Category, all of the above are commerical categories involving sale of the catch to authorized dealers.

The United States quota recommendation from ICCAT is allocated among the various user categories based on historical participation and adjusted biannually to account for underages and overages by category. Commercial quotas are mentioned by using logbook and dealer reports. Recreational landings are monitored by means of data obtained from NMFS Large Pelagics Survey, a survey designed for other purposes but applied as well to this fishery.

Catch limits are also specified within the various users categories described above for the bluefin size categories described in Table 3. These categories (e.g., school, medium and giant) are convenient size increments of a species that can grow to very large size and whose individuals are all vulnerable to capture by various components of the fishery. However, the various size categories have differing fishing mortality rates, which are recognized by existing and past regulations.

Minimum Size for Sale

No bluefin tuna can be legally sold in the United States unless they are at least 185 cm (73 in) curved fork length—a fish of about 235 lb. Such a fish would still be immature and would not spawn for at least two more years at a weight of about 300 lb, if it were a female. Establishing minimum size limits on fish sold commercially provides a disincentive to commercial fishers who might otherwise target juvenile fish, which they are capable of catching in relatively large number using purse seines in the shallow waters of the continental shelf. In so doing, these fish are reserved for the recreational fishery (Angling category) or allowed to live longer and grow larger to the point at which they can be landed legally by both commercial and recreational users.

Measures to Slow the General Category: Monthly Quota and Days Off

The total annual quota is allocated among the user categories, after adjusting for overages and underages from the preceding year, in a manner proportional to previous allocations. In 1995, monthly quotas were also stipulated in regulations published in the Federal Register for the General category. This was done to spread the quota more equitably as the fish migrated along the coast and to allow for a late season fishery. The annual General category quota was apportioned into monthly quotas based on historical records of landings.

Industry groups proposed daily closures as a means of lengthening the fishing season. In 1995, NMFS published regulations that restrict allowable fishing days for vessels permitted in the General category to days other than Sundays and Wednesdays. This management measure serves to prevent overharvest of quota in any period and is tied in part to market closures in Japan (the major export market) to minimize potential negative economic consequences to U.S. fishers.

Observer Requirements

Observer coverage is often the most efficient and effective means of scientifically validating catch, bycatch and discard data, and to collect biological samples. Mandatory observer coverage would improve collection of the scientific data necessary for management decisions affecting the Atlantic tuna fisheries. In 1993, NMFS published regulations requiring observers for any vessel fishing for or incidentally catching Atlantic tunas. Owners of vessels selected for observer coverage are required to notify NMFS prior to a vessel's departure on a fishing trip. If, after notification, NMFS decides to place an observer on board for that fishing trip, the vessel owner would be instructed to wait at the dock until an observer arrives.

Effectiveness of Alternative Regulatory Approaches

Introduction

This section describes a range of alternatives analyzed following the revised (November–December 1994) ICCAT stock assessment and recommendations for the management of Atlantic bluefin tuna. A summary of the consequences of each alternative is also presented. These alternatives are the ones considered and analyzed in detail in the 1995 FEIS for the U.S. Atlantic Bluefin Tuna fishery (NMFS, 1995).

In the case of western Atlantic bluefin tuna, the current situation is an overfished stock and an overcapitalized[6] fishery. In terms of dealing with overfishing, quotas are established by ICCAT and the United States has no authority to deviate. However, it can allocate this quota among its user groups and, depending on the allocation chosen, produce important differences in the balance between economic efficiency and resource conservation (e.g., the ability to rebuild and the potential rate of recovery to MSY levels) as discussed below.

The problem of overcapitalization is becoming particularly acute as a result of increasing numbers of participants entering the fishery to compete for a relatively small, scientific monitoring quota. At present, all but one of the commercial and recreational categories are open access, and the fisheries are managed by fleet category quotas and by varying the timing and length of fishing seasons so as to provide harvest opportunities for all applicable geographic regions, while attempting to ensure that the quotas for each category are not overrun. This has proved extremely difficult and costly to accomplish since the migration patterns and availability of bluefin tuna are not completely predictable. The burden on management, monitoring and enforcement systems has been substantial. In addition, there is widespread belief that current (and historical) allocations between the various commercial and recreational gear categories are not "fair and equitable."

Alternatives that have been considered, as analyzed in the FEIS (NMFS, 1995) are summarized in the matrix presented in Table 1. Matrix columns represent the range of alternatives that were considered as a total western Atlantic quota for bluefin tuna (0, 1,995 mt, 2,200 mt, and 2,600 mt from 1995 onwards). For analysis of the domestic fishery, it was assumed that the United States share of the western Atlantic quota would be 61.90 percent for 1,995 mt (as was the case in 1994), 59.61 percent for 2,200 mt (as was recommended by ICCAT for 1995), then reverts to the historical percentage share of 52.14 percent for 2,660 mt. The rows illustrate alternative domestic allocation schemes, including allocations proportionate to the previous level, reduction/closure of the Purse Seine category, closure of the Gulf of Mexico incidental longline fishery, and closure of the small fish fishery. Some of the alternatives examined are not consistent with statutory authorities for the management of highly migratory species. For example, under ATCA and MFCMA, the United States share of the western Atlantic bluefin tuna quota may not differ from that recommended by ICCAT. However they are included for illustrative purposes.

---

[6] A fishery is considered to be overcapitalized if the fleet size exceeds that necessary to harvest the total allowable catch at the lowest possible cost.

Both the biological and socio-economic consequences have been analyzed and are presented below for each of the five domestic allocation schemes considered, combined with each of the three (non-zero) quota levels.[7] Biological consequences focus primarily on the effects on the bluefin stock. Effects on other species are also considered. The socio-economic effects identify the communities and individuals likely to be affected by each alternative, and the possible impacts on net economic benefits and employment. Three alternatives for controlling access to U.S. fisheries have also been analyzed and are presented below, including a discussion of their predicted effects.

Within each alternative noted in Table 1, a variable number of sub-alternatives have also been identified and analyzed. For example, for the Gulf of Mexico incidental longline fishery, the sub-alternatives considered are: zero quota, prohibition of sale of bluefin tuna during spawning periods, season/area closures, gear restrictions and implementation of an experimental fishery with breakaway gear.

The matrix (Table 1) represents a wide range of reasonable alternatives given the present context of the western Atlantic bluefin tuna fisheries in the United States. The alternatives in the matrix are briefly outlined here and analyzed and discussed in detail in the FEIS (NMFS, 1995).

In the assessment below, consequences of alternatives have been considered in relation to the likely future size of the western Atlantic bluefin tuna stock (both in terms of spawning biomass and total biomass), commercial present value (PV), angler consumer surplus (ACS), employment, and effects on other components of the natural environment. The first three of these were primarily evaluated using a computer model that projected the bluefin stock and associated fisheries into the future (see FEIS for a description of the bioeconomic model, including assumptions). Information from supplemental analyses, scientific papers and other documents, and public hearings and anecdotal reports were also used in the evaluations. Sections 4.4.2, 4.4.3, and 4.4.4 summarize the consequences of the alternatives in terms of their effects on the bluefin tuna resource itself and socio-economic factors. Section 4.5 summarizes the effects on other species and fisheries.

Quota Alternatives

Quotas A–D

For the quota alternatives (columns in Table 1), all projections commence with the 1994 quota level of 1,995 mt.[8] The time period covered is 1994 to the year 2010 (17-year period of analysis). A time horizon of this duration was used to illustrate potential trade-offs between short-term and long-term economic gains; however, estimates of stock size this far in the future are completely dependent on assumptions about incoming recruitment, and should therefore be treated with caution. It is also important to note that neither quotas nor domestic allocations are expected to remain static for 17 years.

There are four quota alternatives in total.

(A) Previous Quota Level (QUOTA A)

The previous quota level (or QUOTA A) would hold the western Atlantic quota constant at 1,995 mt from 1995 onward. The consequences of adopting this quota are analyzed for each of the five alternative domestic allocations.

(B) Current ICCAT Recommendation ("status quo" or QUOTA B)

The current quota is 2,200 mt for 1996 and beyond. The consequences of maintaining this quota indefinitely are analyzed for each of the five alternative domestic allocations. This is also the current ICCAT-recommended quota.

(C) Return to 1991 Quota Levels (QUOTA C)

The effects of increasing the quota to the 1983–1991 level of 2,660 mt are analyzed for each of the five alternative domestic allocations.

(D) No Fishing Alternative (QUOTA D)

The last alternative, QUOTA D, would eliminate all bluefin tuna fisheries (both U.S. and non-U.S.) from 1995 onwards. The biological and socio-economic effects of a complete closure of the U.S. recreational and commercial fishery are analyzed.

In summary, the range of reasonable alternatives considered for the total western Atlantic quota is from zero (a 100 percent reduction through closure of the entire fishery) to 2,660 mt. These alternatives reflect the most recent ICCAT recommendations as well as public comments received on the DEIS, and not necessarily the options available to NMFS under current law (e.g., clauses in ATCA and MFCMA that prevent the imposition of any U.S. quota level other than that recommended and agreed to by ICCAT). The current approach is to adopt the ICCAT-recommended quota of 2,200 mt for the total western Atlantic stock (QUOTA B), as required under ATCA and MFCMA.

Consequences of Quota Alternatives

In terms of spawning biomass and total biomass, the previous quota and the status quo quota (QUOTAS A and B,

---

[7] Since the "no fishing" alternative does not require consideration of domestic allocation, a total of 11 scenarios are analyzed quantitatively.
[8] All quota levels refer to the entire western Atlantic quota, of which the U.S. is assumed to have a constant share of 61.90 percent for a total quota of 1,995 mt, 59.61 percent for a total quota of 2,200 mt, and 52.14 percent for a total quota of 2,660 mt.

respectively) both result in appreciable increases in spawning biomass and total biomass; an increase to 1991 quota levels (QUOTA C) results in modest increases in biomass; and the zero quota (QUOTA D) results in rapid rebuilding. In all cases, there is an unknown probability that the stock will actually continue to decline rather than increase, with the probability of further decline being greater for higher quotas. The long-term rank ordering of the quota alternatives is the same for all domestic allocation alternatives. Redistribution of small fish quota to larger sizes (ALLOCATION E) marginally reduces the rate of rebuilding of the spawning stock in the very short term, but results in a somewhat faster rate of rebuilding of the spawning stock over the long term and the total biomass over the whole time horizon. The other four domestic allocation alternatives have trajectories that are virtually indistinguishable from one another.

Projection model results for the economic indicators are similar across the three non-zero quota alternatives. With the exception of the closure of the small fish fishery, both commercial and recreational net present values increase as the quota increases, under all domestic allocation alternatives. The projection model also indicates that short- and long-term economic gains can be achieved through increasing the quota, but that these are relatively small, particularly when compared to the reductions in the rate of stock rebuilding that would also accrue. Since net benefits in the commercial fishery are calculated as revenues over variable costs only, and fixed costs would have to be covered in the long run, these net income levels should be evaluated relative to fixed costs.

Long-term economic gains, whether from commercial net revenues or from ACS or a combination of the two, can only occur if quotas are held at levels that allow rebuilding. Under the revised (optimistic) SCRS stock assessment, rebuilding is predicted to occur at all four quota levels (albeit at varying rates), so no immediate short-term losses are necessary for long-term economic gain. However, this view is dependent upon optimistic assumptions regarding future (unknown) recruitment and use of a stock assessment model and incomplete data that the scientific community has advised managers is not yet fully reliable.

Domestic Allocations

Allocations A–E

The range of alternatives for domestic allocation of the quota is based on scoping and public hearing comments as well as management experience with the fishery and other considerations. This is essentially an unlimited number of combinations of allocations of the U.S. quota between the five main user categories: General and Harpoon combined, Purse Seine, Incidental North, Incidental South (Gulf of Mexico fishery) and Angling category (fish under 185 cm (73") curved length)[9]. To simplify the analysis, the alternatives examined focus on extremes such as elimination of entire categories. However, the consequences of smaller reductions in these categories can be interpolated or inferred from the results presented.

(A) Increase the Purse Seine Quota by 51 mt (ALLOCATION A)

The first alternative allocates quota based on the 1994 allocations among the five user categories. The actual allocation is based on historical shares of the total quota, and the Purse Seine category would be allocated a quota of 301 mt as it was in 1994.

(B) Status Quo (No Action Alternative, ALLOCATION B)

The second alternative for domestic quota allocation (ALLOCATION B) represents a quota reduction of 51 mt for the purse seine fishery to 250 mt from 301 mt in 1994. NMFS received numerous comments about the need to reallocate quota from the purse seine fleet to other components of the fishery. These comments had been made in scoping meetings, public hearings, written letters, and in a petition received by NMFS.[10] The main reasons given in support of reallocation relate to issues of "fairness and equability," the intended primary use of bluefin tuna quota for "scientific monitoring," and the greater employment generated in the non-Purse Seine categories. The scientific monitoring argument was particularly influential in the selection of ALLOCATION B as the preferred domestic allocation alternative for 1995 and 1996. Both ICCAT and NMFS have stated that, pending stock recovery, the primary role of the western Atlantic bluefin tuna fishery is to provide scientific data for stock monitoring purposes. While all gear categories supply biological information used directly in stock assessments (e.g., landings and size frequency data), the General, Angling and Gulf of Mexico Incidental Longline categories are the only U.S. categories that currently provide continuing catch per unit effort time series used to estimate trends in stock size. NMFS intends to explore other methods for domestic allocation for 1997 and beyond.

(C) Eliminate the Purse Seine Fishery (ALLOCATION C)

This third allocation alternative examines the possible effects of a complete closure of the Purse Seine fishery.

---

[9] These categories differ from the management classification by combining the General and Harpoon categories into one category, and by more explicitly dividing the Incidental category into two separate categories. These modifications are made due to data considerations as well as the need for separate analysis of the Gulf of Mexico incidental fishery, which takes place on known spawning grounds.

[10] On February 28, 1994, NMFS received a petition from the General Category Tuna Association requesting that U.S. Atlantic bluefin tuna regulations on allocation consider impacts on employment as well as the scientific usefulness of the data provided. Copies of this petition are available from NMFS upon request (see names, addresses and phone numbers in Chapter 5).

Again, as for ALLOCATION B, this alternative reflects petitions and comments calling for reallocation of the domestic quota to user groups that generate higher employment and/or provide the most useful scientific data.

(D) Eliminate (or Curtail) the Gulf of Mexico Incidental Fishery (ALLOCATION D)

The fourth allocation alternative, closure of the Gulf of Mexico incidental longline fishery (ALLOCATION D), reflects the ICCAT prohibition on directed fishing in this spawning area. Comments submitted during the scoping process suggested that complete closure of this fishery would remove all economic incentive for bluefin tuna bycatch.[11]

(E) Eliminate the Small Fish (< 185 cm or 73" curved length) Fishery (ALLOCATION E)

The fifth domestic allocation alternative (ALLOCATION E) also stems from scoping meeting comments that point to the potentially higher physical yield possible for the commercial fishing industry from a strictly large-fish fishery. This would call for a closure of the recreational fishery for bluefin tuna (or restriction to catch and release only).

In summary, the reasonable alternatives considered for domestic allocations cover a range from no action (status quo allocation) to elimination of complete fishing sectors. Again, these alternatives represent the range of comments received during the scoping period as well as in the written and verbal comments during the DEIS and SDEIS comment periods. The 1995 regulations reduce the Purse Seine category quota from the previous level of 301 mt to 250 mt, and reallocate the 51 mt difference (ALLOCATION B). The primary motivation for the reallocation was the need to improve the quality of catch per unit effort estimates used for scientific monitoring of trends in stock abundance. NMFS intends to explore other methods for domestic allocation for 1997 and beyond.

Consequences of Allocation Alternatives

In the long term, both spawning biomass and total biomass invariably increase most rapidly for the no small fish fishery alternative (ALLOCATION E), next for either the increased allocation for the Purse Seine category (ALLOCATION A) or the elimination of the Gulf of Mexico fishery (ALLOCATION D), next for the status quo allocation (ALLOCATION B), and finally, for the elimination of the Purse Seine category (ALLOCATION C). This ordering of alternatives reflects (i) the assumption that a reduction in one category is redistributed over the other categories in proportion to its current share, and (ii) differences in yield per recruit (YPR) and spawning per recruit (SPR) estimates that result from the different selectivity patterns exhibited in each fishery. However, in all cases except ALLOCATION E, the differences are so small that the stock trajectories almost coincide. For almost all projections presented here, the 8 percent constraint (no more than 8 percent of any nation's landings may be made up of fish less than 30 kg (66 pounds) in any year) could not be satisfied unless the selectivity pattern of the Angling category was moved towards larger sizes (equivalent to closing down the "school" category partway through the season). Results not presented here indicate that without the 8 percent constraint and/or an increase in the sizes of fish targeted by Anglers, increases in the Angling category quota could further slow the rate of stock rebuilding.

The relative ranking of the allocation alternatives is constant across all quota levels. Net present value in the commercial fishery is highest under closure of the small fish fishery (ALLOCATION E), followed by the increased allocation for the Purse Seine category (ALLOCATION A), and lowest under closure of the Gulf of Mexico fishery (ALLOCATION D). Net present value of ACS is maximized under closure of the Purse Seine Category, is zero under closure of the small fish fishery, and relatively similar for other allocation alternatives.

The economic results also illustrate the tradeoffs between the recreational and commercial fishery in establishing domestic quota allocations. Closure of the small fish fishery results in significant reductions in overall net benefits from the resource due to the loss of ACS (up to $141 million in reduction over the 16-year time horizon from the status quo quota–status quo allocation alternative). These losses are only partially offset by increases in net benefits to the commercial fishery, since the closure of the small fish fishery yields a relatively higher present value of commercial net benefits (for example, net increase of $21 million under QUOTA A). Closure of the Gulf of Mexico fishery reduces net commercial benefits in part due to the biological effect of targeting more heavily on smaller fish, as well as the switch from commercial to recreational benefits. The effects of the allocation alternatives on employment depend upon the sector most affected; some proportion of 1,200 full-time equivalent positions can be affected to some degree by the domestic allocation alternatives.

Access Control Alternatives

Alternatives Considered

The analyses of domestic allocation alternatives covered above all assumed that the fisheries would remain open access; thus, none of the alternatives discussed to this point address the problem of overcapitalization.

---

[11] The General Category Tuna Association petition also requests a prohibition on the sale of incidentally-harvested bluefin tuna.

The three access control alternatives considered below address socio-economic issues such as overcapitalization, access rights, the "race for the fish" and overall economic efficiency. Biological impacts are less relevant since the key issue is one of allocation. The access control systems discussed here assume that the overall quota is pre-set (based primarily on biological criteria). However, it is certainly true that the level of the overall quota would affect the socio-economic impacts of the various alternatives.

The three alternatives for access control (ACCESS CONTROLS A, B, and C) indicated below Table 1 address an increasingly difficult problem in the U.S. Atlantic bluefin tuna fishery; notably, the overcapitalization inherent in this open-access fishery. These access control alternatives can be considered a third dimension to the matrix presented in Table 1, since each cell in the matrix could be applied under one of the three access control systems.

The tree alternatives considered are:

(i) a lottery system
(ii) limited entry with a fleet-wide quota
(iii) an individual transferable quota (ITQ) system.

Each alternative addresses the issue of overcapitalization in the fishery. With less than 10 percent of permitted vessels actually having landed a bluefin tuna in 1993, it is apparent that there is over-investment (overcapitalization) in this fishery. Even though a large number of permittees hold a commercial permit only for the remote possibility that they may land a fish over 185 cm (73") curved fork length, there is considerable competition between commercial vessels for the limited bluefin tuna quota. Through various measures, these three alternative systems attempt to reduce the number of participants and thus, to some extent, the race for the fish.[12]

In the sections below, each alternative is described as it might be implemented. In some cases, a range of options for implementing the allocation scheme is presented. In each case, probable impacts of the alternative are considered.

(A) Lottery

Consider first a lottery system, in which tickets would be provided at a fee not to exceed administrative costs to qualifying individuals (the Magnuson Act expressly limits permit fees to the administrative costs of issuing the permit, thus limiting the possibility of charging for lottery tickets). For the commercial fishery, the number of Atlantic bluefin tuna over 185 cm (73") curved fork length which may be landed in a given year is determined according to the latest stock assessment. As an example, assume that 4,500 bluefin tuna may be caught in a given year, and the number of vessels holding permits is 10,000. Each vessel holding a commercial permit would be eligible to enter the lottery; indeed, the permit number could be used as the "lottery ticket." An independent agent selects a random sample of 4,500 tickets, such that a total of 4,500 vessels would be able to land a bluefin tuna in that year. The sampling would be done without replacement such that no vessel receives the opportunity to harvest more than one fish. Those vessels excluded from this year's fishery would still able to participate in the lottery in subsequent years; however, it may be preferable to limit the pool of lottery participants by applying the lottery in conjunction with a moratorium on the issuance of new permits, particularly if the permits are transferable. Because of the random nature of the selection, the lottery system may be perceived as equitable in the sense that everyone has an equal chance of winning—if not necessarily fair in terms of its impact on the various participants in the fishery, particularly those who are truly dependent on the fishery.

For certain vessels, particularly the five purse seine vessels, it is not possible to operate profitably if only one fish can be landed. There are at least three options for addressing this issue. First, tickets could be made transferable, such that purse seiners, or other vessels operating at a scale of fixed and operating costs that necessitates multiple landings in order to be solvent, may purchase the right to land more than one bluefin tuna. This option would be useful in addressing the problem of a vessel that may be unable to harvest its fish in a given season; transferability therefore would allow them to profit from winning the lottery, despite the inability to land a bluefin tuna.[13] A second possibility would be to assign a separate quota to the purse seine category, without a lottery system, such that there would be a predetermined, multiple-fish quota for each vessel. A third possibility would be for each permit number to be entered into the "lottery pool" at a rate equivalent to the number of bluefin tuna harvested in the previous year or years by that permittee.

Similarly, it may be desirable to have the lottery system modified to have each participant's chances of being selected weighted by his or her dependence on the fishery. For example, a greater number of tickets could be made available for vessel owners for whom a certain percentage of their gross income is derived from fishing (for either bluefin or other species). This would decrease the possibility that those most dependent upon the bluefin tuna fishery would be excluded from the fleet of eligible vessels.

---

[12] To a certain extent, the classic race for the fish has been reduced in the Atlantic bluefin tuna fishery by implementing individual vessel quotas in the Purse Seine category, and a daily landing limit for General category vessels.

[13] Apparently, it is fairly common for full-time, commercial fishermen to have one or two years with no Atlantic bluefin tuna landed, due to weather, bad luck, skill or a combination of factors. Under both ATCA and the MFCMA, NMFS is required to ensure that U.S. fishermen have an opportunity to harvest the ICCAT quota.

Recreational anglers would continue to fish under a separate quota, much as under the present system (bag limits, quota restrictions, no sale of fish under 185 cm (73") curved fork length). Quota allocation could be determined by fishing ability and/or luck (as is presently the case), or by an equivalent, but separate, lottery system. There may be some merit to considering a further division between private and charter vessels.

Whatever the method selected to allocate tickets to land a bluefin tuna, the nature of the fishery would change substantially under a lottery scheme. The number of vessels permitted to actively participate in the fishery targeting bluefin tuna would be more than halved (in this hypothetical case—it could vary in actual implementation). Again, those vessels that were holding a permit "just in case" may continue targeting bluefin tuna in the recreational fishery, or while fishing for other species on a commercial or recreational basis. Those vessels holding a ticket would be able to fish for Atlantic bluefin tuna when, where and how they find it most profitable to do so. For those vessels selected by the lottery, profits would increase, as the vessel would only be used when it is profitable to do so, in contrast with the race for the fish when all vessels are vying for a greater share of the fixed pie. Fishing would tend to occur when costs are lowest and/or prices (and thus gross revenues) are at their highest levels.[14] More importantly, eliminating the race for the fish can reduce potential safety hazards usually associated with derby fishing, particularly during periods when the season is about to close.

Since the initial allocation would be determined by pure luck rather than willingness to pay, the lottery system is not efficient in maximizing net economic benefit from the fishery. To the extent that lottery tickets are transferable, however, economic efficiency would increase. Non-transferability between purse seine and other vessels, and between the recreational and commercial categories, would decrease the economic efficiency of the lottery system (see discussion on ITQ's and transferability below). Since the lottery system introduces another source of risk into the fishery, uncertainty would increase as well. This could hamper the ability of vessel owners to obtain financing for vessel upkeep. Holding a lottery every other year, or even up to every five years, would reduce this source of uncertainty. However, for vessels that are dependent on Atlantic bluefin tuna, those not chosen in a given lottery may not retain their capital investment if the next lottery is held too many years into the future.

As with all forms of limited entry, the lottery system could have the effect of reallocating excluded fishing capacity to other species. In the Atlantic bluefin tuna fishery, some vessels also target other species, including other tunas, groundfish, and lobster (using different gear). Given the limits on these species, it is important to recognize the probable effects of excluding effort from the bluefin tuna fishery. For example, reducing purse seine quota allocations could result in a significant increase in fishing effort on yellowfin and skipjack tuna. The 1994 ICCAT recommendations stated that effort for yellowfin should be capped.

The lottery system would exclude a large number of participants from the commercial fishery, even more so if vessels within the lottery system are allowed to sell their tickets to those already holding a ticket to land bluefin tuna. However, since the overall landings would remain the same with or without the lottery, the reduction in costs in targeting bluefin tuna would not be offset by a reduction in revenues, implying higher net economic benefits. Furthermore, since the vessels holding tickets would be able to fish when costs are at their lowest and prices are expected to be high, net economic benefits to the fishery would be higher under the lottery system than under open access. Direct and indirect employment in the fishery would probably decrease in a lottery system, in line with other excess inputs applied to open access fisheries, although average income of authorized participants might increase, with probable induced economic effects. Since the lottery would not affect the total number of bluefin tuna landed and processed, employment in the processing sector would not be significantly affected.

(B) Limited Entry with Fleet-wide Quota

Limited entry is a generic term referring to any form of control to limit the number of vessels in a fishery—as opposed to open access systems in which no new entrant is excluded by law or by regulation. Thus, a vessel permit moratorium (no new permits issued), an earned income requirement, or individual (non-transferable) vessel quotas (IVQ's) are all examples of limited entry systems. As with all of these allocation schemes, there are numerous options for applying limited entry. For example, the number of vessel permits may be reduced by allocating limited entry permits (LEP's) only to those vessels having demonstrated a "historical dependence" on the fishery. Historical dependence could be defined in many ways. For example, vessels or individuals could be required to have landed at least one bluefin tuna per year over one of the previous three years, or alternatively, to have a minimum percentage of gross income from bluefin tuna fishing. Transfer of the permit to a new owner would be undertaken when the vessel is sold or given away (e.g., to heirs), with the resource rent capitalized in the value of the permit. Regulations would be needed to avoid

---

[14] Although input prices do not vary throughout the season, fishing costs can be lower due to less search time and distance, since vessels can go out when the fish are most available, and in the areas closest to them. Prices of Atlantic bluefin tuna can vary considerably throughout the fishing season (see section 3). However, since the U.S. is a price taker on the Atlantic bluefin tuna world market, it is unlikely that this or other forms of limited entry would affect consumer prices of bluefin tuna.

increased fishing power from a fixed fleet size. "Capital stuffing" would be addressed through controls on vessel upgrades as well as on permit transfers to new vessels of greater fishing capacity. The permit could be transferred to a new vessel if limits on increases in fishing power are respected.[15] Limited entry would be applied only to the commercial fishery, since the net benefits from the recreational fishery would not necessarily be reduced by entry into the fleet (unless congestion becomes problematic).

There is considerable variability in the catch performance of bluefin tuna vessels from year to year. For example, a LEP could be allocated only to vessels that caught at least one bluefin tuna within the past three years. This alternative could result in a greater number of permits than under a lottery system. However, since more than half of the successful vessels land more than one fish, the number of vessels in the fishery under this eligibility scheme may be less than the number of bluefin that can be landed, depending on the quota level.

Under the LEP system with a fleet-wide quota, there would be no limit on the number of bluefin tuna which could be landed by a vessel which has proven its historical dependence. Thus, the race for the fish, with its problems of overcapitalization, higher costs, lower gross revenues and safety, would not necessarily be eliminated under limited entry. The incentive to save costs and to increase revenues by fishing when conditions are optimal would be offset by the fear of not obtaining a profitable share of the fixed global quota. Under a LEP system, division of the total quota into permit categories would remain in effect, since harpoon and hook and line vessels may have difficulty competing with purse seine vessels in a race for the fish.

Despite the continuing derby nature of the fishery, since the number of vessels decreases under this LEP system, the vessels in the fishery would face less competition, less crowding and fewer gear conflicts. In a manner similar to the lottery system, the reduction in inputs applied to the fishery, due to the smaller fleet size, implies higher net economic benefits would accrue. While fishery employment would decrease, the average income of fishermen would rise. Limited entry benefits would be offset by any "capital stuffing" undertaken by vessels which are in the fishery and wish to increase their share of the quota by increasing fishing efficiency. Although the incentive to boost fishing power is also present under open access, there may be higher returns to "capital stuffing" in a limited entry scheme, since competition is only from vessels already in the fleet, rather than from a continued flow of new entrants.

(C) Individual Transferable Quotas

An Individual Transferable Quota (ITQ) system allows each vessel, after the initial allocation process takes place, to determine the optimal quantity of fish to harvest, and to buy or sell quota to achieve that level. With this guaranteed "privilege to fish," each vessel would fish when conditions (costs and prices) are optimal. An ITQ system could be applied to the bluefin tuna fishery with a variety of optional restrictions and additional regulations. Primary restrictions on transferability could include separate quotas for the recreational and commercial fisheries, a fixed total quota for incidental longline bycatch, and perhaps restrictions for individual ownership (e.g., prohibiting non-vessel owners or establishing a maximum percentage of total quota that could be owned by a single entity). As for additional regulations, gear restrictions would still apply (e.g., no directed longline fishery), as would the prohibition on directed fishing in the Gulf of Mexico.

One of the most difficult aspects of establishing an ITQ system is determining the initial allocation of the total allowable catch (or of the U.S. quota, as in the case of Atlantic bluefin tuna). Existing law prohibits the initial sale of such shares, although this may change under reauthorization of the MFCMA. Thus, currently the resource must be "given away," resulting in transfer of the economic rent of the fishery to initial share holders,[16] which may be a source of considerable controversy. An unlimited number of initial allocation schemes are possible for ITQ's. One option would be to divide the overall quota into equal shares for all current permit holders. Alternatively, individual quotas could be established by dividing the quota into percentage shares based on historical bluefin tuna catch over the past five or any other combination of years, such that any commercial vessel that has landed at least one bluefin within the qualifying period would be allocated some initial share. Allocation based on historical catch could be combined with investment considerations by using a formula which includes both catch history and vessel/gear characteristics. Whatever the initial allocation scheme, transferability of the quota shares would assure that the final allocation of the quota would be primarily a function of the relative net economic benefits generated by the various user groups.

The ICCAT-recommended quota for bluefin tuna is expressed in terms of total weight; however, it would be preferable for the domestic ITQ system to "translate" that total weight into a number of fish, possibly in two or more size classes. This would facilitate the determination of the amount of ITQ which must be purchased in order for a vessel to go fishing. Past experience with ITQ systems has pointed to the relative advantages of expressing individual quotas in terms of a percentage of total quota, rather than as a fixed weight (Anderson 1992). Only integer amounts of ITQ's—whole fish—may be initially allocated and

---

[15] The administrative costs of overseeing permit transfers and vessel upgrades could well exceed the benefits of limited entry.

[16] If a limited entry system is used on an interim basis prior to implementation of the ITQ system, the resource access privileges are already "given away", in effect, and the value of the resource is capitalized in the permit prices and/or vessel prices, depending upon the regulations regarding transfer of the access right.

transferred. There are also precedents for defining ITQ's in terms of numbers of fish (Anderson 1992).

As noted above, with transferable quota shares, owners could determine the optimal level of catch associated with the productive capacity of their vessels, and buy or sell shares accordingly. An ITQ system thus allows each vessel to operate more efficiently in terms of costs. Revenues also could be optimized, since timing of fishing effort and/or landings could be sensitive to ex-vessel market prices. An ITQ system could eliminate the race for the fish, and the incentives to overinvest in harvesting capacity. Net economic benefits are likely to be higher under an ITQ system than open access or limited entry with fleet-wide quotas.

Social impacts of an ITQ system can be very significant, although the degree of change would be a function of the initial allocation scheme as well as the restrictions placed on the transferability of quota shares. Quota ownership may be restricted to vessel owners if there is concern among fishermen about shares being held by non-fishing interests. Furthermore, apprehension about concentration of quota ownership could be addressed through ceilings on the percentage that any one entity can own (such as the 1 percent limit placed on halibut/sablefish ITQ's in Alaska)[17]. Despite these restrictions on transferability, the nature of the fleet can change considerably after imposing ITQ's. The number of vessels would most likely decrease as overcapitalization is reduced, with a possible increase in average boat size and horsepower. Employment in the bluefin tuna fishery may decrease due to the reduction in fleet size; however, each crew member remaining in the fishery might work a longer season and net a higher income than under open access. Given the reduction in the number of deck hands hired, it is probable that the more skilled and experienced workers would have a greater probability of being hired than recent entrants into the fishery. Since the quota system would not affect the total number of bluefin tuna landed and processed, it is likely that employment in the processing sector would not be affected.

An ITQ system saves considerable management efforts, most notably in the following areas: geographical and user category allocation, within-season quota adjustments, time/area closures (except for spawning ground/season protection), fish quality, and fleet safety. Furthermore, less research would be necessary in the area of valuation of both commercial and recreational fisheries, since the market would reveal the producer surplus and angler consumer surplus through respective quota prices.

Weak aspects of ITQ systems in general include highgrading of retained catch and potentially higher enforcement/monitoring costs. Highgrading would likely not be a major issue with Atlantic bluefin tuna, since fish are caught one at a time (with the exception of purse seining and the longline incidental catch fishery) and price differentials are difficult to determine at sea. All fish retained for sale are at least 185 cm (73") curved fork length, and price may increase or decrease as the fish size increases, but quality is also an important determinant. The additional weight from a larger fish may be more than offset by a lower price per pound, if the bluefin tuna is a "rubber band," or a low quality, lean fish with poor color and low fat content. Since it is difficult for fishermen to determine market quality—and thus the possible price—of a bluefin tuna at sea, and given the hours of additional effort involved in obtaining a second bluefin (unless stock size increases significantly), highgrading is less likely in this fishery. Indeed, highgrading is not a problem with the ITQ system for bluefin tuna in southern Australia (Anderson 1992:49).

The costs of an ITQ system can be significant, particularly in the initial stages of determining quota allocation and investing in the infrastructure for monitoring the system. However, it is unlikely that the long-run costs would exceed those of the current "micro-managed" system, which includes lengthy public debates over issues such as user group and geographical allocation, and season/area closures. Since country quotas are determined by ICCAT, the greater part of domestic management effort has been directed at allocation issues; thus, management efforts are likely to be substantially reduced under an ITQ system. Because enforcement at the individual vessel level would be necessary for ITQ's, monitoring costs can be high. However, monitoring could be accomplished through issuance of fin tags and from dealer reports, for which a system is already in place for bluefin tuna. Although initially costly, investment in automated reporting systems (such as those involving electronic data recording card systems) would offset these monitoring costs.

Summary of Consequences of Access Control Mechanisms

The three access control alternatives listed below Table 1 address socio-economic issues such as overcapitalization, access rights, the "race for the fish," and overall economic efficiency. Biological impacts are less relevant, and biological consequences are not considered, since it is assumed that an overall quota is pre-set based primarily on biological criteria, and the key issue is therefore one of allocation. The three alternatives considered vary widely in terms of outcomes such as equability, acceptability by commercial and recreational sec-

---

[17] Fishermen may be concerned about non-fishery interests owning quota, and/or a concentration in quota, because of the possible effects on the level of employment and the very nature of the fishery itself. If quota is held by larger economic entities (e.g., corporations) and/or in significant concentrations by sole entities, it would be difficult for smaller participants to win a price war (for both quota and fishery output), with subsequent changes in the nature of the fishery. Furthermore, concentration of the holdings can also lead to market inefficiencies due to monopolistic or monopsonistic effects.

tors, management costs, overall economic efficiency, and employment. It is believed that a correctly-designed ITQ system would maximize economic efficiency, while a lottery system or limited entry system might result in higher employment, although probably lower profitability and less stability.

Effects on Other Species

If fishing opportunities on bluefin decreased (due either to more restrictive quotas, or to further reductions in stock size, or both), it is likely that there would be an increase in effort on associated or alternative fisheries that are also currently managed under open access. Species that would probably be affected include yellowfin tuna, bigeye tuna, albacore, skipjack, other tunas, swordfish, sharks, dolphin, and wahoo. Since the ICCAT recommendations call for no increase in quota in 1996, it is unlikely that effort will switch to other species in the remainder of the year, unless domestic regulations call for reallocation.

Mortality to other species through gear entanglement and incidental catch is not considered a major source of mortality for those species affected. The primary bycatch species of concern include the blue marlin, white marlin, swordfish, and the endangered loggerhead turtle, whose populations are very low.

The U.S. longline and drift gillnet industries' catch of targeted fish, non-targeted fish and turtles, including endangered species, in the western Atlantic, Gulf of Mexico and Caribbean Sea from 1987 through 1995 has been estimated by Cramer (1996) using observer records from representative vessels. Reproduced in Table 15, is the estimated catch for 1995 (as an example), including the number and percentages of animals released alive and discarded dead. Bluefin represent a very small (but very valuable) proportion of the catch. Endangered species are rarely caught by longlines and almost all are released alive. However, drift gillnets do catch significant numbers of both endangered and threatened species as well as marine mammals—most of which are killed before they can be released. Cramer's data also show a substantial incidental catch by longliners of blue marlin, white marlin and other billfish, which by law may not be retained. Unfortunately, the vast majority of these premiere game fish die before they can be released or die shortly thereafter. Their take by the longline industry is a significant problem since their populations have been driven to very low levels by the longline industry. Almost all billfish caught by sport fishing vessels are immediately released alive for conservation purposes.

Conclusions

The bluefin tuna fishery in the United States is conducted on a scientific monitoring quota. While this fishery is of considerable regional economic importance for both commercial and recreational fisheries, the primary purpose of the quota is to obtain biological and CPUE data for monitoring the stocks of western ABT. NMFS has conducted and/or funded research, and implemented domestic regulations, which recognize the scientific information that should be gleaned from the fishery.

Determining the appropriate western Atlantic quota, for which ICCAT is responsible, involves a trade-off between short-term economic gains and long-term rebuilding of the bluefin tuna resource. The higher the quota taken in the short term, the higher the immediate profits from the fishery, and the slower the rate of stock recovery. Short-term profits must also be balanced against increased revenues in the future from a rebuilt stock. The results in terms of stock dynamics are more straightforward: the smaller the quota, the faster the rate of stock rebuilding.

Domestic allocations of the scientific monitoring quota also have an effect on the rate of rebuilding. For a given total quota, large allocations to the small fish fishery are projected to result in somewhat slower rates of stock rebuilding; however, the net economic benefits from the small fish fisheries are substantial. In addition, the CPUE indices generated from the small fish fishery are critical to following the entire range of age classes in the ABT fishery. ICCAT's allowance of 8 percent of the national quota for fish in the 27" to 45" range is recognition of the need for data on these age classes for stock assessment purposes.

With the wide range of user groups, its economic importance, and its many biological features, the bluefin tuna fishery is one of the most complex fisheries managed by NMFS. NMFS continues to strive for research and regulatory mechanisms that will maximize the information that can be generated from the fishery, in an effort to improve our knowledge of this resource, and thus contribute to its recovery.

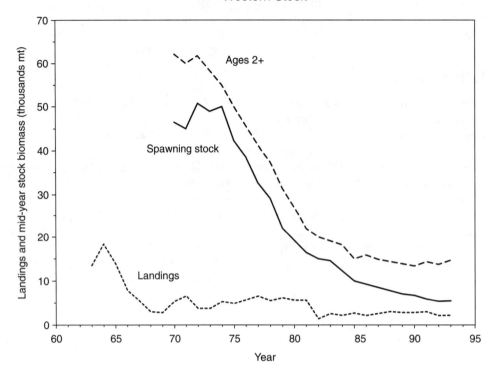

Figure 1. Trends in landings, mid-year spawning biomass and mid-year biomass of ages two and older western Atlantic bluefin tuna. Source: SCRS base case stock assessment from 1994 SCRS Report (ICCAT 1994).

TABLE 1. Summary matrix of major alternatives considered for Atlantic bluefin tuna. Ticks indicate the combinations of quota and domestic allocation alternatives that have been examined analytically, using a computer model. In addition, the consequences of three access control alternatives are examined qualitatively. The access control alternatives essentially represent a third dimension of the matrix, in that they could be applied in combination with any of the individual cells.

Domestic Allocation Alternatives Beginning in 1995 ↓	Quota Alternatives			
	Decrease to 1,995 mt (A)	Status quo: 2,200 mt (B)	Increase to 2,660 mt (C)	No fishing indefinitely (D)
Increase the Purse Seine category by 51 mt (A)	√	√	√	√
Status quo (B)	√	√	√	
Closure of the Purse Seine fishery (C)	√	√	√	
Elimination of the Gulf of Mexico longline fishery (D)	√	√	√	
Elimination of the small fish fishery (E)	√	√	√	

Access Control Alternatives:

Three alternatives for controlling access to the U.S. Atlantic bluefin tuna fishery are also considered in a descriptive format:

    (A) Lottery system
    (B) Limited entry with fleet-wide quota
    (C) Individual transferable quota (ITQ) system.

TABLE 2. U.S. Commercial Atlantic Tuna Landings, 1991–1993

Tuna Species	Commercial Landings (Metric Tons)		
	1991	1992	1993
Yellowfin	5,599	6,195	3,788
Bigeye	889	663	874
Skipjack	786	509	290*
Albacore	229	274	260
Other Tunas	681	587	NA

Source: National Report of the United States: 1994.

TABLE 4. Atlantic Bluefin Tuna Catch by Fishing Area, 1991–1994 (thousands of pounds round weight)								
Fishing Area*	1991		1992		1993		1994	
	Landings	% of Total	Landings	% of Total	Landings	% of Total	Landings	% of Total
1	10.34	0.4%	13.64	0.6%	23.24	1.0%	17.87	0.7%
2	423.48	17.4%	698.32	30.5%	1,033.61	44.7%	745.83	30.8%
3	518.09	21.3%	391.52	17.1%	423.40	18.3%	530.98	21.9%
4	411.04	16.9%	649.19	28.4%	329.42	14.3%	272.63	11.2%
5	622.82	25.6%	204.79	9.0%	317.50	13.7%	639.57	26.4%
6	21.25	0.9%	22.70	1.0%	26.43	1.1%	16.88	0.7%
7	41.19	1.7%	50.11	2.2%	30.82	1.3%	55.29	2.3%
8	9.82	0.4%	10.20	0.4%	10.00	0.4%	36.68	1.5%
9	84.79	3.4%	17.77	0.7%	1.83	0.1%	7.30	0.3%
10	275.90	11.3%	229.04	10.0%	113.91	4.9%	102.35	4.2%
Total	2,436.38		2,287.78		2,310.15		2,425.38	

Source: 1991–1994 NMFS Atlantic bluefin tuna Dealer Report Database.

TABLE 5. 1995 and 1996 Atlantic Bluefin Tuna Quota Allocations by Category

Permit Category	1995 Quota	1996 Quota
General	438	541
Harpoon	47	53
Incidental	125	110
Other	2	1
Longline	123	109
North	23	23
South	100	86
Purse Seine	250	251
Commercial	860	955
Angling	330	243
Lg Schl/Sml Med	178	100
School	148	138
North	78	73
South	70	65
Trophy	4	5
Total Allocated	1190	1198
Reserve	145	108
Total U.S. ICCAT	1335	1306

TABLE 6. Atlantic Bluefin Tuna Landings by Year and Category (metric tons), 1983–95

Category	Year 1983	1984	1985	1986	1987	1988	1989	1990	1991	1992	1993	1994	1995
General	743	642	690	395	401	400	627	645	624	535.6	608.7	642	558
Harpoon	73	68	74	67	56	74	62	39	59	58.4	56.6	59	56.5
Purse Seine	374	398	377	360	367	383	385	384	236	300.0	295.3	301	249
Incidental	116	132	133	130	139	152	112	137	177	136.7	84.9	94.4	72.1
No. LL*	25	37	12	14	8	2	31	3	8	18.4	26.5	27.9	31
So. LL	91	92	120	115	130	149	80	133	168	117.2	56.7	64.13	39.9
Other	0	3	1	1	1	1	1	1	1	1.1	1.7	2.26	1.2
Angling	65	105	149	202	426	277	228	486	431	134.5	297.0	111.9	401.9
TOTAL	1371	1345	1423	1154	1389	1286	1414	1691	1527	1165.2	1342.5	1302.6	1409.6

Sources: Landings data from Northeast Region mandatory dealer report program, except for Angling category landings which are survey-derived. Note that General category figures include school and medium fish sold by General category permit holders (up to July of 1992), and that Angling figures thus reflect school and medium fish caught and/or sold by non-permit holders.

Note: U.S. quota in 1982 was 605 mt. From 1983 to 1991, U.S. quota was 1,387 mt. For 1992–93, U.S. annual quota, excluding angling, is set at 1,022 mt.

*LL indicates longline gear. No north/south split in 1982.

TABLE 7. Summary of Traditional Patterns of Fishing Activities Directed at Atlantic Bluefin Tuna in the United States

Gear	Area	Size of Fish	Traditional Season
Handline, Harpoon, and Rod and Reel	Cape Cod Bay and Gulf of Maine	Giant	June–September
		Medium	August–October
		Small	Summer (unpredictable)
	Cape Hatteras to Cape Cod	Small	June–October
		Medium	June–October
		Giant	August–October
	Gulf of Mexico	Giant	January–June
Purse Seine	Cape Hatteras to Cape Cod	Large Medium and Giant	August–October
	Cape Cod Bay	Large Medium and Giant	August–October

TABLE 8. Atlantic Bluefin Tuna Gross Revenue, 1992–94 (in Thousands of Constant (1994) Dollars)

Category	1992	1993	1994
General	$8,979	$11,946	$12,398
Harpoon	$738	$769	$1,591
Purse Seine	$4,842	$5,141	$5,249
Incidental	$1,122	$956	$1,364
TOTAL	$15,681	$18,812	$20,602

Source: 1992–1994 Dealer Report Database. Figures for 1992 and 1993 are adjusted to 1994 dollars using the Consumer Price Index (Bureau of Labor Statistics) in order to evaluate the impact on fishermen's purchasing power.

TABLE 9. Atlantic Bluefin Tuna: Fleet Size and Success Rate, 1993 and 1994

Category	1993		1994	
	Number of Permitted Vessels	Success Rate (% of Vessels landing at least one fish)	Number of Permitted Vessels	Success Rate (% of Vessels landing at least one fish)
General	9,336	8%	11,643	8%
Harpoon	132	26%	108	34%
Purse Seine	5	100%	5	100%
Incidental	607	27%	2,214	8%
Angling	NA	NA	15,383	NA
TOTAL	10,081*	10%	17,710**	7%

Source: NMFS Permit Office, Gloucester, MA
*Actual landings may differ from the quota due to overages, underages, and in-season transfers.
**Total number of permits less than the sum, since all General category permit holders also hold Angling category permits. Some double-counting still occurs since the Incidental category includes Angling category permit holders.

TABLE 10. Angling Fleet Size Estimates in 1993, by State*

Area	Estimated Fleet Size**
North Carolina	820
Virginia	1,364
Maryland and Delaware	818
New Jersey	1,864
New York	1,394
Connecticut and Rhode Island	709
Massachusetts	857
TOTAL	7,826

Source: NMFS 1993 Large Pelagic Recreational Fishery Survey
*All figures are preliminary
*Estimated with mark-recapture procedures; there is no reported estimate for North Carolina, where exit counts were used.

Questions for Discussion

For Reading 1

1. How many sides are there to this issue? Identify as many agencies and interest groups that contribute to this issue as you can. How do they interact (who are allies, who are antagonists)?

2. Write a brief paragraph describing the positions of the various sides identified in question 1. In doing so, try not to take sides yourself.

3. Based on what you have read in this reading, how well is ICCAT doing in its management of bluefin tuna in the Western Atlantic? What evidence points to that conclusion?

For Reading 2

1. What is the relationship between ICCAT and NMFS? (Summarize each agency's responsibilities.)

2. NMFS sees itself as "caught between a rock and a hard place" with respect to this issue. What contributes to that perception?

3. Briefly describe the various user groups (types of fishers) in the United States who exploit bluefin tuna regulated by NMFS.

4. Briefly describe how bluefin tuna are currently being regulated. How are these regulations imposed on U.S. fishers?

5. As presently managed, what is the relationship between those who catch bluefin tuna (fishers) and those who study them (researchers)? What are the strengths and weaknesses of this relationship?

6. NMFS apparently believes that the Western Atlantic stock of bluefin tuna is improving under its management. What evidence can be cited to support this conclusion? (Can you see any evidence in Figure 1 to support the conclusion?) Can you cite any conflicting evidence?

7. Are the twin goals of "save the fish" and "save the fishery" compatible? Can we do both? If not, which should take precedence?

Going Beyond the Readings

Visit our Web site at <www.harcourtcollege.com/lifesci/envicases2> to investigate wildlife management issues in other regions of the United States and Canada.

unit 12

Northwest: Toxic Wastes and Pollution

Introduction
Readings
Reading 1: The Two Faces of the Exxon Disaster
Reading 2: 1999 Status Report on the *Exxon Valdez* Oil Spill
Reading 3: 2000 Status Report on the *Exxon Valdez* Oil Spill
Questions for Discussion

What If There Were Another Oil Spill in Prince William Sound, Alaska?

Introduction

To many Alaskans, March 23, 1989, was a day of infamy—the day their worst-case scenario became stark, painful reality. On that fateful day, the supertanker *Exxon Valdez* left Valdez harbor and struck an underwater reef, leaking more than 10 million gallons (38 million liters) of crude oil into Prince William Sound. The Sound is a pristine wilderness about twice the size of the state of Delaware. Commercial and sport fishers, kayakers, and sailors plied its waters. Hunters sought trophy bear, sheep, and goats on the Sound's spectacular mountainous islands. Wilderness campers sought solace on isolated beaches. Suddenly a crown jewel of what everyone loved most about Alaska was being defiled by a thick creeping carpet of oil.

How could such a thing happen? What about all those wonderful safeguards promised in the 1970s, before construction started, that made a major spill seem nearly impossible? The oil industry and state and federal governments promised that tankers operating in Prince William Sound would be double hulled; even if a tanker struck something hard enough to penetrate one hull, another hull would protect against spillage. Only qualified, licensed pilots would operate tankers within the Sound. The Coast Guard would monitor tankers from the time they left Valdez harbor until they reached open ocean 40 miles (65 km) to the south. The state of Alaska would test every tanker crew for alcohol and drug use before they left the harbor. A crack spill response team, properly equipped and trained, would be constantly on standby in Valdez.

What went wrong? Basically, it's hard to maintain a state of readiness when nothing seems to be happening. For 12 years, the Trans-Alaska Pipeline System worked flawlessly as more than 8700 loaded tankers left Valdez without serious incident. As the record of successes stretched on and on, vigilance relaxed. One by one, with hardly anyone noticing, safeguards were eased, overlooked, or discarded. Then, on March 23, 1989, a tanker with an experienced crew who had made the trip many times before strayed off course, hit a submerged reef that was well marked on charts, and leaked its oil into the Sound.

What happened next would be comical if it were not so tragic. The crack response team had recently sent their boat off for repairs. An alternate had to be found. Much of their other equipment was being stored in various locations, as were barges on which to put the equipment. By the time arrangements had been made, with gear and crew assembled, nine precious hours had been wasted. Lightering vessels were able to save nearly one million gallons (4 million liters) of oil directly from the tanker. The rest oozed out on the water. Floating booms, deployed to contain the spill, were quickly overwhelmed by wind and waves as weather deteriorated.

Worst of all, no one seemed to be in charge. Was it the responsibility of Exxon Oil Company, who owned and operated the tanker? Or Alyeska Pipeline Company, who was supposed to maintain the spill response team? Or a plethora of state and federal agencies who had various responsibilities in the area? During the first few critical days after the spill there was much saber rattling, but little effective action. Orders given by one agency were countermanded by another. Some wanted to burn the oil, but this would turn water pollution into air pollution. But wasn't that what would become of the oil anyway? Yes, but over a period of time, not all at once. Some wanted to use dispersants to break up the oil, combine with it chemically, and sink it out of sight. That would save beaches and wildlife, but put oil and toxic dispersants into the water column and food chains. Which would be worse, soiled beaches or fouled food chains? No one seemed to know. While these issues were being debated, oil spread throughout the Sound. Eventually, it stretched 750 miles (1200 km) west and south of the spill site, fouling more than 1200 miles (1900 km) of beaches.

Oiled wildlife found alive, mostly sea birds and sea otters, were transported to treatment sites where they were cleaned, nursed back to health, and released. Relatively few were found and even fewer survived. In an area of so few people, most contaminated wildlife died and sank from sight.

As containment failed, attention switched to cleanup. Exxon promised to do whatever was necessary to return the sound to its pre-spill, pristine state. They tried. Their first immediate task was to clean up the beaches. They launched a massive effort. First, crews were sent out to sop up and shovel up as much beached oil as possible. Their next task was to remove the remaining oil that still covered beach rocks, gravel, and driftwood. Nothing like this had ever been tried before. A new technology to treat fouled beaches was literally invented on site. Huge floating boilers heated sea water to near boiling, pumped it through huge booms, and sprayed it on the beaches. Hot water carried oil as it ran off into the sea, there to be contained by booms and skimmed off the surface.

The undertaking was massive. Exxon brought in 1767 tons (1600 metric tons) of gear and hired more than 10,000 people to conduct an extensive cleanup using a technology that had never been field tested, working in an area with few towns, roads, or human amenities. Steam clean beaches in a pristine wilderness? It appeared to work surprisingly well. By winter, most of the soiled beaches had been treated. For the next two years, scaled-down cleanup efforts continued on beaches that had been missed or inadequately treated.

Important policy changes were quickly implemented to lessen the likelihood of a similar spill in the future. Today, tanker crews are isolated and tested for drugs and alcohol before every sailing. The spill response team is extensively trained and their efforts monitored and evaluated by outside experts. Every tanker that leaves Valdez harbor is accompanied by two large tugs. Any tanker that strays off its course is pushed back on.

The total cost of the cleanup to Exxon was in excess of $2.5 billion. That was only the beginning of their expenses. Even before the cleanup was complete, numerous lawsuits were launched against them. Separate suits were initiated by the federal government, the state of Alaska, and several private agencies and citizens groups. Years of litigation ensued.

Meanwhile, as cleanup efforts ended, damage assessment and restoration activities blossomed. For the next several years, teams of scientists, some funded by Exxon, others by federal and state sources, looked at every component of the Prince William Sound ecosystem. They determined the extent to which natural resources had been damaged and monitored how well or poorly they recovered. These efforts included human impacts as well. Prince William Sound supported a commercial and sport fishery and a substantial population of Native Americans. These groups depended extensively on the Sound's natural resources.

Unfortunately, the findings of these studies were not made available to the general public, who were initially eager for information. With so much litigation in the air, lawyers representing all sides sought and received gag orders on all relevant studies until after suits were settled. It was not until the mid-1990s that information gathered in those first critical years after the spill was released to the public. To what extent were the birds, fish, mammals, and waters of the Sound affected by the spill? How well have they recovered? Was the money obtained from Exxon—voluntarily and through fines and litigation—well spent? These are the questions addressed in the readings that follow.

reading 1

"The Two Faces of the Exxon Disaster"
Stephanie Pain

New Scientist, May 22, 1993. Reprinted by permission of the *New Scientist*, London.

Every oil spill is a disaster for wildlife. But the Exxon Valdez spill in March 1989 caused more public anger and dismay than any other. Exxon's supertanker ran aground in one of the world's most beautiful places. Prince William Sound was a remote fiordland surrounded by mountains and glaciers, teeming with sea otters, seals and birds. Its waters provided huge harvests of salmon, halibut and herring.

Exxon's grounded tanker leaked 35,000 tonnes of crude oil into these productive waters. About 40 per cent of it ended up on the beaches of Prince William Sound. Another 10 per cent came ashore along the coast of the Alaska Peninsula and the islands of the Gulf of Alaska. After eight weeks, the oil had travelled 750 kilometres.

For months, newspapers, magazines and television news bulletins were filled with images of oiled birds and dying sea otters. Documentaries and even a TV drama have rerun the events that followed the grounding many times, leaving Exxon with a severe image problem.

Last month, Exxon tried to counter the bad publicity by announcing that its researchers, many of them the country's foremost experts, had found that the damage from the spill lasted only a few months and that Prince William Sound had recovered almost completely. There are some striking differences between this view and the picture painted two months earlier by US government scientists who have spent four years assessing the damage. They said that while most of the shoreline is now clean, and many species are well on the way to recovery, there are still reservoirs of oil and some wildlife continues to suffer.

"There was a massive impact with massive mortalities. It's very clear that long-term recovery is far from complete. In some cases it will take many years," says Doug Wolfe of the National Oceanic and Atmospheric Administration (NOAA), one of the agencies monitoring the spill's effects.

But Exxon attributes this view to the potent images of dying animals. "These images persist in the minds of the public—and in the minds of the scientists working on the spill," says John Wiens, a sea bird expert from the University of Colorado hired by Exxon. "The scientists who went there early on in the spill all thought it would be a disaster—and that impression lasts."

Exxon maintains that its researchers have been more objective. Their studies have been designed according to strict protocols and their data subjected to rigorous statistical testing. "Exxon has programmes that are well designed and statistically sound so that we can demonstrate if an effect is due to hydrocarbons or not," says Al Maki, the company's chief scientist.

But if Alaska's scientists have been influenced by the sights and smells of the early days of the spill, Exxon still has a pack of lawyers at its heels. Although the company paid $1.1 billion in fines and damages to the state and federal governments in March 1991, the legal wrangling is not over yet. Exxon still faces private claims from fishermen for lost earnings and from native Alaskans for the loss of their traditional lifestyle. These claims amount to $2.6 billion, and are due to be heard in 1994. And, if the environmental damage proves to last longer than anticipated when the government settlement was thrashed out, there is a "reopener" clause, which could leave Exxon open to further claims.

Perhaps more important in the long run, the regulations for the Oil spill Pollution Act 1990 are now being drawn up. The Exxon spill promoted the act, and the strictness for the regulations could depend on how damaging the spill eventually turns out to have been.

In February, the Oil Spill Trustees, the six-member board which has the job of overseeing the billion-dollar settlement from Exxon, presented the findings of all the studies carried out by the government's scientists. The oil company sent a posse of observers to the meeting in Anchorage, but turned down the invitation to present its findings on what it considered hostile territory. At the end of April, Exxon finally went public at a meeting of the American Society for Testing and Materials (ASTM), in Atlanta.

No one was surprised that Exxon's perception of what has happened in Prince William Sound since 1989 was rosier than that of the trustees. But speaker after speaker stood up and presented findings that apparently contradicted those described in Anchorage. The "claims" made by Alaskan researchers were countered by "facts" from Exxon.

The researchers working for the trustees accuse Exxon of being selective with its data, ignoring problem areas by sticking doggedly to sites selected at random, while in the case of guillemots, for example, making sweeping statements about the birds' recovery based on findings from a single, small site. "Exxon is picking and choosing the information it is using to assess recovery," said Carol Ann Manen of NOAA.

Manen also accused Exxon of censoring its own researchers. She was to have presented a joint paper in Atlanta with a researcher working under contract to

Exxon. The paper looked at the way the two sides had managed "quality assurance" of their data "and would have provided some balance to the proceedings," she said. The ASTM accepted the paper, said Manen, but "Exxon called my coauthor and said 'withdraw.'" Exxon denies censorship and says it did not have time to look at the paper in detail before the conference.

One of Exxon's main assertions is that despite initial fears that the oil might take years to break down in the icy Alaskan waters, it actually disappeared very quickly. Hydrocarbon specialists from the consulting company Arthur D. Little found that "after a few months" the oil in the water column was no longer toxic to animals. Hardly any of the oil ended up on the sea floor and, by 1990, what remained on the beaches was harmless, they say.

Besides, argues Hans Jahns, one of Exxon's most senior scientists, Prince William Sound was never a pristine, oil-free environment. Natural seeps of oil around the Alaskan coast provide a background of hydrocarbons in sediments on the sea floor. And intensive fishing pollutes the water with diesel fuel. With all this oil about, "the animals of the sea floor are used to living with hydrocarbons," says Jahns. And there is a rich supply of bacteria that feed on hydrocarbons, cleaning up oil much faster than might have been expected.

After the first year, these other sources of oil far outweighed any residues from the Exxon Valdez. "By 1990 the background signal was much higher than Exxon Valdez crude, even next to heavily oiled shores," said David Page of Bowdoin College in Maine. Many samples the trustees said were oil from the Exxon Valdez contained oil from these other sources, claims Exxon.

But in studies carried out ten years ago, NOAA found no trace of oil from the seeps in sediments close to the shore. "Exxon's claims are plausible for deep sediments but not for nearshore, shallow sediments," says Jeff Short of NOAA.

Government researchers agree that in most parts of the sound there is little oil left. But reservoirs of liquid oil remain trapped in the gravel of some beaches or beneath dense beds of mussels. Exxon admits there are a few isolated pockets of oil, but dismisses them as a "specific, limited phenomenon."

In 1992, the total area of sediment that still contained liquid oil from the tanker was less than the size of a basketball court, says Jahns. More controversially, he suggests that the oil is harmless: "It's wrong to call this oil 'fresh' or 'unweathered.' Essentially it's nontoxic—weathered and not hazardous."

NOAA's biologists who have been monitoring the health of mussels since the spill, believe their evidence proves otherwise. They found high concentrations of hydrocarbons in the flesh of mussels a year after the spill, suggesting that they were still taking up oil from

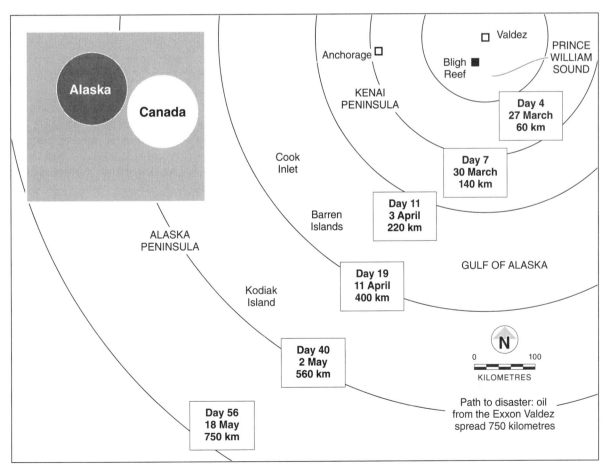

the environment. In 1991, mussels from a clean beach were moved to the oiled site—and within two months had accumulated significant levels of hydrocarbons in their tissues. The transplanted mussels were still accumulating hydrocarbons in 1992.

Exxon's dismissal of the "hot spots" of contamination is one of the main bones of contention between the oil company and the trustees. The company insists that because its studies are rigorously objective and subject to stringent statistical testing, it is reasonable to draw conclusions about the whole region from them.

But the company missed the worst sites, say the trustees. And while these may be small and scattered, they are important locally and to the animals most likely to come into contact with them. According to Malin Babcock of NOAA, even late last summer the action of waves was enough to cause sheens of oil to leak from the mussel beds into the water. These pools could be responsible for problems seen in animals that feed on mussels, particularly the harlequin duck, which has failed to breed every year since the spill.

Exxon complains that people will not believe the ecosystem has recovered until it is restored to the state it was in before the spill. This is not a valid comparison, say many of its scientists. "The need to see things get back to the position prior to the spill is misconceived," says Wiens. "You must assess the perturbation against a backdrop of natural variation. There's a danger of attributing everything that follows the spill to the spill." It is not easy to design a study that shows when a site or a population has reached a stage where it can be said to have recovered. Simply comparing a population before and after the spill does not allow for often large annual differences that are caused by factors such as a food shortage or an unusually warm year. And for most parts of the sound, data collected before the spill are few and far between.

Comparisons of locations that were heavily oiled with places that escaped the spill are also open to criticism. The sites, and the creatures that live there, can be subject to a variety of factors that are not easy to untangle.

Some of Exxon's criteria for recovery failed to convince the trustees. According to Exxon, the communities living in the intertidal zone made a rapid recovery. Rockweed and barnacles for example began to recolonise even the most heavily oiled beaches in the summer of 1989. By 1990, new recruits were arriving thick and fast. But this does not necessarily signal recovery, argues Bob Spies, the trustees' chief scientist. If Exxon had checked the shores later in the year, it would have found that many of the new arrivals had died.

Existing But Not Thriving

Similarly, the return of shore birds to once badly oiled parts of the sound was a sure sign that they regarded these places as habitable, says Robert Day of Alaska Biological Research. Not so, says Michael Fry, a sea bird specialist from the University of California at Davis, who works for the trustees. He points out that birds are so attached to their traditional haunts that they will return under almost any conditions, existing but not thriving.

The animals that did most to harm Exxon's corporate image were the shivering sea otters and sea birds collected by the sackful under the constant eye of the TV cameras. Not surprisingly, the fate of these animals was the subject of some of the fiercest argument in Atlanta.

An estimated 4000 sea otters died in the aftermath of the spill. They died form hypothermia as the oil clogged their fur, destroying their insulation against the cold, from emphysema after breathing in toxic fumes or from poisoning after ingesting oil.

According to the US Fish and Wildlife Service, in 1989 the number of otters in the oiled part of the sound fell by 35 per cent. And only 13 per cent of the pups in the area survived to the following spring, compared with 36 per cent in the unpolluted part of the sound. "This was an alarming difference and did not bode well for the population," says Brenda Ballachey, who works for the service. The numbers of otters remained low in 1990 and 1991.

Ballachey and her colleagues also found that an unusually large number of otters in their prime were dying. Before the spill, about 15 per cent of the otters that died each year were mature animals of breeding age. In 1989, this figure leapt to 44 per cent, and stayed high at 43 per cent in 1991. This could mean that these animals were showing a delayed response to the spill, or that they were showing the effects of chronic exposure.

"By the end of 1991 we didn't feel very comfortable about the status of sea otters in Prince William Sound," says Ballachey. In 1992, however, only 22 per cent of dead otters were mature adults and pup survival increased to 50 per cent, suggesting that recovery may finally have begun.

Exxon's findings bear little resemblance to these. The company's otter expert, David Garshelis, of the Minnesota Department of Natural Resources, says by 1991 there were just as many, if not more, otters than had been counted in surveys in the early 1980s—putting them well within the range of "historical variation." He suggests that the population was expanding through the 1980s and that the spill simply cut the numbers back to where they have had been in the mid-1980s. He also says the death toll was probably much lower than estimated.

While things look more hopeful for sea otters, the trustees are less optimistic about the future of guillemots. Around 20,000 guillemot carcasses were collected after the spill, and government biologists reckon the death toll could have been as high as 300,000.

The oil arrived just as the breeding adults were gathering in the water in "rafts," before heading off to their nesting sites. Some rafts were engulfed by oil. Between 1989 and 1991, the numbers of birds at colonies in the path of the spill were reduced by between 40 and 60 per cent, while those outside it were stable, says the Fish and Wildlife Service.

The service's biologists also found that the guillemots' nesting behaviour changed dramatically. For the past three years, the birds have been laying eggs on average 45 days late. With fewer pairs at the colonies, eggs and chicks are more vulnerable to predators such as gulls. Many chicks that survived were abandoned as winter approached or were swept to their deaths by winter storms. After the spill the number of chicks fledged fell from an average of 0.56 per pair to just 0.09.

The best explanation for this disruption, says Fry, is that because most of the mature breeding birds were killed, the birds returning to the colonies are young and inexperienced. Without the right cues from mature birds their breeding is poorly synchronised. If such a late breeding pattern becomes established the prospects for these colonies are poor, says Fry. If the disruption is temporary, the colonies could take between 20 and 70 years to recover.

Again, Exxon's biologists differ. There are just as many birds as recorded in historical surveys, and they are breeding normally, says Exxon. This assertion is based mainly on a study by Dee Boersma from the University of Washington in Seattle. She studied the guillemots of Light Rock, an outlier of the Barren Islands, which lay directly in the path of the oil. In particular, she monitored a small patch of the rock which had been studied in the 1970s. Recording the birds at their nests with time-lapse photography, she found just as many chicks as two decades before.

Even if the numbers for Light Rock are accurate, they are not necessarily representative of the whole region, says Fry. "I don't really understand why Light Rock should be any different from the rest of the sound—but you can't extrapolate from a 5 by 5 metre square to the whole of the spill area."

While most public concern focused on otters and birds, much of the local and legal interest was in fish, particularly pink salmon. In this case, discrepancies between the two accounts are being analysed in minute detail by the region's fishermen and their legal advisers. Pink salmon form the basis of a multimillion-dollar industry and a large part of the diet of native villagers. The fish spawn at the mouths of streams that run into the sound.

Biologists from Alaska's Department of Fish and Game say as many as 213 salmon streams were contaminated by oil from the Exxon Valdez. Hal Geiger estimates that 1.6 million fish that hatched in 1989 were lost—6 percent of the salmon in the contaminated zone.

Alarm bells began to ring louder when the death toll of pink salmon eggs increased in 1991 and 1992; 40 percent of eggs laid in oiled streams died, twice as many as in clean streams. Pink salmon have a two-year life cycle so these eggs were laid by fish that hatched in 1989 and 1990. Brian Bue, also from the fish and game department, suggests that these survivors may have suffered some form of sublethal damage, "possibly some form of genetic damage." Exxon says there is no evidence for this.

The difference between the two sides is not so much what happened to the fish in 1989 but whether it makes any difference in the long run. Exxon believes not. Salmon returned in record numbers in 1990 and 1991. Most of these were fish from hatcheries which were protected from oil; but the number of wild fish returning was slightly higher than in the previous two years. The Alaska Department of Fish and Game is waiting to see. "If a species lays a lot of eggs, does a 10 per cent extra mortality matter?" asks Spies. The jury is still out on that question.

The Atlanta meeting has left a nasty taste in many scientists' mouths. "What bothers me is everything we've done is being called a 'claim'," says Karen Oakley, a bird biologist with the Fish and Wildlife Service. "They're not claims, we have data to support them." Exxon, on the other hand, feels this was its first chance "to rebut some of the more exaggerated statements" about how the sound continues to suffer.

Privately, many scientists expressed a desire to talk to their opposite numbers to work out why their results differ, without the constant scrutiny of lawyers and company executives. Some say the discrepancies are not really as big as they seem, but while the trustees choose to highlight the worst damage. Exxon and its army of public relations experts ignored it. The problem, says Fry, is that the law requires both sides to carry out their own assessment of damage. "The way the studies are done is not impartial. The science is driven by lawyers, who decide which studies will support claims for damages—or which will help to counter the claims."

reading 2

1999 Status Report on the *Exxon Valdez* Oil Spill: "Legacy of an Oil Spill: 10 Years After *Exxon Valdez*"

Exxon Valdez Oil Spill Trustee Council: 1999 On-line Status Report.* Reprinted by permission.

What Was the Settlement With Exxon?

The settlement among the State of Alaska, the United States government and Exxon was approved by the U.S. District Court on October 9, 1991. It resolved various criminal charges against Exxon as well as civil claims brought by the federal and state governments for recovery of natural resource damages resulting from the oil spill. The settlement had three distinct parts:

Criminal Plea Agreement. Exxon was fined $150 million, the largest fine ever imposed for an environmental crime. The court forgave $125 million of that fine in recognition of Exxon's cooperation in cleaning up the spill and paying certain private claims. Of the remaining $25 million, $12 million went to the North American Wetlands Conservation Fund and $13 million went to the national Victims of Crime Fund.

Criminal Restitution. As restitution for the injuries caused to the fish, wildlife, and lands of the spill region, Exxon agreed to pay $100 million. This money was divided evenly between the federal and state governments.

Civil Settlement. Exxon agreed to pay $900 million with annual payments stretched over a 10-year period. The settlement has a provision allowing the governments to make a claim for up to an additional $100 million to restore resources that suffered a substantial loss, the nature of which could not have been anticipated from data available at the time of the settlement.

What is the Restoration Plan?

The Trustee Council adopted a Restoration Plan in 1994 after an extensive public process. More than 2,000 people participated in the meetings or sent in written comments.

As part of the settlement agreement, $173.2 million went to reimburse the federal and state governments for costs incurred conducting spill response, damage assessment, and litigation. Another $39.9 million went to reimburse Exxon for cleanup work that took place after the civil settlement was reached. The remaining funds were dedicated to implementation of the Restoration Plan, which consists of five [sic] parts:

Research, Monitoring, Restoration $180 million

Surveys and other monitoring of fish and wildlife in the spill region provide basic information to determine population trends, productivity, and health. Research increases our knowledge about the biological needs of individual species and how each contributes to the Gulf of Alaska ecosystem. Research also provides new information and better tools for effective management of fish and wildlife populations.

General Restoration includes projects to protect archaeological resources, improve subsistence resources, enhance salmon streams, reduce marine pollution, and restore damaged habitats.

Habitat Protection $392 million

Protection of habitat helps prevent additional injury to species due to intrusive development or loss of habitat. The Trustee Council accomplishes this by providing funds to government agencies to acquire title or conservation easements on land important for its restoration value.

Restoration Reserve $108 million

This savings account was established in recognition that full recovery from the oil spill would not occur for decades. The reserve fund will support long-term restoration activities after the final payment is received from Exxon in September 2001. The reserve is expected to be worth approximately $140 million by that time.

Science Mgmt., Public Info. & Admin. $31.1 million

This component of the budget includes management of the annual work plan and habitat programs, scientific oversight of research, monitoring and restoration projects, agency coordination, and overall administrative costs. It also includes the cost of public meetings, newsletters and other means of disseminating information to the public.

* The online version of the 1999 Status Report at <www.oilspill.state.ak.us> is a synopsis of the full, printed version of the Status Report.

Uses of Civil Settlement: 10 Years of Restoration
(in millions)

REIMBURSEMENTS FOR ASSESSMENT AND RESPONSE	**213.1**
Governments (includes litigation and cleanup)	173.2 (a)
Exxon (for cleanup after 1/1/92)	39.9
RESEARCH, MONITORING AND GENERAL RESTORATION	**180.0**
FY 1992—FY 1999 Work Plans	108.6
FY 2000—FY 2002 Work Plans (estimate)	35.0
Alutiiq Museum (Kodiak)	1.5
Archaeological Repository/Exhibits (PWS & Kenai Pen)	2.8
Alaska Sealife Center	26.2
Port Graham Hatchery	.8
Reduction of Marine Pollution	5.1
HABITAT PROTECTION	**395.3**
Large Parcel and Small Parcel habitat protection programs (past expenditures, outstanding offers, estimated future commitments and parcel evaluation costs)	
RESTORATION RESERVE	**108.0**
FY 1994—FY 1999	72.0
FY 2000—FY 2002 (anticipated)	36.0
SCIENCE MANAGEMENT, PUBLIC INFORMATION & ADMINISTRATION	**30.9**
FY 1992—FY 1999	24.7
FY 2000—FY 2002 (estimate)	6.2
TOTAL	**927.3**
Exxon Payments	900.0
Interest on Court Registry Investment System (minus fees)	27.3

(a) Reimbursement to governments reduced by $2.7 million included in the FY 1992 Work Plan

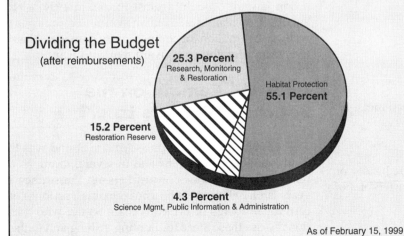

Dividing the Budget (after reimbursements)
- 25.3 Percent — Research, Monitoring & Restoration
- Habitat Protection — 55.1 Percent
- 15.2 Percent — Restoration Reserve
- 4.3 Percent — Science Mgmt, Public Information & Administration

As of February 15, 1999

Uses of Criminal Settlement
(in millions)

FEDERAL (highlights)

The federal government used most of its portion of the criminal settlement to help the Trustee Council fund habitat protection efforts, including:

Akhiok Kaguyak, Inc.	$10.0
Chenega	$10.1
English Bay	$ 1.3
Koniag	$ 7.0
Old Harbor	$ 3.3
Tatitlek	$10.0

(See table on Page 37 for more information)

Small Parcel acquisitions of habitat within:

Kodiak National Wildlife Refuge	$ 3.9
Chugach National Forest	$ 1.5
Kenai National Wildlife Refuge	$ .5

Other federal uses include:

Shoreline Monitoring	$ 3.4
Oil Spill Research	$ 5.6
General Restoration	$ .9

STATE OF ALASKA (highlights)

The State Legislature divided the money among capital improvements benefiting fisheries and research, habitat improvements, subsistence, and new recreational facilities.

Alaska SeaLife Center	$12.5
Kachemak Bay State Park	$ 7.0
Kachemak Bay St. Pk Visitor Center	$ .5
Seward Shellfish Hatchery	$ 3.5
Fort Richardson Hatchery	$ 4.0
State Park Recreational Facilities	$10.9
Kenai River Bank Restoration	$ 3.0
Main Bay Hatchery	$ 2.0
Fishery Industrial Technology Ctr.	$ 3.0
Subsistence Enhancements	$ 6.2
Spill Prevention/Response	$ 2.6
Tatitlek and Chenega Docks	$ .6
PWSAC Hatchery Operations	$ 1.8
Prince William Sound Science Center	$ .3
Shepard Point Road	$ 2.7
Kenai Visitors Center	$ 1.9
Fish Stock Identification	$ 1.0
Port Graham Hatchery	$ .5

Are Injured Species and Resources Recovering?

Ten years after the *Exxon Valdez* oil spill, it is clear that many fish and wildlife species injured by the spill have not fully recovered. It is less clear, however, what role oil plays in the inability of some populations to bounce back.

An ecosystem is dynamic—ever changing—and continues its natural cycles and fluctuations at the same time that it struggles with the impacts of spilled oil. As time passes, separating natural change from oil-spill impacts becomes more and more difficult.

Resources and Services Injured by the Spill

Recovering: Resources showing little or no clear improvement since spill injuries occurred.

Common loon
Cormorants (3 spp.)
Harbor seal
Harlequin duck
Killer whale (AB pod)
Pigeon guillemot

Recovering: Substantive progress is being made toward recovery objective. The amount of progress and time needed to achieve recovery vary depending on the resource.

Archaeological resources
Black oystercatcher
Clams
Common murre
Intertidal communities
Marbled murrelet
Mussels
Pacific herring
Pink salmon
Sea otter
Sediments
Sockeye salmon
Subtidal communities

Recovered: Recovery objectives have been met.

Bald eagle
River otter

Recovery Unknown: Limited data on life history or extent of injury; current research inconclusive or not complete.

Cutthroat trout
Designated Wilderness Areas
Dolly Varden
Kittlitz's murrelet
Rockfish

Human Services: Human services which depend on natural resources were also injured by the oil spill. The services below are each categorized as "recovering" until the resources they depend on are fully recovered.

Commercial fishing
Passive use
Recreation and tourism
Subsistence

Have the People of the Spill Region Fully Recovered?

The lives of the people who live, work, and play in the areas affected by the spill were completely disrupted in the spring and summer of 1989. Commercial fishing families did not fish. Those people who traditionally subsist on the fish, wildlife and plants of the region could no longer trust what they were eating and turned instead to high-priced groceries. Recreational opportunities were mostly shut down and the world-wide image of an attractive and pristine Prince William Sound was tarnished with oil.

Ten years later, a sense of normalcy is returning to the spill region, but residents, fishermen, and the tourism/recreation industry have not fully recovered.

The Trustee Council determined that the "human services" of commercial fishing, subsistence, recreation/tourism, and passive use will have recovered when the injured resources on which they depend are once again healthy and productive. Since that level of recovery has not been achieved, each of these services is considered to be recovering.

On a more personal level, it is clear that many people associated with the spill region have not been able to put the trauma behind them. Reasons for this continuing stress vary, but it is widely accepted that one major obstacle to recovery has been the protracted class action lawsuit brought against Exxon. In 1994, a jury awarded the plaintiffs $5 billion in punitive damages as well as $287 million for compensatory damages. Nearly five years later, this jury award remains on appeal.

Does Oil Remain on the Beaches 10 Years Later?

The western portion of Prince William Sound was the most heavily oiled in 1989 and in several locations oil remains on the surface or just beneath the surface 10 years later. Oil on some beaches remains a serious concern for residents of Prince William Sound who traditionally use these areas for hunting, fishing and gathering. Subsistence activities continue for residents of the sound, but generally they no longer use beaches that contain oil. Recreational users such as campers and kayakers also usually avoid these areas.

Exxon Valdez oil penetrated deeply into cobble and boulder beaches that are common on shorelines throughout the spill area, especially in sheltered habitats that don't receive much winter storm action. Cleaning and natural degradation removed much of the oil from the intertidal zone, but visually identifiable surface and subsurface oil persists at many locations.

A 1989-90 survey of nearly 5,000 miles of shoreline documented oil on approximately 1,300 miles of beach. The oiling was considered heavy or medium on 200 miles of shoreline. The remaining 1,100 miles of oiled shoreline were considered to have light or very light oiling. When crews returned to the beaches in 1993, they found hundreds of sites that contained substantial oil deposits.

What happened to the 10.8 million gallons of oil released into the environment? A 1992 National Oceanic and Atmospheric Administration study provided some insight, estimating that the great majority of the oil either evaporated, dispersed into the water column or degraded naturally. Cleanup crews recovered about 14 percent of the oil and approximately 13 percent sunk to the sea floor. About 2 percent (some 216,000 gallons) remained on the beaches.

In 1997, eight years after the oil spill, villagers from Chenega Bay returned to nearby beaches to clean some of the most heavily-oiled sites. Under the guidance of the Alaska Department of Environmental Conservation, the crew of mostly local residents applied a chemical agent to the weathered oil at five sites, along about one-half mile of beach on LaTouche and Evans islands. They used PES-51, a citrus-based product from the oil of oranges and lemons. PES-51 binds to the oil and floats, allowing both the chemical agent and the oil to be collected through the use of oil-absorbent pads.

One year later, preliminary analysis of targeted sites showed that the cleanup method was largely effective in removing the visible surface oil. But it had little effect on the large deposits of oil beneath rocks and overburden. Winter storms rearranged the beaches, exposing large quantities of oil that never received treatment. NOAA's Auke Bay Lab found no biological injury due to the cleanup.

What Habitat Has Been Protected?

The Trustees established two habitat protection programs. The Large Parcel Program protects blocks of land in excess of 1,000 acres. The Small Parcel Program recognizes the special qualities and strategic value of smaller tracts of land. Small parcels are typically located on coves, along important stretches of river, at the mouths of rivers, adjacent to valuable tidelands, and, often, close to spill-area communities. These properties are acquired for their habitat qualities as well as for their importance for recreational and subsistence use.

The Table on the following page shows the habitat protected through the Large Parcel Program.

What Have We Gained Through the Research, Monitoring, and Restoration Program?

In 1994, the Exxon Valdez Oil Spill Trustee Council set its sights high when it established its mission to restore Prince William Sound and the Gulf of Alaska to the "healthy, productive, world-renowned ecosystem" that existed before the spill. In doing so, the Trustee Council recognized that, in most cases, if protected from harm, injured species will recover on their own. Instead of direct intervention, such as rearing and releasing seabirds, the Trustee Council focused mostly on knowledge and stewardship as the best tools for fostering the long-term health of the marine ecosystem.

Since the Exxon Valdez settlement in 1991, hundreds of research, monitoring, and general restoration projects have been funded with an investment to date of roughly $109 million.

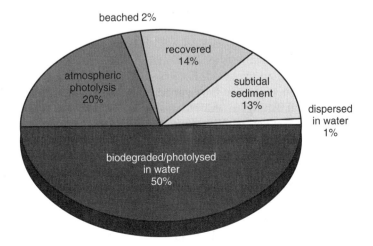

TABLE 1. Large Parcel Status

Parcel Description	Acreage	Coastal Miles	Salmon Rivers	Total Price	Trustee Council's Share
Acquisitions Complete					
Afognak Joint Venture	41,750	99	18	74,133,824	74,133,824
Akhiok-Kaguyak	115,973	202	39	46,000,000	36,000,000
Chenega	59,520	190	45	34,000,000	24,000,000
English Bay	32,537	123	31	15,371,420	14,128,074
Eyak	75,425	189	80	45,100,000	45,100,100
Kachemak Bay State Park	23,800	37	3	22,000,000	7,500,000
Koniag (fee title)	59,674	41	11	26,500,000	19,500,000
Koniag (limited easement)	55,402	—	—	2,000,000	2,000,000
Old Harbor	31,609	183	13	14,500,000	11,250,000
Orca Narrows	2,052	—	2	3,450,000	3,450,000
Seal Bay/ Tonki Cape	41,549	112	5	39,549,333	39,549,333
Shuyak Island	26,665	31	8	42,000,000	42,000,000
Tatitlek	69,814	212	50	34,550,000	24,550,000
TOTAL	634,770	1,419	305	399,324,038	342,580,691

A scientific program of this magnitude has resulted in a leap in knowledge about the marine environment on which all Alaskans depend. A better understanding of the ecosystem, along with significant improvements in the tools fish and wildlife managers use to evaluate populations, means better decisions for the health of those populations and the people who depend on them.

Kenai River Sockeye Salmon Genetics

New research into the genetics of sockeye salmon in the Kenai and Russian rivers provides a good example of how increased knowledge translates into better management tools and decision-making.

In 1998, the return of sockeye salmon to the Kenai River was significantly lower than expected. Fisheries managers reduced the sport catch limit on the river and severely restricted commercial fishing. In order to reach minimum escapement goals for key tributaries of the Kenai River, biologists faced a decision about closing the sport fishery altogether. On a Friday, fisheries managers ordered genetic sampling of the sockeye entering the river. By the following Monday, they had documented that escapement goals to the Russian River would be met and they were able to make a sound decision to keep that popular fishery open. Without the genetic sampling method, developed with funds from the Trustee Council, managers say they would have been forced to close the sport fishery in order to err on the side of caution.

Ecosystem-based Research

Trustee Council-sponsored research is providing more information on fish, marine birds, and mammals than ever imagined. These projects benefit commercial and sport fisheries, aquaculture, subsistence, recreation and tourism. Most prominent among them are three ecosystem-scale projects, known primarily by their acronyms: SEA, NVP, and APEX.

What About the Future of Restoration?

The Trustee Council is funding a balanced restoration effort that will continue long after the last check from Exxon arrives in September 2001.

Two funds are being set up, providing income for a long-term habitat protection program and a multi-decadal research and community-based restoration program.

The original 1994 restoration plan called for the Trustee Council to create a reserve account to fund restoration into the future. By setting aside $12 million annually in an interest-bearing account, the Trustees were able to ensure a $140 million reserve. An additional $30 million is expected to be added to the reserve using unspent funds.

After 18 months of public comment and meetings throughout Alaska, it was clear that the public strongly supported continuation of the Council's two main restoration programs. Trustees decided March 1, 1999, to place $55 million into a long-term habitat protection program and approximately $115 million to fund a research, monitoring and general restoration program.

With this decision, the Trustees will have dedicated in total about 60 percent of available funds or $432 million for habitat protection in the spill region. The program is thus far responsible for acquiring title, conservation easements, or timber easements on about 650,000 acres, including more than 300 salmon streams and 1,400 miles of shoreline.

The remaining 40 percent or $285 million has funded one of the largest marine science efforts in the world, including hundreds of studies to better understand the dynamics of the ecosystem as a whole as well as the roles played by individual fish and wildlife species.

The Trustees emphasized that a balanced approach is necessary for the long-term health of injured fish, wildlife, and other species. Permanent protection of upland habitat is one vital component. The species injured by the spill, however, spend most or all of their lives at sea. Since the sea cannot be protected through acquisition, the best long-term protection is increased knowledge and better tools to support sound management decisions.

The research and restoration fund will start with about $115 million. Earnings from investment of the fund (at the nominal interest rate of 5 percent per year) will provide about $6 million annually to fund a long-term restoration effort.

In addition to research, the fund will promote development of better tools and methods for fish and wildlife management, as well as support community-based projects, including enhancements to subsistence, educational programs, local stewardship of resources, and other projects that have been an on-going part of the current restoration program.

The funding for habitat protection will be flexible enough to be used for a large protection effort on Kodiak Island or elsewhere in the spill region and for protection of key small parcels that are usually located at the mouths of rivers, along salmon spawning and rearing areas, or important coastal areas. The fund is expected to be worth $55 million.

At the rate of 5 percent annual interest, the habitat fund earnings could provide about $2.6 million each year for small parcels. The Trustees could choose, however, to spend the principal on larger protection packages.

The Council's action is based on existing authority and assumes that interest earnings would continue to be approximately 5 percent. The Council has sought Congressional help to expand its investment authority, but has so far been unsuccessful.

Details on how the funds will be established and managed have not yet been worked out. The Trustee Council also has other key issues to decide in the future. Trustees must determine whether to continue supporting the large public involvement process, including the Public Advisory Group, with its associated expenses, or to have a reduced effort. They must also decide whether the Trustee Council should continue to exist as managers of the programs or whether a different oversight entity should be established.

What's Being Done to Prevent Another Spill?

One of the major lessons of the *Exxon Valdez* oil spill was that the spill prevention and response capability in Prince William Sound was fundamentally inadequate.

Ten years ago, nearly 11 million gallons of oil spread slowly over open water during three days of flat calm seas. Despite the opportunity to skim the oil before it hit the shorelines, almost none was scooped up. A response barge maintained by Alyeska Pipeline Service Company was out of service and unavailable for use. Even if it had responded, there were not enough skimmers and boom available to do an effective job.

Dispersants were applied, but were determined to be ineffective because of prevailing conditions. Even if dispersants had been effective, however, there was not enough dispersant on hand to make a dent in the spreading oil slick.

Since that time, several significant improvements have been made in oil spill prevention and response planning.

- The U.S. Coast Guard now monitors fully laden tankers via satellite as they pass through Valdez Narrows, cruise by Bligh Island, and exit Prince William Sound at Hinchinbrook Entrance. In 1989, the Coast Guard watched the tankers only through Valdez Narrows and Valdez Arm.
- Two escort vessels accompany each tanker while passing through the entire sound. They not only watch over the tankers, but are capable of assisting them in the event of an emergency, such as a loss of power or loss of rudder control. Ten years ago, there was only one escort vessel through Valdez Narrows.
- Specially trained marine pilots, with considerable experience in Prince William Sound, are now aboard the ship during its entire voyage through

the Sound. Weather criteria for safe navigation are firmly established.

- Congress enacted legislation requiring that all tankers in Prince William Sound be double-hulled by the year 2015. It is estimated that if the Exxon Valdez had had a double-hull structure, the amount of the spill would have been reduced by more than half. There are presently three double-hulled and twelve double-bottomed tankers moving oil through Prince William Sound. Arco Marine is constructing two new double-hulled tankers, the first of which is expected to be in use in 2000.
- Contingency planning for oil spills in Prince William Sound must now include a scenario for a spill of 12.6 million gallons. Drills are held in the sound each year.
- The combined ability of skimming systems to remove oil from the water is now 10 times greater than it was in 1989, with equipment in place capable of recovering over 300,000 barrels of oil in 72 hours.
- Even if oil could have been skimmed up in 1989, there was no place to put the oil-water mix. Today, seven barges are available with a capacity to hold 818,000 barrels of recovered oil.
- There are now 34 miles of containment boom in Prince William Sound, seven times the amount available at the time of the *Exxon Valdez* spill.
- Dispersants are now stockpiled for use and systems are in place to apply them from helicopters, airplanes, and boats.

The debate continues to rage over whether a spill the size of the *Exxon Valdez* disaster can be contained and removed once it's on the water. But there is little doubt that today the ability of industry and government to respond is considerably strengthened from 10 years ago.

Complacency is still considered one of the greatest threats to oil spill prevention and response. To help combat that threat the Alaska Department of Environmental Conservation (ADEC) conducts both scheduled and unannounced drills and participates in regular training exercises in Prince William Sound each year. Community training programs have been established and local fishing fleets have been trained to respond to spill emergencies.

In addition, the Prince William Sound Regional Citizens' Advisory Council, established by an act of Congress, serves as a citizen watchdog over the Alyeska Terminal, the shipping of oil through the sound, and the government agencies that regulate the industry. A similar citizen's organization watches over oil issues in Cook Inlet.

reading 3

2000 Status Report on the *Exxon Valdez* Oil Spill: "Ecosystem Research"

Exxon Valdez Oil Spill Trustee Council: 2000 Status Report, pp. 2–23 with selected figures and tables. Reprinted by permission.

Looking at the Big Picture

Ecosystem Studies Anchor Trustee Council's Research Program

Since 1993, the Trustee Council has invested tens of millions of dollars in an organized effort to better understand the marine ecosystem that both supports and enchants the people of Southcentral Alaska.

To be good stewards of the resources, we must understand the intricacies of this marine ecosystem to the best of our abilities. What are the changes taking place? What factors limit the productivity of key species? How can we best react to inevitable changes in the sea, caused by both natural fluctuations and human activities?

In the Trustee Council's view, long-term restoration of the spill region will depend on knowledge of natural resources and a better understanding of how populations rise and fall as changes—both natural and human-caused—occur in the sea. It was in the spring of 1993 that the herring crashed in Prince William Sound, its population plummeting from an all-time high the previous year to an all-time low. A few short months later, the return of pink salmon to the sound turned dismal. Something had drastically changed in the ecosystem and commercial fishers demanded action and answers. They took their frustrations to the Valdez Narrows and formed a blockade that for three days kept oil tankers at bay.

The result was an agreement that the Trustee Council would support ecosystem studies focused on herring and pink salmon in Prince William Sound. The six-year, $22 million Sound Ecosystem Assessment (SEA) project was born later that year after months of discussion between commercial fishers, fishery managers, scientists, and others.

This was the genesis of the Trustee Council's emphasis on ecosystem-based studies. The settlement with Exxon was not yet two years old. The previous year, the Trustee Council made the transition from reactive damage assessment studies to a more proactive plan for restoration. A draft of the Restoration Plan (which was adopted in 1994) was already being prepared and in it, the Trustee Council clearly stated its intentions concerning research. "Restoration will take an ecosystem approach to better understand what factors control the populations of injured resources."

In addition to the SEA project, the Trustee Council made a long-term commitment to understand how seabirds are affected by the availability of forage fish. The Alaska Predator Ecosystem Experiment (APEX) received $10.8 million over an eight-year period.

Another ecosystem-based research effort, the Nearshore Vertebrate Predator (NVP) project, began in 1995. That six-year, $6.5 million study used four species injured by the spill—river otters, pigeon guillemots, sea otters, and harlequin ducks—to determine what factors in the nearshore area were limiting recovery.

What have we learned from these major ecosystem projects and dozens of other studies that focused on individual species? This 2000 Annual Report will take a look at some of the major research efforts and the lessons learned.

Sound Ecosystem Assessment

Pink Salmon and Pacific Herring in Prince William Sound

The 1993 collapse of the herring population in Prince William Sound, followed that summer by a poor return of pink salmon, served as a clarion call for a better understanding of the sound ecosystem.

Prince William Sound endured a major ecosystem change after the 1964 earthquake when subtidal areas were lifted above the tide line and community oil tanks were laid open by tsunamis. Twenty-five years later, the northern Gulf of Alaska absorbed the largest oil spill in U.S. history. Along the coast of the northern gulf is where more than half of Alaska's population lives, works, and plays and it is a destination point for many of the one million tourists who vacation in Alaska each summer.

Understanding both the natural changes and the human impacts on the region is vital to maintaining this rich ecosystem on which Alaskans depend. The Sound Ecosystem Assessment (SEA) project is the largest of the ecosystem-scale research projects funded by the Trustee Council. SEA was initiated by commercial fishers and implemented by a team of research scientists coordinated out of Cordova. It included more than a dozen integrated components organized to obtain a clear understanding of the factors that influence productivity of Pacific herring and pink salmon in Prince William Sound.

This project has produced new information about the survival needs of juvenile herring and unraveled mysteries showing the effects of wind and ocean currents on plankton, the very base of the food chain.

Researchers have developed a clear new picture of how pink salmon fry, no more than three inches in length, dodge predators and fatten up on plankton during their first few months in saltwater when their survival is most in doubt.

Pink Salmon

Pink salmon hatch in the streams of Prince William Sound or in the incubators of five hatcheries in the area. During the spring, they emerge from the streams or are released from the hatcheries to begin their migration out of the sound and into the northern Pacific. Research scientists chose to focus their studies at the point where pink salmon appear to be most vulnerable: as fry trying to survive the migration to the open ocean.

Pink salmon fry, in salt water for the first time, must quickly fatten up on animal plankton and at the same time avoid predators that must fatten up on them. Why is it that some years they do this with great success and return as adults in large numbers the following year, while in other years the survival of fry is poor and the adult return dismal?

SEA researchers began their focus on plant and animal plankton. They knew that plankton come from two main sources. Winds and currents can sweep plankton into the sound from the Gulf of Alaska. Or plankton can emerge from the deep recesses of Prince William Sound, including an area reaching 700 meters in depth known locally as the "black hole." But it was not known whether the black hole could produce enough plankton to support the needs of the sound or whether the sound was dependent on production in the Gulf of Alaska.

To determine this, SEA researchers developed a computer-based circulation model that predicts the movement of currents in the sound. A powerful computer was used to simulate circulation using information about the underwater landscape, inflow of saltwater from the Gulf of Alaska, inflow of freshwater from streams, tides, and winds.

Basketball-sized buoys were released at different locations in the sound to help validate the predictions of the model. The buoys, tracked by satellite, measured currents at the surface and at 15 meters depth. The circulation model provided a wealth of information about how water movement impacts biological processes. For instance, the model simulated what would happen as plankton eggs, deep in the black hole, floated to the surface. In this way scientists were able to determine that zooplankton from the black hole were retained in the sound for several months, meaning that the region was not wholly dependent on the gulf for food. The circulation model, using 1996 data, showed that plankton in the southeast portion of the sound were from the gulf while the majority of plankton in the western corridor were from the nearby black hole.

Another SEA model involved the groundbreaking work of predicting when the plankton blooms (the peak of plankton abundance) will occur. The timing and duration of the blooms have enormous implications for survival of salmon fry. Depending on conditions, the onset of the blooms can vary year to year by several weeks. Drs. David Eslinger and Peter McRoy developed a highly accurate model that is based on easily monitored ocean and weather conditions (Figure 1). In general, they found that during stormy, windy years, the onset of plankton blooms are late but prolonged, while those in calm years are early, intense, and short-lived. Other SEA researchers were simultaneously working on another key question: Which is more limiting to salmon fry, food or predation?

SEA's salmon researchers mounted an intensive sampling campaign. Salmon fry and predator fish were caught, counted, weighed, their distributions mapped, and their stomach contents examined to find out their eating habits. Although it was generally known which fish preyed on salmon fry, it was not clearly known how much of an impact different predators had on the salmon population.

Researchers were able to show that pollock are the dominant predator on pink salmon fry, and that knowledge helped confirm an ecological theory. The plankton blooms not only feed the salmon fry, they also help shield them from predators. Scientists determined that, depending on the size and timing of the plankton

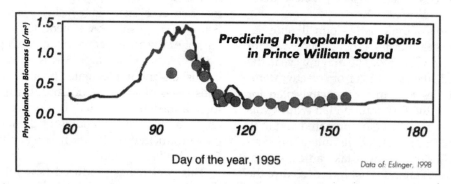

Figure 1. Models generated through the Sound Economic Assessment (SEA) project estimated the timing and size of the animal and plant plankton booms. The lines show the models' predictions compared to the actual measurements, represented by dots.

bloom, pollock may prefer to feed on plankton instead of fry. The opportunistic pollock will feed on whatever prey provides the most energy for the least effort, and this often means they prefer floating plankton over the swimming salmon. When plankton is plentiful, salmon fry are less likely to be preyed upon. To gauge the impact of this discovery, SEA developed another computer-based model. This model was designed to simulate the migration path of the fry. As the fry disperse and move toward the ocean, the computer model predicts where they will go, what food they will encounter, and what predators will encounter them.

The model is more than a fascinating look at predator-prey relationships. It has some very practical applications. In some circumstances, for example, it's better for the survival of fry when hatchery managers release them en masse. But other marine conditions call for a different strategy. When the zooplankton bloom is weak, it can make better sense to release the fry in batches.

Although there are many factors involved in fry survival, SEA has narrowed the field to a subset that can be monitored: light, temperature, fry size at release, fry density and group clustering, plankton bloom timing and abundance, and predator composition and size. These data can be used in a computer model to predict the survival of salmon fry, thereby providing a new tool to more accurately predict the return of adult pink salmon the following year. So what have we learned from SEA?

The basic scientific knowledge gained is almost incalculable and new discoveries will continue as data are analyzed for many years to come. But of more immediate importance are the practical applications for users of the gulf ecosystem.

SEA has developed models that can now tell us where the plankton is coming from, where the currents will take it, when the bloom will occur and how strong it will be. Predator-prey models predict the survival rate of salmon fry. This information is important not only when it comes to planning the release of fry from hatcheries and in forecasting the return of those salmon the following year, but also in understanding how salmon survive and grow.

Pacific Herring

Herring spawn in early April and the hatched larvae spend months drifting around Prince William Sound. Like the plankton, herring larvae are at the mercy of the currents until they metamorphose into juvenile fish in August.

Just as they did with pink salmon, SEA researchers focused primarily on the early stages of life as the most critical for herring survival. What were they feeding on? How did they survive the winter when plankton were practically non-existent? Unlike pink salmon, very little was known about the first year of life for Pacific herring. Researchers were pretty much starting from scratch.

The SEA herring team conducted a painstaking series of surveys to map the distribution of the juvenile herring and document their habitat needs. The groundbreaking information they gathered depended on a highly coordinated series of aerial surveys, hydrocoustic surveys, and intensive net sampling efforts.

SEA learned that young herring begin appearing in small bays in late July and August each year and were feeding on plankton into the fall. However, by late fall their food supply nearly disappears. It turned out that juvenile herring must survive three or four months with very little food. They fast and preserve their energy, or "cut power and float," as some scientists refer to it. If they fail to store up enough energy for the winter, they may die. The energy reserves of the herring, the severity of the winter, and the bay in which they overwinter all play significant roles in their survival.

Dr. A. J. Paul went to eight different Prince William Sound bays in March and measured the energy reserves of juvenile herring found there. He found that the herring in Simpson, Sheep, and Boulder bays had plenty of reserves to survive the winter. Juvenile herring in Jack and Whale bays were low on reserves and those in Eaglek, Paddy, and Drier bays were near the point of starvation. This information confirmed that a particularly cold or stormy winter could cause starvation in many areas and lead to a poor return as adults.

Once again the SEA modelers took this information and built it into a model for herring overwintering survival. The model inputs body protein and energy content measured from a sample of young herring in late fall and, based on expected winter temperatures, estimates the proportion of herring that will survive until spring. This provides another tool for better predicting the survival of a herring year class. Development of the various SEA models is continuing. More fine tuning and testing must be done before they can be used routinely for management purposes.

Alaska Predator Ecosystem Experiment

Forage Fish and Seabirds in the Gulf of Alaska

Common murres, black-legged kittiwakes, harbor seals, and Steller sea lions are examples of apex predators, fish eaters at or near the top of the food chain. Declines in these and other apex populations have occurred in the Gulf of Alaska since the 1970s. At the same time, the gulf has undergone a drastic change in the type and abundance of forage species, such as herring, capelin, sand lance, shrimp, young pollock and juvenile cod. The

Alaska Predator Ecosystem Experiment (APEX) began in 1994 in an effort to determine why some seabird species injured by the oil spill showed no sign of recovery. Such knowledge was seen as essential to undertaking biologically realistic recovery. APEX asked the basic question: How does food availability—the type and abundance of forage fish—limit the ability of seabirds to recover from oil spill injuries?

A shift in the dominant forage fish populations occurred in the last half of the 1970s, likely triggered by a decadal shift in climate. Warming waters resulted in a shift from an ecosystem dominated by shrimp to one dominated by pollock and cod. Small-mesh trawl surveys, conducted annually since 1953, resulted in a strong database with nearly 10,000 individual sampling tows, collected over widely dispersed regions of the Gulf Alaska. These data illustrate a massive change in the marine ecosystem, beginning in 1978. Shrimp and forage fish gave way to pollock and cod, and within two years there was a complete reversal in dominance.

Energetics

If the ecosystem shift forced a change of diet on seabirds, how does that affect egg production and survival of chicks?

APEX researchers measured lipid or fat content of forage fishes because it is the primary factor determining energy available to apex predators. Lipid content of seabird prey ranged from 5 percent of dry mass (Pacific tomcod) to 48 percent of dry mass (eulachon, also known as hooligan). Most of this variation was from species to species, but there were also variations within species related to age, sex, location, and the reproductive status of the fish. Of the main fishes consumed by seabirds, juvenile herring, pre-spawning capelin, and sand lance had the highest energy content (Figure 4). Kittiwake diets were dominated by these high-lipid forage fishes at all study sites and this correlated with good nestling growth and survival. The trend in the 1990s of higher kittiwake productivity associated with increasing availability of sand lance, capelin, and herring was broken in 1997, a poor year at most kittiwake study colonies. The poor productivity of 1997 appears to be associated with fewer herring in their diet.

Controlled laboratory studies were conducted to better understand the nutritional difference between high-lipid and low-lipid fishes in the diets of kittiwakes. Kittiwakes fed a high proportion of sand lance and herring also had high growth rates and productivity (Figure 4). This compared to much lower growth rates and productivity in birds that consumed mostly demersal fishes (pollock) or gadids (cod). Juvenile kittiwakes required roughly twice the amount of pollock and cod to obtain the same growth rates as juveniles fed herring and sand lance. These results support the hypothesis that productivity of kittiwakes and guillemots in the gulf region is strongly linked to the availability of three species of forage fishes: Pacific sand lance, Pacific herring, and capelin. These three species form schools near shore and have high energy densities compared with

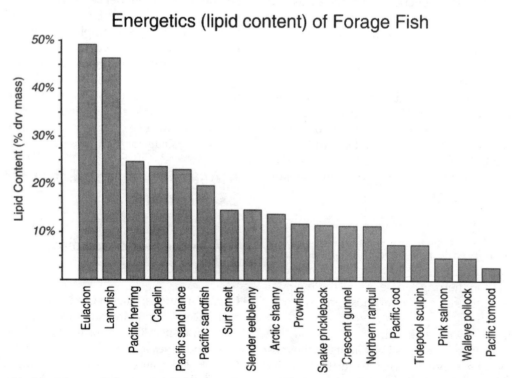

Figure 4. The kinds of forage fish available to seabirds can have a big impact on survival of the chicks. Herring, capelin and sand lance are rich in fats when compared with pollock, cod and pink salmon. Seabird colonies with fatty forage fish in the area do better than those feeding on the lean cuisine of pollock and cod.

most other forage fishes. Recovery of seabird populations injured by the oil spill will likely depend, at least in part, on increases in these key fish stocks.

Feeding Behavior

By observing individual seabirds and schools of fishes the APEX study has found several factors that influence how seabirds forage. In summer, the waters of Prince William Sound are stratified with little mixing and the near-surface fish schools (herring, sand lance, and capelin) are small, occur in low density, and are located close to shore. Seabirds in the sound respond by foraging singly or in small flocks close to shore.

This is in contrast to lower Cook Inlet where there is strong tidal mixing of the water column, the fish are in larger and more dense schools than in the sound, and the fish occur offshore (capelin and pollock) as well as nearshore (capelin and sand lance). When prey are predictable, seabirds learn and remember where prey can be found and individual birds return to the same area repeatedly. They do not always forage on fish that are closest to the colony. Instead, they pass by fish schools to return to the area where they have successfully foraged in the past. This may increase the birds' foraging efficiency.

Seabirds change their foraging strategy in respect to prey abundance: when prey are scarce seabirds generally forage in flocks, but when prey are abundant they often forage alone. This behavior increases their efficiency when food is scarce by using other birds to locate schools.

Modeling

The population dynamics of kittiwakes and other seabirds in Prince William Sound are usually in a state of flux. At any given time, some seabird colonies are growing and some are declining. Although there is strong evidence that variation in food supply underlies much of the fluctuation in colony size, the mechanism by which food supply influences colony dynamics needs to be more clearly defined.

Using detailed data on the movement patterns and foraging behavior of radio-tagged kittiwakes, coupled with extensive concurrent aerial surveys of fish schools, APEX researchers have constructed a computer model designed to mimic the behavior of a foraging kittiwake. This model can be used to simulate the response of a foraging kittiwake to various patterns of food distribution and abundance. These simulated foraging behaviors can then be used to predict the distance that adults must travel in order to forage, and the rate and nature of food deliveries to the chicks. Since chick survivorship is known to be strongly influenced by these factors, it may be possible to make predictions about the performance of individual colonies based on hypothetical fluctuations in the distribution and abundance of the forage fish.

APEX data are also being considered for a model that would help predict the recruitment of juvenile herring to the adult population. Little is known about the survival rate of juvenile herring, so it is difficult to predict how large a group will grow up to join the population of spawning adults each spring. One possible prediction method would use observations of diets in seabirds. By quantifying the amount of juvenile herring in the diet of kittiwakes, for example, a model could extrapolate the strength of the juvenile year classes.

Nearshore Vertebrate Predator Project

Long-term Impacts of Oil in the Nearshore Environment

The plants and animals living along the coast took the brunt of the spilled oil as millions of gallons washed up along hundreds of miles of shoreline. Ten years later, some of the more heavily oiled areas remain polluted with tar, asphalt, and unweathered oil either at or just below the surface. But what about the animals that live there? Of the eight species that remain listed as "not recovering" ten years after the spill, seven use the nearshore environment for nesting, feeding, and resting. Is it oil that is preventing their recovery or are other factors involved, such as food availability, reproductive ability, weather, or predation? How long does it take for populations to rebuild to pre-spill levels?

This central question became the basis of the Nearshore Vertebrate Predator project, a six-year $6.5 million effort conducted by the Alaska Biological Science Center of the U.S. Geological Survey in cooperation with scientists from NOAA and the University of Alaska Fairbanks. This team of scientists sought answers to the fundamental question about oil spills: how long does oil persist in the environment and does it continue to impact wildlife? Researchers narrowed their study to four species injured by the spill: two fish eaters (river otters and pigeon guillemots) and two species that feed on shellfish (sea otters and harlequin ducks). All four species are long-lived and spend most of their time in the nearshore environment.

All four species could face oil contamination in the nearshore waters whenever oil in sediments is released. Other potential avenues of oil exposure, through diet or direct contact, differ for each species. Sea otters, which feed on shellfish and dig through volumes of sediments where residual oil persists, are likely to contact oil while foraging or through their diet. River otters, which catch fish for food, are more likely to be exposed on shore. Pigeon guillemots would have minimal direct exposure to oil, although they do supplement their diets with invertebrates. Harlequin ducks, on the other hand, live on shore and feed on shellfish and could be exposed to oil through preening and diet.

The risk of oil exposure is greater for animals that eat invertebrates, such as clams and mussels, because they concentrate hydrocarbons. Fish, on the other hand, metabolize hydrocarbons quickly and, thus, don't concentrate them.

The NVP research team split up into groups studying each of the four species. They maintained two research sites, one in an area that was heavily oiled in 1989 and the other in an area that saw very little or no oil. In this way they could compare results from oiled areas to non-oiled areas.

Animals were captured, weighed, measured, aged, and had blood samples taken. Before being released some harlequin ducks and river otters had transmitters to allow researchers to follow their movements and to indicate death (should that occur).

The results differed between regions and among species. Harlequin ducks showed signs of continued stress, and populations in the oiled areas had zero growth. Sea otters in the most heavily oiled areas of northern Knight Island have not increased in abundance, despite food supplies capable of supporting population growth. There has, however, been an increase in numbers of sea otters in the comparison group.

Female harlequin ducks in the oiled areas were less likely to survive the winter than those from non-oiled areas (Figure 5). Although the difference was not great (78 percent survival versus 84 percent survival), it is enough to have a significant impact on the population. Researchers calculated that the winter survival rate in oiled areas was enough to cause a 5 percent population decline versus approximately stable populations in the unoiled areas. Survey data are consistent with these data on survival.

Researchers used a biomarker of hydrocarbon exposure to determine whether animals in the oiled area were still being exposed to oil. Based on analysis of blood and tissue samples, both species showed evidence of recent exposure to oil. Researchers say, however, that the exposure appears to be variable, and there is some indication that it is diminishing each year.

For the harlequins, ducks with the highest levels of the biomarker tended to have the lowest body weights, suggesting that oil exposure was negatively affecting their health. In contrast, sea otters from the oiled study area tended to be in better condition than their counterparts in the nonoiled area, perhaps because of greater availability of food, given the low numbers of otters there. However, over the last decade, blood work on sea otters has indicated liver damage in animals residing in the oiled areas, which could be a result of exposure to the oil.

By the end of the NVP study, researchers determined that only a small proportion of the sea otters still exhibited abnormal blood values, which is consistent with declining amounts of residual oil in the environment. It is also important to note that both sea otters and harlequin duck populations appear to be healthy or stable when considering the entire Prince William Sound. Effects of continued exposure to oil are seen primarily in more heavily oiled regions in the western part of the sound. A colony of pigeon guillemots at Naked Island is not rebounding from the impacts of the oil spill, and apparently, both oil and food availability may be affecting recovery. The Jackpot Bay colony is growing dramatically, whereas the Naked Island colony remains depressed. Based on biomarker results, pigeon guillemots at Naked Island continue to be exposed to oil, although perhaps at relatively low levels. However,

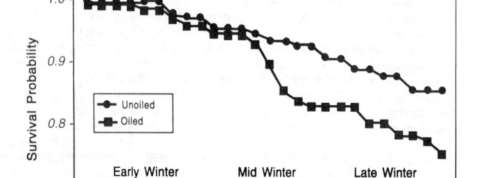

Figure 5. Winter survival of female harlequins was significantly lower in oiled areas versus unoiled areas during the winters of 95–96, 96–97, and 97–98. The difference in survival is enough to cause a 5 percent decline in population in the oiled areas compared to stable populations in the unoiled areas.

blood work on birds from the Naked Island colony suggests that oil may be affecting the function of liver and other organs.

Recovery of the guillemots may also be complicated by diet differences: birds at the healthy Jackpot colony eat twice as much herring and other high-energy fish as do those at the Naked Island colony. A major shift in forage fish (see APEX) has impacted many species and colonies throughout the Gulf of Alaska region. The food available to Naked Island guillemots consisted mostly of pollock and cod, which has about half the amount of fat as herring. In following the trail of river otters, NVP researchers were able to determine that the river otter had recovered from the effects of the spill. Biomarker levels were higher in river otters from the oiled area in 1996, the initial year of the study, but by 1998, no differences were seen between animals in oiled and nonoiled areas. Other measures of health, which indicated that river otters suffered from oil effects in the early 1990's, had also returned to normal by the time of the NVP study. In 1999, the Trustee Council officially declared the river otter "recovered."

The NVP research team concluded in 1999 that continued oil exposure appears to be through a diet of invertebrates and, possibly, by grooming or preening. This can come from mussels left untreated after the oil spill or other intertidal and subtidal invertebrates from the sound's floor. Oil on shorelines or in the nearshore water column may also be a source of contamination, getting onto an animal's fur or feathers.

"The collective evidence supports the hypothesis that patchy, persistent oil in the sound is still being sufficiently mobilized some 10 years post-spill to constrain recovery within the nearshore ecosystem," the NVP final report concludes.

"It is apparent that we are no longer studying populations under acute stress, but rather that components of the invertebrate-based nearshore community are still under chronic, but decreasing levels of stress. This stress is observed not at a regional level where sea otters and harlequin duck populations are stable or growing, but in those areas of the sound most heavily oiled by the 1989 *Exxon Valdez* oil spill and examined under the NVP study."

Research Highlights and Accomplishments

Pink Salmon

Excavation of Port Dick Creek on the Kenai Peninsula opened new spawning habitat for pink and chum salmon.

Improvements to a bypass at Little Waterfall Creek on Kodiak Island increased the number of pinks reaching spawning habitat.

Development of remote video technology may prove a technological breakthrough to economically monitor spawning.

A plankton model is being developed to forecast optimum release times of hatchery salmon smolt.

Genetic work of pink salmon will identify genetic traits such as disease resistance, growth, and timing of the run.

New studies on the sensitivity of pink salmon eggs, larvae and fry to fresh and weathered oil can aid in establishing water quality standards and in contingency planning for future spills.

The study of pristane in mussels could lead to better forecasting of pink salmon runs in PWS.

Pacific Herring

Improved assessment methods have helped fisheries managers better determine herring biomass and set harvest quotas.

Intensive research indicated disease played a major role in the 1993 collapse of the PWS herring population.

A herring model is being developed to help forecast harvest levels by estimating overwinter survival of juveniles.

Changes in the pound fishery were made as a result of questions raised about pounding methods and the possible link to disease.

Genetic research revealed significant differences between the Bering Sea and Gulf of Alaska herring populations.

Studies on the sensitivity of young herring to oil can aid in establishing water quality standards and in spill contingency planning.

Sockeye Salmon

Genetic identification of Kenai River stocks helps protect sockeye salmon by ensuring escapement to vital tributaries.

Cook Inlet sonar provides a better measure of the run strength and better management decisions.

Rearing and survival in nursery lakes are better understood and aid in predicting/managing sockeye runs.

Marine Mammals

Can something be done to arrest the on-going population decline of harbor seals?

Researchers are studying whether the carrying capacity has diminished due to changes in forage fish availability.

Data from wild, healthy seals and sick/injured seals help veterinarians evaluate the health status of individuals and populations.

Genetic data on killer whales will help in interpreting population changes and responding to conservation problems.

Some killer whale populations were found to have high concentrations of contaminants in their blubber.

New information on reproduction is crucial for understanding the effects of harvests on sea otters.

Data on life expectancy of sea otters has led to improved estimates of survival and recovery.

Improved aerial survey techniques are now being used throughout the sea otter's range.

Improved technique allows aging of sea otters using teeth.

Habitat Improvement

Hands-on restoration projects are stabilizing and restoring stream banks which protect rearing sockeye salmon.

A PWS human use model will help predict how increased human uses will impact fish and wildlife habitats and resources. This should help managers reduce or mitigate such impacts.

Seabird/Forage Fish

The availability and quality of forage fish was linked directly to seabird productivity. As a result, the North Pacific Fishery Management Council strictly limited forage fish bycatch and prohibited new commercial fisheries on forage fish species.

Blood samples may provide a simple means of predicting reproductive success of seabirds and serve as a broad indicator of overall environmental stress.

Forage fish schools can be monitored by examination of stomach contents of halibut caught on charter boats.

Non-native foxes that were introduced to some seabird-nesting islands were removed to help increase bird populations.

Surveys of marine bird populations in PWS now provide powerful data on population change, vital for long-term evaluation of the ecosystem.

Researchers identified species-specific indicators for health of seabird populations, such as productivity for kittiwakes and foraging time for murres.

A new method to monitor marbled murrelets on the water was developed.

The first full-scale study of the Kittlitz's murrelet, one of the least known seabird species in the world, was conducted. This seabird is found near tidewater glaciers and may be a victim of global climate change.

Harlequin Duck

New data on populations resulted in curtailment of the sport hunting season in PWS.

New techniques were developed to age and sex ducks, improving population assessments.

First large-scale use of surgically-implanted radios was conducted to study movements and survival.

Information on genetics, site fidelity, and movements will support decisions such as harvest levels and siting of facilities.

Cutthroat Trout and Other Fish

Habitat inventories and in-stream improvements enhanced spawning and survival of cutthroat trout and coho salmon.

Research on cutthroat trout supported changes in regulation of the sport fishery to conserve the resource.

Studies will help determine if PWS pollock should be managed as part of the Gulf of Alaska population.

River Otter

Researchers pioneered a new method for trapping otters for mark-and-recapture population estimates.

Ground-breaking work on oil ingestion will aid in understanding exposure from oil and other contaminants.

Reduction of Marine Pollution

Waste management plans help reduce marine pollution, such as waste oil and hazardous household waste, by providing proper disposal facilities and equipment in PWS, Kodiak Island, and lower Cook Inlet.

Subsistence

Broke new ground by including traditional knowledge of some species as part of the overall research into populations, trends, and health.

Subsistence hunters have been trained to take biosamples from harbor seals for scientific analysis.

Smolt have been released in Boulder Bay near Tatitlek annually since 1994 to create a subsistence fishery.

Sockeye salmon have been stocked in Solf Lake to provide more subsistence resources in the Chenega Bay area.

King salmon fry released near Chenega Bay provide additional subsistence resources.

A clam restoration project is expected to stock littleneck clam populations on subsistence beaches. New techniques to spawn clams could help the shellfish industry as well.

Students take part in ongoing restoration projects, learning the skills and knowledge to take part in restoration activities now and in the future.

Elders/Youth Conferences in 1995 and 1998 brought subsistence users, youth and researchers together to learn and exchange ideas.

Two videos document subsistence uses and traditions involving harbor seals and herring.

A coho salmon project on the Kametolook River near Perryville will strengthen the return to the river and improve subsistence resources.

Ecosystem Synthesis

A dynamic computer model for PWS will help predict effects of changes in the system as variables such as increasing fisheries harvests change.

Maps are being produced bringing together all the new knowledge about species location and seasonal use in PWS.

Gulf Ecosystem Monitoring

A Sentinel Program for the Gulf of Alaska

A health watch for the Gulf of Alaska, in which researchers continually take the pulse of the marine environment and its watershed, is being planned as part of the long-term legacy of oil spill restoration. In March 1999, the Trustee Council established a $120 million fund for a research and monitoring effort now known as GEM—Gulf Ecosystem monitoring—that could span the entire century or more.

In return, the people who live, work, and play in the spill region are expected to gain unprecedented knowledge about the ecosystem they depend on as as well as new tools and strategies for management of fish and wildlife. Such knowledge will provide an early warning system for changes, both natural and human-caused, that occur in the Gulf of Alaska ecosystem, from mountain headwaters to the depths of the open ocean beyond the outer continental shelf.

At the start of the third millennium A. D., the northern Gulf of Alaska remains as pristine an environment as can be found almost anywhere on earth. Environmental signals, especially climate, continue to be the determining factors in the overall health of the ecosystem.

But history in other parts of the world has taught us that eventually the human signals will overtake and overwhelm the natural signals. If anyone doubts this, simply look at the tripling of Alaska's population over the last 40 years, the increased fishing pressure in Southcentral Alaska, the doubling of tourism in the last 10 years, and, soon, the opening of the Whittier Road. And then, consider what the next 40 years might bring. Developed properly, GEM will serve as a sentinel over the gulf, providing an early warning system that will help resource managers, policy makers, and the public to minimize the impacts and better prepare for the inevitable increase in human use.

GEM will cover an area similar to the spill region, including Prince William Sound, lower Cook Inlet, Kodiak Island, and the Alaska Peninsula. Its mission is "to sustain a healthy and biologically diverse marine ecosystem in the northern Gulf of Alaska and the human use of those resources through greater understanding of how productivity is influenced by *human activities and natural changes.*"

The gulf ecosystem is extremely complex, consisting of thousands of interacting species. It will not be possible for GEM to answer all, or even most, of the questions that could be posed about the gulf. Instead, GEM will focus on 1) key species in the system, picked on the basis of perceived ecological importance and human relevance, and 2) on the most telling physical and biological processes responsible for healthy production.

The GEM program will continue to work with resource managers, stakeholders, the scientific community and the public to refine a common set of priorities for research, monitoring and protection in the northern gulf. GEM will coordinate its efforts with other government agencies, universities, and private groups that are already studying individual components of the gulf ecosystem. This will allow GEM to fill in vital gaps without duplicating studies.

The $120 million GEM fund will act as an endowment, providing annual funding of $5 million to $10 million depending on investment earnings. Independent peer review is essential for a high caliber scientific program. Participation in research and monitoring is expected to be completely open to competition. Periodic "State of the Gulf " workshops, public reports, and a GEM website would serve to keep the public, and especially stakeholders, up to date with research.

The draft GEM program will be submitted to the National Research Council in Washington, D.C. for a thorough scientific review before being implemented.

Goals of GEM

Detect: Serve as an early warning system by detecting annual and long-term changes in the marine ecosystem, from coastal watersheds to the central gulf.

Understand: Identify causes of change in the marine ecosystem, including natural variation, human influences, and their interaction.

Predict: Develop the capacity to predict the status and trends of natural resources for use by resource managers and consumers.

Inform: Provide integrated and synthesized information to the public, resource managers, industry and policy makers in order for them to respond to changes in natural resources.

Solve: Develop tools, technologies, and information that can help resource managers and regulators improve management of marine resources and address problems that may arise from human activities.

GEM Implementation Schedule

April 2000

- Submit Draft to the National Research Council for review
- FY 2001 Invitation to seek transition proposals

October 2000

- Initiate FY 2001 transition projects

January 2001

- Receive preliminary NRC feedback
- Begin revisions to GEM plan to address NRC recommendations and use results from transition project

February 2001

- Invite additional transition projects for FY 2002

2001

- Begin FY 2002 transition projects

January 2002

- Trustee Council finalizes GEM Program

February 2002

- Issue GEM invitation for proposals (FY 2003)

October 2002

- Begin GEM monitoring and research program

Habitat Protection

The long-term protection of threatened habitat, considered essential for the well-being of species injured by the oil spill, was one of the earliest goals of the Trustee Council. Even before the Restoration Plan was finalized in 1994, the Trustee Council, in cooperation with the Alaska State Legislature, funded the protection of two key parcels, along Kachemak Bay and on Afognak Island, each under imminent threat of logging.

Six years later, one of the largest habitat protection efforts in the United States is nearly complete. The Council's goal of protecting key habitats throughout the spill region was largely achieved in February of 1999 when a package was finalized with Eyak Corporation protecting 75,452 acres in eastern Prince William Sound.

Through a creative series of conservation easements, timber easements, and fee simple acquisitions, the Trustee Council expanded the portfolio of its Large Parcel program to a total of 635,770 acres protected, including at least 1,400 miles of shoreline and more than 300 salmon streams.

This includes 55,402 acres surrounding the popular Karluk and Sturgeon rivers on Kodiak Island, which is protected by a conservation easement, but only through 2001. The permanent protection of these rivers is the only unfinished business remaining in the Trustee Council's $360 million Large Parcel program. Negotiations with the landowner, Koniag, Inc., are continuing.

An additional $24.4 million has been spent or earmarked to protect at least 11,179 acres through the Small Parcel program, which deals with smaller, more strategically located habitats, usually along rivers, coastal areas, and lagoons. To date, 47 parcels totaling 7,240 acres have been acquired and protected. The Council has set aside funds for another 3,939 acres under consideration for protection.

In March of 1999, the Trustee Council decided that it would use $55 million from its Restoration Reserve account to continue habitat protection efforts into the

Habitat Protection Large Parcel Program

Parcel Description	Acreage	Coastal Miles[3]	Salmon Rivers[4]	Total Price	Trustee Council's Share
Acquisitions Complete					
Afognak Joint Venture	41,750	99	18	$74,133,824	$74,133,824
Akhiok-Kaguyak	115,973	202	39	$46,000,000	$36,000,000
Chenega	59,520	190	45	$34,000,000	$24,000,000
English Bay	32,537	123	31	$15,371,420	$14,128,074
Eyak	75,425	189	80	$45,126,704	$45,126,704
Kachemak Bay State Park inholdings	23,800	37	3	$22,000,000	$7,500,000
Koniag (fee title)	59,674	41	11	$26,500,000	$19,500,000
Koniag (limited easement)	55,402			$2,000,000	$2,000,000
Old Harbor[1]	31,609	183	13	$14,500,000	$11,250,000
Orca Narrows (timber rights)	2,052		2	$3,450,000	$3,450,000
Seal Bay/Tonki cape	41,549	112	5	$39,549,333	$39,549,333
Shuyak Island	26,665	31	8	$42,000,000	$42,000,000
Tatitlek	69,814	212	50	$34,550,000	$24,719,461
TOTAL:	**635,770**	**1,419**	**305**	**$399,353,038**	**$343,357,396**

Negotiations Continuing
Koniag (fee title)[2]

1. As part of the protection package, the Old Harbor Native Corporation agreed to protect an additional 65,000 acres on Sitkalidak Island as a private refuge.
2. Negotiations with Koniag concern fee title to the 55,402 acres that are currently protected under a temporary conservation easement due to expire in December 2001.
3. Approximate miles of coastline.
4. Approximate number of anadromous rivers, streams, and spawning areas.

future. The fund will be used to finance protection of the Karluk and Sturgeon rivers, if negotiations are successful, and to create an account for future parcel acquisitions. How the account will be structured and the types of habitat that will qualify for protection have not yet been determined.

Altogether, the Trustees have dedicated about 60 percent of available settlement funds or $431.4 million for habitat protection in the spill region. Many species injured by the oil spill nest, feed, molt, winter, and seek shelter in the habitat protected through these programs. Several other species live primarily in the nearshore environment and benefit from the protection of the nearby uplands. Habitat protection also supports the restoration of tourism, recreation, commercial fishing and subsistence, all of which are dependent upon healthy productive ecosystems.

Public Participation

While public involvement continues to be at the forefront in the restoration process, it's fair to say that public information efforts peaked in 1999 with the 10th anniversary of the oil spill. "Legacy of an Oil Spill: 10 Years after *Exxon Valdez*" was the theme for a flurry of activities marking a decade since the spill. An all-day public "Report to the Nation" was held in Anchorage on March 23, followed by a three-day scientific symposium to report what has been learned since the spill.

A documentary about restoration activities was released in February, 1999 and provided to biology teachers and every school library in Alaska. It also aired over Alaska's public television stations. In addition, independent documentaries were produced by KTOO-TV in Juneau, KTUU-TV in Anchorage, and by National Geographic Television.

The world's media again focused on the *Exxon Valdez* in large numbers. Executive Director Molly McCammon conducted a standing-room-only press conference and luncheon presentation at the National Press Club in Washington, D.C. in February 1999. In addition, hundreds of newspaper, magazine, television, and radio reporters from 13 countries contacted the Restoration Office as part of their 10th anniversary coverage. Media groups included: CNN, ABC, NBC, 60 Minutes, Discovery, National Geographic, *Newsweek, People, Life, TIME, Christian Science Monitor, Wall Street Journal, USA Today,* and several newspapers from around the country.

A 10-year anniversary exhibit at the Alaska SeaLife Center has been seen by about 400,000 visitors over a two-year period. That exhibit was reproduced as a traveling display, exhibited at the public symposium in March and then shown at the Juneau, Anchorage, and Fairbanks libraries.

The 1999 Annual Report was a special edition answering basic questions about the spill with straightforward answers. That annual report has been widely requested and used by teachers throughout the country who are making the *Exxon Valdez* spill part of their curriculum. The 10-year report has been reproduced on the Trustee Council's web site and can be found at: www.oilspill.state.ak.us.

Alaska Coastal Currents, a series of weekly radio reports and newspaper columns that described the restoration process one piece at a time, provided nearly three years of detailed information to the public leading up to the 10-year event.

New titles were added to the popular Restoration Notebook series, a look at the natural history of injured species and associated restoration activities, written by the scientists in the field. Notebooks on the bald eagle and subsistence were added in 1999 to a series that already included harbor seals, killer whales, marbled murrelets, pigeon guillemots, sea otters, black oystercatchers, and herring. The entire series is available on the web.

The Restoration Update, the newsletter of the Trustee Council and the Restoration Office, continues to be published four times yearly. Newsletters for the last three years are available on the web.

In a scholarly evaluation of public trusts in the United States, the Trustee Council's public involvement process was singled out for its unique and innovative programs. Sally Fairfax, professor of Natural Resources at the University of California Berkeley, noted that the memorandum of understanding that established the Trustee Council specifically required "meaningful public participation in the injury assessment and restoration process."

Fairfax pointed to open meetings, the opportunity for all spill area communities to be included by teleconference at each meeting, public comment periods at each meeting, and the Public Advisory Group as exemplary of a public trust. But it was in regard to the various public involvement projects funded through the Council that she said the Trustees were "charting new territory."

Establishing community facilitators in 10 spill area communities, coordinated through the Chugach Regional Resources Commission, has been an effective way to disseminate information to the villages, get feedback from residents, and help communities apply for funds for local projects, she said. Fairfax also pointed to efforts such as establishment of the Alaska Native Harbor Seal Commission local stewardship of archaeological resources, and recognition of Traditional Ecological Knowledge (TEK) as a tool for restoration. "In the field of TEK, EVOS is truly a pioneer, developing both programs and protocols for addressing the practical problems of sensitive and effective cross-cultural inquiry, and working on the cutting edge of a new field of social science to understand theoretical and social implications of such research," she wrote.

Questions for Discussion

For Reading 1

1. Do all the scientists who studied the impacts of the Exxon oil spill agree on its seriousness and the degree to which the ecosystem recovered?

2. Briefly summarize the differing points of view that exist between experts.

3. What are the motivations of the differing groups of scientists?

For Reading 2

1. Briefly describe the legal settlement that was made with Exxon Oil Company. How is the $900 million from the civil settlement being used?

2. Had the Prince William Sound ecosystem recovered from the oil spill in 1999? What critical species have recovered? Which ones are still recovering? Which ones have not yet recovered? For which species is the recovery status not known?

3. What human activities were affected by the oil spill? Have they recovered?

4. What is the General Restoration Program? Briefly describe what has been gained by the program.

5. In the years since the oil spill, concern has shifted from critical species to habitat protection. How is this being accomplished? Has it been successful?

For Reading 3

1. Briefly describe the ecosystem studies being supported by the Trustee Council. What are the overall objectives of each project? What has each project accomplished?

2. What is your opinion? Exxon Oil Company has provided a huge amount of money to the state and federal governments and to the Trustee Council. Is that money being well spent?

Going Beyond the Readings

Two Web sites are particularly useful for providing the most recent information concerning the Exxon oil spill. One is maintained by the Oil Spill Information Center at <www.alaska.net/~ospic> and the other by the *Exxon Valdez* Oil Spill Trustee Council at <www.oilspill.state.ak.us>.

In addition, visit our Web site at <www.harcourtcollege.com/lifesci/envicases2> to investigate control of toxic waste and pollution issues in additional regions of the United States and Canada.